T0344975

Flow-Induced Vibration Handbook for Nuclear and Process Equipment

Wiley-ASME Press Series

Flow-Induced Vibration Handbook for Nuclear and Process Equipment

Michel J. Pettigrew, Colette E. Taylor, Nigel J. Fisher

This Work is a co-publication between ASME Press and John Wiley & Sons, Inc.

© 2022 ASME

This Work is a co-publication between ASME Press and John Wiley & Sons, Inc.

The right of Michel J. Pettigrew, Colette E. Taylor, Nigel J. Fisher to be identified as the authors of this work has been asserted in accordance with law.

Registered Office
John Wiley & Sons, Inc., 111 River Street, Hoboken, NJ 07030, USA

Editorial Office
111 River Street, Hoboken, NJ 07030, USA

For details of our global editorial offices, customer services, and more information about Wiley products visit us at www.wiley.com.

Wiley also publishes its books in a variety of electronic formats and by print-on-demand. Some content that appears in standard print versions of this book may not be available in other formats.

Library of Congress Cataloging-in-Publication Data applied for:

ISBN: 9781119810964

Cover Design: Wiley
Cover Images: © iStock\olegback

Set in 9.5/12.5pt STIX Two Text by Straive, Pondicherry, India

SKY10031136_110321

Contents

Preface

Excessive flow-induced vibration causing failures by fatigue or fretting wear must be avoided in process and nuclear components. That is the purpose of this handbook. In this book, the term process components is used generally to describe nuclear reactor internals, nuclear fuels, piping systems, and all shell-and-tube heat exchangers, including nuclear steam generators, power plant condensers, boilers and coolers.

There are already a number of good books on flow-induced vibration. So, why another one? This handbook is to help engineers to design, operate, and diagnose heat transfer equipment. The emphasis in this handbook is on two-phase flow-induced vibration. Despite the fact that roughly half of all heat exchanger equipment operates in two-phase flow, previous flow-induced vibration texts have provided limited guidance regarding vibration induced by two-phase flow. The prediction of fretting-wear damage is another important priority. The state of the art is presented in the design guidelines, figures and tables. The use of these design guidelines is illustrated with example calculations. To assist students and new design engineers, the calculations are supplemented and presented with more explanation in an appendix.

Largely, this handbook is the outcome of some 40 years of research and development at the Canadian Nuclear Laboratories. The emphasis of this book is the presentation of design guidelines based on extensive analysis of the literature and, in particular, on experimental data obtained in the field and at the Canadian Nuclear Laboratories in Chalk River.

We believe that this book will be useful to engineering design firms in the nuclear, petrochemical and aerospace industries, graduate schools in mechanical engineering and technical support groups in operating nuclear and petrochemical plants. This handbook is not a textbook, although it could be used as a resource in a graduate course. We also hope that this book will help to stimulate further research in the area of two-phase flow-induced vibration.

Michel J. Pettigrew
Colette E. Taylor
Nigel J. Fisher

Acknowledgments

The authors would like to recognize the institutions and colleagues who have provided permissions, support and inspiration to this project. We begin by recognizing the publishers who have kindly given permission to use copyrighted tables and figures. Rather than add the requested recognition statements to each figure, the following general statements are provided to avoid repetition.

- Figures from the *Congress of the Engineering Institute of Canada (EIC)* reprinted with permission from the EIC.
- Figures from the *International Association for Structural Mechanics in Reactor Technology (IASMiRT) conferences* reprinted courtesy of IASMiRT.
- Figures from *ANL Reports*, copyright by Argonne National Laboratory, managed and operated by the University of Chicago, U.S. Department of Energy, reprinted with permission.
- Figures from the *Nuclear Power Safety*, the *Journal of Fluids and Structures*, the *Journal of Sound and Vibration*, the *Journal of Multiphase Flow* and the *Journal of Nuclear Engineering and Design*, reprinted with permission from Elsevier.
- Figures from *Convective Boiling and Condensation*, Oxford Publishing Limited, Oxford, GB, reproduced with permission of Oxford Publishing Limited through PLSclear.
- Figures from *AERE reports* reproduced with the permission of UKAEA Scientific Publications.
- Figures from *Washington State University Reports* reprinted with permission from the Washington State University Libraries.
- Figures from the *8th International Heat Transfer Conference* reprinted with permission from Begell House Inc.
- Figures from the *Journal of Heat Transfer Engineering* and the *Journal of Nuclear Science and Engineering*, reprinted with permission of the publisher (Taylor & Francis Ltd, https://www.tand-fonline.com/)
- Figures from *Atomic Energy of Canada Ltd. Reports*, used with permission from AECL.

This document is based on some 40 years of research and development conducted at Chalk River Laboratories in the area of flow-induced vibration. This technology development effort was largely supported by Atomic Energy of Canada Limited (AECL) and by the CANDU Owners Group (COG). It also received support from the Heat Transfer and Fluid Flow Service (HTFS), the Centre d'Etudes Nucleaires de Saclay (CEN-Saclay), the Pressure Vessel Research Council (PVRC) and the Washington Public Power Supply System (WPPSS). The support of all these organizations is very gratefully acknowledged. Many people have contributed to this effort including colleagues from industry and universities. Recognizing that we will fail to acknowledge all of our partners, we have decided to name some of the key individuals and institutions.

The authors have benefited from discussions with researchers in other institutions such as F. Axisa and B. Villard, Centre d'Etudes Nucleaires de Saclay; H.G.D. Goyder, UKAEA Harwell; R.T. Hartlen, Ontario Hydro; N.W. Mureithi and many graduate students at École Polytechnique, Montreal; I.G. Currie, University of Toronto; R.J. Rogers, University of New Brunswick and M.P. Païdoussis, McGill University.

Visiting scientists, B.S. Kim, Korea Power Engineering Company Inc., Taejon, Korea; A. Yasuo, Central Research Institute of Electric Power Industry, Japan; and Z.L. Qiao, Xian Jiaotong University, China, contributed to the development of the flow-induced vibration database.

The contributors would also like to recognize the input of colleagues from the Canadian Nuclear Laboratories throughout the past 50 years: J. Albrecht, K.M. Boucher, W.A. Cook, T. Dickinson, P. Feenstra, E.G. Hagberg, Y. Han, G. Knowles, J. Mastorakos, J. McGregor, K. Moore, J.N. Patrick, P.J. Smith, Y. Sylvestre, J. H. Tromp, M.K. Weckwerth, and T. Whan. These individuals ably assisted with construction, instrumentation and installation of various flow loops and test sections, as well as copious data analysis.

Finally, the authors wish to express our gratitude to our understanding partners in life, for allowing us to take the time to write this handbook and for their moral support.

Contributors

Liberat N. Carlucci
Retired, Canadian Nuclear Laboratories
(previously, Atomic Energy of Canada Ltd.)

Nigel J. Fisher
Retired, Canadian Nuclear Laboratories
(previously, Atomic Energy of Canada Ltd.)

Daniel J. Gorman
Professor Emeritus, Ottawa University
(Deceased)

Fabrice M. Guérout
Canadian Nuclear Laboratories

Victor P. Janzen
Retired, Canadian Nuclear Laboratories
(previously, Atomic Energy of Canada Ltd.)

Michel J. Pettigrew
Principal Engineer Emeritus, Canadian
Nuclear Laboratories
Adjunct Professor, Polytechnique Montreal

John M. Pietralik
Retired, Canadian Nuclear Laboratories
(previously, Atomic Energy of Canada Ltd.)

Bruce A. W. Smith
Retired, Canadian Nuclear Laboratories
(previously, Atomic Energy of Canada Ltd.)

Colette E. Taylor
Retired, Canadian Nuclear Laboratories
(previously, Atomic Energy of Canada Ltd.)

David S. Weaver
Professor Emeritus, McMaster University

Metin Yetisir
Canadian Nuclear Laboratories

1

Introduction and Typical Vibration Problems

Michel J. Pettigrew

1.1 Introduction

Excessive flow-induced vibration must be avoided in process and nuclear system components. That is the purpose of this handbook. The term "process components" is used generally here to describe nuclear reactor internals, nuclear fuels, piping systems, and all shell-and-tube heat exchangers, including nuclear steam generators, power plant condensers, boilers, coolers, etc. Higher heat-transfer performance often requires higher flow velocities and more structural supports. On the other hand, additional supports may increase pressure drop and costs. The combination of high flow velocities and inadequate structural support may lead to excessive tube vibration. This vibration can cause failures by fatigue or fretting wear. Failures are very undesirable in terms of repair costs and lost production, particularly for high-capital-cost plants such as nuclear power stations, petroleum refineries and oil exploitation platforms. To prevent these problems at the design stage, a thorough flow-induced vibration analysis is recommended. Such analysis requires good understanding of the dynamic parameters and vibration excitation mechanisms that govern flow-induced vibration.

This handbook covers all relevant aspects of component vibration technology, namely: examples of vibration failures, flow analysis, and vibration excitation and damping mechanisms. The latter includes fluidelastic instability, periodic wake shedding, acoustic resonance, random turbulence, flow-induced vibration analysis and fretting-wear predictions.

Chapter 2 is an overview of flow-induced vibration technology. The reader should start with this chapter. In many cases, Chapter 2 will be sufficient to provide the required information. Each aspect of the technology is covered in detail in the succeeding chapters. Typically, each chapter includes a review of the state of the art, available laboratory data, brief review of theoretical considerations and modeling, parametric analysis, recommendations for design, and sample calculations.

The performance of process components is often limited by excessive vibration in a localized area, e.g., near inlets, outlets, etc. The combination of detailed flow calculations and vibration technology allows the designer to avoid such problems. Flow velocities and support design can be optimized to allow maximum heat transfer in all regions of process components, resulting in higher heat-transfer performance, less corrosion and fouling problems and reduced component size. The latter means capital cost reduction and a more competitive manufacturing industry.

This handbook is for the practicing engineer who is designing or troubleshooting nuclear and process system components. Design guidelines are proposed based on extensive analysis of the

Flow-Induced Vibration Handbook for Nuclear and Process Equipment, First Edition.
Michel J. Pettigrew, Colette E. Taylor, and Nigel J. Fisher.
© 2022 John Wiley & Sons, Inc. This Work is a co-publication between ASME Press and John Wiley & Sons, Inc.

literature and, in particular, on experimental data obtained in the field and at the Canadian Nuclear Laboratories of Atomic Energy of Canada Limited at Chalk River, Ontario, Canada. Although it is not intended as an undergraduate text book, it could be useful as a source of design data and practical examples. To assist students and new design engineers, the example calculations provided throughout this handbook are supplemented and presented with more explanation in Appendix A.

There are already several useful books on flow-induced vibration, e.g. Au-Yang (2001), Blevins (1990), Chen (1987), Kaneko et al (2014), Naudascher and Rockwell (1994), and Païdoussis (1998). So, why another one in the form of a handbook? This book is complementary to the above books for the following reasons. This book has greater emphasis on design guidelines. Much experimental data is presented in the form of comprehensive data bases that include a significant number of two-phase flow results. Particular attention is given to damping in single- and two-phase flow, two-phase flow-induced vibration mechanisms such as fluidelastic instability and random turbulence excitation, and the prediction of fretting-wear damage. Simple examples of calculations are given throughout the handbook.

1.2   Some Typical Component Failures

In heat exchangers, tube failures due to fretting wear may occur at the tube supports or at midspan if the tubes vibrate with sufficient amplitude to contact each other. Figure 1-1 shows an example of tube-to-tube fretting wear. It occurred in the U-bend of an early nuclear steam generator with tubes that were inadequately supported near the outlet in a region of high-velocity two-phase cross flow. Extensive fretting-wear damage was also observed between tube and tube support, as shown in Fig. 1-2. Here, the damage was sufficient to cause a hole in the tube resulting in leakage between tube-side and shell-side. Obviously, this kind of problem must be avoided. An additional support near the outlet region was an easy solution to this problem.

Figure 1-3 shows extensive tube-to-tube fretting-wear damage in the inlet region of a triple segmental liquid-liquid process heat exchanger. The problem was due to the combination of long tube spans (1.45 m) and high flow velocities impinging directly on the tubes in the inlet region. Lacing

Fig. 1-1   Tube-to-Tube Fretting Wear in the U-Bend Region of an Early Nuclear Steam Generator (Pettigrew, 1976).

Fig. 1-2 Tube-to-Support Fretting Wear: Note Hole Through Tube Wall (Pettigrew, 1976).

Fig. 1-3 Fretting Wear in the Inlet Region of a Liquid Process Heat Exchanger: a) Tube-to-Tube Initial Damage, b) Lacing Strip, and c) Damage at Lacing Strip Location (Pettigrew, 1976).

strips were installed to support the tubes near the inlet. Unfortunately, they were excessively loose. Fretting wear occurred between the tube and the lacing strips. Tubes wore through, as shown in Fig. 1-4. Eventually, proper baffle-supports were installed, as shown in Fig. 1-5. No further problems occurred.

In power condensers, very-high-velocity steam may impinge on the tubes, causing excessive vibration. Figure 1-6 shows a fatigue failure of a titanium condenser tube. The vibration amplitude was

Fig. 1-4 Fretting Wear Through Tube Wall at a Lacing Strip Location in a Process Heat Exchanger (Pettigrew et al, 1977).

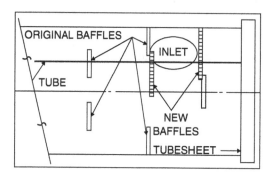

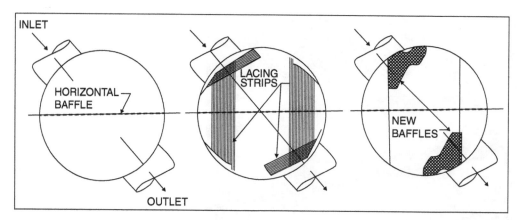

Fig. 1-5 Fretting Wear of Process Heat Exchanger: Repair (Pettigrew and Campagna, 1979 / with permission of Atomic Energy of Canada Limited).

sufficient for the tubes to contact each other at midspan. In this case, the condenser was operated with only four of the six tube bundles, resulting in 150% of design flow velocity. Operation at 100% design flow did not cause any vibration damage.

Figure 1-7 shows tube-to-tube fretting-wear damage in another power condenser. In this case, the damage was sufficient to wear through the tube wall, causing leakage of sea water into the secondary side of a power plant.

An example of fretting-wear damage of a tube located just beyond the baffle cut (window tube) vibrating against a baffle edge is shown in Fig. 1-8. This tube came from a gas-to-gas heat exchanger,

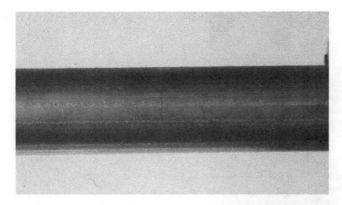

Fig. 1-6 Fatigue Failure of a Titanium Tube in a Nuclear Power Plant Condenser (Pettigrew et al, 1991).

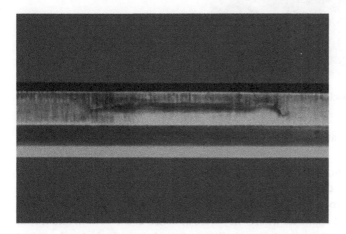

Fig. 1-7 Tube-to-Tube Fretting Wear in a Power Plant Condenser.

Fig. 1-8 Fretting Wear of a Gas Heat Exchanger Tube at a Baffle Edge Location.

Fig. 1-9 Fretting-Wear Damage on Nuclear Fuel (Hot Cell Examination) (Pettigrew, 1976).

which was inadvertently operated at several times above design flow during a commissioning operation.

Figure 1-9 is an example of fretting wear of nuclear fuel as seen through an optical magnifier during "hot cell" examination. The damage occurred at the top of fuel string assemblies in 40% of the high flow fuel channels of a prototype CANDU-BLW[®1] nuclear station. The fuel strings were inserted in upward-flow vertical fuel channels where they were attached at the bottom and free at the top. The flow in each channel became two-phase as boiling occurred along the fuel and reached approximately 16% steam quality near the top. The mass flux was typically 4400 kg/m^2 s. The fretting problem was attributed to transverse flow-induced vibration of the fuel strings. Inadvertently, some of the fuel strings were assembled eccentrically. This caused the strings to be bent and promoted fretting wear. The corrective measures taken were to ensure concentric assembly of the fuel and increase fuel string flexural rigidity to reduce vibration.

One of the most costly vibration related problems took place in the early 1990s at the Darlington Nuclear Power Station, where nuclear fuel bundles were seriously damaged by fatigue and fretting wear (Fig. 1-10). The cost of investigation, repairs and particularly lost production totalled approximately 1 billion dollars Canadian. The problem was caused by acoustic resonance in the inlet headers due to coincidence of the pump pressure pulsation frequency, (30 Hz x 5 vanes = 150 Hz) and the natural acoustic frequency of the headers. The pressure pulsations were transmitted and amplified in the fuel channels, subjecting the fuel bundles to significant pressure fluctuations causing extensive damage. The problem was solved by simply replacing the five-vane pump impellers by seven-vane impellers, thus eliminating the acoustic resonance.

Sometimes vibration problems develop because of changes in operating conditions. For example, pressurized water reactor (PWR) fuel failures occurred in the 1990s due to fretting wear between fuel rods and support grids. The problem was related to longer fuel residence time, which caused increased clearances between the rods and grids due to creep, and deregulation of fuel

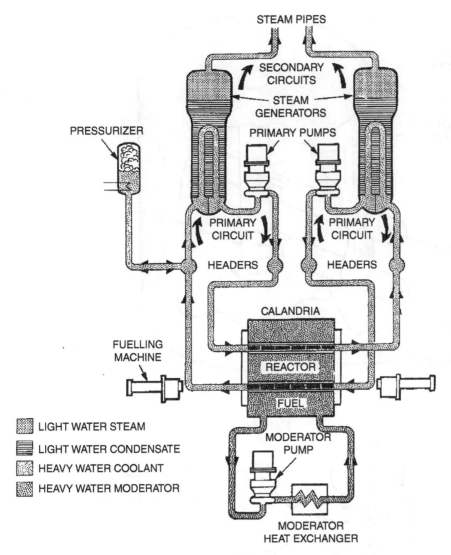

Fig. 1-10 Schematic Drawing of CANDU-PHW®2 Reactor (Pettigrew, 1978 / with permission of Atomic Energy of Canada Limited)

procurement. The latter allowed fuels from different suppliers at the same time in the reactor core. Differences in design caused slight differences in impedance resulting in increased cross flows and, thus, more flow-induced vibration excitation. Changes in support grid design solved this problem.

Vibration problems are not limited to material damage such as fatigue and fretting wear. For example, excessive vibration of control absorber guide tubes due to jet impingement could have caused a serious reactor control problem (see Fig. 1-11a). The problem was avoided by shielding the guide tube with a protective shroud, as shown in Fig. 1-11b.

Many other vibration problems have been encountered, such as fatigue failures of PWR core barrel tie rods and in-core instrumentation nozzles, excessive acoustic noise due to control valve

(a)

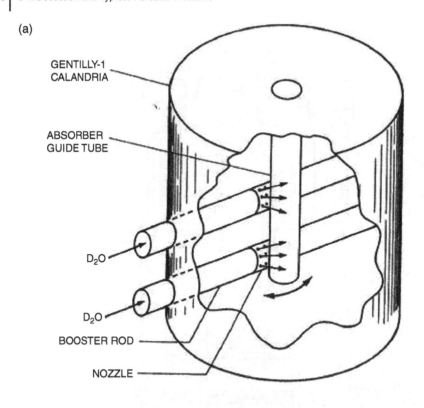

(b)

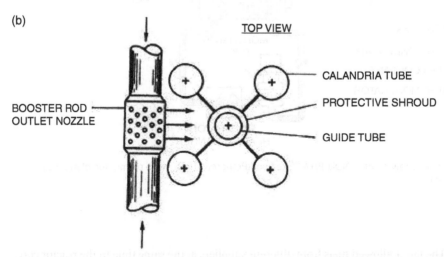

Fig. 1-11 a) Control Absorber Guide Tube Vibration due to Jetting, b) Modification with Protective Shroud (Pettigrew, 1976 / with permission of Atomic Energy of Canada Limited).

dynamics and mechanical damage resulting from acoustic resonance in gas heat exchangers. Although most vibration problems have very costly consequences, they are usually solved by simple design modifications or changes in operating conditions. After the fact, it is easy to see that most problems could have been avoided by proper understanding of flow-induced vibration phenomena.

Thus, it is important to understand flow-induced vibration and damage mechanisms to prevent problems at the design stage and assist plant operators in predicting the life of components. It is hoped that this handbook will help in that direction.

1.3 Dynamics of Process System Components

1.3.1 Multi-Span Heat Exchanger Tubes

From a mechanical dynamics point of view, heat exchanger and steam generator tubes are multi-span beams clamped at the tubesheet and held at the baffle supports with varying degree of constraint (see Fig. 1-12). The degree of constraint depends on support geometry and, particularly, on tube-to-support clearance. Heat exchanger dynamics is inherently a non-linear phenomenon since it depends largely on the interaction (impacting and sliding) between a particular tube and its supports. This non-linearity is particularly important since it governs the evaluation of damping at the supports and the prediction of fretting-wear damage to the tube.

Such analysis requires time domain non-linear simulation of the tube dynamics in which the details of sliding-friction, impact, viscous-shear and squeeze-film forces between tube and tube supports are modelled. Unfortunately, non-linear simulations are difficult and some of the detailed information is lacking. Furthermore, the required non-linear analysis is not yet in the form of a practical design tool. Some progress has been made in this area with the development of codes, such as VIBIC (Fisher et al, 1992), H3DMAP (Sauvé et al, 1987) and EPRI SG FW (Rao et al, 1988) to predict fretting wear of heat exchanger tubes.

In the future, we believe that all vibration analyses will consider non-linear simulation of the dynamic interaction between tubes and supports and will include a fretting-wear damage prediction. This kind of analysis is now done by specialists for very critical or very expensive components, such as nuclear steam generators. Fretting-wear damage prediction is discussed in Chapter 12.

For the time being, quasi-linear vibration analyses are used by the industry for most heat exchangers. Quasi-linear analysis requires the formulation of tube-to-support dynamic interaction forces, such as damping, in terms of equivalent linear values. We have found this approach to be reasonable in practice for the prediction of overall tube vibration response and critical velocities for fluidelastic instability. Such analysis is adequate to eliminate most vibration problems. However, long-term fretting-wear damage and tube life can only be predicted in an approximate manner. Tube vibration measurements in real heat exchangers show generally good agreement between measured and predicted frequencies using the quasi-linear approach.

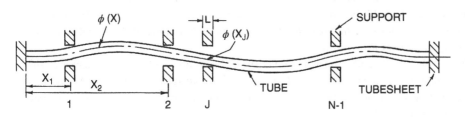

Fig. 1-12   Multi-Span Heat Exchanger Tube with N Spans and N-1 Clearance Supports.

1.3.2 Other Nuclear and Process Components

Other process and nuclear system components, such as nuclear fuels, reactor internals, and piping systems, are often multi-span beams with intermediate clearance-type supports (e.g., piping supports, fuel bearing pads and support grids). Analysis of these components is similar to that of multi-span heat exchanger tubes.

Notes

1 CANDU-BLW = Canada Deuterium Uranium – Boiling Light-Water is a registered trademark of Atomic Energy of Canada Limited.
2 CANDU-PHW = Canada Deuterium Uranium – Pressurized Heavy-Water is a registered trademark of Atomic Energy of Canada Limited.

References

Au-Yang, M.K., 2001, *"Flow-Induced Vibration of Power and Process Plant Components: A Practical Workbook,"* ASME Press, New York, NY, USA.

Blevins, R. D., 1990, *"Flow-Induced Vibration,"* 2nd Edition, Van Nostrand Reinhold Company, New York, NY, USA.

Chen, S. S, 1987, *"Flow-Induced Vibration of Circular Structures,"* Hemisphere Publishing Corporation, New York, NY, USA.

Fisher, N. J., Ing, J. G., Pettigrew, M. J. and Rogers, R. J., 1992, "Tube-to- Support Dynamic Interaction for a Multispan Steam Generator Tube," *Proceedings of ASME International Symposium on Flow-Induced Vibration and Noise,* Anaheim, California, November 8-13, 2, pp. 301–316.

Kaneko, S., Nakamura, T., Inada, F., Kato, M., Ishihara, K., Nishihara, T., 2014, *"Flow-Induced Vibration,"* 2nd Edition, Academic Press, Elsevier, London, UK.

Naudascher, E. and Rockwell, D., 1994, *"Flow-Induced Vibration: An Engineering Guide,"* A.A. Balkma, Rotterdam, Netherlands.

Païdoussis, M. P., 1998, *"Fluid-Structure Interactions: Slender Structures and Axial Flow,"* Vol. 1, Academic Press, Elsevier, London, UK.

Pettigrew, M. J., 1976, "Flow-Induced Vibration of Nuclear Power Station Components," 90th*Annual Congress of the Engineering Institute of Canada,* Halifax, NS, Oct. 4-8, (also *Atomic Energy of Canada Report, AECL-5852*)

Pettigrew, M. J., and Campagna, A. O., 1979, " Heat Exchanger Vibration: Comparison Between Operating Experiences and Vibration Analysis," *Proceedings of the International Conference on Practical Experiences with Flow-Induced Vibrations,* Karlsruhe, Germany, 1979 September, (also *Atomic Energy of Canada Report, AECL-6785*).

Pettigrew, M. J., Sylvestre, Y. and Campagna, A. O., 1977, "Flow-Induced Vibration Analysis of Heat Exchanger and Steam Generator Designs," 4th *International Conference on Structural Mechanics in Reactor Technology,* Paper No. F6/1+, San Francisco, California, Aug. 15-19, (also *Atomic Energy of Canada Report, AECL-5826*).

Pettigrew, M. J., Carlucci, L. N., Taylor, C. E. and Fisher, N. J., 1991, "Flow-Induced Vibration and Related Technologies in Nuclear Components," *Nuclear Engineering and Design,* 131, pp. 81–100.

Rao, M. S. M., Steininger, D. A., and Eisinger, F. L., 1988, "Numerical Simulation of Fluidelastic Vibration and Wear of Multispan Tubes with Clearances at Supports," *Proceedings of the ASME Symposium on Flow-Induced Vibration and Noise-1988*: Vol. 5, *Flow-Induced Vibration in Heat-Transfer Equipment*, pp. 235–250.

Sauvé, R. G. and Teper, W. W., 1987, "Impact Simulation of Process Equipment Tubes and Support Plates-A Numerical Algorithm", *Journal of Pressure Vessel Technology*, 109, pp. 70–79.

2

Flow-Induced Vibration of Nuclear and Process Equipment: An Overview

Michel J. Pettigrew and Colette E. Taylor

2.1 Introduction

Failures due to excessive vibration must be avoided in process equipment, preferably at the design stage. Thus, a comprehensive flow-induced vibration analysis is required before fabrication of process equipment, such as shell-and-tube heat exchangers. It must be shown that tube vibration levels are below allowable levels and that unacceptable resonances and fluidelastic instabilities are avoided.

The purpose of this chapter is to summarize our design guidelines for flow-induced vibration in components operating in gas, liquid, and two-phase flows. This overview chapter can be used by the designer as a guideline for vibration analysis, by the project engineer to get an overall appreciation of flow-induced vibration concerns, or by the plant operator to understand tube failures. This overview pertains to critical regions of shell-and-tube heat exchangers, such as nuclear steam generators (SG), heat exchangers (HX), coolers, condensers and moisture-separator-reheaters (MSR).

2.1.1 Flow-Induced Vibration Overview

The vibration behavior of process system components is governed by vibration excitation mechanisms and by damping mechanisms. Generally, in components such as heat exchangers there are several significant damping mechanisms: 1) friction damping between tube and tube support, 2) squeeze-film damping at the support, 3) viscous damping between tube and shell-side fluid, and 4) damping due to two-phase flow.

Generally, the flow in heat exchanger tube bundles can be parallel (axial flow) or transverse (cross flow) to the tube. In nuclear steam generators, the flow is axial for a large portion of the tube bundle. Vibration excitation forces induced by axial flow are relatively small in heat exchangers. Thus, vibration excitation mechanisms in axial flow may generally be neglected. The vibration behavior is clearly governed by cross-flow vibration excitation mechanisms.

Several vibration excitation mechanisms are normally considered in heat exchange components such as tube bundles in cross flow, namely: 1) fluidelastic instability, 2) periodic wake shedding, 3) turbulence excitation, and 4) acoustic resonance. Fluidelastic instability is by far the most important mechanism and must be avoided in all cases. Periodic-wake-shedding resonance may be of concern in liquid cross flow where the flow is relatively uniform. It is not normally a problem at the entrance region of steam generators because the flow is very non-uniform and quite turbulent (Pettigrew et al, 1973). Turbulence may inhibit periodic wake shedding in tube arrays. Periodic wake shedding is generally not a problem in two-phase flow except at very low void fractions

Flow-Induced Vibration Handbook for Nuclear and Process Equipment, First Edition.
Michel J. Pettigrew, Colette E. Taylor, and Nigel J. Fisher.
© 2022 John Wiley & Sons, Inc. This Work is a co-publication between ASME Press and John Wiley & Sons, Inc.

(i.e., $\varepsilon_g < 15\%$) or under unusual conditions, as discussed in Chapter 11. Turbulence excitation may be important in liquid cross flow. Periodic wake shedding resonance and turbulence excitation are not usually of concern in gas flow since the fluid density is generally low, thereby resulting in relatively small excitation forces. However, both mechanisms should be considered in some gas heat exchangers such as MSRs where relatively high fluid densities exist. Acoustic resonance must be avoided in heat exchangers with shell-side gas flow. However, it is generally not a problem in liquid and two-phase heat exchangers.

2.1.2 Scope of a Vibration Analysis

A heat exchanger vibration analysis consists of the following steps: 1) flow distribution calculations, 2) dynamic parameter evaluation (i.e., damping, effective tube mass, and dynamic stiffness), 3) formulation of vibration excitation mechanisms, 4) vibration response prediction, and 5) resulting damage assessment (i.e., comparison against allowables). The requirements applicable to each step are outlined in this overview. Each step is discussed in more detail in the following chapters of this handbook.

2.2 Flow Calculations

Flow-induced vibration problems usually occur on a small number of vulnerable tubes in specific areas of a component (e.g., piping elements, entrance regions and tube-free lanes in heat exchangers, and U-tubes in nuclear steam generators). Thus, a flow analysis is required to obtain the local flow conditions throughout these heat exchange components. Flow considerations are discussed in detail in Chapter 3.

2.2.1 Flow Parameter Definition

The end results of a flow analysis are the shell-side cross-flow velocity, U_p, and fluid density, ρ, distributions along critical tubes. For flow-induced vibration analyses, flow velocity is defined in terms of the pitch velocity:

$$U_p = U_\infty P/(P-D) \tag{2-1}$$

where U_∞ is the free stream velocity (i.e., the velocity that would prevail if the tubes were removed), P is the pitch between the tubes and D is the tube diameter. For finned tubes, the equivalent or effective diameter, D_{eff}, is used. The pitch velocity is sometimes called the reference gap velocity. The pitch velocity is a convenient definition since it applies to all bundle configurations.

The situation is somewhat more complex in two-phase flow. Another parameter, steam quality or void fraction, is required to define the flow conditions. Two-phase mixtures are rarely homogeneous or uniform across a flow path. However, it is convenient and simple to use homogeneous two-phase mixture properties as they are well defined. This is done consistently here for both specifying vibration guidelines and formulating vibration mechanisms. The homogeneous void fraction, ε_g, is defined in terms of the volume flow rates of gas, $\dot{V}_g$, and liquid, $\dot{V}_\ell$ as:

$$\varepsilon_g = \frac{\dot{V}_g}{\dot{V}_g + \dot{V}_\ell} \tag{2-2}$$

The homogeneous density, ρ, the free stream velocity, U_∞, and the free stream mass flux, $\dot{m}_\infty$, are defined using the homogeneous void fraction:

$$\rho = \rho_\ell(1 - \varepsilon_g) + \rho_g \varepsilon_g \tag{2-3}$$

$$U_\infty = \frac{\rho_\ell \dot{V}_\ell + \rho_g \dot{V}_g}{\rho A} \tag{2-4}$$

$$\dot{m}_\infty = \rho U_\infty \tag{2-5}$$

where ρ_g and ρ_ℓ are the densities of the gas and liquid phase, respectively, and A is the free-stream flow path area.

For both liquid and two-phase cross flow, the pitch velocity, U_p, and the pitch mass flux, $\dot{m}_p$, are similarly defined as:

$$U_p = U_\infty \frac{P}{(P - D)}$$

$$\dot{m}_p = \dot{m}_\infty \frac{P}{(P - D)} = \rho U_p \tag{2-6}$$

2.2.2 Simple Flow Path Approach

For relatively simple components, where the flow paths are reasonably well defined, a flow path approach may be adequate to calculate flow velocities, as illustrated in Fig. 2-1. In the flow path

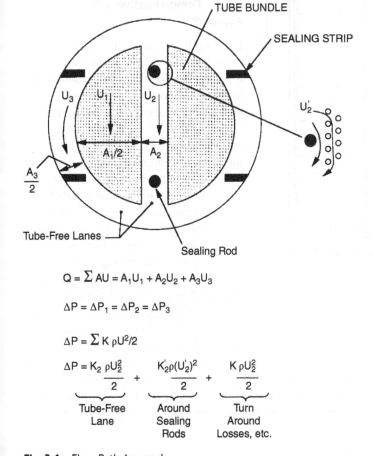

$$Q = \Sigma\, AU = A_1 U_1 + A_2 U_2 + A_3 U_3$$

$$\Delta P = \Delta P_1 = \Delta P_2 = \Delta P_3$$

$$\Delta P = \Sigma\, K\, \rho U^2/2$$

$$\Delta P = \underbrace{\frac{K_2\, \rho U_2^2}{2}}_{\substack{\text{Tube-Free} \\ \text{Lane}}} + \underbrace{\frac{K_2'\rho (U_2')^2}{2}}_{\substack{\text{Around} \\ \text{Sealing} \\ \text{Rods}}} + \underbrace{\frac{K\, \rho U_2^2}{2}}_{\substack{\text{Turn} \\ \text{Around} \\ \text{Losses, etc.}}}$$

Fig. 2-1 Flow-Path Approach.

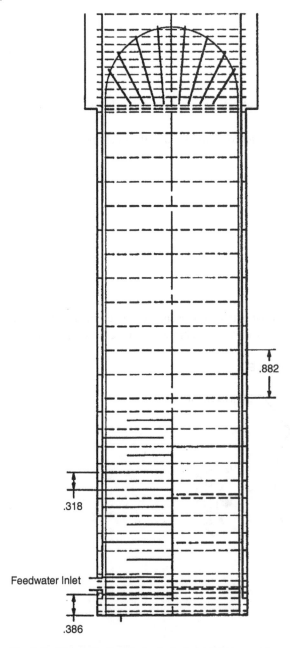

.882

.318

Feedwater Inlet

.386

Fig. 2-2 Thermalhydraulic Analysis: Axial Grid Layout for a Typical Steam Generator with Preheater (Distances in Metres).

analysis approach, characteristic flow paths (i.e., through the tube bundle, between the tube bundle and shell, etc.) between regions of common pressures are identified. Flow impedances (i.e., pressure drop coefficients) are estimated. The flows within each path are then calculated. The resulting flow velocity distributions are then used to estimate vibration excitation mechanisms and predict vibration response.

All typical operating conditions must be considered including the following: 1) as-designed operating conditions, from zero to 100% flow, 2) operating conditions with fouling of the tubes or crudding of the tube supports, and 3) other possible operating conditions (e.g., after chemical cleaning, system testing, etc.).

2.2.3 Comprehensive 3-D Approach

For complex components such as nuclear steam generators and power condensers, a comprehensive three-dimensional thermalhydraulic analysis is required. In such analyses, the component is divided into a large number of control volumes. The equations of energy, momentum and continuity are solved for each control volume. This is done with numerical methods using a computer code such as the THIRST code for steam generators (Pietralik, 1995). The numerical grid outlining the control volumes for the analysis of a typical steam generator is shown in Fig. 2-2. The grid must be sufficiently fine to accurately predict the flow distribution along the tube. Some typical thermalhydraulic analysis results are shown in Fig. 2-3 for the U-bend region of a steam generator. For flow-induced vibration analyses, the results must be in the form of pitch flow velocity and fluid density distributions along a given tube. These distributions constitute the input to a flow-induced vibration analysis of this particular tube. Figure 2-4 shows pitch flow velocity and fluid density distributions for an example condenser tube.

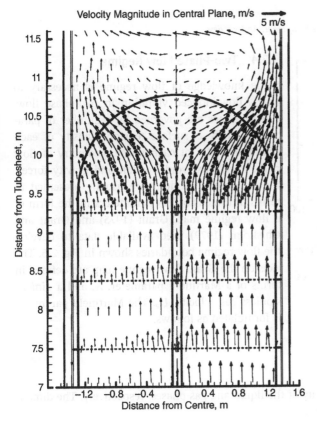

Fig. 2-3 Flow Velocity Vectors in the Central Plane of a Typical Steam Generator U-Bend Region.

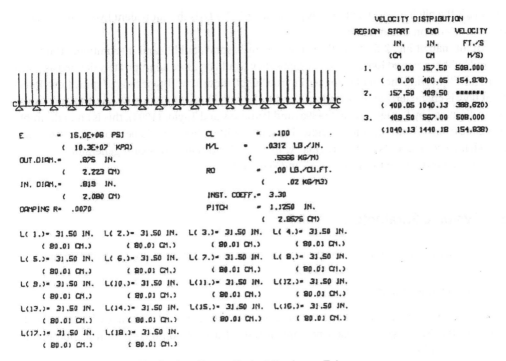

Fig. 2-4 Gap Cross-Flow Distribution Along a Typical Condenser Tube.

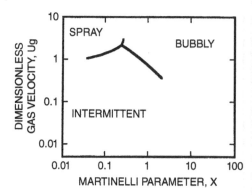

Fig. 2-5 Flow Pattern Map for Two-Phase Flow Across Cylinder Arrays Using Flow Pattern Boundaries (Grant and Chisholm (1979) and Axes Parameters from McNaught (1982)).

2.2.4 Two-Phase Flow Regime

Some knowledge of flow regime is necessary to assess flow-induced vibration in two-phase flow. Flow regimes are governed by a number of parameters, such as surface tension, density of each phase, viscosity of each phase, geometry of flow path, mass flux, void fraction and gravity forces. Flow regime conditions are usually presented in terms of dimensionless parameters in the form of a flow regime map. Grant (1975) and Grant and Chisholm (1979) used available data to develop the flow regime boundaries shown in Fig. 2-5. The axes on the Grant flow regime map are defined in terms of a Martinelli parameter, X, and a dimensionless gas velocity, U_g. The Martinelli parameter is formulated as follows:

$$X = \left(\frac{1-\varepsilon_g}{\varepsilon_g}\right)^{0.9} \left(\frac{\rho_\ell}{\rho_g}\right)^{0.4} \left(\frac{\mu_\ell}{\mu_g}\right)^{0.1} \tag{2-7}$$

where μ_ℓ and μ_g are the dynamic viscosity of the liquid and gas phases, respectively. The dimensionless gas velocity is defined as follows:

$$U_g = \frac{\dot{m}_{pg}}{\left[d_e g \rho_g \left(\rho_\ell - \rho_g\right)\right]^{0.5}} \tag{2-8}$$

where $\dot{m}_{pg}$ is the pitch mass flux of the gas phase, $d_e \cong 2(P - D)$ is the equivalent hydraulic diameter and g is acceleration due to gravity.

The Grant map of Fig. 2-5 shows three flow regimes: spray, bubbly and intermittent. The terms spray and bubbly are used loosely here. Perhaps they would be more appropriately defined as "continuous flow" covering the whole range from true bubbly flow to wall-type flow to spray flow. Intermittent flow is characterized by periods of flooding (mostly liquid) followed by bursts of mostly gas flow. As discussed by Pettigrew et al (1989a) and Pettigrew and Taylor (1994), this is an undesirable flow regime from a vibration point-of-view. Thus, intermittent flow should be avoided in two-phase heat exchange components and, particularly, in the U-bend region of steam generators.

Flow regime is discussed in more detail in Chapter 3.

2.3 Dynamic Parameters

The relevant dynamic parameters for multi-span heat exchanger tubes are mass and damping.

2.3.1 Hydrodynamic Mass

Hydrodynamic mass is the equivalent dynamic mass of external fluid vibrating with the tube. In liquid flow, the hydrodynamic mass per unit length of a tube confined within a tube bundle may be expressed by

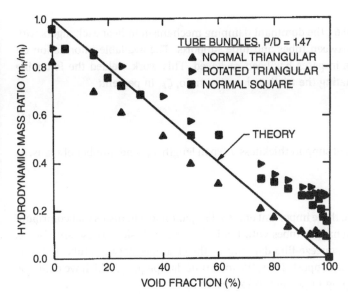

Fig. 2-6 Hydrodynamic Mass in Two-Phase Cross Flow: Comparison Between Theory and Experiments. (Note that m_ℓ is the hydrodynamic mass per unit length in liquid.)

$$m_h = \left(\frac{\pi}{4}\rho D^2\right)\left[\frac{(D_e/D)^2 + 1}{(D_e/D)^2 - 1}\right] \tag{2-9}$$

where D_e is the equivalent diameter of the surrounding tubes and the ratio D_e/D is a measure of confinement. The effect of confinement is formulated by

$$D_e/D = (0.96 + 0.5P/D)P/D \tag{2-10}$$

for a tube inside a triangular tube bundle (Rogers et al, 1984). Similarly, for a square tube bundle (Pettigrew et al, 1989a) confinement may be approximated by

$$D_e/D = (1.07 + 0.56P/D)P/D \tag{2-11}$$

The hydrodynamic mass of tube bundles subjected to two-phase cross flow may be calculated with Eq. (2-9) provided that the homogeneous density of the two-phase mixture, ρ_{TP}, is used in the formulation (Pettigrew et al, 1989a). Figure 2-6 compares Eq. (2-9) to experimental data.

The total dynamic mass of the tube per unit length, m, comprises the hydrodynamic mass per unit length, m_h, the tube mass per unit length, m_t, and the mass per unit length of the fluid inside the tube, m_i:

$$m = m_h + m_t + m_i \tag{2-12}$$

See Chapter 4 for more detail on hydrodynamic mass.

2.3.2 Damping

Vibration energy dissipation (damping) is an important parameter in limiting vibration. Damping in single- and two-phase flows is discussed in detail in Chapters 5 and 6, respectively.

Heat Exchanger Tubes in Gases

As discussed in Pettigrew et al (1986), the dominant damping mechanism in heat exchangers with gas on the shell-side is friction between tubes and tube supports. The available information on damping of heat exchanger tubes in gases has been reviewed. This work yielded the following design recommendation for estimating the friction damping ratio, ζ_F, in percent:

$$\zeta_F = 5\left(\frac{N-1}{N}\right)\left(\frac{L}{\ell_m}\right)^{0.5} \tag{2-13}$$

which takes into account the effect of support thickness, L, span length, ℓ_m, and number of spans, N.

Heat Exchanger Tubes in Liquids

As discussed in Chapter 5, there are three important energy dissipation mechanisms that contribute to damping of multi-span heat exchanger tubes with liquids on the shell-side. These are viscous damping between tube and liquid, squeeze-film damping in the clearance between tube and tube support, and friction damping at the support. Thus, the total tube damping, ζ_T, which we define as the ratio of actual to critical damping in percent, is expressed by

$$\zeta_T = \zeta_V + \zeta_{SF} + \zeta_F \tag{2-14}$$

where, ζ_V, ζ_{SF} and ζ_F are, respectively, the viscous, squeeze-film and friction damping ratios.

Tube-to-fluid viscous damping is related to the Stokes number, $\pi f D^2/2\nu$, and the degree of confinement of the heat exchanger tube within the tube bundle or D/D_e. Rogers et al (1984) developed a simplified formulation for viscous damping, valid for $\pi f D^2/2\nu > 3300$ and $D/D_e < 0.5$ which covers most heat exchangers. Their simplified expression for ζ_V(in percent) is

$$\zeta_V = \frac{100\pi}{\sqrt{8}}\left(\frac{\rho D^2}{m}\right)\left(\frac{2\nu}{\pi f D^2}\right)^{1/2}\left\{\frac{[1+(D/D_e)^3]}{[1-(D/D_e)^2]^2}\right\} \tag{2-15}$$

where ν is the kinematic viscosity of the fluid and f is the natural frequency of the tube for the mode being analysed. Clearly, viscous damping is frequency dependent. Calculated values of damping using Eq. (2-15) are compared against experimental data in Fig. 2-7. The comparison shows reasonable agreement.

Squeeze-film damping, ζ_{SF}, and friction damping, ζ_F, take place at the supports. Semi-empirical expressions were developed, based on available experimental data, to formulate friction and squeeze-film damping as discussed by Pettigrew et al (2011) and in Chapter 5. The available experimental data is outlined in Fig. 2-8. Thus, squeeze-film damping (in percent) may be formulated by

$$\zeta_{SF} = \left(\frac{N-1}{N}\right)\left[\frac{(1460)}{f}\left(\frac{\rho D^2}{m}\right)\left(\frac{L}{\ell_m}\right)^{\frac{1}{2}}\right] \tag{2-16}$$

and friction damping (in percent) by

$$\zeta_F = \left(\frac{N-1}{N}\right)\left[0.5\left(\frac{L}{\ell_m}\right)^{1/2}\right] \tag{2-17}$$

where $(N-1)/N$ takes into account the ratio of the number of supports over the number of spans, L is the support thickness and ℓ_m is a characteristic tube length. The latter is defined as the average of the three longest spans when the lowest modes and the longest spans dominate the vibration response. This is usually the case. When higher modes and shorter spans govern the vibration

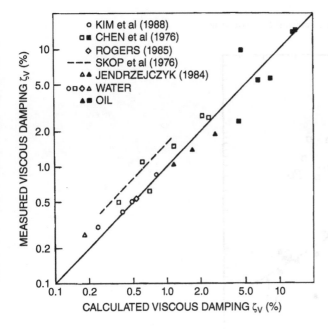

Fig. 2-7 Viscous Damping Data for a Cylinder in Unconfined and Confined Liquids: Comparison Between Theory and Experiment.

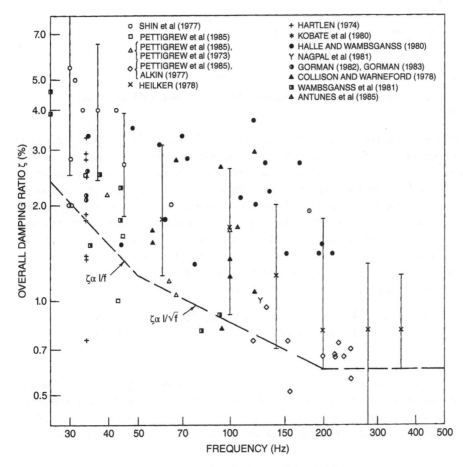

Fig. 2-8 Damping Data for Multi-Span Heat Exchanger Tubes in Water.

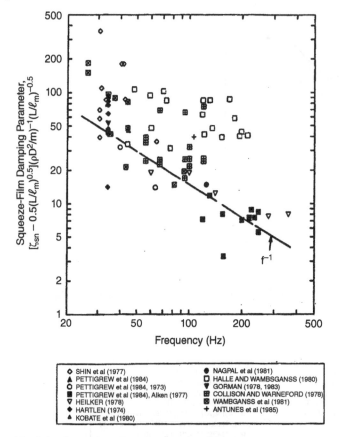

Fig. 2-9 Comparison Between Tube Support Damping Model (Squeeze-Film and Friction) and Experimental Data.

response, then, the characteristic span length should be based on these shorter spans. This could happen when there are high flow velocities locally, such as in entrance or exit regions.

Equations (2-16) and (2-17) are combined to determine the squeeze-film damping parameter, $1460/f$, and compared against available experimental data in Fig. 2-9. There is large scatter in the data, which is mostly due to the nature of the problem. Heat exchanger tube damping depends on parameters that are difficult to control, such as support alignment, tube straightness, relative position of tube within the support, and side loads. These parameters are statistical in nature and probably contribute to much of the scatter. A conservative but realistic criterion for damping is obtained by taking damping values at roughly the lower decile of the available data, as shown in Fig. 2-8.

Substituting Eqs. (2-15), (2-16) and (2-17) in Eq. (2-14):

$$\zeta_T = \frac{100\pi}{\sqrt{8}} \left\{ \frac{\left[1 + (D/D_e)^3\right]}{\left[1 - (D/D_e)^2\right]^2} \right\} \left(\frac{\rho D^2}{m}\right) \left(\frac{2\nu}{\pi f D^2}\right)^{\frac{1}{2}} + \left(\frac{N-1}{N}\right) \left[\frac{(1460)}{f} \left(\frac{\rho D^2}{m}\right) \left(\frac{L}{\ell_m}\right)^{\frac{1}{2}} + 0.5 \left(\frac{L}{\ell_m}\right)^{\frac{1}{2}}\right]$$

(2-18)

Although somewhat speculative, Eq. (2-18) formulates all important energy dissipation mechanisms and fits the data best. Thus, it is our recommendation as a damping criterion for design purposes.

However, if the damping ratio predicted by this equation is less than 0.6%, we recommend taking a minimum value of 0.6%. As shown in Fig. 2-8, a minimum damping of 0.6% appears reasonable.

In Fig. 2-8, there is no experimental data for frequencies below 25 Hz. Rather than extrapolating Eq. 2-18 to lower frequencies, we recommend using a maximum total damping of 5%.

Damping in Two-Phase Flow

The subject of heat exchanger tube damping in two-phase flow was reviewed by Pettigrew and Taylor (1997) as discussed in Chapter 6. The total damping ratio, ζ_T, of a multi-span heat exchanger tube in two-phase flow comprises support damping, ζ_S, viscous damping, ζ_V, and two-phase damping, ζ_{TP}:

$$\zeta_T = \zeta_S + \zeta_V + \zeta_{TP} \tag{2-19}$$

Depending on the thermalhydraulic conditions (i.e., heat flux, void fraction, flow, etc.), the supports may be dry or wet. If the supports are dry, which is more likely for high heat flux and very high void fraction, only friction damping takes place. Support damping in this case is analogous to damping of heat exchanger tubes in gases, Pettigrew et al (1986). Damping due to friction (and impact) in dry supports may be expressed by Eq. (2-13). The above situation is unlikely in well-designed recirculating steam generators. However, it could exist in some areas of once-through steam generators.

When there is liquid between the tube and the support, support damping includes both squeeze-film damping, ζ_{SF}, and friction damping, ζ_F. This situation is analogous to heat exchanger tubes in liquids and the support damping may be evaluated with Eqs. (2.18) and (2.19) above:

$$\zeta_S = \zeta_{SF} + \zeta_F = \frac{(N-1)}{N} \left[\frac{(1460)}{f} \left(\frac{\rho_\ell D^2}{m} \right) \left(\frac{L}{\ell_m} \right)^{\frac{1}{2}} + 0.5 \left(\frac{L}{\ell_m} \right)^{1/2} \right] \tag{2-20}$$

Note that in Eq. (2-20), ρ_ℓ is the density of the liquid within the tube support, whereas m is the total mass per unit length of the tube including the hydrodynamic mass calculated with the two-phase homogeneous density.

Viscous damping in two-phase mixtures is analogous to viscous damping in single-phase fluids (Pettigrew and Taylor, 1997). Homogeneous properties of the two-phase mixture are used in its formulation, as follows:

$$\zeta_V = \frac{100\pi}{\sqrt{8}} \left(\frac{\rho_{TP} D^2}{m} \right) \left(\frac{2v_{TP}}{\pi f D^2} \right)^{1/2} \left\{ \frac{\left[1 + (D/D_e)^3 \right]}{\left[1 - (D/D_e)^2 \right]^2} \right\} \tag{2-21}$$

where v_{TP} is the equivalent two-phase kinematic viscosity as per McAdams et al (1942):

$$v_{TP} = v_\ell \left[1 + \varepsilon_g \left(v_\ell / v_g - 1 \right) \right]^{-1} \tag{2-22}$$

Above 40% void fraction, viscous damping is generally small and could be neglected for the U-bend region of steam generators. However, it is significant for lower void fractions.

There is a two-phase component of damping in addition to viscous damping. As discussed by Pettigrew and Taylor (1997) and in Chapter 6, two-phase damping is strongly dependent on void fraction, fluid properties and flow regimes, directly related to confinement and the ratio of hydrodynamic mass over tube mass, and weakly related to frequency, mass flux or flow velocity, and tube bundle configuration. A semi-empirical expression was developed from the available experimental data to formulate the two-phase component of damping, ζ_{TP}, in percent:

$$\zeta_{TP} = 4.0 \left(\frac{\rho_\ell D^2}{m} \right) \left[f(\varepsilon_g) \right] \left\{ \frac{\left[1 + (D/D_e)^3 \right]}{\left[1 - (D/D_e)^2 \right]^2} \right\} \tag{2-23}$$

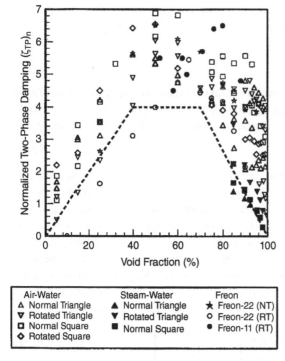

Air-Water	Steam-Water	Freon
△ Normal Triangle	▲ Normal Triangle	★ Freon-22 (NT)
▽ Rotated Triangle	▼ Rotated Triangle	○ Freon-22 (RT)
▢ Normal Square	▪ Normal Square	● Freon-11 (RT)
◇ Rotated Square		

Fig. 2-10 Comparison Between Proposed Design Guideline and Available Damping Data. Normalized Two-Phase Damping Ratio $(\zeta_{TP})_n$:

$$(\zeta_{TP})_n = \zeta_{TP} \left(\frac{\rho_\ell D^2}{m}\right)^{-1} \left\{\frac{[1 + (D/D_e)^3]}{[1 - (D/D_e)^2]^2}\right\}^{-1}$$

As shown in Fig. 2-10, the void-fraction function, $f(\varepsilon_g)$, may be approximated by taking the envelope through the lower decile of the data:

$$f(\varepsilon_g) = 1, \text{for } 40\% \leq \varepsilon_g \leq 70\%$$
$$f(\varepsilon_g) = \varepsilon_g/40, \text{for } \varepsilon_g < 40\% \qquad (2\text{-}24)$$
$$f(\varepsilon_g) = 1 - (\varepsilon_g - 70)/30, \text{for } \varepsilon_g > 70\%$$

The fluid properties/flow regime dependence was difficult to assess in the absence of a broad range of damping data for different two-phase mixtures. Flow regime effects are partly taken care of by the void-fraction function.

Tube fouling is not expected to contribute much to damping. However, support damping is considerably reduced by crudding within the support. At the limit, when tubes are jammed in the support by severe crud deposition, a support damping value of $\zeta_S = 0.2\%$ should be used in analysis.

Dynamic Stiffness and Support Effectiveness The dynamic stiffness of multi-span heat exchanger tubes is simply the flexural rigidity, *EI*, where *E* is the elastic modulus of the tube material and *I* is the area moment of inertia of the tube cross-section. The boundary conditions, that is, the support conditions, are somewhat complex. The tubes are effectively clamped at the tubesheet. To facilitate assembly and to allow for thermal expansion, there is a clearance between tubes and tube supports.

The diametral clearance between tube and intermediate support is typically 0.25 to 0.80 mm (for many nuclear heat exchangers, the diametral clearance is specified to be 0.38 mm or 0.015 in.). Thus, the dynamic interaction between tube and tube support is inherently non-linear. In well-designed heat exchangers, the tube vibration response at mid-span is mostly less than 100 μm root-mean-square (rms) and much less at the supports (usually less than 25 μm rms). This response is significantly less than the available diametral clearance. Thus, the tubes do not generally vibrate back and forth across the available clearance. Instead, most tubes are not centered within the supports and are, therefore, contacting or vibrating very close to one side of the supports. It is difficult to imagine that many tubes of typically one-metre span length, would be located concentrically within a 0.38 mm diametral clearance without touching the support. The chances of a tube not touching a support are probably much less than one percent. Thus, it is reasonable to assume pinned conditions at the supports to allow for a quasi-linear vibration analysis. For the purpose of vibration analysis, drilled hole, broached hole, and egg-crate type supports may be taken as pinned supports.

Anti-vibration bars (AVBs) are often used in the U-bend region of nuclear steam generators. AVBs are reasonably effective in the out-of-plane direction, but offer little restraint in the in-plane (plane of the U-bend) direction (Taylor et al, 1995). Fortunately, the vibration response is usually much less in the in-plane direction than in the out-of-plane direction. Nevertheless, AVBs are less effective than other clearance supports. For steam generator vibration analyses, it is recommended that one (any one) AVB in the U-bend region be assumed ineffective. The possibility that two adjacent AVBs would be ineffective at the same time is considered too improbable to be of concern. For example, if we assume that the probability of one AVB being ineffective is less than one percent, then the probability of two adjacent AVBs being ineffective is less than 0.01% or less than one tube in 10 000, which is less than one tube per steam generator and, thus, insignificant.

In the in-plane direction, on the other hand, recent experience has shown that more than one adjacent AVB may be ineffective and that maybe two or more AVBs should be assumed ineffective in the in-plane direction. In practice, tube-to-support clearances must be small enough to provide an effective support. Thus, pinned support conditions may be assumed provided that the tube-to-support diametral clearance for drilled holes, broached holes, scallop bars, egg crates and lattice bars is equal to or less than 0.4 mm. In the out-of-plane direction, the diametral clearance between tube and AVBs should be sufficiently small to provide effective support (e.g., <0.1 mm).

2.4 Vibration Excitation Mechanisms

As mentioned previously in Section 2.1.1, several vibration mechanisms are normally considered in heat exchange components: fluidelastic instability, random turbulence excitation, periodic wake shedding and acoustic resonance. Formulations for these mechanisms are presented in the following sub-sections (Sections 2.4.1 to 2.4.5) of this chapter. More details are given in Chapters 7 to 11.

2.4.1 Fluidelastic Instability

Fluidelastic instability is the most severe vibration excitation mechanism in heat exchanger tube bundles. Formulations for single- and two-phase cross flow are given below. The topics are discussed in more detail in Chapters 7 and 8 for single- and two-phase flow, respectively.

Single-Phase Cross Flow (Gas or Liquid)

Fluidelastic instability for tube bundles subjected to single-phase cross flow was reviewed by Pettigrew and Taylor (1991). Fluidelastic instability is formulated in terms of a dimensionless flow velocity, U_p/fD and a dimensionless mass-damping parameter, $2\pi\zeta m/\rho D^2$, such that:

$$\frac{U_{pc}}{fD} = K\left(\frac{2\pi\zeta m}{\rho D^2}\right)^{1/2} \tag{2-25}$$

where f is the tube natural frequency, ρ is the fluid density, m is the tube mass per unit length and U_{pc} is the threshold, or critical, flow velocity for fluidelastic instability. A fluidelastic instability constant $K = 3.0$ is recommended for all tube bundle configurations, as shown in Fig. 2-11 (Pettigrew and Taylor, 1991). The damping ratio, ζ, is the total damping ratio as outlined in Section 2.3.2. The above formulation applies to all cross-flow regions such as steam generator preheater regions, heat-exchanger tube bundles between baffle plates, and outlet and inlet regions.

Two-Phase Flow

We have found that fluidelastic instability behavior is somewhat similar in continuous two-phase cross flow, (Pettigrew et al, 1989b). As explained in Chapter 3, continuous flow means two-phase flow regions of a continuous nature such as bubbly, spray, fog and wall flows as opposed to intermittent flow regimes leading to bundle reflooding and large flow oscillations. Such oscillations can lead to much lower critical velocity for fluidelastic instability (Pettigrew et al, 1989b, 1995 and 1994). Thus, intermittent flow regimes should be avoided in two-phase cross flow.

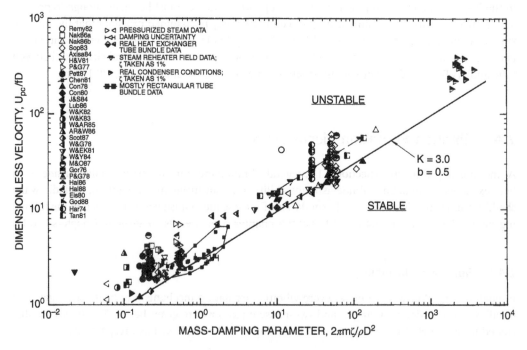

Fig. 2-11 Summary of Fluidelastic Instability Data for Single-Phase Cross Flow: Recommended Design Guidelines.

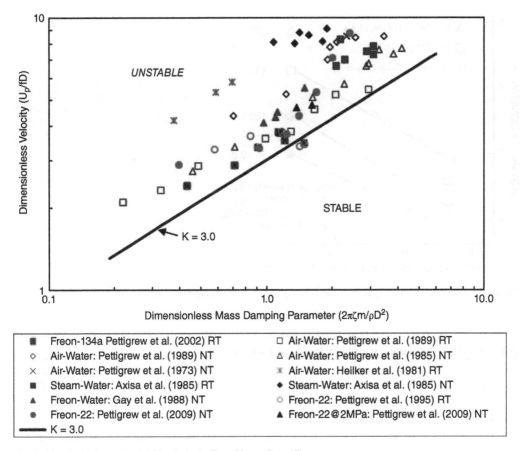

■	Freon-134a Pettigrew et al. (2002) RT	☐	Air-Water: Pettigrew et al. (1989) RT
◇	Air-Water: Pettigrew et al. (1989) NT	△	Air-Water: Pettigrew et al. (1985) NT
✕	Air-Water: Pettigrew et al. (1973) NT	✳	Air-Water: Heilker et al. (1981) RT
■	Steam-Water: Axisa et al. (1985) RT	◆	Steam-Water: Axisa et al. (1985) NT
▲	Freon-Water: Gay et al. (1988) NT	◯	Freon-22: Pettigrew et al. (1995) RT
●	Freon-22: Pettigrew et al. (2009) NT	▲	Freon-22 @ 2MPa: Pettigrew et al. (2009) NT
—	K = 3.0		

Fig. 2-12 Fluidelastic Instability Data in Two-Phase Cross Flow.

Fluidelastic instability in two-phase cross flow can also be formulated using Eq. (2-25). As shown in Fig. 2-12, a fluidelastic instability constant $K = 3.0$ is also recommended for two-phase cross flow, but only for tube bundles of $P/D > 1.47$ which is the case for many steam generators. Lower values of K must be used for bundles of P/D ratios lower than 1.47, as discussed by Pettigrew and Taylor (1994). As shown in Fig. 2-13, the expression

$$K = 4.76(P - D)/D + 0.76 \tag{2-26}$$

would be a reasonable design guideline for $P/D < 1.47$.

2.4.2 Random Turbulence Excitation

Random turbulence excitation is a significant excitation mechanism in both liquid and two-phase cross flow. Formulations for single- and two-phase cross flow are given below. The topics are discussed in more detail in Chapters 9 and 10 for single- and two-phase flow, respectively.

To be able to compare data and find an upper bound, the excitation forces must be presented as a normalized excitation force spectra. Researchers in this field such as Taylor and Pettigrew (1998), and Pettigrew and Gorman (1981) have used various methods of normalizing their results. Therefore, it was necessary to select one means of normalization and apply it to all of the data. The adopted method is the "equivalent power spectral density (EPSD)," first described by Axisa et al (1990).

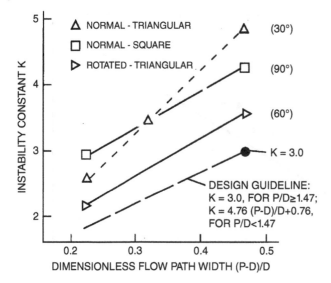

Fig. 2-13 Effect of P/D on Fluidelastic Instability Constant in Two-Phase Cross Flow.

The power spectral density (PSD), $S_F(f)$, can be rendered dimensionless using a pressure scaling factor, p_o, and a frequency scaling factor, f_o, as follows:

$$\widetilde{S}_F(f_R) = \frac{S_F(f)}{(p_o D)^2} f_o \tag{2-27}$$

where, f_R is the reduced frequency, defined as f/f_o, and D is the tube diameter.

A difficulty arises in the calculation of $S_F(f)$ because the correlation length, λ_c, is rarely known. Axisa et al (1990) present a dimensionless EPSD, $\widetilde{S}_F(f_R)_e$, defined as follows:

$$\widetilde{S}_F(f_R)_e = \frac{\lambda_c}{L_e} \widetilde{S}_F(f_R) \tag{2-28}$$

where, L_e is the excited tube length. Using this definition, the dimensionless EPSD for Mode 1 can be defined in terms of the mean square of tube displacement, $\overline{y^2(x)}_1$ as follows:

$$\widetilde{S}_F(f_R)_e = \frac{\overline{y^2(x)}_1 64\pi^3 f_1^3 m^2 \zeta_1}{\phi_1^2 a_1} \frac{1}{(p_o D)^2} f_o \tag{2-29}$$

where, $\phi_1(x_1)$ is the normalized mode shape for the 1st mode, a_1 is the numerical coefficient for the 1st mode, f_1 is the 1st mode tube natural frequency, m is the total tube mass (tube mass + hydrodynamic mass) and ζ_1 is the damping ratio for the 1st mode. Values of ϕ_1^2 and a_1 are 2.0 and 1.1, respectively, for pinned-pinned end conditions.

Using Eq. (2-29), the mean square of tube displacement can be found without knowledge of the correlation length. Instead, a small correlation length is assumed. To correctly compare spectra obtained using experimental rigs with varying geometries, it is necessary to define a dimensionless reference EPSD, $\widetilde{S}_F(f_R)_e^o$, based on a reference excited tube length, L_o, and a reference tube diameter, D_o, as follows:

$$\widetilde{S}_F(f_R)_e^o = \widetilde{S}_F(f_R)_e \times \frac{L_e}{L_o} \times \frac{D_o}{D} \tag{2-30}$$

where, L_e, is the excited tube length. In this chapter, reference lengths of $L_o = 1$ m and $D_o = 0.02$ m are applied.

Single-Phase Cross Flow

In single-phase cross flow, two distinct flow fields are possible. Interior tubes, well within a heat exchanger tube bundle, are excited by turbulence generated within the bundle. This excitation is governed by the tube bundle geometry. On the other hand, upstream or inlet tubes are excited by turbulence generated by upstream components such as inlet nozzles, entrance ports and upstream piping elements. Upstream turbulence levels are governed by the upstream flow path geometry and are very often much larger than those generated within the bundle. Such excitation is often referred to as far-field excitation.

Random turbulence excitation is usually not a problem with gas or vapor cross flow. The pressure fluctuations and resulting excitation forces due to gas cross flow at a given velocity are generally an order of magnitude less than those for a liquid or two-phase mixture at the same velocity. However, gas velocities can be extremely high and at high pressure the densities can be significant. Therefore, some consideration should be given to random excitation in high-pressure gas heat exchangers.

Taylor and Pettigrew (2000) combined data from many sources to arrive at the reference EPSD guideline shown in Fig. 2-14. The lower bound, shown in Fig. 2-14, should be used when the upstream turbulence is less than or equal to the turbulence within the tube bundle. The upper bound, shown in Fig. 2-14, should be used if the upstream turbulence exceeds the turbulence inside the tube bundle. The boundaries are defined as follows:

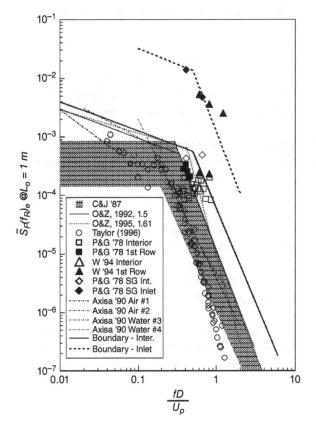

Fig. 2-14 Proposed Guideline for Single-Phase Random Excitation Forces (References Provided in the Legend can be found in Taylor and Pettigrew (2000)).

Interior $\widetilde{S}_F(f_R)_e^0 = 4 \times 10^{-4}(f/f_o)^{-0.5}$, $\quad 0.01 < f/f_o < 0.5$

$\widetilde{S}_F(f_R)_e^0 = 5 \times 10^{-5}(f/f_o)^{-3.5}$, $\quad 0.5 < f/f_o$ $\qquad\qquad$ (2-31)

Inlet $\widetilde{S}_F(f_R)_e^0 = 1 \times 10^{-2}(f/f_o)^{-0.5}$, $\quad 0.01 < f/f_o < 0.5$

$\widetilde{S}_F(f_R)_e^0 = 1.25 \times 10^{-3}(f/f_o)^{-3.5}$, $\quad 0.5 < f/f_o$ $\qquad\qquad$ (2-32)

For single-phase flow $f_o = U_p/D$ and $p_o = \rho U_p^2/2$.

In most cases, the random excitation forces for interior tubes are significantly lower than for upstream tubes. The vibration response of the upstream tubes will be larger. Thus, it may not be necessary to consider the vibration response of interior tubes when they are otherwise identical to the upstream tubes.

Two-Phase Cross Flow

Random turbulence excitation forces can be significant in two-phase cross flow, in particular, in the U-bend region of steam generators. The term turbulence is used loosely in two-phase flows. It describes the dynamics of the two-phase mixture as it flows through a tube bundle.

For void fractions of 10% or less, two-phase flow random forces behave like single-phase flow forces and the single-phase guidelines can be used. At higher void fractions, the effect of the two-phase mixture begins to dominate the single-phase random forces. The physics of these two-phase forces are not well understood but recent study of the effect of void fraction and flow regime have led to reasonable collapse of two-phase data from air-water, Freon liquid-vapor, Freon vapor-water and steam-water experiments.

Preliminary two-phase design guidelines for random excitation of heat exchanger tubes were presented by several authors such as Taylor et al (1989), Axisa et al (1990) and Taylor et al (1996). Most of this data was obtained with air-water mixtures, with some steam-water data. A few years later, de Langre and Villard (1998) introduced a dimensionless scaling that simplifies that suggested by Taylor et al (1996) and more closely follows the principle used in the scaling of single-phase random forces.

De Langre and Villard (1998) show that a two-phase power spectral density, $S_F(f)$, can be rendered dimensionless using Eq. (2-27) with a two-phase pressure scaling factor, $p_o = \rho_\ell g D_B$, and a two-phase frequency scaling factor, $f_o = U_p/D_B$. The length scale, D_B, is defined as follows:

$$D_B = 0.1D/\sqrt{1-\varepsilon_g} \qquad\qquad (2-33)$$

The selection of these scaling factors is described is some detail in de Langre and Villard (1998). Many pressure scaling factors were considered, but the gravity-based factor was the most efficient. The authors point out that gravity in the scaling factor may be related to dynamic pressure caused by the drift velocities between gas and liquid phases. More recent analysis by Taylor and Pettigrew (2019) has shown that a two-phase pressure scaling factor of $p_o = \rho_\ell g D$ is preferred.

Using these scaling factors and the reference EPSD described above in Eq. (2-30), data from many sources were collapsed on a single plot (see Fig. 2-15). The following boundary spectra for two-phase random forces in churn and bubbly flow were defined by Taylor and Pettigrew (2019):

$$\widetilde{S}_F(f_R)_e^0 = 2.0(f/f_o)^{-0.5}, \quad 0.001 \leq f/f_o \leq 0.06$$

$$\widetilde{S}_F(f_R)_e^0 = 4.3 \times 10^{-4}(f/f_o)^{-3.5}, \quad 0.06 \leq f/f_o \leq 1 \qquad\qquad (2-34)$$

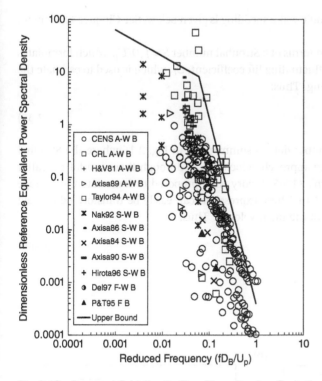

Fig. 2-15 Proposed Guideline for Two-Phase Random Excitation Forces In Churn and Bubbly Flow Regimes (Taylor and Pettigrew, 2019).

2.4.3 Periodic Wake Shedding

Periodic wake shedding, or vortex shedding, may be a problem when the shedding frequency coincides with a tube natural frequency. This coincidence may lead to resonance and large vibration amplitudes. Periodic-wake-shedding resonance in tube bundles subjected to liquid cross flow was observed in early research by Pettigrew and Gorman (1981).

Periodic-wake-shedding resonance may be of concern in liquid cross flow where the flow is relatively uniform. It is not normally a problem at the entrance region of steam generators because the flow is very non-uniform and quite turbulent. Turbulence may inhibit periodic wake shedding in tube arrays. Periodic wake shedding is generally not a problem in two-phase flow except at very low void fractions (i.e., $\varepsilon_g < 15\%$) (Pettigrew and Taylor, 1994). Then, the behavior is similar to liquid flow.

Periodic-wake-shedding resonance is usually not a problem in heat exchangers with shell-side gas flow. The gas density is usually too low to cause significant periodic forces at flow velocities close to resonance. Normal flow velocities in gas heat exchangers are usually much higher than those required for resonance. However, resonance may be possible in high-pressure components such as MSRs with higher density gas on the shell side. It could also happen for higher modes of vibration with higher frequencies corresponding to higher flow velocities. Thus, periodic-wake-shedding resonance cannot always be ignored in gas heat exchangers.

Generally, there is less information on the magnitude of periodic-wake-shedding forces in tube bundles than on the frequency of periodic wake shedding, because acoustic resonance in gas-flow heat exchangers has proven to be a more significant problem in gas-flow heat exchangers than tube vibration due to vortex shedding.

A recommended formulation for periodic wake shedding is given below. See Chapter 11 for more detail.

Periodic wake shedding is described in terms of a Strouhal number $S = f_S D / U_p$, which formulates the wake shedding frequency, f_S, and a fluctuating lift coefficient, C_L, which is used to estimate the periodic forces, F_L, due to wake shedding. Thus:

$$F_L = C_L D \rho U_p^2 / 2 \qquad (2\text{-}35)$$

Periodic-wake-shedding Strouhal number data is summarized in Fig. 2-16 where the Strouhal numbers, S_t, are defined in terms of the approach velocity U_∞. Although there is a lot of scatter in the data, expressions based on Owen's (1965) theory were proposed to formulate the average values, as shown by the curves in Fig. 2-16. These expressions can easily be transformed to yield Strouhal numbers, S, defined in terms of the pitch velocity. Thus,

$$S = (1/1.73)(D/P) \qquad (2\text{-}36a)$$

for normal-triangular bundles,

$$S = (1/1.16)(D/P) \qquad (2\text{-}36b)$$

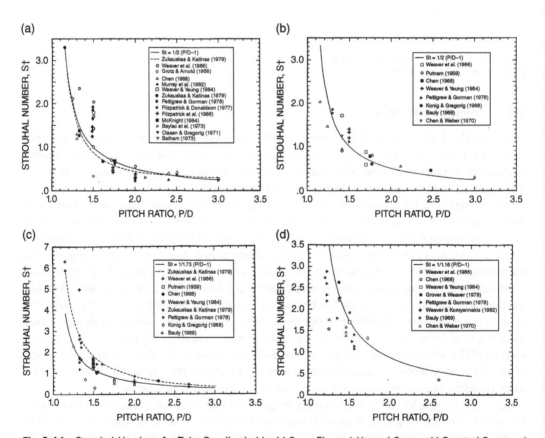

Fig. 2-16 Strouhal Numbers for Tube Bundles in Liquid Cross Flow: a) Normal Square, b) Rotated Square, c) Normal Triangular, and d) Rotated Triangular.

for rotated-triangular bundles, and

$$S = (1/2)(D/P)$$

(2-36c)

for both normal- and rotated-square bundles.

These expressions give Strouhal numbers between about 0.32 and 0.70 for realistic heat exchanger tube bundles of P/D between 1.23 and 1.57. These Strouhal numbers correspond to dimensionless pitch velocities, U_p/fD, between 1.5 and 3.0. Resonance should be assumed possible within this velocity range. In the case of a tube subjected to a non-uniform velocity, only the fluctuating fluid forces corresponding to the region within the above range of dimensionless velocities need to be considered.

A review of the available data showed that fluctuating forces due to periodic wake shedding depend on several parameters such as bundle configuration, location within the bundle, Reynolds number, turbulence level, fluid density and P/D. At the limit when P/D is large, fluctuating force coefficients should approach those for single cylinders. On the other hand, when P/D is very small the force coefficients are small since the fluid mass associated with the formation of periodic wake or vortices is small as there is little space available within the bundle for large vortices.

The available data for fluctuating force coefficients in single- and two-phase flow is compiled in Fig. 2-17 and Fig. 2-18, respectively. In single-phase flow, Fig. 2-17 shows that the fluctuating lift coefficient, C_L, is very dependent on P/D up to $P/D \approx 2.5$. For both single- and two-phase flow within heat exchanger tube bundles of $P/D < 1.6$, a fluctuating force coefficient of $C_L = 0.075$ rms is recommended to calculate periodic-wake-shedding forces.

When resonance is considered possible, the maximum allowable tube vibration amplitude should not exceed 0.02D or 2% of the tube diameter. Below 0.02D, the vibration amplitude is generally not sufficient to correlate periodic wake shedding along the tube, resulting in much lower vibration

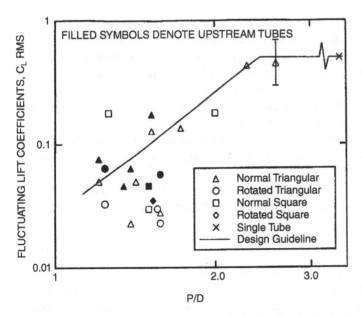

Fig. 2-17 Fluctuating Force Coefficients for Tube Bundles in Single-Phase Cross Flow. (Based on Weaver et al. 1987)

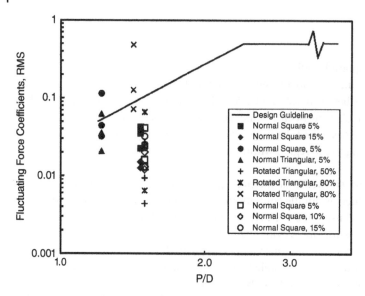

Fig. 2-18 Fluctuating Force Coefficients for Tube Bundles in Two-Phase Cross Flow.

response. If the vibration response exceeds 0.02D, a fatigue analysis and a fretting-wear damage calculation should be done, as outlined in Chapter 12.

2.4.4   Acoustic Resonance

Acoustic resonance may take place in heat exchanger tube bundles when periodic-wake-shedding frequencies coincide with a natural frequency for acoustic standing waves within a heat exchanger cavity. The flow-excited acoustic standing waves are generally transverse to both the axis of the tubes and to the direction of flow. Such resonances normally cause intense acoustic noise with sound pressure amplitudes that can exceed 160 dB. Tube, baffle and duct damage from fatigue may also occur, if nothing is done to eliminate the noise problem. Acoustic resonance is possible in gas heat exchangers with both finned and bare tubes.

Acoustic resonance requires two conditions: 1) coincidence of shedding and acoustic frequencies, and 2) sufficient acoustic energy to overcome the acoustic damping in order that sustained acoustic standing-wave resonance occurs.

An approach to acoustic resonance is given below. For more detail, refer to Chapter 11.

Frequency Estimates: As discussed earlier, periodic wake shedding is governed by the Strouhal number:

$$S = f_s D / U_p \tag{2-37}$$

The Strouhal number may be obtained from Eq. (2-36) or from Fig. 2-16 for different tube bundle configurations and P/D ratios. The range of Strouhal numbers over which acoustic resonance can take place should be extended to allow for possible "lock-in" of the wake. As suggested by Blevins and Bressler (1987), acoustic resonance may be possible between 0.8S and 1.3S, or in terms of periodic-wake-shedding frequency

$$0.8 S U_p / D < f_s < 1.3 S U_p / D \tag{2-38}$$

Acoustic standing wave frequencies, f_{an}, are defined by:

$$f_{an} = n C_e / 2W \tag{2-39}$$

where W is the dimension of the heat exchanger tube bundle cavity in the direction normal to the flow and the tube axes, n is the mode order and C_e is the effective speed of sound (Ziada et al, 1989):

$$C_e \approx C / \sqrt{1 + \sigma} \tag{2-40}$$

The speed of sound, C, is obtained from

$$C = \sqrt{kp/\rho} \tag{2-41}$$

where k is the specific heat ratio ($k = 1.33$ for steam), p is the shell side pressure and ρ is the shell-side fluid density.

The speed of sound is affected by the presence of the tubes. This effect is related to the solidity ratio, σ, which is the ratio of the volume occupied by the tubes over the volume of the tube bundle. For a triangular tube bundle

$$\sigma = \pi D^2 / \left(2\sqrt{3} P^2 \right) \tag{2-42a}$$

and, for a square tube bundle

$$\sigma = \pi D^2 / P^2 \tag{2-42b}$$

The acoustic standing wave frequencies should be calculated for the first few acoustic modes (i.e., the first five modes should suffice). If one or more acoustic mode frequencies fall within the range of periodic-wake-shedding frequency, acoustic resonance conditions are possible.

2.4.5 Susceptibility to Resonance

Coincidence of acoustic and shedding frequencies does not necessarily cause resonance. Resonance also depends on acoustic energy and acoustic damping. Several design criteria have been proposed to evaluate susceptibility to tube bundle resonance. The design guidelines of Blevins and Bressler (1987) and of Ziada et al (1989) are suggested for heat exchanger tube bundles.

Blevins and Bressler (1987) suggest that resonance is less likely for closely packed tube bundles, probably because acoustic damping is higher. Also, closely packed tube bundles probably prevent the formation of larger vortices that are associated with higher acoustic energy. From experimental data for first-mode acoustic resonance, Blevins and Bressler (1987) show that resonance is unlikely for $P/D < 1.6$ and $L/D < 3.0$ for staggered (triangular) tube bundles, and for $L/D < 1.4$ for in-line (square) tube bundles. Here, P is the transverse pitch and L is the longitudinal pitch. However, they do not discuss the applicability of their criterion to higher-order acoustic modes.

Ziada et al (1989) propose a resonance parameter to assess susceptibility to resonance. For staggered arrays (triangular), this parameter has the form

$$G_s = \sqrt{R_{cr}} \left[\frac{\sqrt{(L/D)(P/D - 1)}}{(L/D) - 1} \right] \left(\frac{v}{CD} \right) \tag{2-43a}$$

where R_{cr} is the Reynolds number based on the critical flow velocity at resonance, and v is the kinematic viscosity for the shell-side gas. The resonance parameter, G_s, is correlated versus the flow

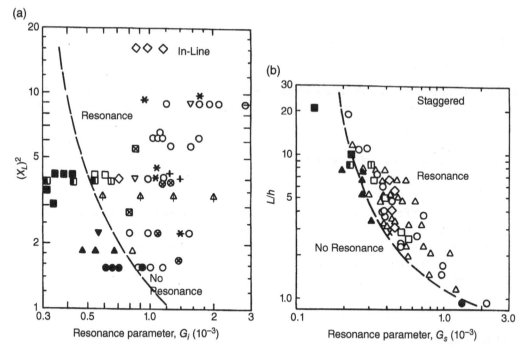

Fig. 2-19 Proposed Damping Criteria (Resonance Parameter) for: a) In-Line (Square) Arrays; b) Staggered (Triangular) Arrays (Ziada et al, 1989 / with permission of Elsevier).

path parameter, L/h, in Fig. 2-19b for both resonance and non-resonance data. The parameter L/h is defined in Ziada et al (1989).

Similarly, for in-line arrays (square)

$$G_i = \sqrt{R_{cr}}(P/D)\left(\frac{v}{CD}\right) \tag{2-43b}$$

The parameter G_i is plotted against $(X_L)^2$ in Fig. 2-19a, where $X_L = L/D$. For a satisfactory design, it must be shown that the heat exchanger tube bundle is in the non-resonance parts of Fig. 2-19.

2.5 Vibration Response Prediction

Due to the complexity of a multi-span steam generator tube, a computer code is used to predict vibration response accurately and to determine the susceptibility of a tube to periodic-wake-shedding resonance and fluidelastic instability. Such a computer code must be capable of calculating mode-shapes and natural frequencies for a number of modes at least equal to the number of tube spans.

Generally, the code will be in two parts: the first part calculates the free-vibration tube characteristics such as mode shapes and natural frequencies, while the second part is a forced vibration analysis that calculates the tube response to turbulence-induced excitation and periodic wake shedding, and the fluidelastic instability ratios. A forced vibration analysis involves the application of the distributed flow velocities and densities that are determined using the methods outlined in Chapter 3.

From a mechanics point-of-view, the tubes are simply multi-span beams clamped at the tube-sheet and held at the supports with varying degrees of constraint. To predict tube response, it is convenient and appropriate to assume that the intermediate supports are hinged (or pinned). With this assumption, the analysis is linear and either a finite-element code or an analytical code can be used, as discussed in Chapter 4.

If the tube-to-support clearances are too large, the tube supports will not be effective and the assumption of hinged supports will not be valid. Tube-to-support diametral clearances of 0.38 mm (0.015 in.) or less are typically used in nuclear heat-exchanger design. Effective supports can become ineffective if a heat exchanger is subjected to significant corrosion or a chemical cleaning technique is used to clean a fouled unit. The effects of future chemical cleaning and corrosion should be considered in the vibration response analysis.

The ultimate goal of vibration analyses is to ensure that neither fretting wear nor fatigue damage occurs in heat exchangers. Although a linear analysis does not predict fretting wear, fretting-wear damage can be estimated, as discussed in Section 2.6.1.

Within vibration prediction codes, the tube response $y(x,t)$ at any point x along the tube and at any time t may be expressed as a normal mode expansion in terms of the generalized coordinates $q_i(t)$ as follows:

$$y(x, t) = \sum_{i=1}^{n} \phi_i(x) q_i(t) \tag{2-44}$$

where $\phi_i(x)$ is the mode shape of the ith mode and n is the number of modes to be used in the analysis. Using Lagrange's equation and assuming that the damping is small and that it does not introduce coupling between modes, the equation of motion for the ith mode is as follows:

$$\ddot{q}_i(t) + 2\omega_i \zeta_i \dot{q}_i(t) + \omega_i^2 q_i(t) = \int_0^\ell g(x, t) \phi_i(x) dx \tag{2-45}$$

where ζ_i is the damping ratio, ℓ is the tube length and ω_i is the angular frequency of the ith natural mode. The natural modes are normalized so that

$$\int_0^\ell m\phi_i^2(x) dx = 1 \tag{2-46}$$

where m is the mass per unit length.

Knowing m, ζ_i, the support locations and the flexural rigidity of the tube, the response to different types of forcing functions, $g(x,t)$, is obtained by solving the family of equations described in Eqs. (2-44) and (2-45). Assumptions that can be used within this type of code include uniform mass distribution and flexural rigidity, negligible effect of shear and rotary inertia, no axial motion of any point along the tube, homogeneous boundary conditions and continuity at the intermediate supports.

2.5.1 Fluidelastic Instability

A fluidelastic instability ratio is determined by calculating an effective velocity based on the velocity and density profiles and the important mode shapes. For the ith mode, the critical velocity, U_{pc}, at which instability occurs, may be expressed as:

$$\frac{U_{pc_i}}{fD} = K \left[\frac{2\pi\zeta_i}{D^2 \rho \int_0^\ell u^2(x) r^2(x) \phi^2(x) dx} \right]^{1/2} \tag{2-47}$$

where non-uniform flow velocities and densities are represented as $U_p(x) = U_p u(x)$ and $\rho(x) = \rho r(x)$ This is a generalized form of Eq. (2-25).

2.5.2 Random Turbulence Excitation

Tube response to turbulence-induced excitation forces must be calculated using random vibration theory, and the response must be summed over all significant modes of vibration. The mean square amplitude response, $\overline{y^2(x)}$, of a continuous uniform cylindrical tube to distributed random forces, $g(x,t)$, may be expressed by:

$$\overline{y^2(x)} = \sum_r \sum_s \frac{\phi_r(x)\phi_s(x)}{16\pi^4 f_r^2 f_s^2} \int_0^\infty H_r^*(f)H_s(f) \int_0^\ell \int_0^\ell \phi_r(x)\phi_s(x)R(x,x',f)dxdx'df \qquad (2\text{-}48)$$

where H_r^* and H_s are complex complementary frequency response functions for the rth and s^{th} modes, $R(x,x',f)$ is the spatial correlation density function and x and x' are points along the length of the tube. More detail is found in Pettigrew et al (1978), Axisa et al (1990) and in Chapters 9 and 10.

2.5.3 Periodic Wake Shedding

The response to periodic wake shedding must be calculated for any modes that are found to have Strouhal numbers in the range defined in Section 2.4.3. Assuming that the damping is small, peak vibration amplitudes, $Y(x)$, at resonance for the ith mode may be expressed as:

$$Y(x) = \frac{\phi_i(x)}{8\pi^2 f_i^2 \zeta_i} \int_0^\ell F_L(x')\phi_i(x')dx' \qquad (2\text{-}49)$$

where $F_L(x')$ is a distributed periodic force formulated by:

$$F_L(x) = C_L \rho r(x)DU_p^2 u^2(x)/2 \qquad (2\text{-}50)$$

where C_L is the dynamic lift coefficient defined in Section 2.4.3. Equation (2-50) is a generalized form of Eq. (2-35).

Periodic wake shedding can result from very localized flow over a single span. If the velocity distribution has large step changes in velocity, an effective velocity approach like that used for calculating fluidelastic instability may not be appropriate for periodic wake shedding. Regions with nozzles and inlets may have to be assessed separately.

2.5.4 Acoustic Resonance

Susceptibility to acoustic resonance is not assessed by calculating the tube response. It is estimated by the method described in Sections 2.4.4 and 2.4.5.

2.5.5 Example of Vibration Analysis

An example of a vibration analysis for a typical heat exchanger U-tube is illustrated in Figs. 2-20 to 2-22. This vibration analysis was done with the heat exchanger vibration analysis code PIPO1.

The tube support geometry and the support locations are shown in Fig. 2-20. The input data required for the PIPO1 code is outlined in Fig. 2-21. This figure shows the thermalhydraulic input

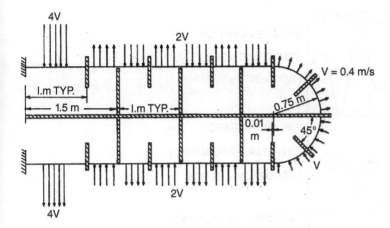

Fig. 2-20 Flow Velocities, Support Locations and Tube Geometry for a Typical Heat Exchanger Tube.

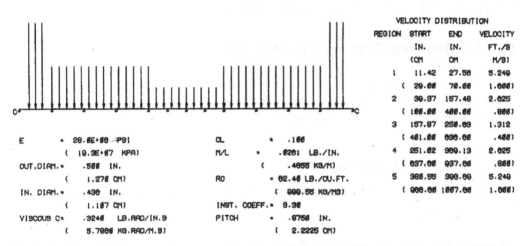

Fig. 2-21 Example of Heat Exchanger Tube Vibration Analysis: Input of Vibration Analysis Code PIPO1.

in the form of pitch flow velocity distribution along the tube. Note the relatively high flow velocity in the inlet region near the tubesheet.

Typical free vibration analysis results are shown in Fig. 2-22 for selected vibration modes. The results include the vibration mode shapes and the natural frequencies.

Figure 2-23 shows the results of a fluidelastic instability analysis for the condenser tube described in Fig. 2-4. For some vibration modes, the ratio of actual to critical flow velocity for fluidelastic instability, U_p/U_{pc}, is greater than one, indicating that fluidelastic instability is possible for this tube, which was subjected to abnormal flow conditions. In reality, fretting-wear and fatigue damage were observed in this condenser.

U-TUBE FREE VIBRATION MODE SHAPES

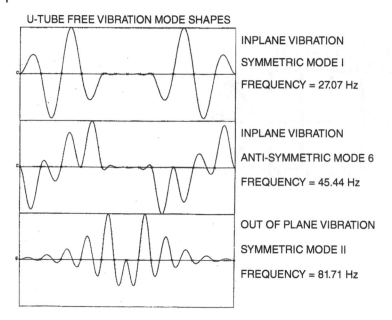

INPLANE VIBRATION

SYMMETRIC MODE I

FREQUENCY = 27.07 Hz

INPLANE VIBRATION

ANTI-SYMMETRIC MODE 6

FREQUENCY = 45.44 Hz

OUT OF PLANE VIBRATION

SYMMETRIC MODE II

FREQUENCY = 81.71 Hz

Fig. 2-22   Heat Exchanger Tube Vibration: Typical Free Vibration Analysis Results.

2.6   Fretting-Wear Damage Considerations

Fretting-wear damage may be assessed using the following methodology, as discussed in detail in Chapter 12. The total fretting-wear damage over the life of the heat exchanger must not exceed the allowable tube wall reduction or wear depth, d_w.

2.6.1   Fretting-Wear Assessment

Fretting-wear damage may be estimated using the following modified Archard equation:

$$\dot{V} = K_{FW}\dot{W}_N \tag{2-51}$$

where $\dot{V}$ is the volume fretting-wear rate, K_{FW} is the fretting-wear coefficient, and $\dot{W}_N$ is the normal work-rate. Work-rate is the available mechanical energy from the dynamic interaction between tube and support and is an appropriate parameter to predict fretting-wear damage. Work-rate may be calculated by performing a non-linear time domain simulation of a multi-span heat exchanger tube vibrating within its supports (Yetisir and Fisher, 1996). Alternately, work-rate may be estimated with the following simplified expression (Yetisir et al, 1998 and Pettigrew et al, 1999):

$$\dot{W}_N = 16\pi^3 f^3 m\ell\overline{y_{\max}^2}\zeta_s \tag{2-52}$$

where m is the total mass of the tube per unit length, and $\left(\overline{y_{\max}^2}\right)$ and f are, respectively, the maximum mean-square vibration amplitude and natural frequency of the tube for the worst mode. The worst mode of a given region is defined as the mode of vibration that has the highest value for the normal work-rate term, $\dot{W}_N$, in Eq. (2-52). The length, ℓ, is that of the span where the vibration amplitude is maximum and ζ_s is the damping ratio attributed to the supports.

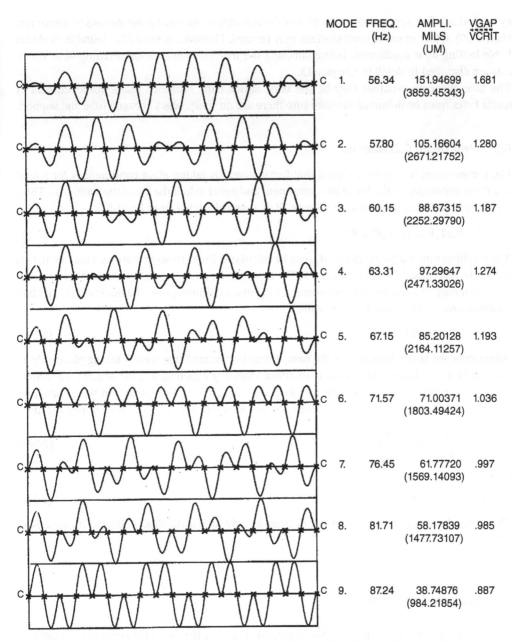

MODE	FREQ. (Hz)	AMPLI. MILS (UM)	VGAP VCRIT
C 1.	56.34	151.94699 (3859.45343)	1.681
C 2.	57.80	105.16604 (2671.21752)	1.280
C 3.	60.15	88.67315 (2252.29790)	1.187
C 4.	63.31	97.29647 (2471.33026)	1.274
C 5.	67.15	85.20128 (2164.11257)	1.193
C 6.	71.57	71.00371 (1803.49424)	1.036
C 7.	76.45	61.77720 (1569.14093)	.997
C 8.	81.71	58.17839 (1477.73107)	.985
C 9.	87.24	38.74876 (984.21854)	.887

Fig. 2-23 Vibration Mode Shapes and Vibration Analysis Results for a Typical Condenser Tube.

2.6.2 Fretting-Wear Coefficients

Tube and tube-support materials should be chosen to minimize fretting-wear damage. Similar materials such as Inconel[®][1]-600 (I600) tubes and I600 supports must be avoided. Incoloy[®]-800 (I800), Inconel[®]-690 (I690) and I600 tubes with 410, 304L, 316L, 321 stainless steel or carbon steel supports are considered acceptable material combinations. A fretting-wear coefficient, K_{FW}, of 20 x 10^{-15} m^2/N may be used as a first approximation for these tube and support material combinations (Guérout and Fisher, 1999). Other material combinations may be used. However, it would be desirable to obtain reliable fretting-wear coefficients before choosing the material combination. Fretting-wear coefficients are discussed in detail in Chapter 13.

The tube supports, whether they be flat bars, lattice bars, broached holes, scalloped bars or circular holes must be deburred to make sure there are no sharp edges between tube and support.

2.6.3 Wear Depth Calculations

To be conservative, it may be assumed that fretting-wear is taking place continuously for a total time, T, corresponding to the life of the component and equal to half the life of the station, T_s. Thus, $T = 0.5T_s$ and the total fretting-wear volume, V, is then calculated from Eq.(2-53):

$$V = 0.5T_s\dot{V} = 0.5T_s K_{FW}\dot{W}_N \tag{2-53}$$

The resulting tube wall wear depth, d_w, can be calculated from the wear volume. This calculation requires the relationship between d_w and V. For example, for a tube within a circular hole or a scalloped bar, it may be assumed that the wear is taking place uniformly over the thickness, L, and half the circumference, D, of the support. Thus:

$$d_w = 2V/(\pi DL) \tag{2-54a}$$

Alternately, for lattice bars and for flat bars, it may be assumed that wear is taking place only on the tubes. Thus, the bars and the wear scars remain flat. For lattice bars in the straight-leg section, wear occurs on both sides of the tube and $M = 4$ in Eq. (2-54b); for flat bars in the U-bend, we assume that wear occurs on one side of the tube and $M = 8$ in Eq. (2-54b). This geometry leads to

$$V = \frac{LD^2}{M}(2\theta - \sin 2\theta) \tag{2-54b}$$

where

$$\theta = arc\ cos[(D - 2d_w)/D] \tag{2-54c}$$

Thus, the tube wall fretting-wear depth, d_w, may be estimated using Eqs. (2-53) and either (2-54a), or (2-54b) and (2-54c).

2.7 Acceptance Criteria

The flow-induced vibration analysis should demonstrate that the heat exchanger is acceptable by satisfying the design acceptance criteria outlined below.

2.7.1 Fluidelastic Instability

The maximum flow velocity in a heat exchanger should be below the critical flow velocity for fluidelastic instability, U_{pc}, based on a fluidelastic instability constant, $K = 3.0$, for liquid and two-phase flows. Thus,

$$U_p/U_{pc} < 1.0 \tag{2-55}$$

2.7.2 Random Turbulence Excitation

The vibration response to random turbulence excitation should be sufficiently low to prevent excessive tube wall reduction due to fretting wear. The tube wall fretting-wear depth, d_w, calculated as per Section 2.6.3 for the entire life of the component should be less than a specified percentage of the nominal design tube wall thickness (e.g., 40%).

2.7.3 Periodic Wake Shedding

Resonance due to coincidence of tube frequency and periodic-wake-shedding frequency should be avoided. If the latter is not possible, the maximum (zero-peak) tube vibration amplitude, Y_{max}, at resonance should be less than two percent of the tube diameter, thus:

$$Y_{max} < 0.02D \qquad (2\text{-}56)$$

Below $0.02D$, the vibration amplitude is generally not sufficient to spatially correlate the formation of vortex shedding with the motion of the tube, thereby resulting in a much-reduced uncorrelated vibration response.

2.7.4 Tube-to-Support Clearance

The tube-to-support clearance must be small enough to provide an effective support. Thus, pinned support conditions may be assumed provided the tube-to-support diametral clearance for drilled holes, broached holes, scallop bars, egg crates and lattice bars is less than or equal to 0.4 mm.

For flat-bar-type U-bend supports, the vibration analysis should satisfy Section 2.7.2 for the out-of-plane direction while assuming that one (any one) support may not be effective in the U-bend region. In the in-plane direction, the diametral clearance between tube and flat bar support should be sufficiently small to provide effective support (e.g., <0.1 mm).

2.7.5 Acoustic Resonance

The analysis should show that acoustic resonance conditions are avoided in the heat exchanger tube bundle. This should be based on at least two generally recognized criteria (e.g., Blevins and Bressler, 1987 and Ziada et al, 1989). The heat exchanger relevant parameter, whether it be shedding frequency, f_s, acoustic frequency, f_{an}, geometry, P/D, L/D or resonance parameter, G_s, or G_i, should be at least 25% away from the resonance criteria or boundary, thus:

$$\text{Heat Exchanger Parameter} < 0.75 \text{ Acoustic Resonance Parameter} \qquad (2\text{-}57)$$

2.7.6 Two-Phase Flow Regimes

Flow conditions leading to intermittent flow regime should be avoided in two-phase cross flow. This applies in particular to the U-bend region of nuclear steam generators.

Note

1 Inconel and Incoloy are registered trademarks of International Nickel Company.

References

Axisa, F., Antunes, J. and Villard, B., 1990, "Random Excitation of Heat Exchanger Tubes by Cross-Flows," *Journal of Fluid and Structures*, 4, No. 3, pp. 321–341.

Blevins, R.D. and Bressler, M.M., 1987, "Acoustic Resonances in Heat Exchangers Part II: Prediction and Suppression of Resonance," *Journal of Pressure Vessel Technology*, 109, pp. 282–288.

de Langre, E. and Villard, B., 1998, "An Upper Bound on Random Buffeting Forces caused by Two-Phase Flows across Tubes," *Journal of Fluids and Structures*, 12, pp. 1005–1023.

Grant, I.D.R., 1975, "Flow and Pressure Drop with Single Phase and Two-Phase Flow in the Shell-Side of Segmentally Baffled Shell-and-Tube Heat Exchangers," *NEL Report No. 590*, National Engineering Laboratory, Glasgow, pp. 1–22.

Grant, I.D.R. and Chisholm, D., 1979, "Two-Phase Flow on the Shell-Side of a Segmentally Baffled Shell-and-Tube Heat Exchanger," *Journal of Heat Transfer*, 101, pp. 38–42.

Guérout, F.M. and Fisher, N.J., 1999, "Steam Generator Fretting-Wear Damage: A Summary of Recent Findings," *ASME Journal of Pressure Vessel Technology*, 121, pp. 304–310.

McAdams, W.H., Woods, W.K., and Herman, L.C., 1942, "Vaporization Inside Horizontal Tubes-II-Benzene-Oil Mixtures," *Trans. ASME*, 64, pp. 193–200.

McNaught, J. M., 1982, "Two-phase Forced Convection Heat Transfer during Condensation on Horizontal Tube Bundles," *7th International Heat Transfer Conference*, 5, Munich, pp. 125–131.

Owen, P.R., 1965, "Buffeting Excitation of Boiler Tube Vibration," *Journal of Mechanical Engineering Science*, 7, pp. 431–439.

Pettigrew, M.J. and Gorman, D.J., 1981, "Vibration of Heat Exchanger Tube Bundles in Liquid and Two-Phase Cross-Flow," *Proceedings of the ASME Pressure Vessel and Piping Conference*, San Francisco, California, 1980 August, Vol. 52, pp. 89–110.

Pettigrew, M.J. and Taylor, C.E., 1991, "Fluidelastic Instability of Heat Exchanger Tube Bundles: Review and Design Recommendations," *ASME Journal of Pressure Vessel Technology*, 113, pp. 242–256.

Pettigrew, M.J. and Taylor, C.E., 1994, "Two-Phase Flow-Induced Vibration: An Overview," *ASME Journal of Pressure Vessel Technology*, 116, pp. 233–253.

Pettigrew, M.J. and Taylor, C.E., 1997, "Damping of Heat Exchanger Tubes in Two-Phase Flow," *Proceedings, 4th International Symposium on FSI, AE & FIV+N, ASME International Congress*, Dallas, Texas, November 16-21, AD-Vol 53.2, pp. 407–418.

Pettigrew, M.J., Platten, J.L. and Sylvestre, Y., 1973, "Experimental Studies on Flow-Induced Vibration to Support Steam Generator Design, Part II: Tube Vibration Induced by Liquid Cross-Flow in the Entrance Region of a Steam Generator," *International Symposium on Vibration Problems in Industry*, Keswick, UK, April.

Pettigrew, M.J., Sylvestre, Y. and Campagna, A.O., 1978, "Vibration Analysis of Heat Exchanger and Steam Generator Designs," *Nuclear Engineering and Design*, 48, pp. 97–115.

Pettigrew, M.J., Goyder, H.G.D., Qiao, Z.L. and Axisa, F., 1986, "Damping of Multispan Heat Exchanger Tubes, Part 1: In Gases," *Symposium on Special Topics of Structural Vibration*, 104, ASME PVP Conference, Chicago, pp. 81–88.

Pettigrew, M.J., Taylor, C.E. and Kim, B.S., 1989a, "Vibration of Tube Bundles in Two-Phase Cross-Flow: Part 1, Hydrodynamic Mass and Damping," *ASME Journal of Pressure Vessel Technology*, 111, pp. 466–477.

Pettigrew, M.J., Tromp, J.H., Taylor, C.E. and Kim, B.S., 1989b, "Vibration of Tube Bundles in Two-Phase Cross-Flow: Part 2, Fluidelastic Instability," *ASME Journal of Pressure Vessel Technology*, 111, pp. 478–487.

Pettigrew, M.J., Carlucci, L.N., Taylor, C.E. and Fisher, N.J., 1991, "Flow-Induced Vibration and Related Technologies in Nuclear Components," *Nuclear Engineering and Design*, 131, pp. 81–100.

Pettigrew, M.J., Taylor, C.E. and Yasuo, A., 1994, "Vibration Damping of Heat Exchanger Tube Bundles in Two-Phase Flow," *Pressure Vessel Research Council Bulletin WRC* 389, pp. 1–41.

Pettigrew, M.J., Taylor, C.E., Jong, J.H. and Currie, I.G., 1995, "Vibration of a Tube Bundle in Two-Phase Freon Cross Flow," *ASME Journal of Pressure Vessel Technology*, 117, pp. 321–329.

Pettigrew, M.J., Yetisir, M., Fisher, N.J., Smith, B.A.W. and Taylor, C.E., 1999, "Prediction of Vibration and Fretting-Wear Damage: An Energy Approach," *Proceedings, ASME-PVP Symposium on Flow-Induced Vibration - 1999*, Boston, USA, August 1-5, PVP-Vol. 389, pp. 283–290.

Pettigrew, M. J., Rogers, R. J. and Axisa, F., 2011, "Damping of Heat Exchanger Tubes in Liquids: Review and Design Guidelines," *Journal of Pressure Vessel Technology*, 133, 014002-1-11.

Pietralik, J.M., 1995, "Thermal-Hydraulic Analysis of Steam Generators: Application of the THIRST Code to Vibration Analysis," Session 3.1, *Proceedings of the 16th Annual Conference of the Canadian Nuclear Society*, Vol. II, Saskatoon, Saskatchewan, June 4-7.

Rogers, R.J., Taylor, C.E. and Pettigrew, M.J., 1984, "Fluid Effects on Multispan Heat Exchanger Tube Vibration," *Proceedings of the ASME Pressure Vessels and Piping Conference*, San Antonio, Texas, June, *ASME Publication H00316: Topics in Fluid Structure Interaction*, pp. 17–26.

Taylor, C.E. and Pettigrew, M.J., 1998, "Guidelines for Random Excitation Forces due to Cross Flow in Steam Generators," *Third International Steam Generator and Heat Exchanger Conference*, Canadian Nuclear Society, Toronto, Canada, June 21-25, pp. 752–763.

Taylor, C.E. and Pettigrew, M.J., 2000, "Random Excitation Forces in Heat Exchanger Tube Bundles," *Journal of Pressure Vessel Technology*, 122, pp. 509–514.

Taylor, C.E., Currie, I.G., Pettigrew, M.J. and Kim, B.S., 1989, "Vibration of Tube Bundles in Two-Phase Cross-Flow: Part 3, Turbulence-Induced Excitation," *ASME Journal of Pressure Vessel Technology*, 111, pp. 488–500.

Taylor, C.E., Boucher, K.M. and Yetisir, M., 1995, "Vibration and Impact Forces Due to Two-Phase Cross Flow in U-Bend Regions of Nuclear Steam Generators," *Proceedings of the 6th International Conference on Flow-Induced Vibration*, London, UK, April 10-12, pp. 401–411.

Taylor, C.E., Pettigrew, M.J., and Currie, I.G., 1996, "Random Excitation Forces Acting on Tube Bundles Subjected to Two-Phase Cross Flow," *Journal of Pressure Vessel Technology*, 118, pp. 265–277.

Weaver, D.S., Fitzpatrick, J.A. and El Kashlan, M., 1987, "Strouhal Numbers for Heat Exchanger Tube Arrays in Cross Flow," *Journal of Pressure Vessel Technology*, 109, pp. 219–223.

Yetisir, M. and Fisher, N.J., 1996, "Fretting-Wear Prediction in Heat Exchanger Tubes: The Effect of Chemical Cleaning and Modelling Ill-Defined Support Conditions," *Proceedings, ASME-PVP Symposium on Flow-Induced Vibration - 1996*, Montreal, Canada, July 21-26, PVP-Vol. 328, pp. 359–368.

Yetisir, M., McKerrow, E. and Pettigrew, M.J., 1998, "Fretting-Wear Damage of Heat Exchanger Tubes: A Proposed Criterion Based on Tube Vibration Response," *Journal of Pressure Vessel Technology*, 120, No. 3, pp. 297–305.

Ziada, S., Oengoren, A. and Buhlmann, E.T., 1989, "On Acoustical Resonances in Tube Arrays Part II: Damping Criteria," *Journal of Fluids and Structures*, 3, pp. 315–324.

3

Flow Considerations

John M. Pietralik, Liberat N. Carlucci, Colette E. Taylor, and Michel J. Pettigrew

3.1 Definition of the Problem

Many flow situations are possible in process systems and nuclear power stations. This fact is illustrated by a few examples.

In the core of nuclear reactors, axial flow generally prevails along the fuel rods. In pressurized-water reactors (PWRs) (Fig. 3-1) and earlier CANDU[1] stations, the flow remains liquid across the fuel. In boiling water reactors (BWRs) (Fig. 3-2) and later CANDU stations (Figs. 3-3a and 3-3b), the coolant is allowed to boil, and the downstream fuel rods and outlet piping are subjected to two-phase flow.

Several flow situations are possible in evaporators, reboilers and nuclear steam generators. For example, in recirculating-type nuclear steam generators (see Fig. 3-4), the tubes are subjected to liquid cross flow in the preheater region and in the recirculated water entrance region near the tubesheet. Within the tube bundle, the shell-side flow is mostly axial. It is liquid at the bottom and gradually becomes two phase as boiling takes place. Two-phase cross flow is predominant in the U-bend region where, depending on the recirculation rate, the steam quality reaches 15-25% or even higher in some designs. This level of quality corresponds to void fractions in excess of 80%.

The high-pressure steam produced by the steam generators flows at high velocity in steam lines to reach the turbine and eventually the condenser, which operates at sub-atmospheric pressures. The condenser is essentially an enormous heat exchanger whose tubes are exposed to high-velocity steam cross flow. There are many other process heat exchangers and reactor internals in nuclear stations which are subjected to cross flow, particularly near inlets and outlets where flow velocities are generally high.

From a flow-induced vibration point of view, process and nuclear components are essentially cylindrical structures (i.e., piping, fuel channels) or bundles of cylinders (i.e., fuel assemblies, heat exchanger tube bundles) subjected to axial or cross flow. The flow may be internal or external to the cylinders. It may be confined to narrow annuli or in close-packed bundles. The flow may be adiabatic or diabatic where boiling or condensing takes place. Single-phase (liquid, vapor or gas) and two-phase (e.g., steam-water) flows are all possible. These cylindrical structures are often multi-span and supported in several places. For example, heat exchanger tubes are supported by baffle plates, fuel rods by bearing pads and support grids, and piping systems by piping supports.

Flow-Induced Vibration Handbook for Nuclear and Process Equipment, First Edition.
Michel J. Pettigrew, Colette E. Taylor, and Nigel J. Fisher.
© 2022 John Wiley & Sons, Inc. This Work is a co-publication between ASME Press and John Wiley & Sons, Inc.

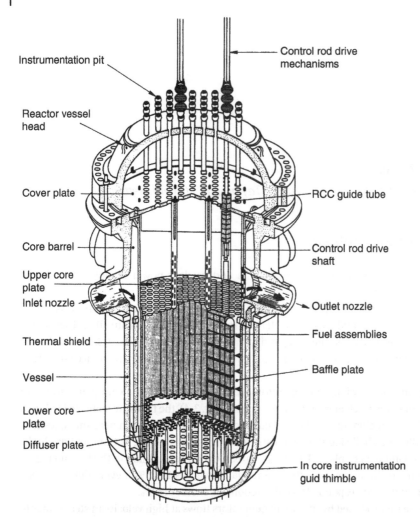

Fig. 3-1 Pressurised Water Reactor Vessel and Internals (Axisa, 1993).

This chapter begins by providing some background on the nature of the flows found in process components (Section 3.2). The following sections provide a simplified approach to flow calculations (Section 3.3) and a multi-dimensional example using thermalhydraulic computer codes (Section 3.4).

3.2 Nature of the Flow

3.2.1 Introduction

Tube failures due to flow-induced vibration have led to an increased awareness of the need to understand the local flow fields in both single- and two-phase shell-side flows in tube bundles. Shell-side flow in two-phase shell-and-tube heat exchangers is generally described by such global parameters as flow rate, pressure drop across the tube bundle, and heat-transfer coefficient.

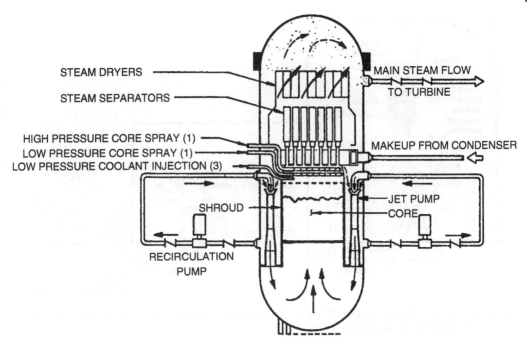

Fig. 3-2 Light-Water Boiling Reactor Pressure Vessel and Internals (Shumway, 1976 / with permission of Elsevier).

(a)

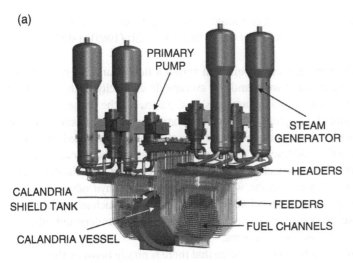

Fig. 3-3a CANDU Heavy Water Reactor (Enhanced CANDU 6 Technical Summary, 2009 / with permission of Candu Energy Inc).

This discussion of the nature of the flow will introduce various flow and geometrical parameters that affect the forces found in single- and two-phase flows (Section 3.2.2). The discussion naturally progresses from unobstructed vertical flow (Section 3.2.3) to flow across a single tube (Section 3.2.4) and then to a tube bundle subjected to vertical cross flow (Section 3.2.5). Sections 3.2.6 and 3.2.7 consider the different properties of air-water and steam-water flows with an emphasis on the effect of nucleate boiling on tube vibration in process equipment. A brief summary (Section 3.2.8) completes Section 3.2.

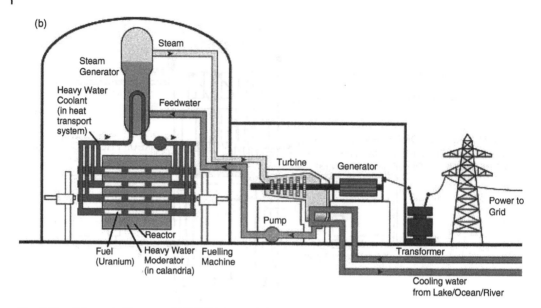

(b)

Fig. 3-3b Schematic Diagram of CANDU Nuclear Power Station (Enhanced CANDU 6 Technical Summary, 2009 / with permission of Candu Energy Inc).

3.2.2 Flow Parameter Definitions

The fundamentals of two-phase flow have been presented in many texts and edited monographs. The texts used frequently by these authors include those written by Wallis (1969) and Collier and Thome (1994).

Two-phase mixtures are rarely homogeneous or uniform across a flow channel. For example, there may be more liquid near the channel walls and the average flow velocity of the gas phase may be somewhat higher than that of the liquid phase causing slip between the phases. However, it is convenient and simple to assume homogeneous two-phase mixture properties. In some cases, it may be desirable to consider slip between phases. For example, Carlucci (1980) used the Smith (1969) slip correlation to calculate the void fraction in his axial-flow experiments to analyze damping and hydrodynamic mass data. This approach may be useful if only one or two vibration parameters are considered. However, a slip correlation which is appropriate for hydrodynamic mass may not be suitable for the formulation of a vibration excitation mechanism such as random excitation. Thus, different two-phase property definitions might be required to understand different dynamic parameters or different vibration mechanisms.

The two-phase definitions within this handbook assume that there is no slip between the gas and liquid phases. This assumption is both convenient and justified, as will be illustrated in Section 3.2.5. The homogeneous void fraction, ε_g, and homogeneous quality, X, are calculated from the known volume flow rates of each phase, gas, $\dot{V}_g$, and liquid, $\dot{V}_\ell$ and the known gas density, ρ_g, and liquid density, ρ_ℓ, as follows:

$$\varepsilon_g = \frac{\dot{V}_g}{\dot{V}_\ell + \dot{V}_g} \quad X = \frac{\rho_g \varepsilon_g}{\rho_g \varepsilon_g + \rho_\ell \left(1 - \varepsilon_g\right)} \tag{3-1}$$

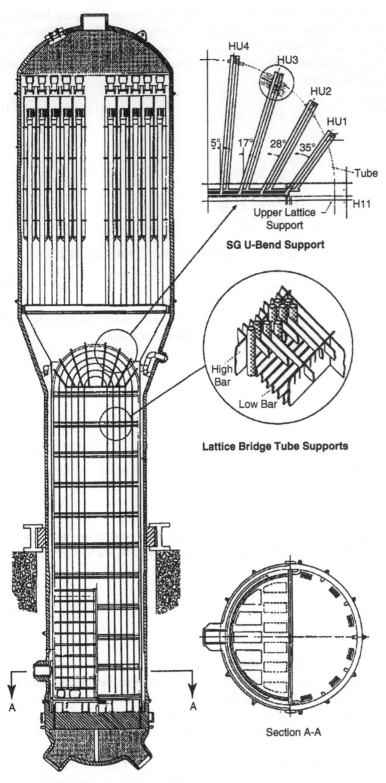

SG U-Bend Support

Lattice Bridge Tube Supports

Section A-A

Fig. 3-4 Schematic Diagram of Recirculating Steam Generator (Sauvé et al, 1999).

An average velocity and homogeneous density can be described by allowing the local velocity and density values to be equal to the average value multiplied by a local spatial distribution function. If the local spatial distributions are assumed to be unity and a homogeneous void fraction is used, the homogeneous density, ρ, free stream velocity, U, free stream mass flux, $\dot{m}$, and superficial velocities, U_g, U_ℓ are defined as follows:

$$\rho = \rho_\ell \left(1 - \varepsilon_g\right) + \rho_g \varepsilon_g \tag{3-2}$$

$$U = \frac{\rho_\ell \dot{V}_\ell + \rho_g \dot{V}_g}{\rho A} \tag{3-3}$$

$$\dot{m} = \rho U = \frac{\rho \dot{V}}{A} \tag{3-4}$$

$$U_g = \frac{\varepsilon_g}{\rho} = \frac{X \dot{m}}{\rho_g}, \quad U_\ell = \frac{\left(1 - \varepsilon_g\right) \dot{m}}{\rho} = \frac{(1 - X) \dot{m}}{\rho_\ell} \tag{3-5}$$

where A is the free-stream cross-sectional area. The pitch velocity, U_p, is defined as follows:

$$U_p = U \frac{P}{(P - D)} \tag{3-6}$$

where P is the pitch or distance between tube centers and D is the tube diameter. Similarly, the pitch mass flux, $\dot{m}_p$, and the tube-bundle cross-flow area, A_p, are defined as follows:

$$\dot{m}_p = \dot{m} \frac{P}{(P - D)} = \rho U_p \tag{3-7}$$

$$A_p = A \frac{(P - D)}{P} \tag{3-8}$$

The Reynolds number is often used in flow calculation simulations within process heat exchangers and steam generators. The cross-flow Reynolds number, Re, is defined as

$$Re = \frac{\rho U_p D}{\mu} = \frac{U_p D}{\nu} \tag{3-9}$$

where μ is the dynamic viscosity and ν is the kinematic viscosity. Many definitions of two-phase viscosity exist (see Ghajar and Bhagwat, 2013). Throughout this handbook, we use the definition from McAdams et al (1942) as follows:

$$\mu = \left(\frac{X}{\mu_g} + \frac{1 - X}{\mu_\ell}\right)^{-1} \tag{3-10}$$

Example 3-1 Flow in a Process Heat Exchanger

In many cases it is possible to do a 'back of the envelope" calculation to get an idea of flow velocities, densities, void fractions, etc. (Fluit and Pettigrew, 2001). It will not be accurate but it could be sufficient to identify components that are comfortably acceptable or definitely not acceptable, or in-between and requiring a more thorough analysis. This calculation and Example 3-2 are both examples of the simple approach.

The principal dimensions of this process heat exchanger are shown in Fig. 3-5. The baffles are 0.5 m apart, the fluid on both tube and shell side is water of density, 1000 kg/m^3, and the shell-side

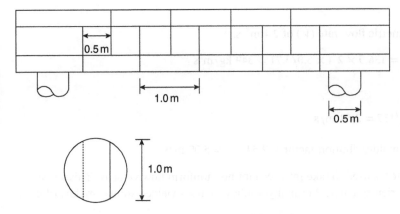

Fig. 3-5 Example 3-1 - Process Heat Exchanger Schematic.

volume flow rate, $\dot{V}$, is 0.25 m³/s. The tubes are in a triangular array with 20 mm outside diameter, 18 mm inside diameter, on a 30 mm pitch for a P/D ratio of 1.5. The minimum flow area, A, is slightly less than

$$A \sim 0.5\text{m} \times 1.0\text{m} = 0.5\,\text{m}^2$$

The flow velocity, U, is

$$U = \dot{V}/A = 0.25/0.5 = 0.5\,\text{m/s}$$

The pitch velocity,

$$U_p = U \times P/(P-D) = 0.5 \times 0.30/(0.30-0.020) = 1.5\,\text{m/s}$$

Example 3-2   Flow in a Nuclear Steam Generator

Figure 3-6 is a sketch of the U-bend region of a typical nuclear steam generator with tube bundle outside diameter, D_{BO}, of 3.0 m. In the U-bend region, the tubes are supported by a number of anti-vibration bars. The tube diameter is 0.02 m and the tube pitch-to-diameter ratio is 1.5. The gas (steam) density is 32 kg/m³ and the liquid (water) density is 760 kg/m³. All the flow in the steam generator tube bundle must exit through the hemispherical area formed by the largest U-tubes. Thus, the flow cross-sectional area, A, is

$$A = \pi/2(D_{BO})^2 = \pi/2(3.0)^2 = 14.14\,\text{m}^2$$

$$A_p = A\left(\frac{P-D}{P}\right) = 4.71\,\text{m}^2$$

Steam Quality: X
   For a circulation ratio (CR = total mass flow/exit steam mass flow) of 5,

$$X = 1/CR = 1/5 = 20\%$$

$$\rho_{TP} = \frac{1}{\dfrac{X}{\rho_g} + \dfrac{(1-X)}{\rho_\ell}} = \frac{1}{\dfrac{0.20}{32} + \dfrac{(1-0.20)}{760}} = 136.9\,\text{kg/m}^3$$

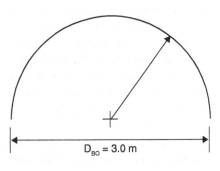

Fig. 3-6 Example 3-2 - U-Bend Schematic.

Pitch Mass Flux: $\dot{m}_p$

For a feedwater volumetric flow rate ($\dot{V}$) of 2.4 m³/s,

$$\dot{m}_p = \rho\dot{V}\cdot CR/A_p = 136.9 \times 2.4 \times 5.0/4.71 = 349 \text{ kg/m}^2\text{s}$$

Flow Velocity: U_p

$$U_p = \frac{\dot{m}_p}{\rho_{TP}} = 349/137 = 2.54 \text{ m/s}$$

$$(U_p)_{max} = U_p \times \text{maldistribution factor} = 2.54 \times 2 = 5.09 \text{ m/s}$$

A maldistribution factor is used to take into account non-uniform flow velocity. This value was originally based on experience. Later, thermalhydraulics analyses yielded similar maldistribution factors.

Void Fraction: ε_g

$$\varepsilon_g = \frac{X/\rho_g}{X/\rho_g + (1-X)/\rho_\ell} = \frac{0.2/32}{0.2/32 + (1-0.2)/760} = 0.856 = 85.6\%$$

The parameters calculated in these examples will be referenced in future chapters as these simple examples are used to illustrate the vibration analysis process. Single-phase and two-phase frequency, damping, vortex shedding, fluidelastic instability, random excitation, work-rate and wear calculations will be provided in the respective chapters.

3.2.3 Vertical Bubbly Flow

Single-phase vertical flow through a tube can be characterized using parameters such as average velocity, velocity profile, Reynolds number and fluid density. The velocity profiles can be parabolic (laminar flow) or relatively flat-topped (turbulent flow). The turbulent nature of the flow can be defined in terms of a turbulence length scale (Hinze, 1975), and the characteristic shape of the spectral density of the turbulent energy for flow in a tube can be assumed to be similar to the shape seen in Fig. 3-7 (Lienhard, 1966).

In two-phase flow, there are many more parameters and profiles to study. Papers by Serizawa et al (1975) and Wang et al (1987) present studies of the turbulence structure and phase distribution in vertical bubbly two-phase flows. They include void-fraction profiles and bubble- and liquid-velocity profiles of both mean- and random-velocity components. Turbulence levels were measured from the fluctuating velocity components.

Serizawa et al (1975) and Wang et al (1987) present very similar results for the effect of void fraction and flow rate on the production of turbulent energy. They found that the turbulence level increased rapidly to a certain value when a small amount of vapor was introduced to a single-phase flow. Thereafter, additional amounts of vapor had less effect. Finally, for higher mass fluxes, the presence of the vapor actually decreased the turbulence levels. The authors suggest that the bubbles of vapor began to act as an energy sink to reduce fluctuations. Energy absorption was especially apparent close to the vertical tube wall, where peaks in the void fraction and turbulent energy production profiles coincided. These observations suggest that the relationship between void fraction and two-phase cross-flow excitation spectra will not be monotonic.

Not only are there many parameters that must be considered in two-phase flow, but these parameters must be defined for each flow pattern or flow regime. Studies of turbulence structure and

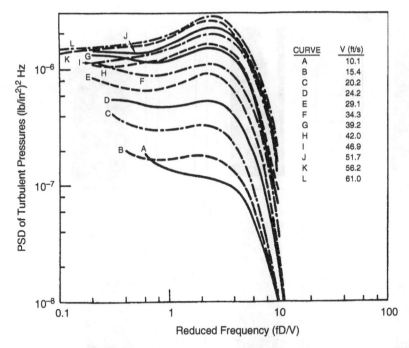

CURVE	V (ft/s)
A	10.1
B	15.4
C	20.2
D	24.2
E	29.1
F	34.3
G	39.2
H	42.0
I	46.9
J	51.7
K	56.2
L	61.0

Fig. 3-7 An Example of Power Spectral Density Curves of Turbulent Pressures as Measured on the Surface of a 1-in. Rod in Parallel Flow in a 2-in. Diameter Tube (Wambsganss and Chen, 1971 / with permission of Argonne National Laboratory).

phase distribution have only dealt with a bubbly flow regime. Although some attempt has been made to define the flow regimes and their respective transition regions quantitatively, recognition and naming of the various regimes is still somewhat subjective. The flow patterns shown in Fig. 3-8 illustrate the most commonly recognized patterns in air-water and steam-water vertical flow. It is important to note that these patterns found in air-water flow are also present in two-phase Freon flow and steam-water flow.

The prediction of flow regimes in two-phase vertical flows has been attempted by many researchers. Most of the experimental work has been done in small diameter tubes containing low-pressure air-water flow and high-pressure steam-water flow. One example of a flow map that was developed from such data is given in Fig. 3-9. Although this map was originally presented by Hewitt and Roberts (1969), Fig. 3-9 is a version with both imperial and SI units found in Collier and Thome (1994). The abscissa and ordinate axes of the flow map define the liquid- and gas-phase superficial momentum fluxes, respectively.

3.2.4 Flow Around Bluff Bodies

The introduction of a bluff body into vertical cross flow can significantly alter the pressure field. In single-phase flow, an example of these changing pressure fields can be seen as vortices being shed from bluff bodies, such as cylinders. Figure 3-10 illustrates the changes in the wake flow of a cylinder with increasing Reynolds number. The shedding frequency, f_S, is defined by the Strouhal number, $f_S D/U$. The Strouhal number is equal to 0.2 for cylinders in a Reynolds number range between 10^2 and 10^5.

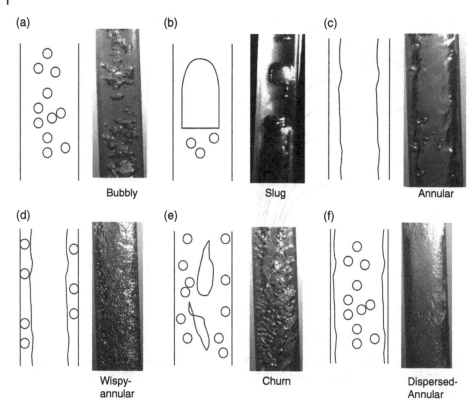

(a)   (b)   (c)

Bubbly   Slug   Annular

(d)   (e)   (f)

Wispy-   Churn   Dispersed-
annular   Annular

Fig. 3-8   Flow Patterns in Vertical Channels.

In two-phase flow across a single bluff body, vortices (periodic forces) have only been observed at very low void fractions (< 15%). In-depth studies by Hulin et al (1982) and Inoue et al (1986) provide descriptions of the nature of the two-phase vortex shedding behind a single cylinder in two-phase cross flow. Hulin et al (1982) state that a reasonably stable vortex street exists in bubbly flows with up to 10% void fraction. At larger void fractions, the amplitude of the Strouhal peak in the force spectrum decreases so that, by 25% void fraction, not even a broad peak in the spectrum can be detected.

3.2.5 Shell-Side Flow in Tube Bundles

The nature of single-phase flow across tube bundles has been an area of some disagreement. Weaver and Fitzpatrick (1987) point out that some periodic excitation does exist in the highly turbulent flow found in tube bundles. The existence of a vortex-type excitation in tube bundles was first verified in flow visualization work reported by Weaver and Abd-Rabbo (1984). Strouhal numbers for tube bundles subjected to fluid cross flow have been summarized by Rae and Wharmby (1987).

The flow regimes introduced in Section 3.2.3 are altered by the presence of a tube bundle. The first flow patterns for shell-side air-water flow across a tube bundle were published in Grant (1975), and Grant and Chisholm (1979). Figure 3-11 shows these flow patterns for vertical shell-side flow as

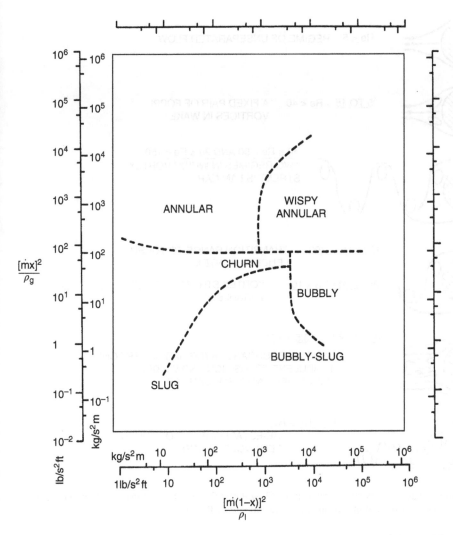

Fig. 3-9 Flow Pattern Map for Vertical Flow in Tubes with 1 to 3 cm Diameters (Collier and Thome, 1994, from Hewitt and Roberts, 1969 / with permission of Oxford Publishing Limited).

they were presented in a paper by Collier (1979). Forty years later, these flow patterns are still being referenced in flow-pattern review papers such as Cheng et al (2008).

Prediction of the flow pattern for a given flow condition is most easily done through the use of a flow pattern map. A flow map by Grant (1975) (also found in Grant and Chisholm (1979)) was the first cross-flow map. In Fig. 3-12, Grant's map has been positioned on axes used by McNaught (1982). Physical property corrections were applied to generalize the air-water results for application to other fluids. The abscissa is the Martinelli parameter, X, which, in this case, is defined assuming turbulent homogeneous flow and assuming that the total mass flux, $\dot{m}$, is equal to $(\dot{m}_\ell + \dot{m}_g)$. The Martinelli parameter is calculated using the following equation:

$$X = \left(\frac{1-\varepsilon_g}{\varepsilon_g}\right)^{0.9} \left(\frac{\rho_\ell}{\rho_g}\right)^{0.4} \left(\frac{\mu_\ell}{\mu_g}\right)^{0.1} \tag{3-11}$$

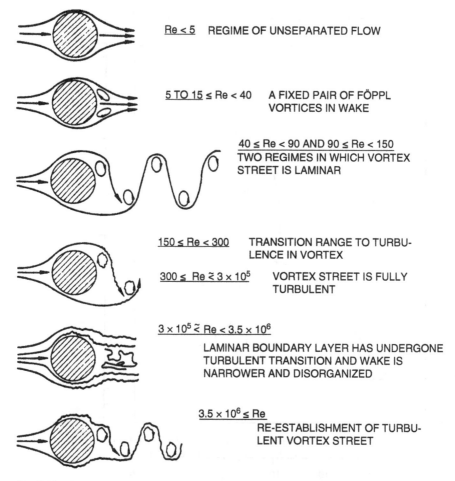

Fig. 3-10 Boundary-Layer and Wake Changes in the Flow Past a Cylinder as the Reynolds Number Increases (Lienhard, 1966 / with permission of Washington State University Press).

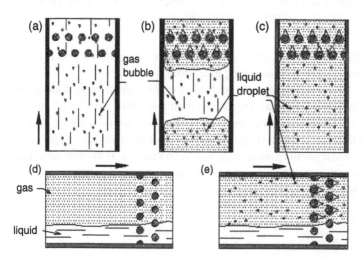

Fig. 3-11 Two-Phase Flow Patterns in Vertically Upwards and Horizontal Cross Flow: a) Bubbly, b) Intermittent, c) Annular, d) Stratified, and e) Stratified-Spray (Grant and Chisholm, 1979 from Collier, 1979).

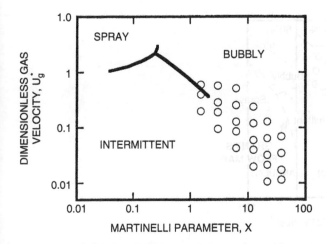

Fig. 3-12 Flow Pattern Map for Two-Phase Flow Across Cylinder Arrays Using Flow Pattern Boundaries (Grant and Chisholm (1979) and McNaught (1982)). (Note that circular symbols indicate air-water flow conditions from Taylor et al, 1996).

where μ_ℓ and μ_g are the dynamic viscosities of liquid and gas phases, respectively. The dimensionless gas velocity, U_g^*, is defined as follows:

$$U_g^* = \frac{\dot{m}_{pg}}{\left[D_e g \rho_g \left(\rho_\ell - \rho_g\right)\right]^{0.5}} \tag{3-12}$$

where $\dot{m}_{pg}$ is the pitch mass flux of the gas phase, $D_e \cong 2(P - D)$ is the equivalent hydraulic diameter and g is acceleration due to gravity.

The flow map provided by Grant does not cover the full range of flow conditions used in industry. Most of the indicated flow conditions in Fig. 3-12, corresponding to air-water cross-flow data from Taylor et al (1996), fall outside of the flow map boundaries provided by Grant. Other flow pattern maps can be examined to assist in understanding the possible flow patterns within a tube bundle. In Pettigrew and Taylor (1994), the Grant flow map was superimposed on a flow map taken from McQuillan and Whalley (1985), see Fig. 3-13. McQuillan and Whalley developed a flow map for two-phase flow inside a vertical tube that is based on flow-pattern transition equations. The position of each transition line is defined by an expression involving flow properties and geometric parameters. The McQuillan and Whalley map provides significantly more flow pattern information in the range of flow conditions studied in the Taylor et al (1996) tests.

Although the Grant map was developed based on observations of flow across a tube bundle and should, therefore, be more relevant, the McQuillan and Whalley map provides further insight into the possible flow regimes occurring in the gaps between tubes. A comparison of the two maps shows that Grant's spray flow corresponds to lower-mass-flux annular flow in a vertical tube. In a tube bundle, spray flow rather than annular flow may occur at these lower flow rates. Grant's bubbly regime is probably more accurately referred to as a region of continuous flow that changes in the McQuillan and Whalley map from true bubbly flow to churn flow and then to annular flow as the superficial gas velocity increases. The intermittent flow regime of Grant's map corresponds to the high-void-fraction churn region of the McQuillan and Whalley map.

Almost 20 years after Grant's map was first published, Ulbrich and Mewes (1994) performed a comprehensive analysis of available flow regime data resulting in shell-side cross-flow flow regime

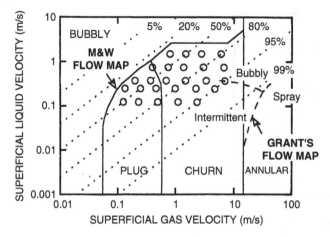

Fig. 3-13 Flow Regime Map for Vertical Two-Phase Flow Using McQuillan and Whalley (1985) Type Map with Flow Condition Data from Taylor et al (1996).

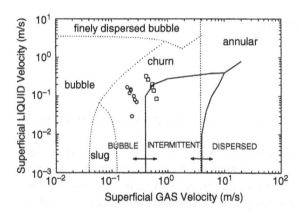

Fig. 3-14 Ulbrich and Mewes Flow Regime Map (Solid Line) for Vertically Upward Two-Phase Flow (Modified from Feenstra et al (1996)). Data Symbols: Square - Fluidelastic Threshold Conditions and Circle - Conditions for Damping Measurements.

boundaries that cover a larger range of flow rates. They found that their new transition lines had an 86% agreement with available data. Their flow map, with sample data is shown in Fig. 3-14 (Feenstra et al, 1996). The flow regime boundary transitions are solid lines and the flow regimes are identified with upper-case text. The dotted lines outline a previous flow regime map based on Freon 11 flow in a vertical tube (Taitel et al, 1980). Once again, we see that this two-phase cross-flow map is similar to a map for flow in a vertical tube.

Almost every study of flow regimes in tube bundles has concluded that three distinct flow regimes exist. In fact, several studies have shown that these regimes can easily be identified by measuring the probability density function (PDF) of the gas component of the flow (Ulbrich et al, 1997; Noghrehkar et al, 1995; and Lian et al, 1997). These researchers observed that the bubbly regime has a single low-frequency peak, annular-dispersed flow has a single high-frequency peak and intermittent flow has either a broadband PDF (Ulbrich et al, 1997) or it shows both peaks (Lian et al, 1997).

More recently, flow regime maps for vertical cross flow in tube bundles have been developed by Mao and Hibiki (2017) and Kanizawa and Ribatski (2016). Both teams have developed flow maps with flow regime transitions that are defined using expressions that include fluid property and

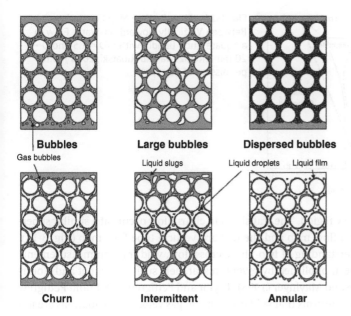

Fig. 3-15 Flow Patterns in Vertical Two-Phase Cross Flow from Kanizawa and Ribatski (2016) / with permission of Elsevier.

geometric considerations. Flow in tube bundles is beginning to reach a level of understanding similar to that for flow in vertical tubes.

Kanizawa and Ribatski (2016) provide an updated diagram of flow patterns in two-phase cross flow based on their observations (see Fig. 3-15). These flow patterns were defined in their paper as follows:

- Bubbles: continuum liquid phase with gas bubbles smaller than tube spacing
- Large Bubbles: characterized by amorphous bubbles with dimensions larger than the tube spacing
- Dispersed bubbles: characterized by a high number of gas bubbles smaller than observed for bubble flow pattern
- Churn: characterized by chaotic movement of the phases, with small portions of liquid moving up and down
- Intermittent: characterized by intermittent passage of liquid slugs and large gas portions, both at high velocity
- Annular: characterized by continuum liquid film over tube and shell walls, with gas flow in the core of the tube bundle

The Kanizawa and Ribatski (2016) flow regime map (Fig. 3-16) was developed using differential pressure transducers with a k-means clustering method. Viewing through a window was also done but they found that the pressures transducers gave clarity to the flow regime transitions. The transition lines in this flow regime map were created based on the fluid properties and tube bundle geometry used by Ulbrich and Mewes (1994). The agreement is good with the exception that Kanizawa and Ribatski (2016) have identified both churn and intermittent flow within the Ulbrich and Mewes (1994) intermittent regime. It should also be noted that these two-phase cross-flow flow

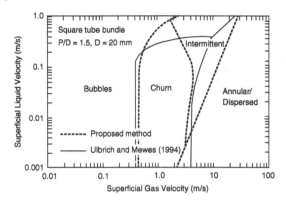

Fig. 3-16 Flow Regime Map Comparison Between Vertically Upward Air-Water Flow in a Square Tube Bundle with P/D = 1.5 and D = 20 mm. (Kanizawa and Ribatski, 2016 / with permission of Elsevier).

regime maps do not have a plug flow region. The transition to churn flow occurs at a similar superficial gas velocity in both this cross-flow map and the upward vertical tube flow map (Fig. 3-13).

In an effort to improve the understanding of two-phase cross flow, void fraction measurements taken within tube bundles have been published by several researchers such as Haquet and Gouirand (1995), Lian et al (1997), Mann and Mayinger (1995), Taylor and Pettigrew (2001) and Pettigrew et al (2004). This collection of work confirms that void fraction and gas velocity distributions are not uniform within a subchannel. Haquet and Gouirand (1995) and Mann and Mayinger (1995) found that the highest void fractions occurred behind tubes. They also found high gas velocities but lower void fractions in the center of the channel. On the contrary, measurements by Lian et al (1997), Taylor and Pettigrew (2001) and Pettigrew et al (2004) found that void fraction is much higher in between tubes than at locations behind tubes. This discrepancy may be explained through an understanding of the flow regimes that these researchers encountered. It is possible that measurements of lower void fraction behind tubes were taken within the bubbly and intermittent flow regimes, while the flow rates quoted in Mann and Mayinger (1995) suggest that these tests were largely performed in the annular-dispersed regime.

While measuring flow within a bundle, these researchers also measured the slip between the liquid and gaseous phases in their experiments. All researchers found that some slip occurred and two groups indicated that a drift-flux type model seemed to reasonably predict the slip (Taylor and Pettigrew, 2001; Haquet and Gouirand, 1995). However, none of the researchers showed that the use of slip models was useful in collapsing vibration data such as two-phase damping or hydrodynamic mass. Pettigrew et al (2004) concluded that two-phase flow within the stream is fairly homogeneous and the slip between phases appears to be relatively small. At this time, two-phase vibration parameters continue to be presented using the assumption of homogeneous flow.

An understanding of the local void fraction around a typical tube may be very important in determining the appropriate model for vibration parameters such as damping and hydrodyamic mass. Obviously, further research in this area is required.

Since the tubes within a bundle act as turbulence generators, the position of the tube within the bundle (upstream or interior) and the bundle geometry will affect the turbulence-induced excitation. Bundle-configuration effects such as pitch-to-diameter ratio and bundle orientation are discussed in Chapters 9 and 10, for single-phase and two-phase cross flow, respectively.

The existence of periodic wake shedding (or vortex shedding) in two-phase flow has been debated for many years. Vortex shedding was not expected to occur in a tube bundle at void fractions higher than 10%, since significant vortices were not shed from a single cylinder at higher void fractions (Hara, 1982). In contrast, Haquet and Gouirand (1995) and Pettigrew et al (2004) have noted the existence of vortex shedding or at least quasi-periodic forces within a tube bundle in two-phase cross flow. Periodic wake-shedding forces in single-phase and two-phase flow are discussed in Chapter 11.

3.2.6 Air-Water versus Steam-Water Flows

Air-water is the most common two-phase mixture used in the experimental analysis of two-phase flows, due to economic and safety factors. The high costs of running tests in steam-water require the experimenter to seek other options. Some two-phase testing has been done using other fluids, such as Freon, that more closely match one or another of the steam-water properties. In all cases, one is left with the task of determining the proper methods for scaling the two-phase flow-induced vibration data. Table 3-1 provides two-phase properties for air-water, steam-water and Freon.

The effects of nucleate boiling on pressure fluctuations in a steam-water mixture have been considered in the scaling of air-water data. Again, due to the expense of building a rig and running steam-water tests, limited research has been done to consider either subcooled or saturated boiling effects. What has been done is covered in some detail in the next section.

Even less effort has gone into the study of the effect of changing the two-phase mixture fluid properties. Dallavalle et al (1973) suggested that the two most important mixture properties with respect to flow-induced vibration are surface tension, σ, and vapor-liquid density ratio, (ρ_g/ρ_l). They concluded that a decrease in the magnitude of one or both of these properties would lead to a decrease in the excitation force levels.

3.2.7 Effect of Nucleate Boiling Noise

Before any experiments were performed, it was thought that nucleate boiling at the surface of steam generator tubes or along nuclear fuel rods would be a significant vibration excitation mechanism. Several laboratories performed experimental programs to test that hypothesis.

Vibration of an Electrically Heated Cylinder in Axial Steam-Water Two-Phase Flow. An exploratory vibration experiment on a heated cylinder in axial flow (see Fig 3-17) was conducted in a steam-water test loop. Vibration measurements were taken under various flow conditions while

Table 3-1 Physical Properties of Freon-22, Freon-134a, Air-Water, and Steam-Water.

Property	Phase	Unit	Freon-22		Freon-134a		Air-Water	Steam-Water
Temperature		°C	23	30	26	39	20	257
Pressure		MPa	1	1.2	0.69	1.0	0.1	4.5
Density	Gas	kg/m^3	42.2	50.7	33.3	48.7	1.2	22.6
	Liquid	kg/m^3	1200	1174.0	1202	1150	998.5	789.0
	Gas/Liquid	ratio	0.035	0.043	0.028	0.042	0. 0012	0.029
Dynamic	Gas	Pa·s	1.3×10^{-5}	1.4×10^{-5}	1.27×10^{-5}	1.34×10^{-5}	1.8×10^{-5}	1.8×10^{-5}
Viscosity	Liquid	Pa·s	2.0×10^{-4}	1.95×10^{-4}	2.06×10^{-4}	1.75×10^{-4}	1.0×10^{-3}	1.0×10^{-4}
	Gas Liquid	ratio	0.065	0.072	0.062	0.076	0.018	0.174
Kinematic	Gas	m^2/s	3.1×10^{-7}	2.8×10^{-7}	3.8×10^{-7}	2.7×10^{-7}	1.5×10^{-5}	7.9×10^{-7}
Viscosity	Liquid	m^2/s	1.7×10^{-7}	1.6×10^{-7}	1.7×10^{-7}	1.5×10^{-7}	1.0×10^{-6}	1.3×10^{-7}
	Gas/Liquid	ratio	1.848	1.740	2.224	1.808	15.00	6.10
Surface Tension	Liquid	N/m	0.0082	0.0072	0.0079	0.0063	0.0727	0.0246

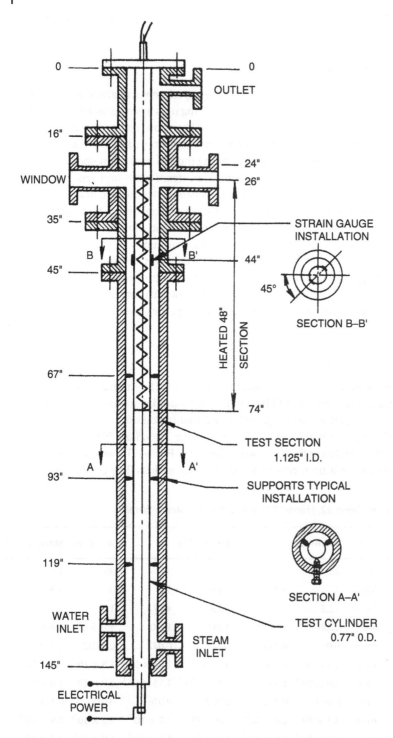

Fig. 3-17 Sketch of Axial-Flow Heated-Cylinder Test Section (Pettigrew and Gorman, 1973 / with permission of Atomic Energy of Canada Limited).

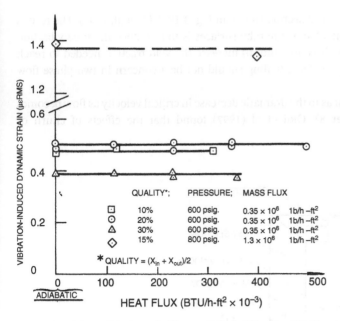

Fig. 3-18 Effect of Nucleate Boiling on Cylinder Vibration (Pettigrew and Gorman, 1973 / with permission of Atomic Energy of Canada Limited).

steam was either added at the inlet of the test section or generated by the heated cylinder (Pettigrew and Gorman, 1973). The contribution of subcooled and bulk nucleate boiling noise to vibration amplitude was investigated. The effect of other flow parameters such as pressure, mass flux and steam quality was also studied. The results showed that the vibration response is highly dependent on flow regime. As shown in Fig. 3-18, nucleate boiling noise did not cause significant vibration in this experiment. The effects of nucleate boiling noise are negligible for void fractions above 10%. Below 10%, two-phase effects are known to be very small, and nucleate boiling effects were noticeable but not large.

In reactor, fuel vibration tests were carried out in the National Research Universal reactor at the Chalk River Laboratories, Chalk River, Ontario, Canada (Pettigrew, 1993). These tests also found that nucleate boiling effects were not significant.

The limited effect of subcooled nucleate boiling in axial flow was also noted by El-Hawary (1975). El-Hawary quotes from authors who stated that the largest effects from subcooled boiling would be at frequencies around 10 Hz. In keeping with their hypothesis, he used a tube with a natural frequency of 7 Hz so that the bubbles would cause the tube to resonate. However, the subcooled boiling effects that he found were not substantial. There is no doubt that subcooled nucleate boiling produces pressure fluctuations, but tube response to these fluctuations has, thus far, proved to be negligible in comparison to the other two-phase mechanisms at void fractions above 10%.

All of the nucleate boiling tests described above compare nucleate boiling forces to adiabatic excitation forces in two-phase axial flow. Adiabatic two-phase excitation forces are several times higher in cross flow than in axial flow. Therefore, nucleate boiling forces in cross flow will be even smaller relative to cross-flow adiabatic forces.

Two-Phase Cross-Flow-Induced Vibration of Tube Bundles with Tube Surface Boiling. In a study of two-phase cross-flow excitation of a tube array comprising heated tubes in Freon 11, vapor was generated both upstream of the test array and at the tube surfaces themselves. The experiments were conducted in the small Freon 11 test section shown in Fig. 3-19 (Gidi et al, 1997). The results showed that the effect of vapor generated at the tube surfaces is to a) reduce the tube vibration amplitude, mainly in the transverse direction and, b) increase the void fraction needed to reach fluidelastic instability. In summary, nucleate boiling should not be a concern in two-phase flow components.

This work may also provide insight as to the dramatic decrease in critical velocity as flow becomes churn or intermittent (see Chapter 8). Gidi et al (1997) found that the effects of churn or

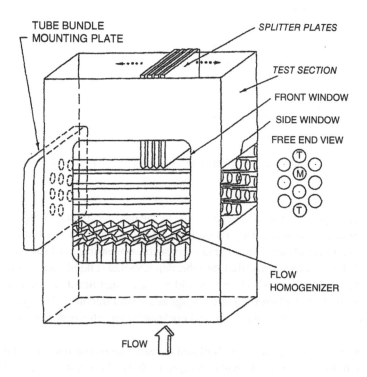

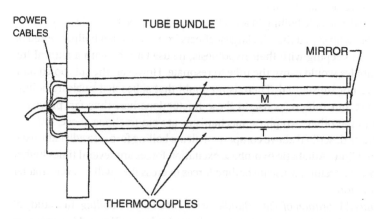

Fig. 3-19 Cross-Flow Tube Bundle Test Section with Heated Tubes (M and T) in Freon-11 Flow Loop (Gidi et al, 1997).

intermittent flow could be suppressed through tube surface boiling. The delaying of critical velocity by nucleate boiling was confirmed in low-pressure steam water tests reported by Mitra et al (2009).

3.2.8 Summary

In this section, approaches to defining and mapping flows in tubes and tube bundles have been examined. Flow parameter definitions have been introduced. A brief description of the development of two-phase flow regime flow maps has been given so the reader appreciates that flow regime maps were being developed at the same time as two-phase vibration experimental programs were underway. As a result, researchers simply used the best flow regime information available at the time the experimental programs were conducted.

3.3 Simplified Flow Calculation

Obviously, the first step in a flow-induced vibration analysis is to evaluate operating conditions and flow velocities everywhere in the component. In cases where flow paths are well defined, flow velocities are simply calculated from the flow areas. This is the case between segmental baffle supports in a heat exchanger. The flow may simply be assumed uniform between baffles. For a conservative approach, leakages across baffle-supports and other internal seals may be neglected, particularly if the leakage paths are likely to be plugged by sludge deposition.

Often the flow is divided between several parallel flow paths. For example, in a heat exchanger, the flow may go through the tube bundle, between the bundle and shell, and sometimes through a central tube-free lane (Fig. 3-20). In such cases, flow calculations may be based on equal pressure

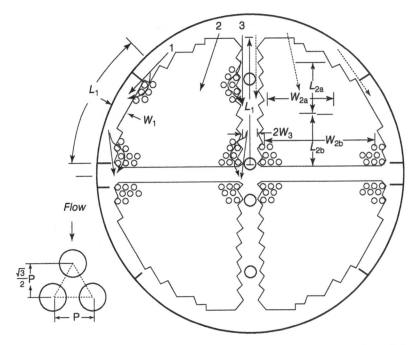

Fig. 3-20 Example 3-3 - Cross-Section of Heat Exchanger Showing Flow Paths 1) Between Tube Bundle and Shell; 2) Through Tube Bundle; 3) in Tube-Free Lane.

drop between regions common to all flow paths. In some cases, particularly when flow impedances are difficult to evaluate, we also calculate an average flow velocity based on the total flow areas of the paths. The highest velocity estimate is then used in any subsequent vibration analysis.

Inlets and outlets are often critical regions. For example, in heat exchangers, the flow may be jetting directly towards the tube bundle or may be deflected by an impingement plate around which secondary jetting occurs. Either way, we assume that little expansion of the jets takes place before hitting the tubes. To be conservative, we use the jet velocity in our vibration analysis.

The approach above shows that simple rules and common sense can lead to conservative flow calculations for many process components.

Example 3-3   Flow in a Process Heat Exchanger Using Flow Paths

This example uses a simple spreadsheet-based method for estimating flow velocities in the shell-side of a hypothetical, four-pass heat exchanger with single segmental baffles. For this example, details of geometry and flow conditions are given in Table 3-2 and Fig. 3-20 shows a simplified cross-section of the heat exchanger.

Equations and Assumptions The total cross flow is assumed to be single-phase and equally split between the top two passes; hence, we will consider only the flow across the top-left pass.

The flow across this pass is assumed to comprise three main paths: the first between the outer edge of the tubes and the shell, the second across the interior of the tube bundle, and the third along the tube-free lane. The total mass flow rate is, therefore, given by

Table 3-2   Example 3-3 - Shell-Side Process Conditions and Geometry.

Fluid	Water
Temperature, T (°C)	50
Density (kg/m³), ρ	988
Viscosity, μ (kg/m/s)	5.440E-04
Total mass flow rate in cross-flow, $G_{tot} = 2G$ (kg/s)	80
Shell inside diameter, D_s (m)	0.935
Number of passes	4
Number of tubes per pass	222
Tube arrangement	equilateral triangular
Tube diameter, D (m)	0.019
Tube pitch, P (m)	0.025
Baffle type	single segmental
Baffle spacing, B_s (m)	0.468
Tube-to-shell spacing, W_1 (m)	0.04
1/2 Width of tube-free lane, W_3 (m)	0.02
Number of sealing strips	8
Sealing strip-to-tube gap, G_{ss} (m)	0.006
Number of sealing rods	2
Sealing rod-to-tube gap, G_{sr} (m)	0.003

$$G = \sum_i G_i \qquad (3\text{-}13)$$

Path 2 is further subdivided into two serial segments to account for the non-uniform flow area of the tube bundle. The tube-free regions at the beginning and end of each path are assumed to act as plenums with the pressure drop across each path, comprising one or more segments, being equal and given by

$$\Delta p = \frac{\rho}{2}\left(\sum_j K_j U_j^2\right)_i = \frac{G_i^2}{2\rho}\left(\sum_j \frac{K_j}{A_j^2}\right)_i \qquad (3\text{-}14)$$

where K is the hydraulic loss factor.

Equation (3-14) can be used to express the mass flow rate of each path in terms of the pressure drop:

$$G_i = \sqrt{\frac{2\rho\Delta p}{\left(\sum_j \frac{K_j}{A_j^2}\right)_i}} \qquad (3\text{-}15)$$

Substituting for G_i in Eq. (3-14), the pressure drop can be expressed in terms of the total mass flow rate and the individual path flow areas and corresponding loss factors:

$$\Delta p = \frac{G^2}{2\rho}\left[\sum_i \left(\sum_j \frac{K_j}{A_j^2}\right)_i^{-0.5}\right]^{-2} \qquad (3\text{-}16)$$

The total loss factor $K_{i,j}$ for segment j of path i, is given by the following:

$$K = \begin{cases} \left(\sum_n k_n + \dfrac{f_a L}{W}\right)_{i,j} & \text{for Paths 1 and 3 : bundle bypass flow} \\[3mm] \left(\sum_n k_n + N_r f_c\right)_{i,j} & \text{for Path 2 : bundle cross flow} \end{cases} \qquad (3\text{-}17)$$

where f_a and f_c are the friction factors for axial and cross flow, L is the path length, and W is the path width.

The form loss factor, $k_{n,j}$ accounts for pressure losses resulting from the contraction and expansion of the flow as it enters and leaves the tube bundle for Path 2, and as it flows past the sealing strips in Path 1 and the sealing rods in Path 3.

In the above, flows along tube-free regions are idealized as channel-type, axial flows. Hence for these, f_a is the friction factor for axial flow in smooth pipes or channels, estimated from the Blasius relationship for $Re_a > 10^5$ (Heat Exchanger Design Handbook, 1984):

$$f_a = 0.184\, Re_a^{-0.2} \qquad (3\text{-}18)$$

where the axial flow Reynolds number is based on the path segment velocity and channel width (note that to calculate the Reynolds number, the channel width for Path 3 $= 2W_3$, see Fig. 3-20).

For cross flow in the bundle, the friction factor is given by (Heat Exchanger Design Handbook, 1984):

$$f_c = 5.2\left(\frac{1}{P/D-1}\right) Re_c^{-0.29} \qquad (3\text{-}19)$$

where the cross-flow Reynolds number is based on the tube gap velocity and tube diameter. Equation (3-19) is valid for equilateral triangular bundles over the Reynolds number range $7 \times 10^3 < Re_c < 2 \times 10^5$. For equilateral triangular arrangements, N_r, the number of tube rows crossed is estimated from:

$$N_r \cong Int \left[\frac{2L}{\sqrt{3}P} \right] \tag{3-20}$$

The components of the form loss factor for Path 2 are estimated from (Idelchik, 1994):

$$k_{con} = 0.5 \left(1 - \frac{A_{ds}}{A_{us}} \right)^{0.75} \cong 0.5 \left(1 - \frac{P - D}{P} \right)^{0.75} \tag{3-21}$$

for the flow entering the tube bundle at the top, and

$$k_{exp} = \left(1 - \frac{A_{us}}{A_{ds}} \right) \cong \left(1 - \frac{P - D}{P} \right) \tag{3-22}$$

as it leaves the tube bundle at the bottom.

To develop expressions for the sealing strip and sealing rod loss factors, we consider that there are secondary leakage paths between tubes near the sealing device. This "porous" effect is approximated by two parallel secondary flow paths, one through the gap between the sealing strip or rod and the closest tube, and the other between neighboring individual tubes (Fig. 3-20). The local pressure drop across each secondary path is the same and can be expressed in terms of the loss factor, mass flow rate and flow area for each secondary path, resulting in

$$\Delta p_{loc} = \frac{k_{dev,g}}{2\rho} \left(\frac{G_{dev,g}}{A_{dev,g}} \right)^2 \tag{3-23}$$

for the gap between sealing strip or sealing rod and nearest tube, and

$$\Delta p_{loc} = \frac{k_{dev,nt}}{2\rho} \left(\frac{G_{dev,nt}}{A_{dev,nt}} \right)^2 \tag{3-24}$$

for the path between the neighboring tubes.

The form loss factor for the flow in the gap between the device and the nearest tube is calculated as the sum of contraction and expansion losses as the flow enters and leaves the gap (Idelchik, 1994):

$$k_{dev,g} = C \left[0.5 \left(1 - \frac{W_{dev,g}}{W_i} \right)^{0.75} + \left(1 - \frac{W_{dev,g}}{W_i} \right) \right] \tag{3-25}$$

where C equals 1.0 for sealing strips and 0.5 for sealing rods to reflect lower losses because of the roundness of the rod.

The form loss factor for the flow between neighboring tubes is defined assuming tube bundle entrance (contraction) and exit (expansion) losses and frictional losses resulting from the crossing of N_g tube-to-tube gaps. Frictional losses are estimated from the tube bundle cross-flow friction factor based on the Reynolds number of the segment of Path 2 nearest the sealing device. Thus, we have

$$k_{dev,nt} = 0.5 \left(1 - \frac{P - D}{P} \right)^{0.75} + N_g f_{2j} + \left(1 - \frac{P - D}{P} \right) \tag{3-26}$$

The local pressure drop can be represented in terms of the mass flow and corresponding area of that path and an effective form loss factor for the device:

$$\Delta p_{loc} = \frac{k_{dev,eff}}{2\rho} \left(\frac{G_i}{A_{i,j}}\right)^2 \tag{3-27}$$

where

$$G_i = G_{dev,g} + G_{dev,nt} \tag{3-28}$$

Thus, from Eqs. (3-23), (3-24), (3-27) and (3-28), the effective form loss factor of device for Path Segment j of Path i is given by

$$k_{dev,eff} = \frac{A_{i,j}^2 k_{dev,g} k_{dev,nt}}{\left(A_{dev,nt} k_{dev,g}^{0.5} + A_{dev,g} k_{dev,nt}^{0.5}\right)^2} \tag{3-29}$$

The corresponding velocity in the gap between the device and nearest tube is given by

$$U_{dev,g} = \left(\frac{k_{dev,eff}}{k_{dev,g}}\right)^{0.5} \frac{G_i}{\rho A_{i,j}} \tag{3-30}$$

while the velocity in the gap of neighboring tubes is calculated from

$$U_{dev,nt} = \left(\frac{k_{dev,eff}}{k_{dev,nt}}\right)^{0.5} \frac{G_i}{\rho A_{i,j}} \tag{3-31}$$

Depending on the level of detail required, the above approach can be extended to include any number of parallel paths with each path subdivided into any number of segments.

Calculation Procedure

Step 1: Examine the heat exchanger cross-section to identify the number of flow paths along with their corresponding lengths and widths.

Step 2: Use segment path lengths in the bundle and the longitudinal pitch to calculate the corresponding number of tube rows crossed. Use the path widths and distance between baffles to calculate the path areas. Path lengths, flow areas and number of tube rows crossed are listed in Table 3-3, along with formulas used.

Step 3: Calculate all bundle entrance and exit, and device form loss factors, making sure to use the factor of 0.5 for the sealing rods (listed in Table 3-4).

Step 4: Calculate the Reynolds number for each path. Iteration is required because friction factors are functions of path velocities. For the first iteration, all path velocities are equal and estimated from the total mass flow rate across the bundle and the flow area near the shell centerline. For subsequent iterations, path velocities are those from the previous iteration.

Step 5: Calculate the total loss factor for each path.

Step 6: Calculate the path pressure drop.

Step 7: Calculate the mass flow rate for each path and the corresponding velocities.

Step 8: Calculate representative gap velocities between sealing devices and the nearest tube as well as between neighboring tubes.

Step 9: Return to Step 4, until an acceptable level of convergence is reached. For this example, all parameters converge to within 1% in three iterations.

Table 3-3 Example 3-3 – Path Lengths and Flow Areas.

Length of Path 1, L_1 (~ 1/6 of circumference) (m)	0.490
Width of Path 1, W_1 (m)	0.040
Flow area of Path 1, $A_1 = W_1 B_s$ (m^2)	0.019
Length of Path 2a, L_{2a} (m)	0.150
Width of Path 2a, W_{2a} (m)	0.222
Flow area of Path 2a, $A_{2a} = ((P\text{-}D)/P)W_{2a}B_s$ (m^2)	0.025
Length of Path 2b, L_{2b} (m)	0.150
Width of Path 2b, W_{2b} (m)	0.330
Flow area of Path 2b, $A_{2b} = ((P\text{-}D)/P)W_{2b}B_s$ (m^2)	0.037
Number of rows crossed along Path 2a, $N_{r,2a} = Int[L_{2a}/((3^{0.5}/2)P)]$ (-)	6
Number of rows crossed along Path 2b, $N_{r,2b} = Int[L_{2b}/((3^{0.5}/2)P)]$ (-)	6
Length of Path 3, L_3 (m)	0.408
Width of Path 3, W_3 (m)	0.020
Flow area of Path 3, $A_3 = W_3 B_s$ (m^2)	0.009
Tube gaps crossed by flow near sealing strip and sealing rod, N_g	3

Table 3-4 Example 3-3 - Tube Bundle and Device Form Loss Factors.

Tube bundle (reference gap velocity):	
Tube bundle entrance, $k_{2b,en} = 0.5(1 - (P - D)/P)^{0.75}$	0.41
Tube bundle exit, $k_{2b,ex} = 1\text{-}(P\text{-}D)/P$	0.76
Sealing strip and rod (reference gap velocity):	
Top sealing strip (ss), $k_{t,ss} = 0.5(1 - G_{ss}/W_1)^{0.75} + (1 - G_{ss}/W_1)$	1.29
Bottom sealing strip, $k_{b,ss} = 0.5(1\text{-}G_{ss}/W_1)^{0.75} + 1$	1.44
Top sealing rod (sr), $k_{t,sr} = [0.5(1 - G_{sr}/W_3)^{0.75} + (1 - G_{sr}/W_3)]/2$	0.65
Bottom sealing rod, $k_{b,sr} = [0.5(1 - G_{sr}/W_3)^{0.75} + 1]/2$	0.72

Results

Table 3-5 lists relevant calculated parameters for the base case (Case 1) as well as three additional cases conducted to examine geometry and modeling changes. These include path Reynolds numbers, loss factors, overall pressure drop, and flow rates, along with bundle gap velocities and the highest gap velocities between sealing devices and the nearest tube and between neighboring tubes.

As expected, bundle gap velocities in the upper part of the bundle are higher than those at the lower part because of the reduced flow area. The gap velocities between the upper sealing rod and nearest tube are about 40% higher than those between the upper sealing strip and nearest tube because of the lower hydraulic resistance of the latter. As well, their magnitudes are up to 2.5 times higher than in the tube bundle. Gap velocities between tubes near the top sealing strip and rod are somewhat lower than those in the upper part of the tube bundle.

Two other cases were calculated to examine the effect of a 25% reduction of the sealing-rod-to-tube gap width, and the addition of a third sealing rod, half-way between the existing two. The reduction of the gap (Case 2) has a negligible effect on the maximum velocity. It results in a higher

Table 3-5 Example 3-3 – Results.

Case #	Porous			Non-porous
	1	2	3	4
n_{sr}	2	2	3	2
G_{sr} (m)	0.004	0.003	0.004	0.004
Re_1	95,082	97,048	99,229	108,405
Re_{2a}	38,855	39,763	40,772	45,283
Re_{2b}	26,139	26,750	27,429	30,463
Re_3	137,512	137,949	115,889	155,613
$k_{1t,ss,nt}$	2.14	2.14	2.13	-
$k_{1t,ss,eff}$	18.34	18.31	18.29	-
$k_{1b,ss,nt}$	2.26	2.26	2.25	-
$k_{1b,ss,eff}$	19.97	19.94	19.91	-
K_1	38.53	38.48	38.42	121.79
K_{2a}	2.36	2.34	2.33	2.27
K_{2b}	2.95	2.93	2.92	2.85
$k_{3t,sr,nt}$	2.26	2.26	2.25	-
$k_{3t,sr,eff}$	4.86	6.76	6.75	-
$k_{3b,sr,nt}$	2.26	2.26	2.25	-
$k_{3b,sr,eff}$	5.29	7.12	7.11	-
K_3	10.32	14.05	20.78	33.25
Δp (Pa)	2,299	2,395	2,503	3,014
G_1 (kg/s)	6.42	6.56	6.71	4.14
G_2 (kg/s)	27.38	28.02	28.73	31.91
G_3 (kg/s)	6.20	5.43	4.56	3.96
$U_{1t,ss,g}$ (m/s)	1.31	1.34	1.37	1.49
$U_{1t,ss,nt}$ (m/s)	1.02	1.04	1.06	-
U_{2a} (m/s)	1.12	1.15	1.18	1.31
U_{2b} (m/s)	0.76	0.77	0.79	0.88
$U_{3t,sr,g}$ (m/s)	1.89	1.90	1.60	2.14
$U_{3t,sr,nt}$ (m/s)	0.98	1.02	0.86	-

Notes:
Case 1: Base Case.
Case 2: 25% reduction in sealing-rod-to-tube gap width.
Case 3: Addition of a third sealing rod, half-way between existing two rods.
Case 4: All flow in Paths 1 and 3 assumed to flow in sealing-strip-to-tube and sealing-rod-to-tube gaps.

rod loss factor for the sealing rod, which on its own would lead to a higher local pressure drop and, hence, less flow. However, this change is almost entirely offset by the lower flow area which results in a higher velocity for a given flow rate. The addition of a third sealing rod (Case 3) has a significant effect in reducing the rod-to-tube gap velocity by about 20% from the reference case.

A "non-porous" case (Case 4) was calculated where all flows in Paths 1 and 3 were assumed to pass between the sealing-strip-to-tube and sealing-rod-to-tube gaps, respectively. As expected, this design case resulted in higher velocities and is, therefore, considered conservative when compared to the more realistic base case, which accounts for leakage of some of the flow in Paths 1 and 3 into the tube bundle near the sealing devices.

3.4 Multi-Dimensional Thermalhydraulic Analysis

In situations where components are geometrically complex, analyses that are more detailed are required to calculate thermalhydraulic parameters such as pressure, velocity and temperature under single-phase conditions, and additionally, vapor quality, void fraction, flow regime, and mixture density under two-phase conditions. For this purpose, a number of specialized computer codes have been developed (largely by the nuclear power industry) for multi-dimensional fluid flow and heat transfer modeling of heat transfer equipment such as steam generators, power plant condensers, and shell-and-tube heat exchangers. These codes are based on porous-media concepts whereby the effects of tube bundles, baffles and other obstacles are modeled by distributed hydraulic resistances and the effects of the tube bundle, such as flow volume reduction, by an isotropic porosity factor (Patankar and Spalding, 1974).

3.4.1 Steam Generator

This section describes an example of a steam generator thermalhydraulic code. Other codes are similar in terms of equations solved, boundary conditions, treatment of the computational domain, solution techniques, results, and capabilities. One of the leading steam generator thermalhydraulic computer codes is THIRST[2]. It was developed for the three-dimensional thermalhydraulic analysis of vertical recirculating steam generators (RSGs).

Steam generators in pressurized water reactors transfer heat from primary to secondary coolants. Their design is similar to that of shell-and-tube heat exchangers with vertical U-tubes and buoyancy-driven two-phase flow on the secondary side. A typical design is shown schematically in Fig. 3-4. The regions of interest from a thermalhydraulic point of view include the tube bundle, separators, annular downcomer, flow entrance region with windows, and tube-free regions close to the tube bundle. The primary hot coolant flows inside the tubes, and the secondary boiling water circulates in the shell region and downcomer. The section of the RSG above the separator deck separates the two-phase mixture and returns the secondary water through the downcomer to the riser.

THIRST is a finite-volume three-dimensional computer code that provides detailed descriptions of all necessary thermalhydraulic parameters in the shell side of RSGs. It solves the energy balance for the primary-side fluid, and the primary heat flux is the source for the secondary-side energy balance. The computational domain stretches from the tubesheet to the primary separators and includes both the boiling section and the downcomer.

THIRST assumes that the flow is steady and incompressible, the shell and shroud walls are adiabatic, the downcomer flow is two-dimensional (i.e., calculated flow velocity, density, etc. are constant in the radial direction), and laminar and turbulent shear stress forces are replaced by distributed hydraulic resistances. The tube supports and preheater baffles are treated as hydraulic resistances in the axial direction only, and there is no carry-over (water droplets in the outlet steam). Carry-under (steam bubbles in the downcomer flow) is allowed, at typically less than

0.25% of steam mass flow rate, and is caused by the primary separators, whose efficiency is high but less than 100%.

The computational domain is treated as a continuum, which means that the tube bundle can be modelled using the porous-media concept. Using this concept, the tube bundle is idealized as a continuum with an equivalent fluid flow area and fluid volume, and hydraulic resistance. The local fluid flow area and fluid volume are based on the corresponding geometric values but are decreased by a local isotropic porosity factor, defined as the ratio of flow volume to geometric volume. Hydraulic resistances of the structural elements are expressed by distributed hydraulic friction factor relationships for the tube bundle and hydraulic loss factor relationships for tube support plates and other internal structures.

In analyses, it is typically assumed that the flow is homogeneous and the tube bundle resistance is anisotropic. Homogeneous flow is commonly used in vibration analyses, although slip velocity models give a more realistic approximation for the void-fraction-steam-quality relationship. Because vibration analyses, correlations, and recommendations are typically based on the homogeneous model, the thermalhydraulic results given in this chapter use this model. Anisotropic tube bundle resistance is used because the resistance in the direction perpendicular to the tube is about 20 to 50 times larger than that in the direction parallel to the tube. This phenomenon is especially important in the U-bend region, where tubes constantly change direction, thus affecting the hydraulic resistance.

THIRST solves the conservation equations of continuity, momentum, and energy in three-dimensional cylindrical coordinates in the shell side of the RSG. The equations are written for the time- and volume-averaged mean conditions:

$$\frac{\partial}{\partial z}(\beta\rho u) + \frac{1}{r}\frac{\partial}{\partial r}(\beta\rho rv) + \frac{1}{r}\frac{\partial}{\partial\theta}(\beta\rho w) = 0 \tag{3-32}$$

$$\frac{\partial}{\partial z}(\beta\rho uu) + \frac{1}{r}\frac{\partial}{\partial r}(\beta\rho ruv) + \frac{1}{r}\frac{\partial}{\partial\theta}(\beta\rho uw) = \beta\cdot\left(-\frac{\partial p}{\partial z} - \rho g - R_z\right) \tag{3-33}$$

$$\frac{\partial}{\partial z}(\beta\rho uv) + \frac{1}{r}\frac{\partial}{\partial r}(\beta\rho rvv) + \frac{1}{r}\frac{\partial}{\partial\theta}(\beta\rho vw) = \beta\cdot\left(-\frac{\partial p}{\partial r} + \frac{\rho\cdot w^2}{r} - R_r\right) \tag{3-34}$$

$$\frac{\partial}{\partial z}(\beta\rho uw) + \frac{1}{r}\frac{\partial}{\partial r}(\beta\rho rvw) + \frac{1}{r}\frac{\partial}{\partial\theta}(\beta\rho ww) = \beta\cdot\left(-\frac{\partial p}{\partial\theta} - \frac{\rho\cdot v\cdot w}{r} - R_\theta\right) \tag{3-35}$$

$$\frac{\partial}{\partial z}(\beta\rho uh) + \frac{1}{r}\frac{\partial}{\partial r}(\beta\rho rvh) + \frac{1}{r}\frac{\partial}{\partial\theta}(\beta\rho wh) = \beta\cdot S_h \tag{3-36}$$

The symbols are explained in the Nomenclature.

In the momentum equations, the three terms on the left-hand side describe convection, while the terms on the right-hand side describe the pressure, gravity, centrifugal and Coriolis forces, and distributed resistance terms. The distributed resistance term, R, describes all important fluid-solid interaction resistances (i.e., that of the tube bundle, entrance windows, downcomer orifice, tube support plates, U-bend supports, separators, preheater baffles, partition plate, and thermal plate). The energy equation includes the convection terms and the heat source term resulting from the primary-side heat transfer.

The required boundary conditions are the primary inlet enthalpy and mass flow rate, the feedwater enthalpy, and initial estimates of the feedwater mass flow rate, reheater drains enthalpy and flow rate, primary inlet and secondary-side pressures, blowdown mass flow rate, downcomer water level, and carry-under ratio.

The closure relations include formulations for the volume porosity for each control volume in the region under consideration, solid-fluid resistances, primary-to-secondary heat transfer, and fluid properties. For the equilateral triangular tube layout, the volume porosity is described by

$$\beta_t = 1 - \frac{\pi}{2\sqrt{3}} \cdot \left(\frac{D}{P}\right)^2 \tag{3-37}$$

The porosity is set to β_t in control volumes completely filled with tubes, to unity in control volumes without tubes and to a weighted value for control volumes partially filled with tubes.

The local hydraulic resistance term is calculated from

$$R_i = \Phi^2 \cdot \left(\frac{\Delta p}{\Delta l}\right)_i = \Phi^2 \frac{\cdot k_i}{\Delta l_i} \cdot |U \cdot \rho|_i \cdot U_i \tag{3-38}$$

where the two-phase flow pressure drop multiplier, Φ, is calculated from empirical correlations in general, and can be derived analytically for the homogeneous model. The hydraulic resistance definition is based on a locally one-dimensional flow and is used for all hydraulic resistances except for the tube bundle resistance. For the tube bundle, two hydraulic-resistance approaches are possible: (1) a more sophisticated, fully three-dimensional anisotropic approach, and (2) a simpler, one-dimensional anisotropic approach. In both approaches, the interaction term is written in vectorial form as:

$$\vec{R} = \overline{\overline{K}} \vec{U} \tag{3-39}$$

where, $\overline{\overline{K}}$ is a 3×3 matrix describing the hydraulic resistance in three directions including cross-coupling terms. In the fully anisotropic approach, the interaction force in a given direction is calculated as a function of all velocity components. In the locally anisotropic approach, the interaction force in a given direction is a function of one velocity component, the component in that direction. In both approaches, the interaction force vector is not parallel to the velocity vector at a given point. As expected, the resulting velocity distributions are different for the two approaches.

The local secondary-side heat transfer source is calculated from

$$S_h = \alpha \cdot A_v \cdot (T_p - T_s) \tag{3-40}$$

where the overall heat transfer coefficient, α, includes the primary and secondary heat-transfer coefficients, the tube fouling, and the tube wall heat transfer resistance. This heat source in the secondary fluid is equal to the heat sink in the primary fluid:

$$S_h = -(\rho \cdot u)_p \cdot \frac{\Delta h_p}{\Delta l} \cdot \frac{A_v \cdot D_{in}^2}{4 \cdot D} \tag{3-41}$$

The above differential equations are transformed into algebraic equations using the finite-volume concept. The computational domain is divided into numerous control volumes, in the order of hundreds of thousands, and the algebraic equations with boundary conditions are solved iteratively for each control volume. The iteration process ends when flow and heat transfer differences between two successive iterations are smaller than the input criteria.

The results of an analysis of a hypothetical steam generator are shown in Figs. 3-21 to 3-23. Figure 3-21 illustrates the distributions of the vector velocity, steam quality, steam-water mixture

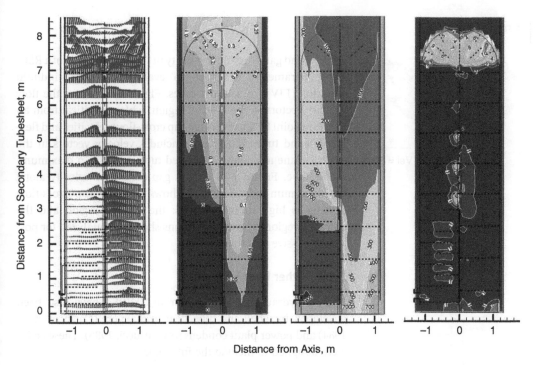

Fig. 3-21 Predicted Velocity Vectors (Left), Steam Quality (Second from the Left), Mixture Density (Third from the Left), and Gap Cross-Flow Velocity (Right) in the Central Plane of a Recirculating Steam Generator Using the THIRST Code.

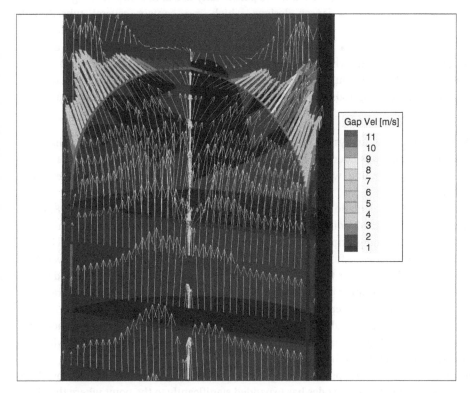

Fig. 3-22 Vector Fluid Velocity for a Maximum-Radius Tube Shown by Thick-Line Vectors. Thin-Line Vectors Show Fluid Velocity at the Central Plane. They are Colored by Gap Cross-Flow Velocity with the Color Scale Included. Also Shown are U-Bend Supports and Tube Support Plates Inside the Shroud.

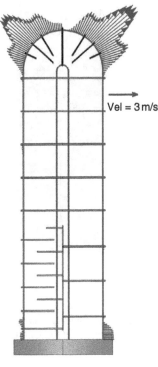

Vel = 3 m/s

Fig. 3-23 Gap Cross-Flow Velocity for the Maximum-Radius Tube.

density, and gap cross-flow velocity in the central plane of a RSG. These parameters are used in evaluation of flow-induced vibration (FIV) for selected tubes. Figure 3-22 shows the flow velocity vectors in the U-bend region, which is critical from an FIV viewpoint because of large gap cross-flow velocities and flexible U-bend tubes. The figure includes velocity vectors in the central plane and along a selected tube, which is a maximum-radius tube. Figure 3-23 depicts gap cross-flow velocity along the maximum-radius tube; it shows that the gap cross-flow velocity is high in two regions: the U-bend region and the entrance region. These two regions should be analyzed for possible excessive vibrations.

3.4.2 Other Heat Exchangers

Thermalhydraulics codes similar to THIRST have also been developed for shell-and-tube heat exchangers (Carlucci et al, 1984) and power plant condensers (Carlucci, 1986). These codes were designed to include the following:

- single-phase and homogeneous two-phase flow with provisions for boiling and/or condensation heat transfer,
- tube bundle layouts of arbitrary arrangement and boundaries (this feature is particularly useful in characterizing some condenser designs which can contain several tube bundle arrangements, each having a highly irregular boundary),
- inlet nozzles with and without impingement plates, and
- sealing strips/rods located anywhere in the modeled region. Sealing strips are often used in designs having "pull-through" bundles where significant tube-free gaps exist between the tube bundle and the shell. Sealing rods are generally used in the central tube-free lane of a U-tube bundle.

Figure 3-24 shows a flow analysis from the power plant condenser thermalhydraulics code (Carlucci, 1986). The code was later extended to include a coupled multi-axial-slice capability to account for changes in cooling water temperature on the tube-side along the tube length (Frisina et al, 1989).

Figure 3-25a shows flow calculation results for a cooling heat exchanger (Frisina et al, 1987). This heat exchanger was limited to 67% of its shell-side design flow because of vibration due to high flow velocities near the upper sealing strip. An improvement to 82% of design flow was achieved by relocating the sealing strips that were further away from the centerline. However, the rating was then limited by high flow velocities near the sealing rods. The addition of a sealing rod allowed 98% of design flow rating without vibration problems (Fig. 3-25b). In the above calculations, all parameters were kept at design conditions except for the shell-side flow.

Other examples of specialized multi-dimensional codes based on the porous media concept are described by Ramón and Prieto Gonzálaz (2001) for power plant condensers, and Prithiviraj and Andrews (1998) for shell-and-tube heat exchangers.

Since the 1980s, several general purpose computational fluid dynamics (CFD) codes have been developed. The application of these codes has expanded significantly to the point where they have

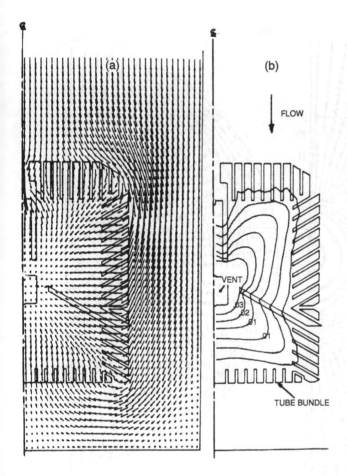

Fig. 3-24 Predicted (a) Velocity and (b) Air Concentration Distributions in a Power Plant Condenser Configuration (Carlucci, 1986 / with permission of Begell House, Inc.).

become accepted analysis and design tools in nearly all industries. In the power, process and petrochemical industries, CFD tools are routinely used for detailed flow and heat transfer analysis of a wide range of components. These components include nuclear reactors and heat transfer equipment such as shell-and-tube heat exchangers, steam generators and power plant condensers.

The ability of CFD codes to model heat transfer equipment with flexible, unstructured grids at scales ranging from a porous-media treatment to detailed flow around individual tubes and other devices makes them ideally suited to the following:

- evaluating different design concepts by optimizing the placement of devices such as baffles, including type, orientation and spacing;
- modeling impingement plates and sealing strips or rods to maximize performance, minimize fouling, and avoid flow-induced vibration damage; and
- evaluating operating units and identifying changes in operating conditions or, where feasible, design to avoid flow-induced vibration damage and improve thermal performance.

Figure 3-26 shows CFD flow and heat transfer predictions (in the form of path lines and color-coded temperature distributions) in a shell-and-tube heat exchanger (Mohammadi et al, 2009).

(a)   (b)

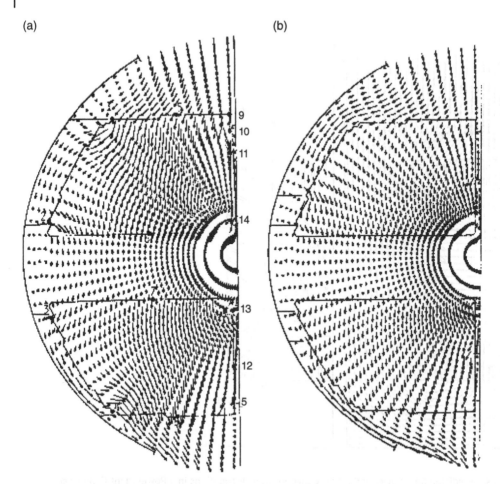

Fig. 3-25 Process Heat Exchanger Predictions of Velocity Distribution for (a) Original Design Configuration and (b) with Sealing Strips Moved Closer Together and Sealing Rods on Heat Exchanger Centerline (Frisina et al, 1987).

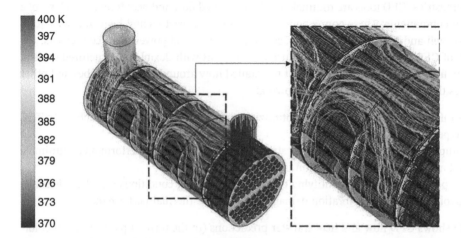

Fig. 3-26 CFD Predictions of Path Lines in a Shell-and-Tube Heat Exchanger with Horizontal Baffles, Colored According to Temperature (Heating Prandtl Number = 1.6, Nozzle Reynolds Number $\approx$ 105) (Mohammadi et al, 2009 / with permission of Taylor & Francis).

The study was carried out to assess the effect of baffle cut and orientation (horizontal and vertical) with respect to the inlet and outlet nozzles on heat transfer and pressure drop performance. The detailed flow around individual tubes was simulated with about 1.2 million control volumes using the re-normalization group k-ε turbulent model.

Acronyms

BWR	Boiling Water Reactor
CANDU	Canada Deuterium Uranium
CFD	Computational Fluid Dynamics
FIV	Flow-Induced Vibration
PDF	Probability Density Function
PWR	Pressurized Water Reactor
RSG	Recirculating Steam Generator
THIRST	Thermalhydraulics in Recirculating Steam Generators

Nomenclature

A	= cross-sectional area, m^2
A_v	= tube surface area per fluid volume
a	= Reynolds number coefficient
B_s	= baffle spacing, m
b	= Reynolds number exponent
C	= factor = 1 for sealing strips and 0.5 for sealing rod
D_s	= shell inside diameter, m
D	= tube outside diameter, m
D_i	= tube inside diameter, m
f	= cylinder vibration frequency, Hz
f	= friction factor
g	= gravity constant, i.e., 9.81 m/s^2
G	= mass flow rate, kg/s
G	= gap width between sealing device and nearest tube, m
h	= enthalpy, J/kg
K	= form loss factor (for sealing strips and rods, bundle entrance and exit)
K	= overall loss factor (includes friction and form losses)
$\overline{\overline{K}}$	= hydraulic resistance coefficient matrix
k	= hydraulic resistance coefficient
l	= distance, m
L	= path length, m
$\dot{m}$	= mass flux, kg/(m^{2}s)
N_g	= number of gaps crossed by flow between tubes neighboring sealing devices, set to 3 for Example 3-3
N_r	= number of tube rows crossed for cross flow in tube bundle

p = pressure, Pa
P = tube pitch, m
r = radial coordinate, m
R^- = resistance force per unit volume, N/m^3
Re = Reynolds number, $Re = \frac{\rho U W}{\mu}$ for channel flow and $Re = \frac{\rho U_p D}{\mu}$ for cross flow
S = Strouhal number
S_h = heat source from the primary side, W/m^3
T = temperature, °C
U = fluid velocity, m/s
U_g^* = dimensionless gas velocity
u = axial velocity, m/s
$\dot{V}$ = volumetric flow rate, m^3/s
v = radial velocity, m/s
w = circumferential velocity, m/s
W = path width, m
X = steam quality
X = Martinelli parameter
z = axial coordinate, m
α = overall heat transfer coefficient, W/(m^2 K)
β = volumetric porosity
Δp = pressure drop, Pa
ε_g = void fraction
θ = circumferential coordinate
μ = dynamic viscosity (assumed constant over cross-section). kg/(m·s)
ν = kinematic viscosity (assumed constant over cross-section), m^2/s
ρ = fluid density (assumed constant over cross-section), kg/m^3
σ = surface tension, N/m
Φ = two-phase flow multiplier

Subscripts

1,2,3 = numbers for Paths 1 to 3
2a, 2b = segment numbers for Path 2
a = axial flow
b = bottom sealing strip or rod
BO = bundle outside diameter, m
c = cross flow
con = contraction
dev = device (sealing strip or sealing rod)
ds = downstream
eff = effective
en = entrance
ex = exit

exp	= expansion
g	= gas
i	= path number
j	= path segment number
ℓ	= liquid
loc	= local
n	= form loss number
nt	= neighboring tubes
p	= pitch
r	= radial
S	= Strouhal or shedding
sr	= sealing rod
ss	= sealing strip
t	= top sealing strip or rod
tot	= total
us	= upstream
z	= axial
θ	= circumferential

Notes

1 CANDU = CANada Deuterium Uranium is a registered trademark of Atomic Energy of Canada Limited.
2 ThermalHydraulics In Recirculating STeam generators, a code developed by Atomic Energy of Canada Limited.

References

Axisa, F., 1993, "Chapter 9: Flow-Induced Vibration of Nuclear System Components," *Technology for the '90s*, (Principal Editor M. K. Au-Yang), ASME, New York, pp. 897–956.

Carlucci, L. N., 1980, "Damping and Hydrodynamic Mass of a Cylinder in Simulated Two-Phase Flow," *ASME Journal of Mechanical Design*, 102, pp. 597–602.

Carlucci, L. N., 1986, "Computation of Flow and Heat Transfer in Power Plant Condensers," *8th Int. Heat Transfer Conf.*, San Francisco, pp. 2541–2546.

Carlucci, L. N., Galpin, P. F., and Brown, J. D., 1984, "Numerical Predictions of Shell-Side Heat Exchanger Flows," 22nd Heat Transfer Conference, Niagara Falls, N.Y., Paper in ASME Booklet, *A Reappraisal of Shellside Flow in Heat Exchangers*, Edited by W.J. Marner and J.M. Chenoweth.

Cheng, L., Ribatski, G. and Thome, J. R., 2008, "Two-Phase Flow Patterns and Flow-Pattern Maps: Fundamentals and Applications," *Applied Mechanics Reviews*, 61, September.

Collier, J. G., 1979, "Two-Phase Gas-Liquid Flows within Rod Bundles," *Turbulent Forced Convection in Channels and Bundles: Theory and Applications to Heat Exchangers and Nuclear Reactors*, Vol. 2, Washington, Hemisphere Publishing Corp., pp. 1041–1055.

Collier, J. G. and Thome, J. R., 1994, *Convective Boiling and Condensation*, 3rd Edition, Oxford University Press, New York.

Dallavalle, F., Rossini, T. and Vanoli, E., 1973, "Vibrations Induced by a Two-Phase Parallel Flow to a Rod Bundle: Preliminary Experiments on the Surface Tension and Gas Density Effects," *Proceedings from UKAEA/NPL International Symposium on Vibration Problems in Industry*, Paper 525, Keswick.

El-Hawary, M., 1975, "Experimental Investigations of the Vibratory Response of a Heated Rod to Subcooled Boiling," *Westinghouse Company Ltd. Report CWAPD-264*.

Enhanced CANDU 6 Technical Summary, 2009, AECL Publication.

Feenstra, P. A., Weaver, D. S. and Judd, R. L., 1996, "Damping and Fluidelastic Instability of a Tube Array in Two-Phase R-11 Cross Flow," *ASME PVP Conference, PVP-Vol. 328, Flow-Induced Vibration*, July 21-26, Montreal, pp. 89–102.

Frisina, V. C., Carlucci, L. N., Campagna, A. O., Pettigrew, M. J. and Crawford, R. M., 1987, "Heat Exchanger Flow Analysis to Resolve Potential Vibration Problems," Paper in ASME Booklet, *Maldistribution of Flow and Its Effect on Heat Exchanger Performance*, ASME HTD 75, Edited by J.B. Kitto Jr. and J.M. Robertson.

Frisina, V. C., Carlucci, L. N., McNaught, J. M. and Mussalli, Y., 1989, "Advanced Predictive Performance Modelling for Condensers and Feedwater Heaters," *EPRI Conf. on Heat Rate Improvement*, Knoxville, September 26-28.

Fluit, S. M. and Pettigrew, M. J., 2001, "Simplified Method for Predicting Vibration and Fretting-Wear in Nuclear Steam Generators," *Proceedings, ASME-PVP Symposium on Flow-Induced Vibration*, Atlanta, Georgia, USA, July 22-26, PVP 420-1, pp. 7–16.

Gidi, A., Weaver, D. S. and Judd, R. L., 1997, "Two-Phase Flow Induced Vibrations of Tube Bundles with Tube Surface Boiling," *ASME Fluid-Structure Interaction, Aeroelasticity, Flow-Induced Vibration and Noise*, AD 53-2. pp. 381–389.

Grant, I. D. R., 1975, "Flow and Pressure Drop with Single-Phase and Two-Phase Flow in the Shell-Side of Segmentally Baffled Shell-and-Tube Heat Exchangers," *NEL Report 590*, National Engineering Laboratory, Glasgow, Scotland, pp. 1–22.

Grant, I.D.R. and Chisholm, D., 1979, "Two-Phase Flow on the Shell-Side of a Segmentally Baffled Shell-and-Tube Heat Exchanger," *Journal of Heat Transfer*, 101, pp. 38–42.

Haquet, J. F. and Gouriand, J. M., 1995, "Local Two-Phase Flow Measurements in a Cross-Flow Steam-Generator Tube Bundle Geometry: The Minnie II XF Program," *Advances in Multiphase Flow 1995*, Elsevier Science B.V.

Hara, F., 1982, "Two-Phase Cross Flow-Induced Forces Acting on a Circular Cylinder," ASME Symposium on *Flow-Induced Vibration of Circular Structures*, PVP-Vol. 63, Orlando, pp. 9–17.

Heat Exchanger Design Handbook, 1984, Chapter 2: *Fluid Mechanics and Heat Transfer*, Hemisphere Publishing Corporation, Rev. 1.

Hewitt, G. F. and Roberts, D. N., 1969, "Studies of Two-Phase Flow Patterns by Simultaneous X-Ray and Flash Photography," *AERE-M2159*, Harwell, England.

Hinze, J. O., 1975, *Turbulence*, 2nd Edition, McGraw-Hill Book Company, New York.

Hulin, J-P., Fierfort, C. and Coudol, R., 1982, "Experimental Study of Vortex Emission Behind Bluff Obstacles in a Gas Liquid Vertical Two-Phase Flow," *International Journal of Multiphase Flow*, 8(5), pp. 475–490.

Idelchik, I. E., 1994, *Handbook of Hydraulic Resistance*, 3rd Edition, CRC Press, Begell House.

Inoue, A., Kozawa, Y., Yokosawa, M. and Aoki, S., 1986, "Studies in Two-Phase Cross Flow, Part 1: Flow Characteristics Around a Cylinder," *International Journal of Multiphase Flow*, 12(2), pp. 149–167.

Kanizawa, F. T. and Ribatski, G., 2016, "Two-Phase Flow Patterns Across Triangular Tube Bundles for Air-Water Upward Flow," *International Journal of Multiphase Flow*, 80, pp. 43–56.

Lian, H. Y., Noghrehkar, G., Chan, A. M. C. and Kawaji, M., 1997, "Effect of Void Fraction on Vibrational Behaviour of Tubes in Tube Bundle under Two-Phase Cross Flow," *Journal of Vibration and Acoustics*, 119, pp. 457–463.

Lienhard, J.H., 1966, "Synopsis of Lift, Drag and Vortex Frequency Data for Rigid Circular Cylinders," *Washington State University, College of Engineering, Research Division Bulletin 300.*

Mann, W. and Mayinger, F., 1995, "Flow-Induced Vibration of Tube Bundles Subjected to Single- and Two-Phase Cross Flow," *Advances in Multiphase Flow 1995*, Elsevier Science B.V.

Mao, K. and Hibiki, T., 2017, "Flow Regime Transition Criteria for Upward Two-Phase Cross-Flow in Horizontal Tube Bundles," *Journal of Applied Thermal Engineering*, 112, pp. 1533–1546.

McAdams, W. H., Woods, W. K., and Heroman, L. V., 1942, "Vaporization Inside Horizontal Tubes II. Benzene Oil Mixtures," *Transactions ASME*, 64, pp. 193–200.

McNaught, J. M., 1982, "Two-Phase Forced Convection Heat Transfer during Condensation on Horizontal Tube Bundles," *7th International Heat Transfer Conference*, 5, Munich, pp. 125–131.

McQuillan, K. W. and Whalley, P. B., 1985, "Flow Patterns in Vertical Two-Phase Flow," *International Journal of Multiphase Flow*, 11(2), pp. 161–175.

Mitra, D., Dhir, V. K. and Catton, I., 2009, "Fluid-Elastic Instability in Tube Arrays Subjected to Air-Water and Steam-Water Cross-Flow," *Journal of Fluids and Structures*, 25, pp. 1213–1235.

Mohammadi, K., Heidemann, W. and Muller-Steinhagenn, H., 2009, "Numerical Investigation of the Effect of Baffle Orientation on Heat Transfer and Pressure Drop in a Shell and Tube Heat Exchanger With Leakage Flows," *Journal of Heat Transfer Engineering*, 30(14), pp. 1123–1135.

Noghrehkar, R., Kawaji, M., and Chan, A. M. C., 1995, "An Experimental Study of Local Two-Phase Parameters in Cross Flow Induced Vibration in Tube Bundles," *Flow-Induced Vibration*, Bearman (ed.), Rotterdam: Balkema, pp. 373–382

Patankar, S. and Spalding, D. B., 1974, "A Calculation Procedure for the Transient and Steady-State Behaviour of Shell-and-Tube Heat Exchangers," Paper in *Heat Exchangers: Design and Theory Sourcebook*, N. Afgan and E.U. Schlunder, eds., pp. 155-176, Hemisphere, Washington, D.C.

Pettigrew, M. J. and Gorman, D. J., 1973, "Experimental Studies on Flow-Induced Vibration to Support Steam Generator Design, Part 1: Vibration of a Heated Cylinder in Two-Phase Axial Flow," *Proceedings of the UKAEA/NPL International Symposium on Vibration Problems in Industry*, Keswick, UK, Paper 424 (also *Atomic Energy of Canada Report, AECL-4514*).

Pettigrew, M. J. and Gorman, D. J., 1993, "The Vibration Behaviour of Nuclear Fuel Under Reactor Conditions," *Nuclear Science and Engineering*, 114, pp. 179–189.

Pettigrew, M. J. and Taylor C. E., 1994, "Two-Phase Flow-Induced Vibration: An Overview," *Journal of Pressure Vessel Technology*, 116, pp. 233–253.

Pettigrew, M. J., Zhang, C., Mureithi, N. W. and Pamfil, D., 2004, "Detailed Flow Measurements in a Rotated Triangular Tube Bundle Subjected to Two-Phase Cross Flow," *Proceedings of Flow-Induced Vibration Symposium* (de Langre and Axisa, editors), École Polytechnique, Paris, July.

Prithiviraj, M. and Andrews, M. J., 1998, "Three Dimensional Numerical Simulation of Shell-and-Tube Heat Exchangers, Part I: Foundation and Fluid Mechanics," *Numerical Heat Transfer*, 33, Issue 8, pp. 799–816, June.

Rae, G. J. and Wharmby, J. S., 1987, "Strouhal Numbers for In-line Tube Arrays," *Proceedings of the International Conference on Flow-Induced Vibrations*, Bowness-on-Windermere, England.

Ramón, I. R., and Prieto Gonzálaz, M., 2001, "Numerical Study of the Performance of Church Window Tube Bundle Condenser," *International Journal of Thermal Science*, 40, pp. 195–204.

Sauvé, R. G., Savoia, D., Morandin, G. and Kozluk, M. J., 1999, "Monte Carlo Simulation of Fatigue Crack Growth in Steam Generator Tubes Under Flow Induced Vibration," *ASME PVP Conference Proceedings, PVP-Vol. 389, Flow-Induced Vibration*, Boston, USA, pp. 255–263.

Serizawa, A., Kataoka, I. and Michiyoshi, I., 1975, "Turbulence Structure of Air-Water Bubbly Flow - II. Local Properties," *International Journal of Multiphase Flow*, 2, pp. 235–246.

Shumway, R.W., 1976, "General Features of Emergency Core Cooling Systems," *Nuclear Power Safety*, Elsevier, New York.

Smith, S., 1968, "Void Fractions in Two-Phase Flow: A Correlation Based Upon an Equal Velocity Head Model," *Proceedings of the Institute of Mechanical Engineers*, 184, pp. 647–664.

Taitel, Y., Barnea, D., and Dukler, A., 1980, "Modelling Flow Pattern Transitions for Steady Upward Gas-Liquid Flow in Vertical Tubes," *AI ChE Jl*, 26, pp. 345–354.

Taylor, C. E. and Pettigrew, M. J., 2001, "Effect of Flow Regime and Void Fraction on Tube Bundle Vibration," *ASME J. Pressure Vessel Technology*, 123, pp. 407–413.

Taylor, C. E., Pettigrew, M. J. and Currie, I. G., 1996, "Random Excitation Forces in Tube Arrays Subjected to Two-Phase Cross Flow," *ASME J. Pressure Vessel Technology*, 118, pp. 265–277.

Ulbrich, R. and Mewes, D., 1994, "Vertical, Upward Gas-Liquid Two-Phase Flow Across a Tube Bundle," *Int. J. Multiphase Flow.* 20, pp. 249–272.

Ulbrich, R., Reinecke, N., and Mewes, D., 1997, "Recognition of Flow Pattern for Two-Phase Flow Across Tube Bundle," *4th World Conference on Experimental Heat Transfer, Fluid Mechanics and Thermodynamics*, Brussels, June 2-6.

Wallis, G. B., 1969, *One-Dimensional Two-Phase Flow*, McGraw-Hill Book Co., New York.

Wambsganss, M.W. and Chen, S.S., 1971, "Tentative Design Guide for Calculating the Vibration Response of Flexible Cylindrical Elements in Axial Flow," *Argonne National Laboratory Report ANL-ETD-71-07*.

Wang, S. K., Lee, S. J., Jones Jr., O. C. and Lahey Jr., R.T., 1987, "3-D Turbulence Structure and Phase Distribution Measurements in Bubbly Two-Phase Flows," *International Journal of Multiphase Flow*, 13, pp. 327–343.

Weaver, D. S. and Abd-Rabbo, A., 1984, "A Flow Visualization Study of a Square Array of Tubes in Water Cross Flow," *ASME-WAM Symposium on Flow-Induced Vibrations*, New Orleans, pp. 165–177.

Weaver, D. S. and Fitzpatrick, J. A., 1987, "A Review of Flow-Induced Vibrations in Heat Exchangers," *Proceedings of the International Conference on Flow-Induced Vibrations*, Bowness-on-Windermere, England.

4

Hydrodynamic Mass, Natural Frequencies and Mode Shapes

Daniel J. Gorman, Colette E. Taylor, and Michel J. Pettigrew

4.1 Introduction

One of the most important limitations on the efficiency with which a nuclear electric power plant can be operated is dictated by the volumetric flow rates with which fluids can be circulated through associated heat transfer equipment such as heat exchangers and steam generators. Failure to maintain fluid flow velocities in this equipment below certain critical levels will result in unacceptable levels of vibration induced in tubes of the tube bundles in this equipment. This sustained vibration will result in fretting wear on tube surfaces as the tubes impact on tube-support surfaces or, in extreme cases, when tubes impact on neighboring tubes of the same bundle. It is of critical importance, therefore, that design criteria be developed and employed to predict the frequency and amplitude of tube vibration induced by proposed flow rates so that tube vibration, and associated fretting wear, can be limited to acceptable levels.

In this chapter, we will review and examine in detail the analytical procedures available to predict the added (or hydrodynamic) mass, frequencies, mode shapes and amplitudes of tube response in tube bundles subjected to external parallel or transverse fluid flow. This may be single-phase flow (e.g. liquid or vapor) or it may be two-phase flow (a combination of liquid and vapor).

There are two major steps involved in carrying out the analysis of flow-induced vibration of tubes. First, we must conduct what is called a free vibration analysis of tube behavior. Later, in a second step, we conduct a theoretical analysis that permits us to predict the actual response of the tubes to the proposed conditions. In this chapter, we present the free vibration analysis techniques. In later chapters, actual response corresponding to various vibration mechanisms (e.g., fluidelastic instability, random excitation and vortex shedding) is predicted.

There are two distinct types of tube bundles employed in heat exchangers and steam generators. The first is the straight tube bundle. The second type differs from the first in that it consists of a bundle of U-shaped tubes, referred to as U-tubes, i.e., tubes fabricated in the shape of a U. Such tubes are sometimes described as "Hair Pin" tubes. Analysis of the free vibration of tubes of U-tube bundles is considerably more involved than that related to tubes of straight tube bundles. For this reason, we will begin the free vibration sections by describing analytical techniques employed in the free vibration analysis of tubes in straight tube bundles.

In this chapter, the calculation of tube mass is discussed in Section 4.2, while Section 4.3 outlines the calculation of frequencies, mode shapes and vibration amplitudes for straight tubes found in typical heat exchangers, including examples. Section 4.4 provides the basic theory for curved tubes,

Flow-Induced Vibration Handbook for Nuclear and Process Equipment, First Edition.
Michel J. Pettigrew, Colette E. Taylor, and Nigel J. Fisher.
© 2022 John Wiley & Sons, Inc. This Work is a co-publication between ASME Press and John Wiley & Sons, Inc.

and Section 4.5 outlines the free vibration analysis of steam generator U-tubes, with one example. Concluding remarks are provided in Section 4.6.

4.2 Total Tube Mass

While a tube vibrating in air has an accelerating mass that is, for practical purposes, equivalent to the tube mass, a tube vibrating in a higher density fluid has an effective accelerating mass that is greater than the tube alone. The liquid does not actually act as a lumped mass about the cylinder; rather, fluid forces in phase with the tube increase the dynamic forces. This effective increase in mass is known as the added mass or hydrodynamic mass, m_h. Therefore, the total tube mass per unit length of the tube is as follows:

$$m = m_t + m_i + m_h \tag{4-1}$$

where $m_t = \dfrac{\pi}{4}\rho_t(D^2 - D_i^2)$ and $m_i = \dfrac{\pi}{4}\rho_i D_i^2$ $\tag{4-2}$

where m_t is the tube mass per unit length, m_i is the mass of the fluid inside the tube per unit length, D is the tube diameter, D_i is the tube inside diameter, ρ_t is the tube density and ρ_i is the density of the fluid inside the tube.

For an infinitely long, unconfined, single cylinder, the theoretical added mass is equal to the mass of the fluid displaced by the cylinder. In real three-dimensional applications, a confinement coefficient, C_h, is introduced to account for the boundary conditions.

$$m_h = \frac{C_h \rho \pi D^2}{4} \tag{4-3}$$

where ρ is the density of the fluid surrounding the tube.

For the case of a tube vibrating in a fluid region bounded by a circular cylinder, the added mass confinement coefficient can be defined as follows:

$$C_h = \frac{(D_{ci}/D)^2 + 1}{(D_{ci}/D)^2 - 1} \tag{4-4}$$

where D_{ci} is the inner diameter of the cylinder. Definitions of C_h for single-phase and two-phase flows across a tube bundle are provided in Sections 4.2.1 and 4.2.2, respectively.

Added mass can also be calculated from measured frequencies in air and the fluid medium. The frequencies in air, f_a, and in the fluid, f, can be defined as follows:

$$f_a = \frac{\lambda^2}{2\pi L^2}\left(\frac{EI}{m_t}\right)^{0.5} \tag{4-5}$$

$$f = \frac{\lambda^2}{2\pi L^2}\left(\frac{EI}{m_t + m_h}\right)^{0.5} \tag{4-6}$$

where λ is the first-mode eigenvalue, L is the tube length, E is the Young's modulus of elasticity, I is the area moment of inertia $\left(I = \pi/64\left(D^4 - D_i^4\right)\right)$ and EI is known as the tube stiffness. By combining these equations, one obtains an expression for the hydrodynamic mass in terms of the measured frequencies:

$$m_h = m_t \left[\left(\frac{f_a}{f} \right)^2 - 1 \right] \tag{4-7}$$

This definition of hydrodynamic mass, like the previous definition, depends upon the three-dimensional characteristics (confinement) of the surrounding geometry. In this expression, the confinement is taken into account by using the response frequencies that were measured in configurations similar to those found in actual steam generators.

Since the hydrodynamic mass is a mass per unit length, some care must be taken in the case where fluid excitation does not take place over the full length of the tube. In this case, the generalized mass must be calculated using appropriate integration limits.

4.2.1 Single-Phase Flow

Values of C_h have been developed in single-phase flow for various tube geometries by authors such as Moretti and Lowry (1976), Rogers et al (1984) and Chen (1987). Their results produce similar confinement coefficients. They all approached the problem by defining an effective or hydraulic diameter, D_e, in place of the inside cylinder diameter, D_{ci}, in Eq. (4-4), as follows:

$$m_h = \left(\frac{\rho \pi D^2}{4} \right) \left[\frac{(D_e/D)^2 + 1}{(D_e/D)^2 - 1} \right] \tag{4-8}$$

In this chapter, we will use the triangular array prediction developed by Rogers et al (1984) to define the effective diameter:

$$\frac{D_e}{D} = \left(0.96 + 0.50 \frac{P}{D} \right) \frac{P}{D} \tag{4-9}$$

where, P is the tube pitch. When this definition was applied to a square array in two-phase flow, Pettigrew et al (1985) found that it was not a good fit and later, in Pettigrew et al (1989), they proposed the following for square tube arrays:

$$\frac{D_e}{D} = \left(1.07 + 0.56 \frac{P}{D} \right) \frac{P}{D} \tag{4-10}$$

With these definitions, added mass can be calculated for single-phase flow.

Example 4-1 Single-Phase Total Tube Mass in a Process Heat Exchanger

This calculation continues from Example 3-1 in Chapter 3, a single-phase heat exchanger shown in Figure 3.5.

Mass Parameters:

$\rho = \rho_i = 1000 \ \text{kg/m}^3$, $\rho_t = 8000 \ \text{kg/m}^3$, $D = 0.02 \ \text{m}$, $D_i = 0.018 \ \text{m}$, $P/D = 1.5$, triangular array

Mass Calculation: m, kg/m
Tube mass per unit length using Eq. (4-2),

$$m_t = \frac{\pi}{4} \rho_t \left(D^2 - D_i^2 \right) = \frac{\pi}{4} \times 8000 \times \left(0.02^2 - 0.018^2 \right) = 0.478 \ \text{kg/m}$$

Inside fluid mass per unit length using Eq. (4-2),

$$m_i = \frac{\pi}{4}\rho_i D_i^2 = \frac{\pi}{4} \times 1000 \times 0.018^2 = 0.254 \text{ kg/m}$$

Effective-diameter-to-tube-diameter ratio, for a triangular tube array, using Eq. (4-9), for a triangular tube array,

$$D_e/D = (0.96 + 0.5P/D)P/D = (0.96 + 0.5 \times 1.5) \times 1.5 = 2.565$$

Hydrodynamic mass per unit length from Eq. (4-8),

$$m_h = \left(\frac{\pi}{4}\rho D^2\right)\left[\frac{(D_e/D)^2 + 1}{(D_e/D)^2 - 1}\right]$$

$$= \left(\frac{\pi}{4} \times 1000 \times 0.02^2\right)\left[\frac{2.565^2 + 1}{2.565^2 - 1}\right]$$

$$= 0.3142 \times 1.36 = 0.427 \text{ kg/m}$$

Total tube mass from Eq. (4-1),

$$m = m_t + m_i + m_h = 0.478 + 0.254 + 0.427 = 1.16 \text{ kg/m}$$

4.2.2 Two-Phase Flow

Use of the single-phase equations for hydrodynamic mass has been extended to two-phase flow through the use of the two-phase homogenous density defined previously in Eq. 3-2:

$$\rho = \rho_\ell(1 - \varepsilon_g) + \rho_g \varepsilon_g$$

where ρ_ℓ and ρ_g are the liquid- and gas-phase densities, respectively, and ε_g is the void fraction.

The applicability of this approach to two-phase hydrodynamic mass has been studied by several researchers. In two-phase flow, it is convenient to present measured added-mass data as a ratio of the measured added mass in a two-phase fluid divided by the predicted added mass in water. Pettigrew et al (1989) determined the measured values for two-phase hydrodynamic mass in air-water flow using Eq. (4-7). The measured vibration frequencies were recorded at velocities of about one-half fluidelastic instability. The authors point out that the tube frequency is relatively independent of mass flux below instability. However, significant shifts in tube frequency are often observed at instability.

In Fig. 4-1, the measured hydrodynamic mass ratio in a normal-square tube bundle is plotted against mass flux for a range of void fractions. Hydrodynamic mass is practically independent of mass flux below 90% void fraction. However, above 90% void fraction, the hydrodynamic mass tends to decrease with mass flux. It is believed that this high-void effect is probably due to the presence of an intermittent flow regime.

Pettigrew et al (2001) presented measured two-phase hydrodynamic mass ratio plotted against void fraction, for a range of bundle geometries, as shown in Fig. 4-2. They found that the agreement between experimental results is good except at void fractions above 80%. For air-water, this discrepancy may be related to the effect of the intermittent flow regime. Fig. 4-2. also shows that the comparison to theory is good over the range of P/D and bundle geometry that is likely to be found in the process heat exchanger field.

The effect of void fraction slip model on the collapse of two-phase mass ratios was presented in Taylor and Pettigrew (2001) (see Fig. 4-3). They used air-water, Freon and steam-water data to illustrate that none of the slip models are obviously better than the homogeneous model in collapsing

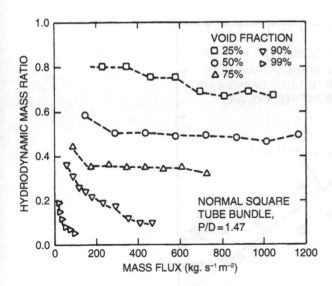

Fig. 4-1 Effect of Void Fraction and Mass Flux on Hydrodynamic Mass Ratio (Pettigrew et al, 1989).

the measured hydrodynamic mass ratios. In Fig. 4-3, a comparison is made between the drift-flux model and the homogeneous model. In addition, the data clearly shows that the theoretical model works equally well for all fluid mixtures.

Example 4-2 Two-Phase Total Tube Mass in a Nuclear Steam Generator

This calculation continues from Example 3-2 in Chapter 3, a two-phase heat exchanger U-bend with triangular geometry shown in Fig. 3-6

Mass Parameters:

$$\rho_i = 1000 \text{ kg/m}^3, \quad \rho_t = 8000 \text{ kg/m}^3, \quad D = 0.02 \text{ m}, \quad D_i = 0.018 \text{ m}$$

$$P/D = 1.5, \quad \rho_{TP} = 137 \text{ kg/m}^3$$

Mass Calculation: m, kg/m
Tube mass per unit length using Eq. (4-2),

$$m_t = \frac{\pi}{4}\rho_t\left(D^2 - D_i^2\right) = \frac{\pi}{4} \times 8000 \times \left(0.02^2 - 0.018^2\right) = 0.478 \text{ kg/m}$$

Inside fluid mass per unit length using Eq. (4-2),

$$m_i = \frac{\pi}{4}\rho_i D_i^2 = \frac{\pi}{4} \times 1000 \times 0.018^2 = 0.254 \text{ kg/m}$$

Effective-diameter-to-tube-diameter ratio using Eq. (4-9),

$$D_e/D = (0.96 + 0.5P/D)P/D = (0.96 + 0.5 \times 1.5) \times 1.5 = 2.565$$

Hydrodynamic mass per unit length using Eq. (4-8),

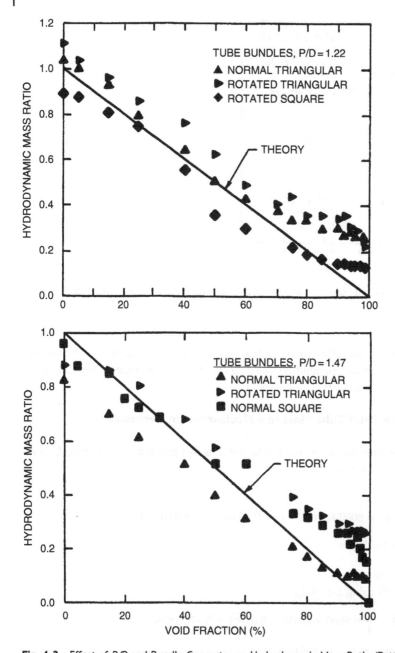

Fig. 4-2 Effect of *P/D* and Bundle Geometry on Hydrodynamic Mass Ratio (Pettigrew et al, 2001).

$$m_h = \left(\frac{\pi}{4}\rho_{TP}D^2\right)\left[\frac{(D_e/D)^2 + 1}{(D_e/D)^2 - 1}\right]$$

$$= \left(\frac{\pi}{4} \times 137 \times 0.02^2\right)\left[\frac{2.565^2 + 1}{2.565^2 - 1}\right]$$

$$= 0.0430 \times 1.36 = 0.0585 \text{ kg/m}$$

Total mass per unit length using Eq. (4-1),

$$m = m_t + m_i + m_h = 0.478 + 0.254 + 0.0585 = 0.79 \text{ kg/m}$$

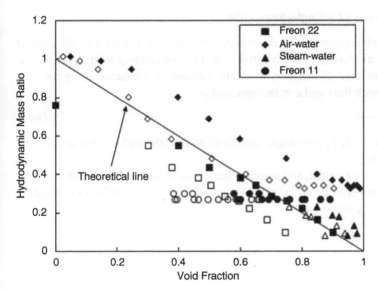

Fig. 4-3   Effect of Slip Ratio and Fluid Mixture on Hydrodynamic Mass Ratio Comparing Homogeneous Model (Solid Symbols) and Drift-Flux Slip Model (Open Symbols) (Taylor and Pettigrew, 2001).

4.3   Free Vibration Analysis of Straight Tubes

Historically, the straight-tube bundle is the type most commonly used in heat transfer equipment. In such a bundle, the tubes are usually rigidly attached at each extremity to a heavy circular perforated plate known as a tubesheet. Tube-to-tubesheet joints are traditionally achieved by means of a "tube rolling process". The tubesheets form the end enclosures for a circular cylindrical tank or shell. The tubes pass through a series of perforated circular plates, known as baffle plates, located at various locations along the cylindrical tank or shell. Each tube receives lateral support from the baffle plate through which it passes.

Exact analytical solutions for the free vibration of straight multi-span beams have been available in the literature for many years. Some examples include Darnley (1921), Gorman (1975), Blevins (1991) and Au Yang (2001). These authors and others have shown that slender tubes employed in steam generators and heat exchangers may be treated as slender beams. It is also known from classical beam theory that, provided certain boundary conditions, as well as tube intermediate support conditions (baffle plate support conditions), are satisfied, the tubes will have certain natural (resonant) frequencies at which they will vibrate if disturbed. Associated with each natural frequency there will be a particular shape (mode shape) in which the tube will vibrate. We refer to these as tube natural frequencies and mode shapes.

The straight tube solutions introduced in this chapter will be the ones with end conditions that are typically used in the free vibration analysis of steam generator U-tubes. In this section, the basic equations used in developing the exact solution for straight-tube free vibration have been extracted from Gorman (1975). While we will be interested in free vibration analysis (obtaining the natural frequencies and mode shapes) of multi-span tubes (tubes with numerous intermediate supports), we will begin, for illustrative purposes, with the analysis of a single-span tube.

4.3.1 Free Vibration Analysis of a Single-Span Tube

A single-span tube, with simple support at each end is depicted in Fig. 4-4. The term simple support implies that no tube lateral displacement is permitted and the tube bending moment is zero. It is known that lateral displacement of the tube, w, at any distance, x, measured along the tube (Fig. 4-4) will be harmonic with time and can be expressed as,

$$w(x,t) = w(x) \cos (\omega t - \alpha) \tag{4-11}$$

where t is the symbol for time, α is a phase angle (usually of no interest) and ω is the circular frequency of tube free vibration.

Focusing on a differential tube element, and equating the inertia and net transverse shear forces acting on the element, one arrives at the ordinary homogeneous fourth-order differential equation governing tube lateral vibratory motion as,

$$\frac{d^4 w(x)}{dx^4} - \beta^4 w(x) = 0 \tag{4-12}$$

where, $\beta^4 = \dfrac{m\omega^2 L^4}{EI}$

Here β is the eigenvalue, m is the effective total mass of the tube per unit length, E is the Young's modulus of the tube material, and I is the area moment of inertia of the tube cross-section in bending. For the present, we will consider the effective mass per unit length as being equal to the mass per unit length of the actual tube, only. As discussed in Section 4.2, eventually, we will have to include added mass due to the presence of internal and external fluids.

The solution for Eq. (4-12) is well known and is written as,

$$w(x) = A \sin \beta x + B \cos \beta x + C \sinh \beta x + D \cosh \beta x \tag{4-13}$$

where coefficients A, B, etc., are evaluated through enforcement of prescribed boundary conditions at each end of the tube. Returning to the tube of Fig. 4-4, it is seen that in order to satisfy the simple support condition at the end, $x = 0$, and recalling that bending moment in the tube is expressed as,

$$M = EI \frac{d^2 w}{dx^2} \tag{4-14}$$

we formulate the boundary conditions as,

$$w(x)|_{x=0} = 0 \tag{4-15}$$

and,

$$\left. \frac{d^2 w}{dx^2} \right|_{x=0} = 0 \tag{4-16}$$

Substituting Eq. (4-13) in Eq. (4-15), and Eq. (4-16), respectively, we obtain

$$B + D = 0 \tag{4-17}$$

and,

$$B - D = 0 \tag{4-18}$$

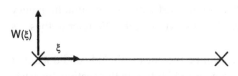

Fig. 4-4 Straight Tube with Simple Support at Each End.

These conditions imply that $B = D = 0$. Enforcing the simple support condition at the other end of the tube we obtain,

$$A \sin \beta L + C \sinh \beta L = 0 \tag{4-19}$$

and,

$$-A \sin \beta L + C \sinh \beta L = 0 \tag{4-20}$$

where L equals the tube length.

It is known that for a non-trivial solution to exist for the constants A and C, the determinant of the coefficient matrix of Eq. (4-19) and Eq. (4-20) must equal zero.

$$\sin \beta L \sinh \beta L = 0 \tag{4-21}$$

Equation (4-21) is satisfied for values of βL equal to $n\pi$, where n takes on a positive integer value greater than zero. Solutions with n equal zero, or negative, are of no interest. Returning to the definition of β (Eq. (4-12)), it is seen that for each value of this parameter, $n\pi$, i.e., for each value of n, there is associated a natural, or resonant, circular frequency of beam free vibration. Focusing on any one of these frequencies and returning to Eq. (4-19) we see that the constant C must equal zero. We conclude that the beam free vibration mode associated with any value of β becomes,

$$w(x) = A \sin \beta x \tag{4-22}$$

Since we are interested in the shape of the mode only, we can assign any non-zero value to the constant A. It is convenient to set it equal to unity. It should be noted that while, from a mathematical point of view, there are an infinite number of free vibration frequencies and mode shapes, only the first few modes will be of practical interest. As we go to the higher modes, the number of crossings of the axis of the mode shape, i.e., the number of node points, will increase. However, the governing differential equation, (Eq. (4-12)) is only valid so long as the inter-node point distances are large in comparison to the tube outer diameter. This is in keeping with the Bernoulli-Euler assumptions upon which the slender beam theory utilized here is based.

It is highly advantageous to work with dimensionless quantities when conducting beam or tube vibration studies. In keeping with the rules of differential calculus, it is shown that we may write the governing differential equation for beam free vibration as,

$$\frac{d^4 w(x/L)}{L^3 d(x/L)^4} - \beta^4 L w(x/L) = 0 \tag{4-23}$$

Introducing the symbol $\xi = x/L$, and defining $W(\xi)$ as the tube lateral displacement divided by L, Eq. (4-23) is written as,

$$\frac{d^4 W(\xi)}{d\xi^4} - \beta^4 W(\xi) = 0 \tag{4-24}$$

where,

$$\beta^4 = \frac{m\omega^2 L^4}{EI} \tag{4-25}$$

The natural frequencies of the tube are then written as,

$$f = \frac{\beta^2}{2\pi L^2} \sqrt{EI/m} \tag{4-26}$$

The solution of Eq. (4-24) is now written as,

$$W(\xi) = A \sin \beta\xi + B \cos \beta\xi + C \sinh \beta\xi + D \cosh \beta\xi \tag{4-27}$$

where $0 \le \xi \le 1$

Returning now to the tube with simple support at each end, the new formulation of the boundary conditions becomes,

$$W(\xi) = \frac{d^2 W(\xi)}{d\xi^2} = 0.|_{\xi = 0 \text{ and } \xi = 1} \tag{4-28}$$

with the shape,

$$W(\xi) = \sin \beta\xi$$

Example 4-3 Calculation of Frequency in a Process Heat Exchanger

This calculation continues from Example 3-1 in Chapter 3, a single-phase heat exchanger shown in Fig. 3.5. The total tube mass was calculated in Example 4-1, earlier in this chapter.

Frequency Parameters:

$$E = 200 \times 10^9 \text{ N/m}^2, \quad \ell = 0.5 \text{ m}, \quad \ell_{window} = 1.0 \text{ m}$$

$$D = 0.02 \text{ m}, \quad D_i = 0.018 \text{ m}, \quad m = 1.16 \text{ kg/m}$$

For first-mode, pinned-pinned boundary conditions, $\beta_1 = \pi$

Frequency Calculation: f, Hz

From Eq. (4-26), $f_i = \dfrac{\beta_i^2}{2\pi} \left(\dfrac{EI}{m\ell^4} \right)^{\frac{1}{2}}$

$$I = \frac{\pi}{64} \left(D^4 - D_i^4 \right) = \frac{\pi}{64} \left(0.02^4 - 0.018^4 \right) = 2.7 \times 10^{-9} \text{ m}^4$$

$$f_1 = \frac{\pi^2}{2\pi} \left(\frac{200 \times 10^9 \times 2.7 \times 10^{-9}}{1.16 \times 0.5^4} \right)^{\frac{1}{2}} = 135.6 \text{ Hz}$$

$$f_{1window} = \frac{\pi^2}{2\pi} \left(\frac{200 \times 10^9 \times 2.7 \times 10^{-9}}{1.16 \times 1.0^4} \right)^{\frac{1}{2}} = 33.9 \text{ Hz}$$

Example 4-4 Calculation of Frequency in a Nuclear Steam Generator U-Bend

This calculation continues from Example 3.2 in Chapter 3, a two-phase nuclear steam generator U-bend shown in Fig. 3.6. The tube mass was calculated in Example 4-2, earlier in this chapter.

U-Bend Frequency Parameters:

$$E = 200 \times 10^9 \text{ N/m}^2, \quad \ell = 0.59 \text{ m}$$

$$D = 0.02 \text{ m}, \quad D_i = 0.018 \text{ m}$$

$$m = 0.788 \text{ kg/m}$$

Assuming that the U-tube curved spans can be approximated as straight spans, the first-mode, pinned-pinned boundary conditions eigenvalue can be defined as $\beta_1 = \pi$

Frequency Calculation; f, Hz

From Eq. (4-26), $f_i = \dfrac{\beta_i^2}{2\pi} \left(\dfrac{EI}{m\ell^4} \right)^{\frac{1}{2}}$

$$I = \frac{\pi}{64}\left(D^4 - D_i^4\right) = \frac{\pi}{64}\left(0.02^4 - 0.018^4\right) = 2.7 \times 10^{-9} \text{ m}^4$$

$$f_1 = \frac{\pi^2}{2\pi}\left(\frac{200 \times 10^9 \times 2.7 \times 10^{-9}}{0.79 \times 0.59^4}\right)^{\frac{1}{2}} = 118 \text{ Hz}$$

If every second support is not effective, then $\ell_{ineffective} = 2 \times 0.59 = 1.18$ m

$$f_{1\,ineffective} = \frac{\pi^2}{2\pi}\left(\frac{200 \times 10^9 \times 2.7 \times 10^{-9}}{0.79 \times 1.18^4}\right)^{\frac{1}{2}} = 29.5 \text{ Hz}$$

4.3.2   Free Vibration Analysis of a Two-Span Tube

While tube bundles of most practical heat exchangers and steam generators will be of more than two spans in length, i.e., there will be more than one interior baffle plate, we will consider here for illustrative purposes the free vibration analysis of tubes (or beams) of two spans in overall length. Since most tubes will be clamped at each extremity, we will consider here a two-span tube clamped at each end. Such a tube is depicted schematically in Fig. 4-5. Let the actual span lengths be designated as ℓ_1 and ℓ_2, respectively, where subscripts 1 and 2 refer to quantities related to the left- and right-hand spans, respectively.

We introduce the quantities μ_1 and μ_2 to represent the dimensionless lengths of the spans, where $\mu_1 = \ell_1/L$, and $\mu_2 = \ell_2/L$, L being, we recall, the tube overall length. It will be obvious that μ_1 and μ_2 will each be < 1.0 and their sum will equal unity. Dimensionless distance along the left-hand span is denoted by the symbol ξ, where ξ equals distance along the tube divided by tube length L. It will be observed that the parameter ξ, will vary between zero and μ_1. Similar rules apply to the dimensionless distance ξ measured along the right-hand span except that it is convenient to measure ξ from the right-hand extremity of the tubes, as indicated in the figure. Amplitudes of free harmonic vibration of the left- and right-hand spans are designated as $W_1(\xi)$ and $W_2(\xi)$, respectively, where we can now write,

$$W_1(\xi) = A_1 \sin\beta\xi + B_1 \cos\beta\xi + C_1 \sinh\beta\xi + D_1 \cosh\beta\xi \qquad (4\text{-}29)$$

and

$$W_2(\xi) = A_2 \sin\beta\xi + B_2 \cos\beta\xi + C_2 \sinh\beta\xi + D_2 \cosh\beta\xi \qquad (4\text{-}30)$$

To construct the eigenvalue matrix from which the eigenvalue β is extracted, we begin by enforcing the appropriate boundary conditions for the problem under consideration. Focusing on the left-end boundary of the tube, we have two boundary conditions to enforce, i.e., the tube lateral displacement and slope must each equal zero at this location. Setting the displacement equal to zero at the origin, it will be obvious that D_1 must equal the negative of B_1. It further follows that upon setting the first derivative of the function $W_1(\xi)$ with respect to ξ equal to zero at the origin, we obtain the relationship, $C_1 = -A_1$. Due to symmetry, identical relationships will apply to the constants A_2, B_2, etc. related to the right-hand span of the tube. We, therefore, obtain

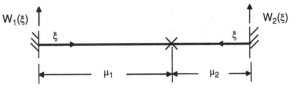

Fig. 4-5   Two-Span Beam with Outer Ends Clamped.

$$W_1(\xi) = A_1[\sin\beta\xi - \sinh\beta\xi] + B_1[\cos\beta\xi - \cosh\beta\xi] \tag{4-31}$$

and

$$W_2(\xi) = A_2[\sin\beta\xi - \sinh\beta\xi] + B_2[\cos\beta\xi - \cosh\beta\xi]. \tag{4-32}$$

Four of the unknown constants have thereby been eliminated from the displacement expressions. We next enforce the conditions that lateral displacement must equal zero at the common junction of the two spans (the internal simple support). Two more constants are thereby eliminated from the displacement expressions and we arrive at the following expressions for span displacements,

$$W_1(\xi) = A_1\{[\sin\beta\xi - \sinh\beta\xi] - \gamma_1[\cos\beta\xi - \cosh\beta\xi]\} \tag{4-33}$$

and

$$W_2(\xi) = A_2\{[\sin\beta\xi - \sinh\beta\xi] - \gamma_2[\cos\beta\xi - \cosh\beta\xi]\} \tag{4-34}$$

$$\text{where,}\ \gamma_1 = \frac{\sin\beta\mu_1 - \sinh\beta\mu_1}{\cos\beta\mu_1 - \cosh\beta\mu_1}\ \text{and}\ \gamma_2 = \frac{\sin\beta\mu_2 - \sinh\beta\mu_2}{\cos\beta\mu_2 - \cosh\beta\mu_2} \tag{4-35}$$

There remain two constraints to be imposed on the above displacement functions as follows. First, there must be continuity of slope of the two segments of the tube where they join at the internal support. Secondly, there must also be continuity of bending moment in the tube segments at the same location. This gives rise to the following two homogeneous equations relating the unknown constants A_1 and B_1 as follows:

$$\begin{aligned} A_1\{\cos\beta\xi - \cosh\beta\xi + \gamma_1[\sin\beta\xi + d\sinh\beta\xi]\}|_{\xi=\mu_1} \\ + A_2\{\cos\beta\xi - \cosh\beta\xi + \gamma_2[\sin\beta\xi + \sinh\beta\xi]\}|_{\xi=\mu_2} = 0 \end{aligned} \tag{4-36}$$

and,

$$\begin{aligned} A_1\{\sin\beta\xi + \sinh\beta\xi - \gamma_1[\cos\beta\xi + \cosh\beta\xi]\}|_{\xi=\mu_1} \\ - A_2\{\sin\beta\xi + \sinh\beta\xi - \gamma_2[\cos\beta\xi + \cosh\beta\xi]\}|_{\xi=\mu_2} = 0 \end{aligned} \tag{4-37}$$

Eigenvalues for this problem are obtained by searching out these values of β for which the determinant of the coefficient matrix relating the constants A_1 and A_2 of the above equations is caused to vanish. These values of β are the eigenvalues. This is a simple two by two matrix and it is easy to plot the determinant of the matrix versus the parameter β for any range of interest. Standard computer routines are available for evaluating the determinant of the coefficient matrix. Any value of β which causes the determinant plot to cross the axis in going from positive to negative, or vice-versa indicates the existence of an eigenvalue in the immediate area. With the eigenvalue obtained, we can substitute it in the matrix and set one of the coefficients, A_1 or A_2 equal to unity. We thus have a set of non-homogeneous algebraic equations available (here there will be two equations) and we can solve for the other unknown, A_1 or A_2, using either one of these equations. With values assigned to A_1 and A_2, we now have available the mode shape associated with any eigenvalue β through Eq. (4-33) and Eq. (4-34). We will find that the mode shapes become more complicated as we move upwards from the lowest (fundamental) eigenvalue. There are certain verification checks which can be made upon obtaining the eigenvalues. It will be appreciated, for example, that as we let the parameter μ_1 approach zero for the present problem, the computed eigenvalues should approach those of the single-span beam with clamped conditions imposed at each end.

4.3.3 Free Vibration Analysis of a Multi-Span Tube

We now have all of the fundamental theory necessary to perform a free vibration analysis of a straight tube of any number of spans. We will concentrate on the analysis of tubes with more than two spans. For illustrative purposes, we will consider the multi-span tube of N spans, as depicted in Fig. 4-6, where N is any positive number greater than two. Here we will consider the tube outer ends to be clamped as is usually the case for heat exchangers or steam generators. It will be seen that any of the other two classical boundary conditions, i.e., simple support or free end conditions could be enforced at the tube outer ends.

Focusing attention on the first span at the left end of the tube, it will be recalled that three of the four constants appearing in the displacement expression for this span (Eq. (4-27)) are immediately evaluated upon enforcing the clamped outer end condition and the condition of zero lateral displacement at the inner end of the span. Utilizing the subscript 1 to indicate this span, we see that mode shape lateral deflection is expressed by our previously developed Eq. (4-33). In view of symmetry, we see that deflection in the Nth span is given by Eq. (4-34), provided we replace the subscript 2 of the earlier equation with the subscript N.

We next focus attention on lateral displacement related to the interior spans. We use the subscript n to designate any one of the internal spans where n can vary from two to N-1. It will be obvious that lateral displacement must equal zero at each end of the span. We, therefore, write (Eq. (4-27))

$$A_n \sin \beta\xi + B_n \cos \beta\xi + C_n \sinh \beta\xi + D_n \cosh \beta\xi|_{\xi = 0.} = 0 \tag{4-38}$$

and,

$$A_n \sin \beta\mu_n + B_n \cos \beta\mu_n + C_n \sinh \beta\mu_n + D_n \cosh \beta\mu_n = 0 \tag{4-39}$$

Utilizing the above two equations to eliminate the constants C_n and D_n, we obtain the expression for the intermediate span displacement as,

$$W_n(\xi) = A_n(\sin \beta\xi - \theta_n \sinh \beta\xi) + B_n(\cos \beta\xi - \cosh \beta\xi + \phi_n \sinh \beta\xi) \tag{4-40}$$

where $\theta_n = \dfrac{\sin \beta\mu_n}{\sinh \beta\mu_n}$; and $\phi_n = \dfrac{\cosh \beta\mu_n - \cos \beta\mu_n}{\sinh \beta\mu_n}$

There are, therefore, two unknown constants associated with each internal span of the tube and a total of $2(N-1)$ unknown constants associated with the entire tube. It will be obvious that we require $2(N-1)$ homogeneous algebraic equations to set up the eigenvalue matrix for the free vibration problem under consideration. Setting up the two equations enforcing continuity of slope and bending moment of the tube at each end of the N-1 internal tube supports provides us with these equations.

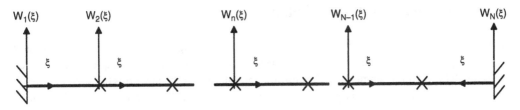

Fig. 4-6 Multi-Span Beam with Outer Ends Clamped.

Table 4-1 Schematic Representation of Eigenvalue Matrix Elements for Four Span Tube.

A_1	A_2	B_2	A_3	B_3	A_4
α_{11}	α_{12}	α_{13}			
α_{21}	α_{22}	α_{23}			
	α_{32}	α_{33}	α_{34}	α_{35}	
	α_{42}	α_{43}	α_{44}	α_{45}	
			α_{54}	α_{55}	α_{56}
			α_{64}	α_{65}	α_{66}

For illustrative purposes, we consider the free vibration analysis of a tube of four spans. There will of course be three internal support points. The non-zero elements of the eigenvalue matrix associated with this problem are represented schematically in Table 4-1. The first and second rows of elements express continuity of tube slope and bending moment at the first tube support. These elements are available from Eq. (4-33) and (4-40). Elements of the third and fourth row express the same conditions related to the second internal support and are obtainable from the same equations using the appropriate subscripts. Finally, the fifth and sixth row elements arise from enforcing the above continuity conditions at the third internal support. Here Eq. (4-33) and Eq. (4-40) are employed, again using the correct subscripts.

Eigenvalues are, of course, obtained by searching out those values of the parameter β which cause the determinant of the above eigenvalue matrix to vanish. The matrix is easily generated and the determinant evaluated using modern digital computer routines. Finally, one of the non-zero coefficients, let us say A_N, is arbitrarily set equal to unity, thereby converting the set of equations associated with the matrix into a non-homogeneous set. A solution for the remaining constants A_1, A_2, etc., is then readily obtained, again, utilizing modern computer routines. With values assigned to all of the constants, the exact mode shape associated with any eigenvalue is available. These constants may be stored for further reference at any time.

It will be appreciated that, by following the above procedure, the free vibration analysis for straight tubes of any number of spans is readily achieved. Each additional span will add two more rows and columns to the eigenvalue matrix. It will also be evident that computer routines are easily set up to generate the basic eigenvalue matrix for tubes of any number of spans. It will only be necessary to provide, as input to the computer, the number of spans and span lengths, for any heat exchanger or steam generator tube of interest.

4.4 Basic Theory for Curved Tubes

A U-tube is composed of three distinct segments as shown in Fig. 4-7. These are the two straight segments (legs) and the curved segment. We will be looking for free vibration frequencies and mode shapes involving all three segments with continuity maintained at the junctions between the curved and straight segments. We already have explored free vibration theory related to straight tubes. Before continuing with the general U-tube free vibration analysis, we must first present the theory of free vibration analysis as it relates to curved tubes. A typical segment of curved tube is depicted in Fig. 4-8.

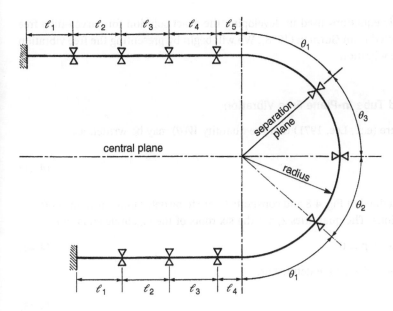

Fig. 4-7 View of Typical Steam Generator U-Tube with Support Points Located Along the Curved Tube Section.

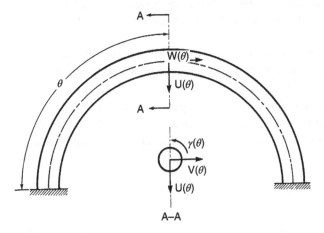

Fig. 4-8 Nomenclature Utilized in Representing Lateral and Rotational Displacement Along Curved Tube Region.

A curved tube segment, as shown in Fig. 4-8, has two distinct mode families in which it can undergo free vibration:

1) In-plane modes - In this mode family, all tube motion occurs in the plane of the tube. Two displacement functions are required to describe this motion. They are displacements $W(\theta)$ and $U(\theta)$, as depicted in Fig. 4-8.
2) Out-of-plane modes - In this mode family, two displacements are also required to describe the tube motion. They are tube rotation, $\gamma(\theta)$, and tube lateral displacement, normal to the plane of the tube, $V(\theta)$, as depicted in Fig. 4-8.

In this section, the basic equations used in developing the exact solution for curved-tube free vibration have been extracted from Gorman (1988). We will begin by presenting the free vibration theory related to in-plane vibration.

4.4.1 Theory of Curved Tube In-Plane Free Vibration

It is shown in the literature (e.g., Lee, 1971) that the quantity $W(\theta)$ may be written as,

$$W(\theta) = \sum_{J=1}^{6} C_J e^{\lambda_J \theta} \tag{4-41}$$

where coordinate θ is as defined in Fig. 4-8 and constants C_J are to be determined through enforcement of boundary conditions. The quantities λ_J are the six roots of the algebraic equation

$$\lambda^6 + 2\lambda^4 + (1+p)\lambda^2 + p = 0 \tag{4-42}$$

where p is the free vibration eigenvalue defined as,

$$p = \frac{mR^4\omega^2}{EI} \tag{4-43}$$

and where parameters m, ω, E, and I are as defined earlier in the straight tube analysis. Here we render the eigenvalue dimensionless by means of the dimension R, the radius of curvature of the curved tube (Fig. 4-7). The symbol p is used here to designate the eigenvalue for curved and U-tube analysis in order to remain consistent with symbols used in the literature.

Equation (4-42) is seen to be cubic in the quantity λ^2.

In conducting an in-plane free vibration analysis of a curved tube, several computational steps are required. First, we must choose a trial eigenvalue, p, which is known to be lower than the actual first-mode eigenvalue. We then proceed to solve for those values of λ^2, ie., λ_1^2, λ_2^2, and λ_3^2 which satisfy Eq. (4-42). One of the values of λ^2, let us say λ_1^2, will always be negative and real. This quantity can always be readily evaluated by means of a simple computer program, i.e., we search for this value of λ^2 which satisfies Eq. (4-42). Equation (4-42) is thereby reduced to a quadratic equation and we can write,

$$\lambda_{2,3}^2 = \frac{-(\lambda_1^2 + 2) \pm \sqrt{(\lambda_1^2 + 2)^2 + 4p/\lambda_1^2}}{2} \tag{4-44}$$

Given the positive root values p_1 and p_2 where $p_1 = 0.11340$, and $p_2 = 17.637$, it is shown that three possible forms of solution exist for the quantity $W(\theta)$ as follows:

1) For $p \geq p_2$, λ_2^2 and λ_3^2 will each be real and positive. $W(\theta)$ will then be written as,

$$W(\theta) = C_1 \sin(G_1\theta) + C_2 \cos(G_2\theta) + C_3 \sinh(G_3\theta)$$
$$+ C_4 \cosh(G_4\theta) + C_5 \sinh(G_5\theta) + C_6 \cosh(G_6\theta) \tag{4-45}$$

where,

$$G_1 = G_2 = +\sqrt{|\lambda_1^2|}; G_3 = G_4 = +\sqrt{|\lambda_2^2|}; G_5 = G_6 = +\sqrt{|\lambda_3^2|}$$

and C_1, C_2, etc., are constants to be determined.

For small tube displacements (a condition which we assume to be satisfied), extensions of the tube centerline will be small and we are entitled to write,

$$U(\theta) = \frac{dW(\theta)}{d(\theta)} \tag{4-46}$$

From Eq. (4-45) we, therefore, obtain,

$$
\begin{aligned}
U(\theta) = {} & C_1 G_1 \cos(G_1\theta) - C_2 G_2 \sin(G_2\theta) + C_3 G_3 \cosh(G_3\theta) \\
& + C_4 G_4 \sinh(G_4\theta) + C_5 G_5 \cosh(G_5\theta) + C_6 G_6 \sinh(G_6\theta)
\end{aligned} \tag{4-47}
$$

2) For $p_1 < p < p_2$, the roots λ_2^2 and λ_3^2 will form a complex conjugate pair and we may write,

$$\lambda_{1,2}^2 = \alpha \pm i\gamma \tag{4-48}$$

where,

$\alpha = -(\lambda_1^2 + 2)/2$ and $\gamma = \frac{1}{2}\sqrt{-4p/\lambda_1^2 - (\lambda_1^2 + 2)^2}$

It is convenient now to introduce the following quantities,

$$\psi = \tan^{-1}(\gamma/\alpha), r = (\alpha^2 + \gamma^2)^{1/4} \sin(\psi/2)$$
$$\text{and } s = (\alpha^2 + \gamma^2)^{1/4} \cos(\psi/2)$$

$W(\theta)$ may then be written as,

$$
\begin{aligned}
W(\theta) = {} & C_1 \sin(\lambda_1\theta) + C_2 \cos(\lambda_1\theta) + C_3 \sin(r\theta)\cos(s\theta) \\
& + C_4 \sin(r\theta)\cosh(s\theta) + C_5 \cos(r\theta)\sinh(s\theta) + C_6 \cos(r\theta)\cosh(s\theta)
\end{aligned} \tag{4-49}
$$

and, upon differentiating, we have,

$$
\begin{aligned}
U(\theta) = {} & C_1\lambda_1 \cos(\lambda_1\theta) - C_2\lambda_1 \sin(\lambda_1\theta) \\
& + C_3\{r\cos(r\theta)\sinh(s\theta) + s\sin(r\theta)\cosh(s\theta)\} \\
& + C_4\{r\cos(r\theta)\cosh(s\theta) + s\sin(r\theta)\sinh(s\theta)\} \\
& + C_5\{s\cosh(s\theta)\cos(r\theta) - r\sin(r\theta)\sinh(s\theta)\} \\
& + C_6\{s\sinh(s\theta)\cos(r\theta) - r\sin(r\theta)\cosh(s\theta)\}
\end{aligned} \tag{4-50}
$$

3) For $p \le p_1$, λ_2^2 and λ_3^2 will each be real and negative. When we introduce the quantities

$$G_1 = G_2 = +\sqrt{|\lambda_1^2|}, \quad G_3 = G_4 = +\sqrt{|\lambda_2^2|}, \text{and } G_3 = G_4 = +\sqrt{|\lambda_3^2|}$$

we obtain,

$$
\begin{aligned}
W(\theta) = {} & C_1 \sin(G_1\theta) + C_2 \cos(G_2\theta) + C_3 \sin(G_3\theta) \\
& + C_4 \cos(G_4\theta) + C_5 \sin(G_5\theta) + C_6 \cos(G_6\theta)
\end{aligned} \tag{4-51}
$$

and upon differentiating we obtain,

$$
\begin{aligned}
U(\theta) = {} & C_1 G_1 \cos(G_1\theta) - C_2 G_2 \sin(G_2\theta) + C_3 G_3 \cos(G_3\theta) \\
& - C_4 G_4 \sin(G_4\theta) + C_5 G_5 \cos(G_5\theta) - C_6 G_6 \sin(G_6\theta)
\end{aligned} \tag{4-52}
$$

We now have available all possible solution forms for curved tube segments undergoing in-plane free vibration.

4.4.2 Theory of Curved Tube Out-of-Plane Free Vibration

For this free vibration mode family, only the tube transverse lateral motion $V(\theta)$ and rotational motion $\gamma(\theta)$ (Fig. 4-8) will be non-zero.

Introducing the quantity $B(\theta)$, where $B(\theta) = R\gamma(\theta)$, it is shown in the literature (e.g., Lee, 1971) that we may write,

$$B(\theta) = \sum_{J=1}^{6} C_J e^{\lambda_J \theta} \tag{4-53}$$

where the C_J are constants to be determined based on the tube boundary conditions and the λ_j^2 are the six roots of the algebraic equation,

$$\lambda^6 + 2\lambda^4 + (1 - pk)\lambda^2 + p = 0 \tag{4-54}$$

which is cubic in the quantity λ^2.

Here $k = GI_0/EI$, where G is the modulus of elasticity in shear of the tube material and I_0 is the polar moment of inertia of the tube cross-section. Here p is used to represent the free vibration eigenvalue of tubes in out-of-plane vibration. For this mode family, p is defined as follows,

$$p = \frac{mR^4\omega^2}{kEI} \tag{4-55}$$

There is a strong similarity between the above Eq. (4-54) and Eq. (4-55) and the corresponding Eq. (4-42) and (4-43) related to in-plane vibration. Again the first root, λ_1^2, of Eq. (4-54) will be real and negative. It is easily obtained by a numerical search once a trial value for the parameter p is selected. The other two roots λ_2^2 and λ_3^2 are again obtained by means of Eq. (4-54).

The form of solution for the quantity $B(\theta)$ will depend on the trial value p and the quantities p_1 and p_2 in a manner identical to that described for the in-plane modes. However, now p_1 and p_2 take on the values as given immediately below,

$$p_1 = \frac{(2k^2 + 9k + 27/4) - \sqrt{(2k^2 + 9k + 27/4)^2 - 4(k^4 + k^3)}}{2k^3} \tag{4-56}$$

and,

$$p_2 = \frac{(2k^2 + 9k + 27/4) + \sqrt{(2k^2 + 9k + 27/4)^2 - 4(k^4 + k^3)}}{2k^3} \tag{4-57}$$

For each of the possible ranges of the parameter p, i.e., $p \geq p_2$, $p_1 \leq p \leq p_2$, and $p \leq p_1$, the solutions for $B(\theta)$ are identical in form to corresponding solutions provided for the quantity $W(\theta)$ in Eq. (4-45), Eq. (4-49) and Eq. (4-51).

We next turn our attention to the quantity $V(\theta)$. It is shown in the literature (e.g., Lee, 1971) that the quantities $B(\theta)$ and $V(\theta)$ are related as follows,

$$V(\theta) = \frac{1}{(1 + k)} \left\{ \int \left(\int \left(\int B(\theta) d\theta \right) d\theta - kB(\theta) \right\} \tag{4-58}$$

where the constants of integration vanish for all possibilities of interest here.

Utilizing the appropriate expressions for $B(\theta)$ and performing the integration indicated above, one obtains the expressions for $V(\theta)$ as follows,

For $p \geq p_2$,

$$V(\theta) = -C_1 G_{1k} \sin(G_1\theta) - C_2 G_{2k} \cos(G_2\theta) + C_3 G_{3k} \sinh(G_3\theta)$$
$$+ C_4 G_{4k} \cosh(G_4\theta) + C_5 G_{5k} \sinh(G_5\theta) + C_6 G_{6k} \cosh(G_6\theta) \tag{4-59}$$

where,

$$G_{1k} = \{k + 1/G_1^2\}/(1+k), G_{2k} = \{k + 1/G_2^2\}/(1+k)$$
$$G_{3k} = \{1/G_3^2 - k\}/(1+k), G_{4k} = \{1/G_4^2 - k\}/(1+k)$$
$$G_{5k} = \{1/G_5^2 - k\}/(1+k), \text{ and } G_{6k} = \{1/G_6^2 - k\}/(1+k)$$

and G_1, G_2 etc., are as defined for Eq. (4-51).

For $p_1 \leq p \leq p_2$,

$$V(\theta) = CEE(1)\cos(G_1\theta) - CEE(2)\sin(G_2\theta)$$
$$+ CEE(3)\cos(r\theta)\sinh(s\theta) + CEE(4)\sin(r\theta)\cosh(s\theta)$$
$$+ CEE(5)\cos(r\theta)\cosh(s\theta)$$
$$+ CEE(6)\sin(r\theta)\sinh(s\theta) \tag{4-60}$$

where, $G_1 = G_2 = +\sqrt{|\lambda_1^2|}$ and r and s are as defined for Eq. (4-49).

$$CEE(1) = -C_2 G_{2k}, CEE(2) = C_1 G_{1k}, CEE(3) = C_5 G_{5k} + C_4 H_{4k}$$
$$CEE(4) = C_4 G_{4k} + C_5 H_{5k}, CEE(5) = C_6 G_{6k} + C_3 H_{3k}$$

and $\quad CEE(6) = C_3 G_{3k} + C_6 H_{6k}$

$$G_{1k} = \frac{1}{k+1}\{k + 1/G_1^2\}, G_{2k} = G_{1k}$$

$$G_{3k} = \frac{1}{k+1}\left\{\frac{s^2 - r^2}{(s^2 + r^2)^2} - k\right\}, G_{4k} = G_{3k}$$

$$H_{3k} = \frac{1}{k+1}\left\{\frac{-2rs}{(r^2 + s^2)^2}\right\}, H_{4k} = H_{3k}$$

For $p \leq p_1$

$$V(\theta) = -C_1 G_{1k} \sin(G_1\theta) - C_2 G_{2k} \cos(G_2\theta) + C_3 G_{3k} \sin(G_3\theta)$$
$$+ C_4 G_{4k} \cos(G_4\theta) + C_5 G_{5k} \sin(G_5\theta) + C_6 G_{6k} \cos(G_6\theta) \tag{4-61}$$

We now have available all possible solutions for the displacement and rotation functions related to out-of-plane free vibration of curved tube segments.

4.5   Free Vibration Analysis of U-Tubes

We are now in a position to compute free vibration frequencies and associated mode shapes of any practical steam generator U-tubes. The main task we face is to develop the eigenvalue matrix associated with the tube of interest. As will be seen, this involves enforcing prescribed boundary conditions at the extremities of each straight tube and curved tube span of the U-tube as well as enforcing appropriate continuity conditions at each inter-span location. All the free vibration mode shapes of such tubes will be symmetric or anti-symmetric with respect to the uppermost point on

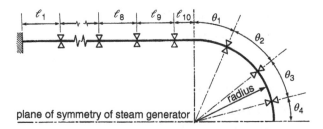

Fig. 4-9 View of Half of Steam Generator U-Tube, Referred to as a Half-Tube, for Tube Possessing Geometric Symmetry. Distances are Measured from Left End of Spans.

the curved portion of the tube. In fact, it will become clear that we need analyze only one-half of such U-tubes, as shown schematically in Fig. 4-9. We will begin by setting up the equations necessary for analyzing such tubes.

In U-tube bundles, the tube extremities usually pass through the same circular tubesheet during assembly. Each tube will have one extremity pass through a perforation on one-half of the tubesheet while the other extremity will pass through a matching perforation on the other half of the tubesheet.

Similar to straight tube bundles, the tube ends are rigidly attached to the tubesheet, usually by means of the tube rolling process. Again, as in the case of straight tube bundles, the U-tubes receive lateral support from baffle plates located at certain locations along the circular tank or shell. Conventional baffle plates cannot be employed in the U-bend region of the bundle. Instead, in modern U-tube bundles, this support is provided by what are called "flat bar restraints" or "anti-vibration bars", bars inserted radially into the bundle between the tubes in the curved tube region, or by some other means.

A schematic view of a typical steam generator U-tube is depicted in Fig. 4-7. For the moment, we will consider a U-tube which, unlike the one shown in the figure, possesses complete symmetry with respect to a central vertical plane running perpendicular to the plane containing the U-tube. The symmetry involves not only the geometry of the straight and curved portions of the tube but also the distribution of the tube interior supports.

Shell-side flow in a U-tube steam generator generally enters at the tubesheet in cross flow with the tube bundle and then turns to go vertically upwards in parallel with the straight-tube section. Shell-side fluids will normally exit the shell at the end of the shell opposite from the tubesheet passing in cross flow across the U-bend section of the U-tube.

4.5.1 Setting Boundary Conditions for the In-Plane Free Vibration Analysis of U-Tubes Possessing Geometric Symmetry

In the work to follow, all tube displacements, eigenvalues, etc., will be rendered dimensionless through division by the curved tube radius, R. Accordingly, the tube bending moments and transverse shear forces for straight and curved tubes, respectively, are written as:

For straight tube segments,

$$M(\xi) = \frac{EI}{R^2}\frac{d^2r(\alpha\xi)}{d\xi^2}, Q(\xi) = \frac{EI}{R^3}\frac{d^3r(\xi)}{d\xi^3}$$

and for curved tube segments (in-plane modes)

$$M(\theta) = \frac{EI}{R^2}\left\{\frac{d^2U(\theta)}{d\theta^2} + \frac{dW(\theta)}{d\theta}\right\},$$

$$\text{and } Q(\theta) = \frac{EI}{R^3}\left\{\frac{d^3U(\theta)}{d\theta^3} + \frac{d^2W(\theta)}{d\theta^2}\right\},$$

For curved tube segments (out-of-plane modes)

$$M(\theta) = \frac{EI}{R^2}\left\{B(\theta) - \frac{d^2V(\theta)}{d\theta^2}\right\}$$

$$Q(\theta) = \frac{EI}{R^3}\left\{(1 + k)\frac{dB(\theta)}{d\theta} + k\frac{dV(\theta)}{d\theta} - \frac{d^3V(\theta)}{d\theta^3}\right\}$$

The twisting moment in the curved section equals M_Z where,

$$M_Z(\theta) = \frac{kEI}{R^2}\left\{\frac{dB(\theta)}{d\theta} + \frac{dV(\theta)}{d\theta}\right\}$$

The parameters β and ξ, when conducting U-tube analysis are as follows,

$$\beta^4 = m\omega^2 R^4/EI$$

and ξ equals displacement along the straight tube divided by R.

We begin the analysis by considering the first (lower) span of the straight section of the U-tube, Fig. 4-9. Assuming the lower end to be clamped and enforcing conditions of zero lateral displacement at each extremity of this span, it is readily shown that we may write,

$$r_1(\xi) = A_1\left\{\sin(\beta\xi) - \sinh(\beta\xi) + \frac{\sin(\beta\mu_1) - \sinh(\beta\mu_1)}{\cosh(\beta\mu_1) - \cos(\beta\mu_1)}(\cos(\beta\xi) - \cosh(\beta\xi))\right\} \quad (4\text{-}62)$$

where the subscript 1 indicates the first span and μ_1 equals the span length divided by R. Only minor changes would be required in the above equation should the tube lower end be simply supported or free (see Gorman, 1975).

We next examine an intermediate span along the straight tube section, i.e., a straight span lying between the first and last straight span of the straight tube section. It will have zero lateral displacement at each extremity and it is easily shown that its lateral displacement along the span can be expressed as,

$$r_n(\xi) = A_n\{\sin(\beta\xi) - \theta_n\sinh(\beta\xi)\}$$
$$+ B_n\{\cos(\beta\xi) - \cosh(\beta\xi) + \phi_n\sinh(\beta\xi)\} \quad (4\text{-}63)$$

where,

$$\theta_n = \frac{\sin(\beta\mu_n)}{\sinh(\beta\mu_n)} \text{ and } \phi_n = \frac{\cosh(\beta\mu_n) - \cos(\beta\mu_n)}{\sinh(\beta\mu_n)}$$

Subscript n refers to the nth span along the straight tube section. There will be two unknown constants, A_n and B_n, associated with each intermediate span.

We utilize subscript N to designate the final straight span in the straight tube segment. Enforcing a condition of zero lateral displacement at the supported end of this span. it is easily shown that we may write,

$$r_N(\xi) = A_N \sinh (\beta\xi) + B_N \{ \cos (\beta\xi) - \cosh (\beta\xi) \} + C_N \sinh (\beta\xi) \tag{4-64}$$

where the three unknowns, A_N, B_N, C_N, are to be evaluated later.

It will be obvious that, at each intermediate support point along the straight tube segment, two homogeneous algebraic equations can be written relating the unknowns A_1, A_2, B_2, etc. This is accomplished through enforcing conditions of continuity of tube slope and bending moment at the point of support. We turn next to conditions to be enforced at the junction of the last straight span and the first curved span. Having selected a trial eigenvalue for the tube free vibration mode under study, we find that expressions for the tube in-plane displacements, $U(\theta)$ and $W(\theta)$ in the curved spans are already available with their unknown constants.

Focusing on the straight-tube-curved-tube junction, we must enforce conditions of continuity of lateral displacement, slope, bending moment and transverse shear force. We also enforce the condition that $W(\theta)$ equals zero at the junction. These continuity and boundary conditions are depicted schematically at the left end of Fig. 4-10.

To complete construction of the eigenvalue matrix, we must look at conditions to be enforced at the junctions between curved spans as well as conditions to be enforced at the upper terminus of the half-tube upon which attention is focused. At curved-span junctions, we must enforce conditions of zero axial and lateral displacement for each of the adjacent curved spans. We must also enforce conditions of continuity of slope and bending moment at the junction. Again, these conditions are represented schematically in Fig. 4-10.

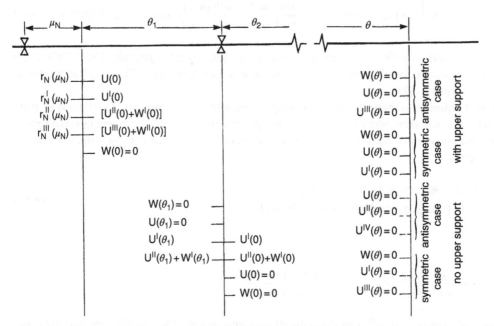

Fig. 4-10 Schematic Representation of Boundary and Continuity Conditions to be Enforced at Straight-Tube-Curved-Tube Junction, Curved-Tube-Curved-Tube Junction, and at Upper Extremity of Half-Tube for In-Plane Vibration Analysis of U-Tube Possessing Geometric Symmetry.

We next examine conditions to be enforced at the upper terminus of the half-tube. It is necessary to provide appropriate equations for the case where a support point exists at the upper terminus as well as the case where such a support does not exist. We must also, for each of these cases, provide appropriate equations for the analysis of symmetric modes as well as for anti-symmetric modes.

Consider first the case where a support exists at the upper extremity. Then for symmetric mode analysis we have $W(\theta) = U(\theta) = U^{I}(\theta) = 0$, where primes indicate differentiation with respect to θ. For the anti-symmetric case, we will have $W(\theta) = U(\theta) = U^{III}(\theta) = 0$.

Next, consider the case where no support exists at the upper extremity. For the symmetric mode analysis, we will have $W(\theta) = U^{I}(\theta) = U^{III}(\theta) = 0$; while, for the anti-symmetric modes, we will have $U(\theta) = U^{II}(\theta) = U^{IV}(\theta) = 0$. All of these conditions are represented schematically at the right side of Fig. 4-10.

4.5.2 Development of the In-Plane Eigenvalue Matrix for a Symmetric U-Tube

It is now considered appropriate to describe preparation of the eigenvalue matrix for the above in-plane free vibration problem. This will serve as a model for all other U-tube free vibration problems to be investigated.

The eigenvalue matrix is the coefficient matrix for the entire set of homogeneous algebraic equations discussed above. These homogeneous equations relate the unknown constants appearing in the displacement expressions related to the set of straight and curved spans of which the half-tube is composed.

For illustrative purposes, we will consider the case where the half-tube is composed of four straight ($N = 4$) and three curved spans ($m = 3$). The total number of straight span unknowns, $A_1, A_2,....A_N, B_N, C_N$, will equal $2N$. In the curved tube section, there will be six unknowns per span for a total of 18 with the combined number of unknowns equal to 26. A review of the number of homogeneous algebraic equations relating these unknowns, as discussed above, reveals that there are a total of 26 such equations available for the system under consideration. As indicated earlier, the eigenvalue matrix for the problem under investigation is, in fact, the coefficient matrix of the above set of algebraic equations.

In the free vibration analysis of such a problem, we begin by selecting a trial eigenvalue which we know to be lower than the actual first-mode eigenvalue. We then generate the eigenvalue matrix based on the trial eigenvalue. The next step is to obtain the determinant of the matrix and store it. We then augment the trial eigenvalue and repeat the above process until we find a trial eigenvalue for which the sign of the determinant changes. We then search for a trial eigenvalue, in between the last two, for which the determinant of the associated matrix is as close to zero as we choose. This latter eigenvalue will be the first-mode eigenvalue for the problem under study. By setting one of the non-zero unknowns, A_1, A_2, B_2, etc., equal to unity, we convert the original set of homogeneous equations to a non-homogeneous set. Upon solving this latter set of equations, and thereby assigning values to the unknowns, we can generate the associated mode shape.

We can then search for as many of the higher eigenvalues and associated mode shapes as we wish.

4.5.3 Generation of Eigenvalue Matrices for Out-of-Plane Free Vibration Analysis of U-Tubes Possessing Geometric Symmetry

Development of the eigenvalue matrix for this family of problems follows closely that described for the in-plane analysis. Only the differences will be elaborated upon here. Parameters β and ξ, are as defined for the in-plane mode analysis. Expressions for the straight-span displacements remain unchanged, as are the equations enforced at straight-tube internal supports. It is only at the

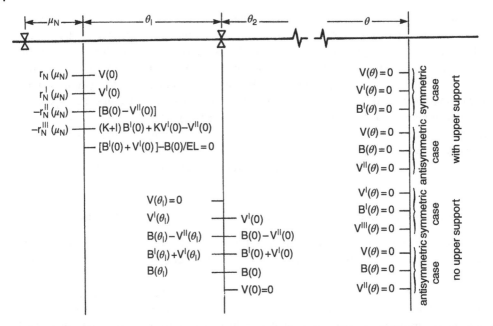

Fig. 4-11 Schematic Representation of Boundary and Continuity Conditions to be Enforced at Straight-Tube-Curved-Tube Junction, Curved-Tube-Curved-Tube Junction, and at Upper Extremity of Half-Tube for Out-of-Plane Vibration Analysis of U-Tube Possessing Geometric Symmetry.

straight-tube-curved-tube junction, as well as at all curved-tube supports and the curved-tube upper terminus, that the equations differ. Conditions to be enforced at the locations discussed immediately above are represented schematically in Fig. 4-11.

At the straight-tube-curved-tube junction, we must enforce continuity of tube lateral displacement and slope as well as bending moment and transverse shear force. The fifth equation is based on the assumption that the twisting moment in the curved span is equal to that in the straight span. The equation is based on the assumption that no rotation of the straight tube is permitted at the lower end and effects of rotary inertia of the straight tube may be neglected.

At the curved-span junctions, in addition to setting lateral displacement equal to zero, we must enforce continuity of slope, bending moment, twisting moment, and tube rotation. Again, conditions to be enforced at the tube upper terminus depend on whether symmetric or anti-symmetric modes are under study and whether or not a support exists at the tube upper terminus.

Example 4-5 Symmetric Steam Generator U-Tube Free Vibration Analysis

We consider an actual steam generator U-tube possessing geometric symmetry. The tube is of stainless steel with inner and outer diameters of 0.436 in. (11.07 mm) and 0.500 in. (12.7 mm), respectively. The straight-tube segments are composed of nine spans, each of 41.34 in. (1.05 m) length. The lower ends of the straight sections are clamped. The curved region of the tube begins immediately above the last straight-span support. Its radius of curvature, R, equals 63.9 in. (1.62 m) and the total subtended angle of the region equals 180 degrees. There are no supports in the curved region.

The entire tube is considered to be submerged in water and the interior of the tube is considered to be filled with water. It is necessary, therefore, to add the interior mass of liquid per unit length of tube as well as the mass of water per unit length displaced by the tube, to the tube mass per unit length in order to arrive at the total effective mass per unit length of the tube (see Section 4.2). All free vibration frequency and mode shape computations were conducted using the appropriate equations described herein. Here there is no straight span between the last straight tube support and the beginning of the curved tube section. We handle this problem following the analytical methods described earlier, by introducing a short straight span between the last straight-span tube support and the beginning of the curved-tube segment and letting the length of this short span become vanishingly small. Alternatively, we can neglect this short span and introduce special boundary and continuity conditions at the junction of the last straight span and the curved tube segment.

In-plane modes. The first two mode shapes for in-plane U-tube free vibration are plotted in Fig. 4-12. Support points are rotated into the plane of the paper so that lateral tube motion may be better viewed. It is seen that the first (lowest frequency) in-plane mode is anti-symmetric with respect to the tube mid-point. There exists only one node point in the curved tube region. This node point is necessary to satisfy continuity requirements as the tube is considered in-extensible. The second free vibration mode is obviously a symmetric one. For both of these modes, almost all the lateral motion occurs in the curved tube region because there is a fairly dense distribution of supports along the straight-tube segments, while there are no supports along the curved-tube segment.

Out-of-plane modes. The first two out-of-plane modes are plotted in Fig. 4-13. The first mode is a symmetric one and has no node points in the curved tube region. Its frequency is lower than the first in-plane mode frequency and it may be referred to, therefore, as the fundamental mode. The second out-of-plane mode is of the anti-symmetric type. Note that the tube natural frequencies are much lower than those normally encountered in steam generators or heat exchangers. This is because we have no tube support in the curved-tube region to better illustrate the mode shapes. In practically all real systems, tube supports would be installed in the curved-tube region and fundamental frequencies of tube vibration would be much higher.

The U-tube free vibration modes in Figs. 4-12 and 4-13 were chosen for illustrative purposes. The same free vibration analysis of these steam generator tubes could be conducted using the general

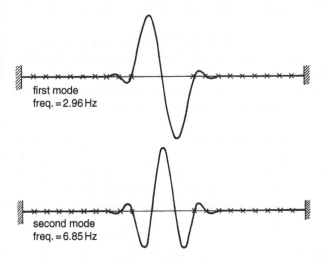

Fig. 4-12 Schematic Representation of First and Second Free Vibration In-Plane Modes of Illustrative Example.

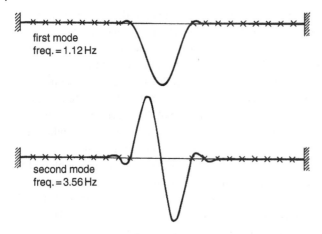

Fig. 4-13 Schematic Representation of First and Second Free Vibration Out-of-Plane Modes of Illustrative Example.

approach where advantage is not taken of geometric symmetry. The same frequencies and mode shapes are found; however, the eigenvalue matrices employed are necessarily twice as large

4.5.4 Free Vibration Analysis of U-Tubes Which Do Not Possess Geometric Similarity

U-tubes lacking geometric symmetry can be analyzed as follows. We can begin at one straight-tube extremity and develop the eigenvalue matrix by working along the straight tube, continuing along the curved-tube segment, and then continuing down along the other straight-tube leg. It would, of course, be necessary to impose appropriate displacement constraints and continuity conditions at the junctions of each pair of adjacent spans as well as appropriate boundary conditions at the U-tube lower ends. In what is to follow, we describe an approach which takes advantage of the analytical techniques already developed.

To begin, we divide the U-tube into two components, the division being made at a desirable location on the curved tube section. A system of equations is prepared for each component in a manner identical to that already described for the half-segment of a U-tube possessing geometric symmetry. The same computer routine can be employed for each segment. The major difference here is that a system of boundary and continuity conditions must be enforced at the junction of the two straight-tube-curved-tube segments.

Two types of U-tube may be encountered. In one type, there will be at least one support point in the curved-tube region. In this case, it is highly preferable for purposes of vibration analysis to divide the tube at one of these support points. If there are no support points in the curved-tube region, then it is preferable to divide the U-tube at the half-way point along the curved region.

U-tube with one or more support points located in the curved tube region. A typical arrangement of such a U-tube is depicted in Fig. 4-7. Let us first consider such a tube to be undergoing in-plane free vibration. It will be evident that at the point of separation (junction) of the two U-tube components, the separation plane in Fig. 4-7, we must require the lateral and axial displacements of each curved span adjacent to the junction to equal zero. Furthermore, conditions of

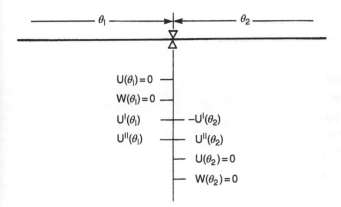

Fig. 4-14 Schematic Representation of Boundary and Continuity Conditions to be Enforced at Junction of U-Tube Components for In-Plane Vibration Analysis of U-Tubes Lacking Geometry Symmetry and with Supports in Curved Region.

continuity of slope and bending moment must also be enforced across the junction. These conditions are depicted schematically in Fig. 4-14.

Consider next the case of out-of-plane free vibration for the same tube. At the junction of the two components, as discussed above, we must require the lateral displacement of each curved span adjacent to the junction to equal zero. Conditions of continuity of slope, bending moment, twisting moment and tube rotation must also be enforced across the junction. These conditions are depicted schematically in Fig. 4-15.

U-tube with no support points located in the curved tube region. It is convenient to analyze such tubes in a manner identical to the U-tubes with one or more supports, above. The two U-tube components are considered to be joined at the extreme uppermost point on the U-tube i.e., midway along the curved section. Complete continuity must be enforced at the common junction between the two tube segments. It is shown that this is achieved for in-plane mode analysis by equating the quantities which appear opposite each other in Fig. 4-16.

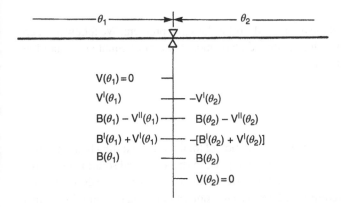

Fig. 4-15 Schematic Representation of Boundary and Continuity Conditions to be Enforced at Junction of U-Tube Components for Out-of-Plane Vibration Analysis of U-Tubes Lacking Geometry Symmetry and with Supports in Curved Region.

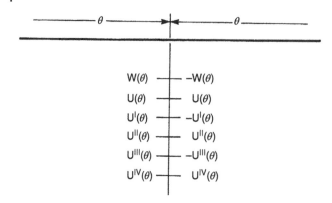

Fig. 4-16 Schematic Representation of Boundary and Continuity Conditions to be Enforced at Junction of U-Tube Components for In-Plane Vibration Analysis of U-Tubes Lacking Geometric Symmetry and without Supports in Curved Region.

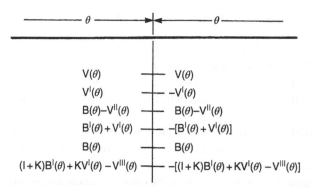

Fig. 4-17 Schematic Representation of Boundary and Continuity Conditions to be Enforced at Junction of U-Tube Components for Out-of-Plane Vibration Analysis of U-Tubes Lacking Geometric Symmetry and without Supports in Curved Region.

We turn next to analysis of out-of-plane mode analysis for the same problem. We follow the steps described immediately above; however, for the out-of-plane modes, we must equate the quantities which appear opposite each other in Fig. 4-17.

4.6 Concluding Remarks

This chapter has provided an introduction to the calculation of total mass, including hydrodynamic mass and inside fluid mass. This knowledge was then applied to the analytical solution of the free vibration of straight tubes and U-tubes. This chapter provides the reader with an orderly and concise presentation of solutions to the required differential equations and boundary conditions which must be enforced to arrive at the appropriate eigenvalue matrix. The chapter showed how advantage could be taken of symmetry in the vibration modes when symmetry exists. With the aid of this chapter, the reader should be able to prepare the eigenvalue matrix for any U-tube of interest and have the formulae necessary to develop a digital computer program to conduct the associated free vibration analysis.

Knowledgeable readers will appreciate that the analytical analysis can be modified to satisfy conditions other than those described in this chapter. For example, one may introduce elastic supports at the extremities of the U-tubes. Furthermore, one is free to employ different tube properties or mass distributions along different tube spans. Tube added mass may be considered to be lower in the curved tube region as rising liquid is converted to steam.

Nowadays, engineering problems are often solved with numerical methods such as finite element analysis for solid mechanics problems or computational fluid dynamics for fluid dynamics problems. Such methods are practical, efficient, and no longer limited by computer power. Alternately, algebraic or analytical methods such as the approach presented in this chapter may be used. There are cases, such as time domain simulation, where a combination of numerical and analytical approaches is advantageous. In any case, an understanding of the basic equations, as presented in this chapter, is essential.

Nomenclature

A, B, C, D	= coefficients
$B(\theta)$	= $R\gamma(\theta)$
C_h	= confinement coefficient
C_J	= six constants defined by Eq. (4-41) and boundary conditions
$CEE(J)$	= six constants defined by Eq. (4-60)
D	= tube outside diameter, m
D_{ci}	= surrounding-cylinder inner diameter, m
D_i	= tube inside diameter, m
D_e	= effective or hydraulic diameter, m
E	= Young's modulus of tube material, Pa
f	= tube vibration frequency, Hz
f_a	= tube vibration frequency in air, Hz
G	= modulus of elasticity in shear of tube material, Pa
G_J	= absolute values of λ_J
G_{Jk}, H_{Jk}	= constants defined by Eq. (4-60) and Eq. (4-61)
I	= area moment of inertia of tube cross-section, m^4
I_o	= polar moment of inertia of tube cross-section, m^4
k	= GI_o/EI
$\ell_1, \ell_2, \ell_3...$	= straight span length, m
L	= tube length, m
M	= bending moment in the tube, N · m
M_z	= twisting moment in the tube, N · m
m	= total (effective) mass per unit length of tube, kg/m
m_h	= hydrodynamic mass per unit length of tube, kg/m
m_i	= mass of the internal fluid per unit length of tube, kg/m
m_t	= tube mass per unit length of tube, kg/m
N	= total number of spans
n	= span number for internal spans from 2 to N-1

P	= tube pitch, m
p	= eigenvalue symbol used in curved tube spans, m
p	= free vibration eigenvalue for curved tube
p_1, p_2	= intermediate limits in range of eigenvalues
Q	= transverse shear force in tube, N
R	= radius of curvature of curved tube region, m
$r(\xi)$	= lateral displacement in straight tube spans, m
r, s, ψ	= constants defined by Eq. (4-49)
t	= time, s
$U(\theta)$	= in-plane lateral displacement in curved tube spans, m
$V(\theta)$	= out-of-plane lateral displacement in curved tube spans, m
$W(\theta)$	= in-plane longitudinal displacement in curved tube spans, m
$W(\xi)$	= lateral displacement of straight tube span divided by L
$w(x,t)$	= lateral displacement of straight tube span, m
x	= distance measured along the tube, m
α	= phase angle, degrees
λ_i	= i^{th} mode eigenvalue
λ_J	= six roots of the algebraic equation defined by Eq. (4-42)
ξ	= distance along straight span divided by L
θ	= angular coordinate measured along curved spans, degrees
θ_n, ϕ_n	= coefficients defined in Eq. (4-40)
ω	= circular frequency of tube free vibration, rad/s
β	= eigenvalue symbol used in straight tube spans
$\gamma(\theta)$	= angular rotation of curved tube in out-of-plane vibration, rad/s
$\mu_1, \mu_2, \mu_3...$	= dimensionless straight span length
ρ	= external-fluid density, kg/m^3
ρ_g	= gas-phase density, kg/m^3
ρ_i	= internal-fluid density, kg/m^3
ρ_ℓ	= liquid-phase density, kg/m^3
ρ_t	= tube density, kg/m^3

References

Au-Yang, M. K., 2001, *Flow-Induced Vibration of Power and Process Plant Components*, Professional Engineering Publishing Ltd, New York, NY.

Blevins, R. D., 1991, *Flow-Induced Vibration*, 2nd Ed., Van Nostrand Reinhold, New York.

Chen, S. S., 1987, *Flow-Induced Vibration of Circular Cylindrical Structures*, Hemisphere Publishing Corporation, Washington.

Darnley, E. R., 1921, "The Transverse Vibration of Beams and the Whirling of Shafts Supported at Intermediate Points," *Phil. Mag, S6*, 41(241), pp. 81–97.

Gorman, D. J., 1975, *Free Vibration Analysis of Beams and Shafts*, J. Wiley and Sons, Inc., Toronto.

Gorman, D. J., 1988, "Exact Analytical Solutions for the Free Vibration of Steam Generator U-Tubes," *ASME Journal of Pressure Vessel Technology*, 110, pp. 422–429.

Lee, L. S., 1971, "Vibration of U-Bend Segments of Heat Exchanger Tubes," *Atomic Energy of Canada Ltd. Report, AECL-3735.*

Moretti, P. M. and Lowery, R. L., 1976, "Hydrodynamic Inertia Coefficients for a Tube Surrounded by Rigid Tubes," *ASME Journal of Pressure Vessel Technology,* 98, pp. 190–193.

Pettigrew, M. J., Tromp, J. H., and Mastorakos, J., 1985, "Vibration of Tube Bundles Subjected to Two-Phase Cross-Flow," *ASME Journal of Pressure Vessel Technology,* 107, pp. 335–343.

Pettigrew, M. J., Taylor, C. E. and Kim, B. S., 1989, "Vibration of Tube Bundles in Two-Phase Cross Flow Part 1: Hydrodynamic Mass and Damping, *ASME Journal of Pressure Vessel Technology,* 111, pp. 466–477.

Pettigrew, M. J., Taylor, C. E. and Kim, B. S., 2001, "The Effects of Bundle Geometry on Heat Exchanger Tube Vibration in Two-Phase Cross Flow," *ASME Journal of Pressure Vessel Technology,* 123, pp. 414–420.

Rogers, R. J., Taylor, C. E. and Pettigrew, M. J., 1984, "Fluid Effects on Multispan Heat Exchanger Tube Vibration," *ASME PVP Conference,* San Antonio, Texas, June 17-21, H00316.

Taylor, C. E. and Pettigrew, M. J., 2001, "Effect of Flow Regime and Void Fraction on Tube Bundle Vibration," *ASME Journal of Pressure Vessel Technology,* 123, pp. 407–413.

5

Damping of Cylindrical Structures in Single-Phase Fluids

Michel J. Pettigrew

5.1 Introduction

Several flow-induced vibration excitation mechanisms may lead to severe vibration in heat exchangers and other process equipment. These mechanisms are fluidelastic instability, periodic wake shedding, turbulence excitation, and acoustic resonance. Excessive flow-induced vibration may cause failures either by fatigue or fretting wear. To avoid such problems, a thorough flow-induced vibration analysis considering all the above vibration excitation mechanisms should be carried out at the design stage. Such analyses require information about damping.

This chapter addresses the question of damping of cylindrical structures such as heat exchanger tubes, nuclear fuels, and piping elements. The different energy dissipation mechanisms that contribute to damping are discussed. The available experimental data from the literature and from our own measurements is reviewed and analyzed. Three important energy dissipation mechanisms emerge.

In liquids, these are: viscous damping between tube and liquid, squeeze-film damping in the clearance between tube and support, and friction damping at the support. Viscous damping only accounts for approximately 25% of the total damping of a typical heat exchanger tube. Thus, about 75% of the damping energy is dissipated at the support. Squeeze-film damping appears to be the most important energy dissipation mechanism. Squeeze-film damping is related to support thickness and is inversely proportional to tube frequency.

In gases, friction damping at the supports is dominant and is related to support thickness.

5.2 Energy Dissipation Mechanisms

Generally, there are several possible energy dissipation mechanisms that could contribute to damping, namely:

Tube related:

- internal or material damping,
- viscous damping between structure and fluid,
- flow-dependent damping due to fluid flow around tubes,
- damping due to fluid flow inside piping elements, and
- damping due to two-phase flow.

Flow-Induced Vibration Handbook for Nuclear and Process Equipment, First Edition.
Michel J. Pettigrew, Colette E. Taylor, and Nigel J. Fisher.
© 2022 John Wiley & Sons, Inc. This Work is a co-publication between ASME Press and John Wiley & Sons, Inc.

Support related:

- friction at the tube-to-tubesheet joints,
- sliding friction between tubes and tube supports,
- energy dissipated by tube-to-support impact and resulting traveling waves,
- squeeze-film damping in the clearance between tubes and supports, and
- viscous shear damping between tubes and supports.

As observed by Goyder (1982), there are two principal types of tube motion at the support: rocking motion and lateral motion, as shown in Fig. 5-1. The motion at any given support may be a combination of rocking motion and lateral motion. Rocking-type motion is predominant in the lower vibration modes. Damping due to rocking motion is likely to be less than for lateral motion and, thus, may be more relevant in practice.

The dynamic interaction between tubes and tube supports may be characterized as three main types: sliding, impacting and scuffing. Scuffing is impacting at an angle followed by sliding (Fig. 5-2). In practice, the dynamic interaction between a tube and each tube support is likely to be a combination of the above. The energy dissipated by sliding is due to friction and is related to the product of contact force and displacement, as discussed further in Chapter 12. The energy dissipated by impact is in the form of local deformation of the support followed by stress wave propagation in the support and local deformation of the tube followed by high-frequency traveling waves in the tube.

Material damping is very small in heat exchanger tubing. Haslinger and Thompson (1980) measured damping ratios of roughly 0.01% on tubes welded or brazed at both ends with no intermediate support. For such welded tubes, material damping is dominant since the contribution from other energy dissipation mechanisms is minimal. Thus, generally, the contribution of material damping may be neglected.

Since viscosity and density are very small in air and light gases, tube-to-fluid viscous damping, squeeze-film damping and viscous shear damping are not significant in heat exchangers with gas on the shell side. However, these damping mechanisms may be significant in heavy gases (e.g., high pressure gas) and are very important in liquids and two-phase flows.

Fluid flow through a tube bundle will affect damping because of flow-dependent damping and fluidelastic forces. However, it is very difficult to separate energy dissipating fluid forces from fluidelastic excitation forces in experiments with flexible tube bundles. There is limited information on flow-dependent damping in tube arrays. Chen and Jendrzejczyk (1980) show that damping is not

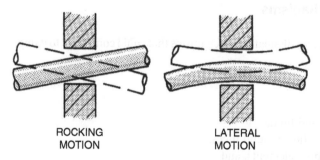

ROCKING
MOTION

LATERAL
MOTION

Fig. 5-1 Types of Tube Motion at Support Location.

(a)

SLIDING

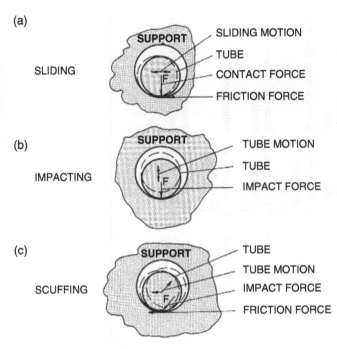

(b)

IMPACTING

(c)

SCUFFING

Fig. 5-2 Types of Dynamic Interaction between Tube and Tube Support.

very dependent on flow velocity below the critical velocity for fluidelastic instability. In this chapter, flow-dependent damping forces are considered together with fluidelastic instability forces.

Some researchers describe the decrease in damping at flow velocities close to critical flow velocities for fluidelastic instability as negative damping. Instead, we consider that the apparent reduction in damping is part of the fluidelastic instability phenomenon.

Fluid flow inside a tube is unlikely to affect damping unless the flow velocity is very high (i.e., approximately that required for axial-flow fluidelastic instability). Such internal flow velocities are not normally encountered in process equipment. However, two-phase flow or sloshing due to partially filled tubes would contribute to damping. Scriven and Hopley (1980), Charreton et al (2015), and Gravelle et al (2007) have studied this phenomenon. This source of damping depends very much on flow regime and void fraction. It would be risky to rely on the contribution of this energy dissipation mechanism unless we have specific data for the situation being analyzed.

Some energy may be dissipated due to friction at the tube-to-tubesheet joints. Such dissipation will obviously depend on the type of tube joint. Friction is insignificant in welded joints as explained earlier. For a tube clamped at both ends but without intermediate support, we have measured damping ratios of roughly 0.07% over a frequency range of 40 to 150 Hz (Pettigrew et al, 1985a). In a different experiment on a cantilever tube hydraulically expanded in one tubesheet, damping was roughly 0.2% (Pettigrew et al, 1989). Haslinger and Thompson (1980) measured somewhat less damping, 0.02%, on large single-span tubes with rolled joints at both ends. Measurements on bent tubes (i.e., large bends for thermal expansion) welded at both ends with an intermediate rolled joint, yielded damping values of 0.4-0.8% (Aita et al, 1983). The above damping data is outlined in Table 5-1. Thus, friction at the tube joints may contribute some damping, depending on the type of joint. However, since it is relatively small, it is prudent not to consider it unless it is reasonably well known for the heat exchanger design being considered.

Table 5-1 Damping in Tube-to-Tubesheet Joints (Pettigrew et al, 1986a).

f (Hz)	Diam. (mm)	Wall (mm)	Span Length Configuration	Material Tube/Tubesheet	Amp. (μm)	Method	Mode No.	Damping ζ(%)	Reference	Comments
40-150	15.9	1.59	0.50 – 1.52 m	Incoloy 800/ Brass	100	Log Dec	1	0.07	Pettigrew et al (1985c)	Clamped joints
33	13.0	1.07	600 mm	St. St. 304/ Carbon Steel		Log Dec Random	1	0.20	Pettigrew et al (1989)	Hydraulic expansion joint
762	16.9	3.45	325 mm	St. St. 304		Log Dec Sweep	1	0.011	Haslinger and Thompson (1980)	Brazed joint, single tube
770-798	16.9	3.45	325 mm	St. St. 304		Log Dec Sweep	1	0.004 - 0.009	Haslinger and Thompson (1980)	Brazed joint, tube bundles
123-129	64	13	1.5 m	St. St. 304		Log Dec Sweep	1	0.026 (0.020 - 0.034)	Haslinger and Thompson (1980)	Rolled joint
14.5	25		1 m 1.1 m 1 m	Incoloy 800	500	Log Dec Random	1	0.04-0.08	Aita et al (1983)	Welded joints intermediate rolled joint

Friction due to relative motion between a tube and its tube supports is a significant energy dissipation mechanism. In liquids, there may be some lubrication between tubes and tube supports, preventing metal-to-metal contact. Such lubrication could alter the nature of the friction forces. On the other hand, there should be some viscous shear damping forces due to sliding between tubes and tube supports. The exact nature of sliding forces in liquids is not yet well understood.

At present, we do not know how to separate the energy dissipated by tube impact and the resulting traveling waves from the other damping mechanisms. Impact damping may be much less in liquids than in gases. Impact forces are probably attenuated and their contact rise time lengthened by squeeze-film forces, which would tend to reduce higher-frequency traveling waves. Thus, we do not consider impact damping as a separate energy dissipation mechanism. It may be considered together with friction damping or squeeze-film damping.

In summary, tube-to-fluid viscous damping, squeeze-film damping, and friction damping are the most important energy dissipation mechanisms in liquids. As will be seen later, squeeze-film damping in the clearance between tube and support is particularly important. The formulation of these mechanisms is developed in Section 5.6.

Damping due to a two-phase mixture surrounding the tubes can be very significant (Carlucci and Brown, 1983; Pettigrew et al, 1985a, 1989; and Pettigrew and Taylor, 2004). This situation exists in many heat exchangers such as boilers, condensers, and nuclear steam generators. Two-phase damping depends on such parameters as mass flux, void fraction, flow geometry, and flow regime. This topic is discussed in detail in Chapter 6.

5.3 Approach

Heat exchanger tube dynamics is inherently a non-linear phenomenon since it depends largely on the dynamic interaction between tubes and tube supports, particularly so for squeeze-film and friction-type damping that depend entirely on the relative motion between tubes and tube supports. The analysis of this phenomenon would require a time domain non-linear simulation of the tube dynamics in which the details of sliding and impacting between tubes and tube supports are modeled. Unfortunately, we are still lacking some of the detailed information required to model the dynamic interaction between tubes and tube supports. Furthermore, the required non-linear analysis is difficult and is not yet generally available to the designer. Some progress has been made in this area with the development of codes such as VIBIC (Fisher et al, 2005), H3DMAP (Morandin and Sauvé, 2003) and SGFW-PC (Rao et al, 1977), to predict fretting wear of heat exchanger tubes. These codes could be adapted to predict damping of heat exchanger tubes.

However, for the time being, most designers are limited to quasi-linear vibration analyses for which they have to obtain an equivalent linear damping value corresponding to realistic vibration amplitudes (Pettigrew et al, 1985b and Gorman, 1988). At first sight, this approach may appear contradictory since the diametral clearance between tubes and supports is in the range of 0.25-0.8 mm. However, in well-designed heat exchangers, the tube vibration response at mid-span is mostly less than 100 μm root-mean-square (rms) and much less at the supports (usually less than 25 μm rms), which is significantly less than the available diametral clearance. Thus, the tubes do not generally vibrate back and forth across the available diametral clearance. Instead, most tubes are not centered within the supports and are, therefore, contacting or vibrating very close to one side of the supports. Given some tube out-of-straightness and tube support misalignment, it is difficult to imagine that many tubes of typically one-meter span length would be located concentrically within, say a

0.40 mm diametral clearance without touching the support. The chances of a tube not touching a support are very small. In summary, the vibration response is significantly larger than the motion at the supports and the system may be modeled as a quasi-linear system with equivalent linear damping over the small range of amplitude of interest. Thus, it is reasonable to assume pinned conditions to allow for a quasi-linear analysis to calculate the vibration response.

The objective of this chapter is to recommend appropriate damping values based on available data. When the development in the understanding of damping described in this chapter took place, the following heat exchanger tube damping references were available: Pettigrew et al (1986a and 1986b), Pettigrew et al (1985c and 1985d), Carlucci and Brown (1983), Taylor et al (1998), Hartlen (1974), Chen (1983), Collinson and Warneford (1978), Mulcahy (1980), Jendrzejcyk (1984), Rogers et al (1984), Shin et al (1977), Kim et al (1988), Chen et al (1976), Rogers and Ahn (1986), Collinson and Taylor (1982), Lowery and Moretti (1975), Blevins (1975), Goyder (1982a, 1982b). In addition, there is some useful damping information buried in a number of other publications related to heat exchanger vibration, as given in the list of references. Most of this information was qualified as being tentative or preliminary, which is not surprising considering the non-linear nature of the problem and the difficulties inherent in measuring small damping values. Our approach was to review the available data and formulate a conservative but realistic minimum damping criterion based on the main energy dissipation mechanisms discussed earlier. We take the lower decile of all the available data as a minimum damping level. The lower decile means that 10% of the data is below the minimum damping criterion. This is reasonable since the smaller damping values measured are usually the more reliable and since vibration problems are usually associated with lower damping values. The higher damping values reported may be due to significant preloads at the supports caused by slightly bent tubes or misaligned supports. In any case, the higher damping values are of no concern here since they do not lead to vibration problems.

A sample of the liquid damping data extracted from the available references is outlined in Table 5-2. There is field data on real multi-span heat exchanger tube bundles and laboratory data on single tubes in the references listed above. The tube configurations included straight, curved, U-tube, single-span through to more than twenty spans. The end conditions included cantilever, clamped-clamped and pinned-pinned. The methods used by the researchers to analyse the damping data included log decrement, sine sweep, energy sweep and random approaches.

This damping data expressed in terms of the damping ratio, ζ, is generally higher than 0.6%. The damping ratio is the ratio of the damping coefficient, C, over the critical damping coefficient, C_c. The data was analysed to find trends and to formulate a damping criterion in terms of tube and tube-support parameters.

5.4 Damping in Gases

In gases, the damping data expressed as the damping ratio, ζ, varied from 0.2 to 8%, which is not very useful information from a design viewpoint. Thus, the data was analyzed to find trends and to outline the more relevant parameters. Damping values and important parameters such as tube vibration frequency, f, diameter, D, support thickness, L, span length, ℓ, diametrical clearance between tube and support, D_S - D, wall thickness, t, tube and support materials, and number of spans, N, were tabulated in the form of a database to facilitate the analysis. Our approach was to establish a conservative but realistic minimum damping criterion based on analysis of the above data. As discussed earlier, we propose to use the lower decile of the available data as a minimum

Table 5-2 Sample Database Page on Heat Exchanger Tube Damping in Liquids (Mostly Water).

Data No.	Freq. f (Hz)	Diam. D (mm)	Wall t (mm)	Tube Mass m_t (kg/m)	Total Mass m (kg/m)	Span Length ℓ_1 (mm)	ℓ_2 (mm)	ℓ_3 (mm)	Clear. $D_s - D$ (mm)	Thick L (mm)	P/D^*	Damping Total ζ_T (%)	Viscous ζ_V (%)	Support ζ_S (%)	Spans N	Mode No.
1	30.5	15.9	3.2	0.92	1.2	1170	1070	558	0.25	12.7	∞	2	0.21	1.79	6	1
2	30.5	15.9	3.2	0.92	1.2	1170	1070	558	0.25	19	∞	11	0.21	10.79	6	1
3	30.5	15.9	3.2	0.92	1.2	1170	1070	558	0.25	38.1	∞	2.8	0.21	2.59	6	1
4	40.3	15.9	3.2	0.92	1.2	1170	1070	558	0.76	38.1	∞	8	0.18	7.82	6	
5	31.2	15.9	3.2	0.92	1.2	1170	1070	558	0.25	38.1	∞	5	0.21	4.79	6	1
6	30.6	15.9	3.2	0.92	1.2	1170	1070	558	0.51	38.1	∞	2	0.21	1.79	6	1
7	33.1	15.9	3.2	0.92	1.2	1170	1070	558	0.76	38.1	∞	4	0.20	3.80	6	
8	41.5	15.9	3.2	0.92	1.2	1170	1070	558	0.25	38.1	∞	8	0.18	7.82	6	
9	42.5	15.9	3.2	0.92	1.2	1170	1070	558	0.51	38.1	∞	4	0.18	3.82	6	
10	64.2	15.9	3.2	0.92	1.2	734	734	734	0.25	38.1	∞	2	0.15	1.85	7	
11	30	15.9	3.2	0.92	1.2						∞	2.6	0.21	2.39		
12	37	15.9	3.2	0.92	1.2						∞	2.4	0.19	2.21		
13	45	15.9	3.2	0.92	1.2						∞	1.7	0.18	1.52		
14	44.4	19	1.57	0.755	1.202	1450	1450	1450	0.8	12.7	1.25	1.6	0.43	1.34	8	
15	33.7	19	1.57	0.755	1.202	1450	1450	1450	0.8	12.7	1.25	2.5	0.49	2.30	8	
16	39.8	13	1.04	0.39	0.566	890	890	0	0.4	25.4	1.57	2.16	0.45	1.71	2	
17	63.5	13	1.04	0.39	0.566	890	890	0	0.4	25.4	1.57	1.15	0.36	0.79	2	
18	119	13	1.1	0.326	0.508	483	483	483	0.4	12.7	1.47	0.74	0.31	0.47	10	2
19	154	13	1.1	0.326	0.508	483	483	483	0.4	12.7	1.47	0.74	0.28	0.52	10	
20	156	13	1.1	0.326	0.508	483	483	483	0.4	12.7	1.47	0.51	0.27	0.26	10	

(Continued)

Table 5-2 (Continued)

Data No.	Freq. f (Hz)	Diam. D (mm)	Wall t (mm)	Tube Mass m_t (kg/m)	Total Mass m (kg/m)	Span Length ℓ_1 (mm)	Span Length ℓ_2 (mm)	Span Length ℓ_3 (mm)	Clear. $D_s - D$ (mm)	Thick L (mm)	P/D^*	Damping Total ζ_T (%)	Damping Viscous ζ_V (%)	Damping Support ζ_S (%)	Spans N	Mode No.
21	199	13	1.1	0.326	0.508	483	483	483	0.4	12.7	1.47	0.66	0.24	0.46	10	
22	225	13	1.1	0.326	0.508	483	483	483	0.4	12.7	1.47	0.73	0.23	0.56	10	
23	234	13	1.1	0.326	0.508	483	483	483	0.4	12.7	1.47	0.66	0.22	0.48	10	
24	219	13	1.1	0.326	0.508	483	483	483	0.4	12.7	1.47	0.66	0.23	0.48	10	
25	246	13	1.1	0.326	0.508	483	483	483	0.4	12.7	1.47	0.56	0.22	0.38	10	
26	130	13	1.1	0.326	0.508	483	483	483	0.4	12.7	1.47	0.95	0.30	0.72	10	
27	219	13	1.1	0.326	0.508	483	483	483	0.4	12.7	1.47	0.67	0.23	0.49	10	
28	246	13	1.1	0.326	0.508	483	483	483	0.4	12.7	1.47	0.7	0.22	0.53	10	
29	60	19	1.09	0.503	0.925	1099	819	610	0.63	25.4	1.33	1.8	0.42	1.38	5	
30	100	19	1.09	0.503	0.925	1099	819	610		25.4	1.33	1.7	0.32	1.38	5	

*For a single tube (without surrounding tubes), $P/D = \infty$

damping level. This approach is reasonable since the smaller damping values measured are usually the more reliable and since vibration problems are usually associated with lower damping values. On the other hand, taking the lower decile is not unduly conservative. The higher damping values reported may have been due to anomalies such as crooked tubes or misaligned supports and are probably not representative of normal heat exchanger tubes.

The database was reviewed in an attempt to find trends or significant parameters as follows.

5.4.1 Effect of Number of Supports

Assuming that span length and all other tube parameters are kept constant, the total vibration energy in a tube is proportional to the number of spans. The energy dissipated by friction at the support is obviously related to the number of supports.

In a two-span heat exchanger tube with one support, there is the vibration energy of two spans but only one support to dissipate energy. Thus, damping should be less than for a tube with a large number of spans and a large number of supports. There appeared to be such a trend in the available damping data. Accordingly, the data was normalized such that:

$$\zeta_n = \zeta N/(N-1) \tag{5-1}$$

Where ζ_n is the normalized damping ratio and N is the number of spans. Normalized damping ratios are shown in Fig. 5-3 versus tube vibration frequency.

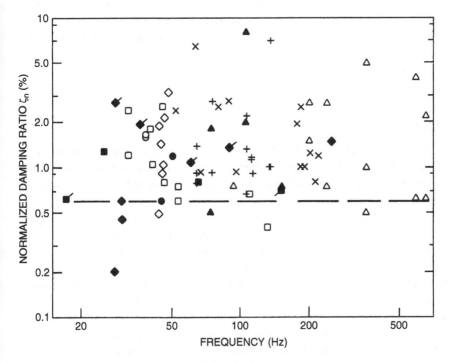

Fig. 5-3 Damping of Heat Exchanger Tubes in Air.

5.4.2 Effect of Frequency

From the data of Fig. 5-3, it is not possible to establish a trend with frequency. There is no obvious reason why there should be a trend with frequency. Thus, frequency does not appear to be a significant parameter.

5.4.3 Vibration Amplitude

There is no conclusive trend of damping as a function of amplitude. Very often, the amplitude is not given with the damping measurements. Some researchers say that damping decreases with vibration amplitude. For sliding-type damping, a decreasing trend would make sense since friction forces do not increase with amplitude, as would be expected for linear damping. Thus, it would be reasonable to expect damping to decrease with amplitude.

Other researchers have found damping to increase with amplitude, which may be explained as follows. At low amplitude, the tube is in contact with the support and sliding-type friction dominates. At higher amplitude, the tube may start rattling within the tube support. This rattling would cause impact-type damping and non-linearities which would tend to increase damping. In practice, both sliding and impacting are possible and a definite trend of damping as a function of amplitude has not been established.

5.4.4 Effect of Diameter or Mass

Large and massive tubes should experience large friction forces and the energy dissipated should be large. However, the potential energy in the tube would also be proportionally greater in more massive tubes. Thus, the damping ratio, which is related to the ratio of energy dissipated per cycle to the potential energy in the tube, should be independent of tube size or mass. No meaningful trend over the range of diameters 13 to 25 mm was observed, although this parameter was not specifically studied in any of the publications reviewed.

5.4.5 Effect of Side Loads

In real heat exchangers, side loads, also called preloads, are possible due to misalignment of the tube supports or due to fluid drag forces. Side loads may increase or reduce damping. Small side loads may prevent impacting and, thus, reduce damping, whereas large side loads may increase damping by increasing friction.

Goyder (1982) measured damping ratios as low as 0.2% in an experiment on a carefully aligned tube with a single support. When the support was misaligned, the damping increased to 0.8%. In both cases, the tube was impacting against the support in a rocking-type motion.

With lateral-type motion, which is uncommon for low frequency modes of vibration, the reverse was observed. Damping increased from 1% for high side load to 8% for zero side load. This trend is because tube impacting with lateral motion tends to decrease with side load.

In practice and in the majority of experiments reviewed here, misalignment and side load are not a controlled parameter, which probably explains the large scatter in the damping data. Unfortunately, the designer cannot take advantage of this parameter to increase damping. Therefore, realistic minimum damping values should be used in design.

5.4.6 Effect of Higher Modes

Damping appears to decrease with mode order for mode orders higher than the number of spans (Blevins, 1990, and Halle and Wambsganss, 1980). This trend is not surprising since these higher order modes involve relatively less interaction between tube and tube support. However, these higher modes are very rarely the cause of problems in practice.

5.4.7 Effect of Support Thickness

There is clear indication in the reviewed data that support thickness is a dominant parameter. Blevins (1990) and Hartlen (1974) found in laboratory experiments that damping is roughly proportional to support thickness (see Fig. 5-4). The available data is plotted against support thickness in Fig. 5-5. Beyond 15 mm thickness, it is not clear that thicker supports increase damping.

The relationship with support thickness may be explained. The energy dissipated by friction is proportional to the product of contact force and displacement. The contact force should be independent of support thickness. On the other hand, the displacement should be proportional to support thickness for rocking-type motion with sliding interaction with the tube support, as shown in Fig. 5-6a. Damping due to rocking-type motion with impact interaction should also increase with support thickness, because impact forces are expected to increase with support thickness, as shown in Fig. 5-6b.

Damping due to lateral-type motion, however, should not be affected by support thickness. Thus, friction damping due to the combination of rocking and lateral motion should be somewhat less dependent on support thickness. Consequently, damping in realistic heat exchanger configurations should also be less dependent on support thickness than the controlled laboratory experiments of Blevins (1990) and Hartlen (1974), as discussed below.

We tried several approaches to correlate damping with support thickness. We know that the total vibrating energy of a tube is related to its length, ℓ_t. On the other hand, the total energy dissipated at the supports is related to their thickness, L. Therefore, damping should be a function of the

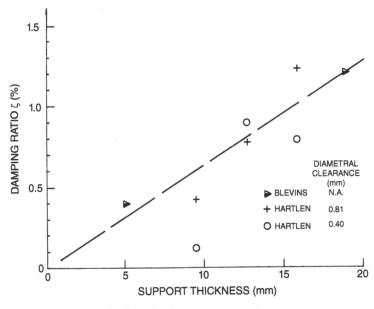

Fig. 5-4 Effect of Tube Support Thickness on Damping in Gases (Air).

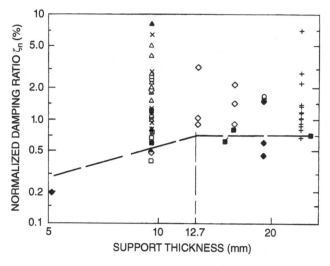

Fig. 5-5   Effect of Tube Support Thickness on Normalized Damping Ratio in Gases (Air).

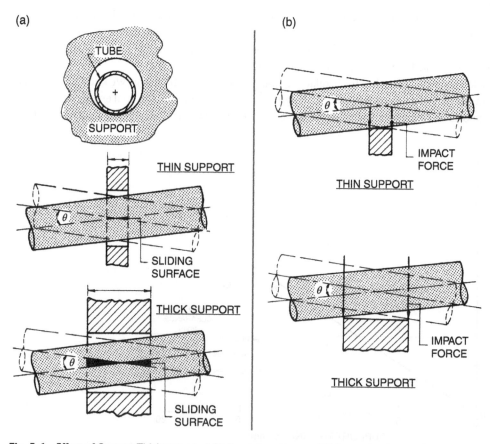

Fig. 5-6   Effect of Support Thickness on a) Sliding Interaction, and b) Impacting Interaction.

dimensionless number (L/ℓ_t). For convenience, we use a characteristic span length, ℓ_m, to represent tube length. We take ℓ_m as the average of the three longest spans. This approach is based on the assumption that lower modes, which involve primarily longest spans, are more vulnerable to vibration. Modes involving shorter spans are less vulnerable since the ratio of energy dissipated at the support to the vibration energy in the tube (i.e., the damping) is larger. This is a crude attempt at weighting the effect of mode shape in the damping analyses.

The normalized damping ratio is presented as a function of (L/ℓ_m) in Fig. 5-7. It appears that an exponent of 0.5, i.e., $(L/\ell_m)^{0.5}$, best fits the minimum damping values. This somewhat-less-than-unity exponent is probably due to the combination of rocking and lateral motion, as discussed earlier. However, the trend is not very clear. It is obvious that additional data is required to establish the effect of support thickness with certainty. Figure 5-8 shows the term $\zeta_n(L/\ell_m)^{-0.5}$, versus tube vibration frequency. It suggests a minimum reasonable damping level.

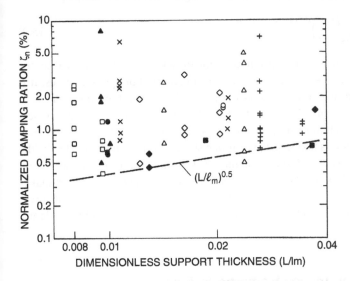

Fig. 5-7 Effect of Dimensionless Support Thickness (L / ℓ_m) on Normalized Damping Ratio.

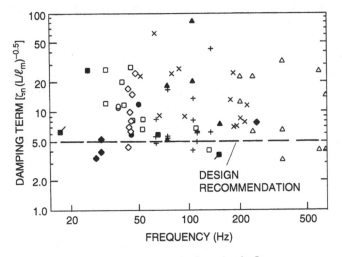

Fig. 5-8 Design Recommendation for Damping in Gases.

5.4.8 Effect of Clearance

At first glance, it is difficult to find a correlation between damping and tube-to-support clearance. For the normal range of tube-to-support diametrical clearances (i.e., 0.40 to 0.80 mm), there is no conclusive trend in the reviewed damping data. For very small clearances, in the order of 0.20 mm, damping appears to be larger, Hartlen (1974).

We conducted further parametric analysis of the data in an attempt to find some correlation between damping and tube-to-support clearance. Both the dimensionless thickness, (L/ℓ_m), and the dimensionless clearance, defined as $(D_S-D)/D$, were used in the analysis. We found that damping appears to increase slightly with smaller clearances. The minimum damping line roughly fits the dimensionless clearance parameter to the exponent -0.2. If the effect of clearance is taken into account, it appears that the exponent of (L/ℓ_m) is around 0.4 rather than 0.5. Using the term $\zeta_n/[D/(D_S-D)]^{0.2}$, which is damping normalized for clearance, leads to the following formulation of damping:

$$\zeta = 1.6\left(\frac{N-1}{N}\right)\left(\frac{L}{\ell_m}\right)^{0.4}\left(\frac{D}{D_S-D}\right)^{0.2} \tag{5-2}$$

The above interpretation of the data is highly speculative, due to lack of data. For this reason, the effect of diametral clearance will not be considered in the formulation of design guidelines. It is included here for the purpose of discussion and to encourage further research in this area.

Although surprising at first, the relatively weak effect of clearance may be explained. A heat exchanger tube would normally touch most supports on one side or the other. Thus, the dynamic interaction between tube and support is taking place near one side and is not much affected by the proximity of the opposite side, which depends on the diametral clearance. This is true in most cases, except when the vibration amplitude is very large or the clearance is very small. Therefore, it makes sense that damping in gases should not be much affected by clearance. It is possible that side forces in the support may be larger for very small clearances, thus explaining the somewhat higher damping in the latter case.

5.5 Design Recommendations for Damping in Gases

Following the above discussions, we have derived some design guidelines to evaluate damping. Our approach is to obtain a reasonable minimum damping value. This is achieved by taking the damping level at roughly the lower decile of the available damping data, as already explained.

A simple recommendation emerges from Fig. 5-3. Most damping values are above 0.6%. The lowest value on this figure is probably not relevant since it was obtained with an unusual tube configuration comprising a single span within two clearance supports (Pettigrew et al, 1986a). Thus, a practical design guideline would be to take 0.6% as the damping ratio for heat exchanger tubes with gases on the shell side.

A more physically-based expression may be formulated taking into account the dominant effect of support thickness. From Figs. 5-7 and 5-8, we may derive,

$$\zeta = 5.0\left(\frac{N-1}{N}\right)\left(\frac{L}{\ell_m}\right)^{0.5} \tag{5-3}$$

where ζ is the damping ratio in percent.

This expression fits the minimum damping data reasonably well. We recommend this expression for design applications since it is more rigorous and less conservative in some cases. The above guidelines apply to the normal range of heat exchanger tubes, i.e.: diameters from 12 to 25 mm, support thicknesses from 6 to 25 mm, diametrical clearances from 0.4 to 0.8 mm, and frequencies from 20 to 600 Hz. This range of parameters corresponds to the available damping data.

The above guidelines are somewhat tentative. Fundamental work is required to establish the trends with more certainty and to confirm the form of the expressions. For example, the effect of clearance, the effect of tube and support materials, the contribution of impact damping and the effect of support thickness, in particular narrow supports, should be studied. Additional measurements on realistic heat exchanger tube configurations are necessary to validate the damping expressions from a statistical point-of-view. However, the above guidelines provide a practical and reasonable minimum damping criterion based on the currently available data.

In summary, design guidelines are recommended to evaluate damping of heat exchanger tubes with gases on the shell-side. Support thickness is a dominant parameter, whereas diametric clearance between tube and support is much less important. Other parameters such as frequency, mass and diameter do not appear relevant.

Example 5-1   Calculation of Damping in a Gas Heat Exchanger

Suppose the hypothetical Tube No. 1 for which $N = 5$, $L = 15$ mm, and $\ell_m = 0.6$ m. For this tube, the damping ratio is calculated using Eq. (5-3) to be $\zeta = 0.63\%$

$$\zeta = 5.0 \left(\frac{N-1}{N}\right) \left(\frac{L}{\ell_m}\right)^{0.5} = 5.0 \left(\frac{5-1}{5}\right) \left(\frac{0.015}{0.6}\right)^{0.5} = 0.63\%$$

Suppose now Tube No. 2 for which $N = 10$, $L = 10$ mm, and $\ell_m = 0.6$ m. For this second tube $\zeta = 0.58\%$.

5.6   Damping in Liquids

The damping data extracted from the available references is plotted in Fig. 5-9. This data is for multi-span heat exchanger tubes in water. The data was tabulated in a database to facilitate analysis.

This damping data expressed in terms of the damping ratio, ζ, is generally higher than 0.6% and appears to be dependent on frequency. This data was analyzed to find trends and to formulate a damping criterion in terms of tube and tube-support parameters.

5.6.1   Tube-to-Fluid Viscous Damping

Of the three main energy dissipation mechanisms, viscous damping between tube and fluid is the easiest to formulate. In still fluid, the contribution of viscous fluid forces to damping is related to the Stokes Number, S or $\pi f D^2 / 2v$, where v is the kinematic viscosity of the liquid, and to the degree of confinement of the heat exchanger tube within the tube bundle, as shown in Eq. (5-4):

$$\zeta_V \, \alpha \left(\frac{\pi f D^2}{2v}\right)^{-0.5} \tag{5-4}$$

where ζ_V is the viscous damping ratio, f is the tube vibration frequency, and D is the tube diameter. This implies that viscous damping is inversely proportional to the square root of frequency. Chen

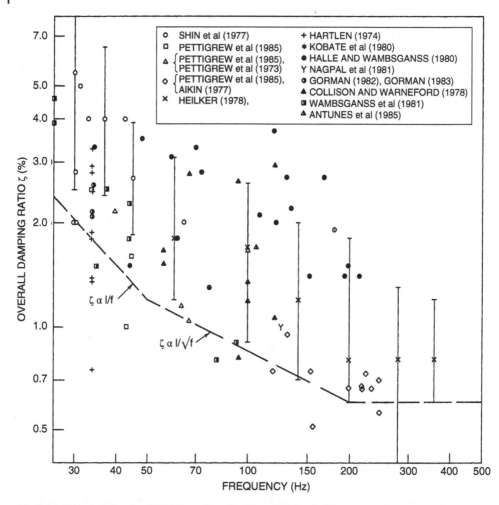

Fig. 5-9 Damping Data for Multi-Span Heat Exchanger Tubes in Water.

(1983) developed a comprehensive theory for viscous damping of a cylinder in confined viscous fluid. Rogers et al (1984) developed a simplified version, valid for $S > 3300$ and $D/D_e < 0.5$, which covers most heat exchangers. Their simplified expression for ζ_V (in percent) is:

$$\zeta_V = \frac{100\pi}{\sqrt{8}} \left(\frac{\rho D^2}{m}\right)\left(\frac{2v}{\pi f D^2}\right)^{0.5} \left\{ \frac{\left[1 + \left(\frac{D}{D_e}\right)^3\right]}{\left[1 - \left(\frac{D}{D_e}\right)^2\right]^2} \right\} \tag{5-5}$$

where ρ is the density of the fluid, m is the total mass of the tube per unit length and includes the mass of fluid inside the tube, m_i, the tube mass, m_t, and the hydrodynamic mass, m_h (also called added mass), and D_e is the effective diameter of the surrounding tubes or the inside diameter of the surrounding cylinder if the tube is in an annulus.

The effective diameter is derived from data for a tube surrounded by rigid tubes. For a triangular tube bundle, Rogers et al (1984) found:

$$D_e/D = (0.96 + 0.50P/D)P/D \qquad (5\text{-}6)$$

Similarly, for a square tube bundle, we found (Pettigrew et al, 1989):

$$D_e/D = (1.07 + 0.56P/D)P/D \qquad (5\text{-}7)$$

where P is the pitch between tubes. The hydrodynamic mass $m_h = C_m(\pi\rho D^2/4)$ where C_m is a coefficient to take into account the effect of confinement, i.e.,

$$C_m = \frac{\left[1 + \left(\frac{D}{D_e}\right)^2\right]}{\left[1 - \left(\frac{D}{D_e}\right)^2\right]} \qquad (5\text{-}8)$$

The assumption of a cylinder surrounded by rigid cylinders was made for simplicity. The dynamics of a cylinder surrounded by flexible cylinders is much more complicated because the hydrodynamic coupling between cylinders makes many coupled modes of vibration possible. Evaluation of the damping for each hydrodynamically coupled mode is beyond the state-of-the-art. At flow velocities below that required for fluidelastic instability, several modes of vibration are generally present at the same time. We assume that the average damping of these modes is similar to the damping of a cylinder surrounded by rigid cylinders. Obviously, this question requires further study.

Equation (5-5) was verified against available data found in several references: Kim et al (1988), Skop et al (1976), Rogers (1985), Jendrzejczyk (1984), and Chen et al (1976). The agreement between experiment and theory is generally good, as shown in Fig. 5-10. The experimental values are slightly lower than calculated for more viscous liquids such as oils. The lower damping values measured in oils are probably due to the approximate nature of Eq. (5-5) at lower Stokes Numbers,

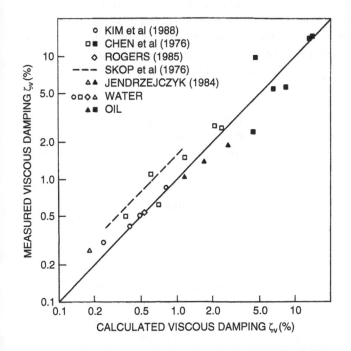

Fig. 5-10 Viscous Damping Data for a Cylinder in Confined (Chen et al, 1976) and Unconfined Liquids (All Other Data): Comparison Between Theory and Experiment.

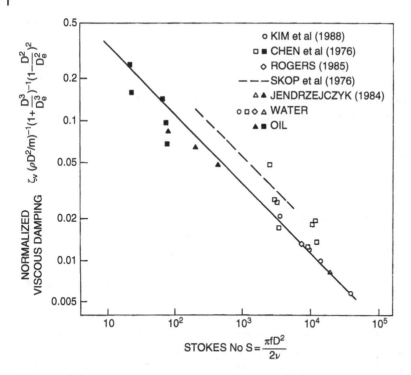

Fig. 5-11 Viscous Damping of Cylinders in Liquids Versus Stokes Number.

as shown in Fig. 5-11. However, this effect is not very significant from a practical point-of-view. Thus, we recommend the use of Eq. (5-5) to evaluate tube-to-fluid viscous damping.

For every data point of Fig. 5-9, we calculated the viscous damping, and subtracted it from the overall damping, ζ_T. The results, which represent damping at the supports, ζ_S, are presented in Fig. 5-12. They include squeeze-film damping, ζ_{SF}, and friction damping, ζ_F, thus:

$$\zeta_T - \zeta_V = \zeta_S = \zeta_{SF} + \zeta_F \tag{5-9}$$

Interestingly, the viscous damping between tube-and-fluid is relatively small, and is generally between 10% and 30% of the overall damping. Thus, most of the energy dissipation takes place at the supports. We will see later that friction damping appears to be less important, meaning that squeeze-film damping is dominant.

The above finding is particularly important for crudded heat exchangers. If the clearance gap between tubes and tube supports is solidly jammed with crud, support damping becomes very small. The tubes would then be more prone to excessive vibration and perhaps fatigue problems.

5.6.2 Damping at the Supports

The principal energy dissipation mechanisms at the supports, squeeze-film damping and friction damping, obviously depend on the dynamic interaction between tubes and tube supports as discussed in Section 5.2. In real heat exchangers, the dynamic interaction between tubes and supports is generally a combination of sliding, impacting, and scuffing and involves both lateral and rocking motion. Damping due to rocking motion should be less than that due to lateral motion since it

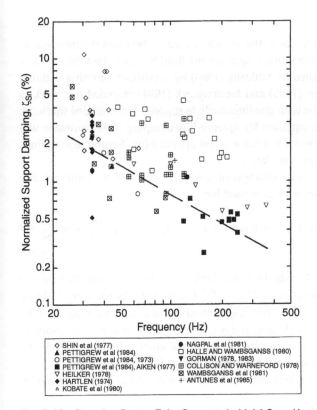

Fig. 5-12 Damping Due to Tube Supports in Multi-Span Heat Exchanger Tubes.

involves less relative motion at the supports. Rocking motion is predominant in the lower modes, which are generally more vulnerable to vibration problems. Thus, rocking motion may be more important in practice.

The impacting category of dynamic interaction involves both squeeze-film and impact damping. As discussed in Section 5.2, we cannot at this stage treat damping due to impacting and the resulting traveling waves separately from the other damping mechanisms.

Also, the energy dissipated by sliding may be due to metal-to-metal contact friction forces or to viscous shear forces or a combination of the two. These considerations apply to both lateral and rocking-type motion.

The total vibration energy in a tube is proportional to the number of spans, N, assuming that span length and all other tube parameters are kept constant. The energy dissipated at the clearance supports by squeeze-film damping and friction damping is obviously related to the number of clearance supports, $(N - 1)$. In a two-span heat exchanger tube with one support, there is the vibration energy of two spans but only one support to dissipate energy. Thus, damping should be less than for a tube with a large number of spans and supports. Accordingly, the data was normalized such that

$$\zeta_{Sn} = \zeta_S N/(N - 1) = (\zeta_T - \zeta_V)N/(N - 1) \tag{5-10}$$

where ζ_{Sn} is the normalized damping related to the supports. Figure 5-12 shows ζ_{Sn} as a function of frequency.

5.6.3 Squeeze-Film Damping

Squeeze-film damping is due to the fluid forces in the annular clearance between the tube and its supports. Fluid forces are generated as tube motion squeezes the fluid between the tube and tube support. This phenomenon was first studied by Mulcahy (1980) for a cylinder vibrating within a concentric cylinder of finite length. Chen (1983) and Jendrzejczyk (1984) extended this work to the case of a two-span heat exchanger tube with one intermediate support. We extended this work further to a multi-span tube to develop an equation for squeeze-film damping. This expression will be subsequently reviewed in terms of relevance, accuracy and applicability. Several assumptions and linearizations will be required, as explained later.

The equation of motion of a multi-span heat exchanger tube with N spans and N - 1 intermediate concentric tube supports (see Fig. 5-13) may be expressed by:

$$EI\frac{\partial^4 u}{\partial x^4} + \left[C_V + \overline{C}_{SF}\delta(x_1) + \overline{C}_{SF}\delta(x_2) + \dots + \overline{C}_{SF}\delta(x_J) + \dots + \overline{C}_{SF}\delta(x_{N-1})\right]\frac{\partial u}{\partial t} + m\frac{\partial^2 u}{\partial t^2} = 0$$

$$(5\text{-}11)$$

where EI is the flexural rigidity; u is the tube vibration amplitude at any point x along the tube and at any time t; $x_1, x_2, x_J,$ and $x_{(N-1)}$ are the locations of the $1^{st}, 2^{nd}, J^{th}$, and last tube supports; C_V is the tube-to-fluid viscous damping coefficient; $\overline{C}_{SF}$ is the integration of C_{SF} over the thickness, L, of the support; C_{SF} is the squeeze-film damping coefficient, and $\delta(x_J)$ is a delta function corresponding to the location of each support. This function is required to relate the energy dissipated locally by squeeze-film damping at the Jth tube support to the vibrating energy in the whole tube.

It is convenient to express the vibration response in terms of generalized coordinates, $q_i(t)$, such that:

$$u(x,t) = \sum_{i=1}^{n} q_i(t)\phi_i(x) \tag{5-12}$$

where $\phi_i(x)$ is the mass normalized mode shape of the ith mode defined as

$$\int_0^\ell m\phi_i^2(x)dx = 1 \tag{5-13}$$

It may be shown from Eqs. (5-11) and (5-12) that

$$\frac{d^2 q_i}{dt^2} + 2\zeta_i\omega_i\frac{dq_i}{dt} + \omega_i^2 q_i = 0 \tag{5-14}$$

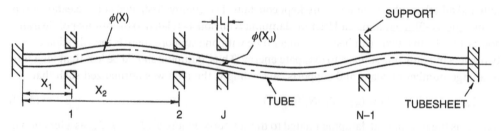

Fig. 5-13 Heat Exchanger Tube with *N* Spans and (*N*-1) Intermediate Supports.

where

$$\zeta_i = \frac{C_V}{2m\omega_i} + \frac{1}{2\omega_i}\sum_{J=1}^{N-1}\overline{C}_{SF}\phi_{Ji}^2 \tag{5-15}$$

Since

$$\zeta_i = \zeta_{Vi} + \zeta_{SFi}$$

$$\zeta_{SFi} = \frac{1}{2\omega_i}\sum_{J=1}^{N-1}\overline{C}_{SF}\phi_{Ji}^2 \tag{5-16}$$

where ϕ_{J_i} represents the ith normalized modal amplitude at the Jth support location.

It is convenient to define a characteristic tube length, ℓ_m, which we take as the average of the three longest spans. This definition is based on the assumption that the lower modes, which involve primarily the longer spans, dominate the vibration response. Modes involving the longest spans tend to be more vulnerable because the ratio of energy dissipated at the support to vibration energy in the tube is lower. As well, the relative motion at the support tends to be more the rocking type than the lateral type. When higher modes and shorter spans govern the vibration response, the characteristic span length should be based on these shorter spans. This situation could happen when there are high flow velocities locally such as in entrance or exit regions or over U-bend regions of nuclear steam generators. This approach is a crude attempt at weighting the effect of mode shape in the squeeze-film damping predictions. Thus, for a given mode:

$$\sum_{J=1}^{N-1}\phi_{Ji}^2 = \frac{A_i^*}{m\ell_m} \tag{5-17}$$

where A_i^* is simply a factor integrating the effect of relative motion between tube and support for all the supports.

From Chen (1983) and Jendrzejczyk (1984):

$$C_{SF} = \frac{\pi}{4}\rho D\left(\frac{D}{D_S - D}\right)\omega_i[-Im(H)]\left[1 - \frac{cosh\left(\dfrac{2z}{D_S}\right)}{cosh\left(\dfrac{L}{D_S}\right)}\right] \tag{5-18}$$

Thus,

$$\overline{C}_{SF} = \int_{-\frac{L}{2}}^{+\frac{L}{2}} C_{SF}dz = \frac{\pi}{4}L\rho D^2 K\left(\frac{D}{D_S - D}\right)\omega_i[-Im(H)] \tag{5-19}$$

where z is a point along the thickness of the support, D_S is the support hole diameter, and K is a factor to take into account three-dimensional flow effects (i.e., side-leakage due to the finite length of the support). The factor K may be expressed as:

$$K = 1 - \frac{D}{L}\tanh\left(\frac{L}{D}\right) \tag{5-20}$$

As shown in Fig. 5-14, K may be linearized as

$$K = 0.225(L/D)^{1.3} \tag{5-21}$$

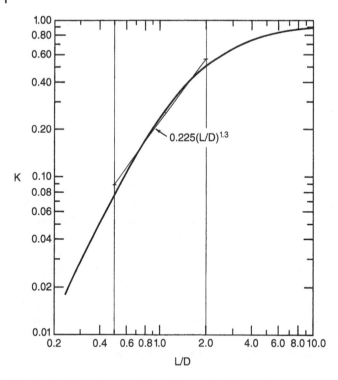

Fig. 5-14 Linearization of Three-Dimensional Factor K.

for L/D between 0.5 and 2.0, which is a realistic range for heat exchanger tube supports.

The function $Im(H)$ derived by Mulcahy (1980) is very complicated. However, it is practically linear for values of $\sqrt{8v/\omega}/(D_S - D)$ greater than 0.3 as shown in Fig. 5-15, such that

$$-Im(H) = (48v/\omega)/(D_S - D)^2 \tag{5-22}$$

where $\omega = 2\pi f$. For many realistic heat exchangers, $\sqrt{8v/\omega}/(D_S - D)$ is near or greater than 0.3, thus the linearization formulated in Eq. (5-22) is acceptable. For $\sqrt{8v/\omega}/(D_S - D)$ smaller than 0.3, Eq. (5-22) would yield conservative damping predictions.

Inserting Eqs. (5-17), (5-19), (5-21), and (5-22) into Eq. (5-16) gives:

$$\zeta_{SFi} = \frac{\pi\rho D^2}{8m\omega_i}\left(\frac{D}{D_S - D}\right)\left(\frac{A_i^* L}{m\ell_m}\right) 0.225 \left[\frac{L}{D}\right]^{1.3} \omega_i \left[\frac{48v}{\omega_i(D_S - D)^2}\right] \tag{5-23}$$

Rearranging:

$$\zeta_{SFi} = A_i^*\left(\frac{\rho D^2}{m}\right)\left(\frac{L}{\ell_m}\right)\left(\frac{L}{D}\right)^{1.3}\left(\frac{D}{D_S - D}\right)\left[\frac{v}{f(D_S - D)^2}\right] \tag{5-24}$$

We now have to estimate constant A_i^* and validate the form of Eq. (5-24) with the experimental data. These steps will be discussed below.

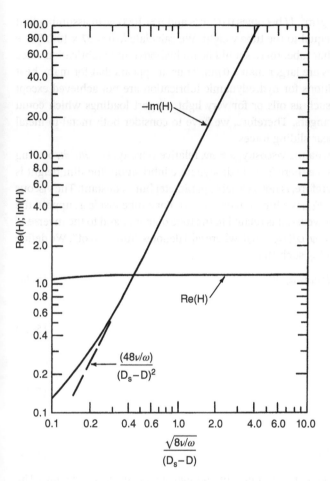

Fig. 5-15 Damping and Hydrodynamic Mass Functions, Im(h) and Re(h), (Mulcahy, 1980).

5.6.4 Damping due to Sliding

Sliding of the tube within the support may generate metal-to-metal contact friction forces and viscous shear forces. If the conditions are such that hydrodynamic lubrication occurs, the tube becomes supported entirely by a film of liquid and metal-to-metal contact is prevented. Under these conditions, the liquid film between the tube and the support is thicker than the surface asperities of the tube and the support. The problem may be studied by using hydrodynamic lubrication theory. According to Ouziaux and Perrier (1966), the thickness of the liquid film, h, is expressed in dimensionless form as

$$h/b = K_p \sqrt{\rho v U/p} \tag{5-25}$$

where b is the effective bearing length, K_P is a dimensionless coefficient having a value of roughly 0.40 for hydrodynamic lubrication, U is the relative velocity between the two load bearing surfaces, and p is the bearing load per unit thickness. For a heat exchanger tube within a support, b is much less than D, and U can be taken as ωu. For a typical tube of $D = 20$ mm, $\ell_m = 1.0$ m, $m = 1.0$ kg/m, $\omega = 200$ rad/s, vibrating at an amplitude $u = 100\,\mu$m within a support of thickness $L = 10$ mm, in water of $\nu = 10^{-6}$ m^2/s, and $\rho = 1000$ kg/m^3, U becomes 0.02 m/s and $p = 1000$ N/m assuming that the loading at the support is roughly equal to the tube weight. We calculate $h/b = 57$ x 10^{-6}. Since $b \ll D = 20$ mm, $h \ll 1.14$ μm. For that tube, there should be no hydrodynamic lubrication since the tube and support surface asperities are larger than 1.0 μm. Thus, it appears that for many heat exchanger tubes in liquids, the conditions for hydrodynamic lubrication are not achieved except perhaps for higher viscosity liquids such as oils or for very light support loadings which could be encountered in vertical heat exchangers. Therefore, we have to consider both metal-to-metal contact friction forces and viscous shear sliding forces.

Viscous shear forces are related to dynamic viscosity, $\mu = \rho\nu$, relative velocity, $U = \partial u/\partial t$, bearing length, b, and liquid film height, h. Since there is no hydrodynamic lubrication, the film height is governed by surface asperities and, therefore, is not a variable parameter but a constant. The viscous shear force per unit support thickness, F_{VS}, is a function of $\rho\nu \, \partial u/\partial t$. For a tube inside a support, the bearing length is not easily defined. However, it is related to the tube diameter and to the inverse of the tube-to-support clearance; thus, $b = \psi[D/(D_S - D)]$, where $\psi[\ ]$ denotes "function of". We define a viscous shear damping coefficient, C_{VS}, such that:

$$F_{VS} = C_{VS}\partial u/\partial t = \psi[D/(D_S - D)]\rho\nu\partial u/\partial t \tag{5-26}$$

from which,

$$C_{VS} = \nu\rho\psi[D/(D_S - D)] \tag{5-27}$$

and the integrated coefficient, $\overline{C}_{VS}$:

$$\overline{C}_{VS} = \int_{-\frac{\ell}{2}}^{+\frac{\ell}{2}} C_{VS}dz \tag{5-28}$$

Thus,

$$\overline{C}_{VS} = \nu\rho L\psi[D/(D_S - D)]\psi(L/D) \tag{5-29}$$

where the function of L/D is added to take into account three-dimensional effects or side flow. The corresponding viscous shear damping ratio, ζ_{VSi}, for a multi-support heat exchanger tube is similarly expressed by

$$\zeta_{VSi} = \frac{1}{2\omega_i}\sum_{j=1}^{N-1}\overline{C}_{VS}\phi_{ji}^2 = \frac{\overline{C}_{VS}A_i^*}{2m\omega_i\ell_m} \tag{5-30}$$

where A_i^* is a factor integrating the effect of the relative sliding motion between tube and support for all the supports.

Inserting Eqs. (5-17) and (5-29) in (5-30), and rearranging, we get:

$$\zeta_{VSi} = A_i^*\left(\frac{\nu\rho}{mf}\right)\left(\frac{L}{\ell_m}\right)\psi\left(\frac{D}{D_S - D}\right)\psi\left(\frac{L}{D}\right) \tag{5-31}$$

The form of Eq. (5-31) is similar to Eq. (5-24). Also, we have no way of separating viscous shear damping from squeeze-film damping in the data of Fig. 5-12. Since their formulation is similar, we shall consider them together as squeeze-film damping and assume they both follow the same formulation.

Damping of heat exchanger tubes in gases due to metal-to-metal contact friction forces is discussed in Section 5.4. Friction damping appears independent of most heat exchanger parameters except for support thickness and the number of supports when there are very few supports. There is no reason to believe that metal-to-metal friction damping should depend on different parameters in liquids.

5.6.5 Semi-Empirical Formulation of Tube-Support Damping

The problem now is to compare the damping formulation of Eqs. (5-24) and (5-31) to the data of Fig. 5-12. The objective is to derive semi-empirical expressions to formulate tube damping in liquids. Several assumptions and simplifications are necessary since the available data is not sufficient to study all the relevant parameters. Furthermore, the large scatter in the data makes this task very difficult. Therefore, we shall proceed in stages. Starting from the simplest formulation, we shall develop a more physically-based expression.

The data of Fig. 5-12 shows that damping at the supports decreases with frequency. Since the theory of Eq. (5-24) also implies a decrease in squeeze-film damping proportional to the inverse of frequency, we accept that this damping mechanism follows the relationship $\zeta_{SF} \propto 1/f$. On the other hand, the component of support damping due to metal-to-metal friction is considered independent of frequency, as discussed in Section 5.4.

We know that tube-support damping, ζ_S, comprises two damping mechanisms, i.e.,

$$\zeta_S = \zeta_{SF} + \zeta_F \quad \text{or} \quad \zeta_{Sn} = \zeta_{SFn} + \zeta_{Fn} \tag{5-32}$$

As a first attempt to correlate the data of Fig. 5-12, we simply subtracted a constant value for friction damping, ζ_{Fn}, so that the remaining damping satisfies the relationship $\zeta_{SF} \propto 1/f$. This required a friction damping value of roughly 0.2%. The remaining squeeze-film damping, ζ_{SFn}, satisfies the relationship $\zeta_{Sn} = 50/f$ as shown in Fig. 5-16. Thus, we get the following expression for tube-support damping, in percent:

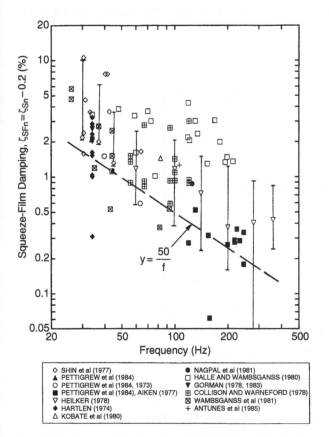

Fig. 5-16 Squeeze-Film Damping of a Multi-Span Heat Exchanger Tube in Water.

$$\zeta_{Sn} = 50/f + 0.2 \qquad\qquad (5\text{-}33)$$

To obtain conservative but realistic damping guidelines, we took the damping level corresponding to the lower decile of the available damping data. This approach means that 90% of the data is above the suggested relationship and 10% below. This is termed the minimum realistic damping level.

As a second step, we attempted to develop a more physically based model for squeeze-film damping. We started by assuming that friction damping is negligible since it is relatively small, as shown above. Then, we correlated systematically the parameters of Eq. (5-24) with the data of Fig. 5-12 to obtain the best possible fit.

The dimensionless thickness, L/ℓ_m, appears to be an important parameter as discussed by Hartlen (1974) and shown in Fig. 5-17. Although there is a lot of scatter in the data, it appears that L/ℓ_m exponent 0.5 best fits the data. The parameter $(\rho D^2/m)$ was not studied systematically since it only varied over a narrow range. However, the data appeared to follow the theory and the parameter $(\rho D^2/m)$ was accepted as correct.

We also studied the effect of the parameter $D/(D_S - D)$. Although Hartlen (1974) and Shin et al (1977) have observed an increase in damping with $D/(D_S - D)$, we found no significant trend over the range of available data. Most of the data was for diametrical clearances between 0.40 and 0.80 mm, which is normal for conventional heat exchangers. Of course, this trend would not necessarily apply for very tight clearances, such as the case of crudded supports, or for very large clearances.

Consequently, the parameter $D/(D_S - D)$ was removed from the formulation. This also indicated that the term $(D_S - D)$ in the parameter $v/[f(D_S - D)^2]$, is not correct. Thus, $(D_S - D)$ was eliminated since we do not, as yet, know of a better term. Clearly, a more appropriate term must be found to represent confinement in the dimensionless parameter $v/[f(D_S - D)^2]$, which is the equivalent of the Stokes number.

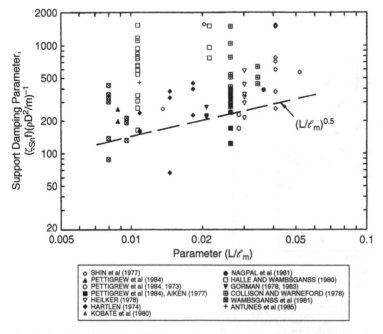

Fig. 5-17 Effect of Support Thickness Parameter L/ℓ_m on Damping due to Tube Supports.

The effect of viscosity could not easily be studied since most of the data was for water at room temperature. Taylor et al (1998) studied damping of multi-span tubes at 25, 60 and 90°C and found little effect of temperature. Nagpal et al (1981) also measured damping at 2 and 65°C. Although there was a lot of scatter in the data, they found surprisingly little effect of temperature on damping. The effect of viscosity on squeeze-film damping obviously needs to be studied further.

We did not see much effect of the parameter (L/D), indicating that side leakage or three-dimensional effects are not significant in the data studied. Thus, we are left with a simplified expression for squeeze-film damping ζ_{SFn}, that is,

$$\zeta_{SFn} = \left(\frac{1600}{f}\right)\left(\frac{\rho D^2}{m}\right)\left(\frac{L}{\ell_m}\right)^{0.5} \tag{5-34}$$

This expression is a reasonable (but not final) fit of the minimum damping data available, as shown in Fig. 5-18. However, it is quite different from the theoretical expression of Eq. (5-24). The most likely explanation is that the theory was developed for a tube concentric within its support (Mulcahy, 1980), whereas in practice, the tube is very likely to touch the support on one side. The hydrodynamic forces and, hence, the damping forces are much larger near the contact area. This phenomenon is controlled by the local proximity, h, of the tube to the support, not by the average

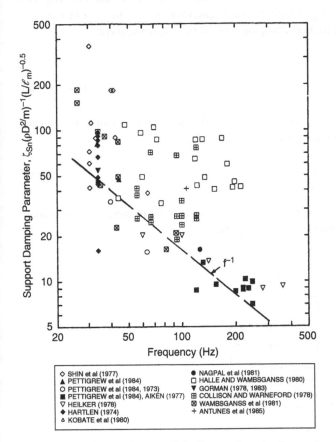

Fig. 5-18 Comparison between Tube Support Damping Parameter and Experimental Data for Heat Exchanger Tube Damping in Water (Squeeze-Film Model Only).

clearance, $D_S - D$, between tube and support. This reasoning explains why the parameter $D/(D_S - D)$ is not significant. Since the tube is generally very near the support, the ratio L/h is large, the impedance to side flow is large and three-dimensional effects are minimized. Thus, the effect of the parameter (L/D) should be small.

There is some data for support types other than drilled holes. Most of this data is for vertical tubes with supports, such a broached holes and egg crates. This data is related to nuclear steam generator studies. Except for one data point, all the other data is for horizontal tubes with drilled-hole-type supports. Interestingly, the vertical tube data with other support types is not significantly different than the horizontal tube (drilled-hole) data. This observation also suggests that damping is controlled by local contact and not by the overall geometry of a support. In this respect, the local contact of a tube on a lattice bar is not significantly different than the local contact of a tube with a drilled hole.

In reality, the tube will be at a slight angle with the axis of the support most of the time, particularly for rocking-type motion, where the tube and the axis of the support are parallel only for a small portion of the time. Thus, contact between tube and support takes place on a point or corner rather than on a line, as shown in Fig. 5-19. This behavior would tend to reduce the effect of support thickness on the dimensionless parameter, L/ℓ_m, and may explain the less than unity exponent, that is, 0.5, for the parameter L/ℓ_m.

The latter model, formulated by Eq. (5-34), neglects the contribution of friction damping. This tends to slightly underestimate damping at higher tube frequencies, as may be seen in Fig. 5-18. Thus, as a third step, we shall add a friction damping term of the following form:

$$\zeta_{Fn} = B(L/\ell_m)^{0.5} \tag{5-35}$$

as suggested in Pettigrew et al (1986b), where $B = 0.5$ was determined to be a best fit to the available data.

Combining Eq. (5-35) for friction damping and Eq. (5-34) for squeeze-film damping, we get a complete formulation for support damping (i.e., $\zeta_{S_n} = \zeta_{SFn} + \zeta_{Fn}$):

$$\zeta_{Sn} = \frac{1460}{f}\left(\frac{\rho D^2}{m}\right)\left(\frac{L}{\ell_m}\right)^{0.5} + 0.5\left(\frac{L}{\ell_m}\right)^{0.5} \tag{5-36}$$

as shown in Fig. 5-20. This model is more physically based since it contains most of the important parameters which influence heat exchanger tube damping.

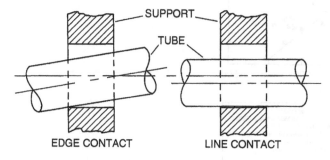

Fig. 5-19 Type of Contact Between Tube and Support.

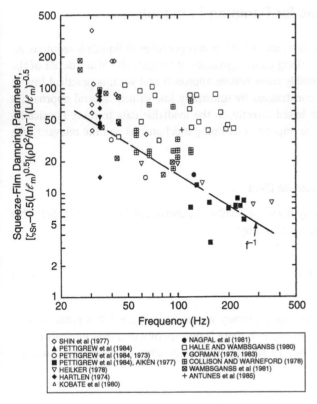

Fig. 5-20 Comparison between Tube Support Damping Model (Squeeze-Film and Friction) and Experimental Data.

5.7 Discussion

The current situation is far from ideal, as illustrated in Fig. 5-20. There is large scatter in the data and the trends are not always clear. This scatter is due to the rather variable quality of the data and, more likely, to the very nature of the problem. Heat exchanger tube damping depends on both deterministic parameters and statistical parameters. Support thickness, clearance between tube and support, tube diameter, spacing between supports, and tube frequency are deterministic parameters. Their effect should be predictable. On the other hand, many parameters are difficult to control, such as support alignment, tube straightness, side loads, and relative position and motion of the tube within the support. These parameters are statistical in nature and probably contribute to much of the scatter in the damping data, explaining why heat exchanger tube damping has been so elusive thus far.

Further damping measurements on many realistic heat exchanger tube configurations are required to provide the statistical basis for a validated design criterion. However, we cannot wait until all results are available. "Best-effort" design recommendations are needed now.

5.8 Design Recommendations for Damping in Liquids

A conservative but realistic criterion for damping of heat exchanger tubes in liquids is required. As already mentioned, this is achieved by taking damping values at roughly the lower decile of the available damping data. This is a reasonably conservative approach and yet it is practical from a design point of view. It is not unduly constrained by unusually low values. Several approaches are possible from the simplest criterion based directly on the available data to more physically-based expressions taking into account the important damping mechanisms and the relevant tube and support parameters.

5.8.1 Simple Criterion Based on Available Data

A simple criterion can be derived directly from the available experimental data of Fig. 5-9. This would yield a damping ratio in percent, expressed by:

$$\zeta = 60/f \quad \text{for } f < 50 \text{ Hz}$$
$$\zeta = 8.5/\sqrt{f} \quad \text{for } 50 < f < 200 \text{ Hz} \tag{5-37}$$
$$\zeta = 0.6\% \quad \text{for } f > 200 \text{ Hz}$$

This criterion takes into account the effect of frequency, as shown in Fig. 5-9. It is probably adequate for many normal heat exchangers. In Fig. 5-9, there is no experimental data for frequencies below 25 Hz. Rather than extrapolating Eq. (5-37) to lower frequencies, we recommend using a maximum damping of 5%.

5.8.2 Criterion Based on the Formulation of Energy Dissipation Mechanisms

A better criterion takes into account the three most important energy dissipation mechanisms, i.e.: viscous damping, ζ_V, squeeze-film damping, ζ_{SF}, and friction damping ζ_F. Thus,

$$\zeta_T = \zeta_V + \zeta_{SF} + \zeta_F \tag{5-38}$$

Substituting Eqs. (5-5) and (5-36) in (5-38), we get:

$$\zeta_T = \frac{100\pi}{\sqrt{8}} \left\{ \frac{\left[1 + \left(\frac{D}{D_e}\right)^3\right]}{\left[1 - \left(\frac{D}{D_e}\right)^2\right]^2} \right\} \left(\frac{\rho D^2}{m}\right) \left(\frac{2\nu}{\pi f D^2}\right)^{1/2} + \left(\frac{N-1}{N}\right) \left[\frac{(1460)}{f} \left(\frac{\rho D^2}{m}\right) \left(\frac{L}{\ell_m}\right)^{1/2} + 0.5 \left(\frac{L}{\ell_m}\right)^{1/2}\right] \tag{5-39}$$

where the support damping mechanisms are adjusted by $(N - 1)/N$ to take into account the ratio of number of supports over number of spans.

Although somewhat speculative, Eq. (5-39) formulates all important energy dissipation mechanisms and fits the data best. Thus, it is our current recommendation as a damping criterion for design purposes. However, if the damping ratio predicted by this equation is less than 0.6%, we recommend taking a minimum value of 0.6%. As shown in Fig. 5-9, a minimum damping of 0.6% appears reasonable. In Fig. 5-20, there is no experimental data for frequencies below 25 Hz. Rather than extrapolating Eq. (5-39) to lower frequencies, we recommend using a maximum damping of 5%.

Example 5-2   Calculation of Damping in a Process Heat Exchanger

To illustrate the application of Eq. (5-39), we use the typical multi-span heat exchanger shown in Fig. 3-5. This calculation continues from Chapters 3 and 4 with a tube of diameter $D = 20$ mm, mass including hydrodynamic mass of $m = 1.16$ kg/m, frequency within the window of $f = 33.9$ Hz, and a triangular tube bundle of $P/D = 1.5$. Further bundle geometry and flow parameters required for damping include $N - 1 = 5$ supports of thickness $L = 15$ mm, of characteristic span length $\ell_m = 1.0$ m, operating in water of viscosity $v = 10^{-6}$ m^2/s and of density $\rho = 1000$ kg/m^3.

$$D_e/D = (0.96 + 0.50\, P/D)\, P/D = 2.57$$

Eq. (5-39) gives:

$$\zeta_{SF} = \left(\frac{5}{6}\right)\left[\left(\frac{1460}{33.9}\right)\left(\frac{1000 \times 0.02^2}{1.16}\right)\left(\frac{0.015}{1.0}\right)^{1/2}\right] = 1.52\%$$

$$\zeta_V = \frac{100\pi}{\sqrt{8}}\left\{\frac{1 + \left(\frac{1}{2.57}\right)^3}{\left[1 - \left(\frac{1}{2.57}\right)^2\right]^2}\right\}\left(\frac{1000 \times 0.02^2}{1.16}\right)\left(\frac{2 \times 10^{-6}}{\pi \times 33.9 \times 0.02^2}\right)^{1/2}$$

$$= 0.39\%$$

$$\zeta_F = 0.5\left(\frac{5}{6}\right)\left(\frac{0.015}{1.0}\right)^{1/2} = 0.05\%$$

Thus, the total tube damping, ζ_T, is

$$\zeta_T = \zeta_V + \zeta_{SF} + \zeta_F$$

$$\zeta_{T\,window} = 0.39 + 1.52 + 0.05 = 1.96\%$$

This is a realistic damping value for a heat exchanger tube in water.
Damping away from the window, with a tube frequency of 135.6 Hz can be similarly calculated.

$$\zeta_T = 0.193 + 0.579 + 0.078 = 0.85\%$$

Nomenclature

A^*	= modal factor
b	= effective bearing length, m
C_m	= confinement coefficient
C	= damping coefficient
$\overline{C}$	= damping coefficient integrated over thickness
D	= tube diameter, m
D_e	= effective diameter of surrounding tubes, m
D_S	= support diameter, m
EI	= flexural rigidity, Pa $\cdot$ m^4 or N $\cdot$ m^2
f	= tube vibration frequency, Hz

F_{VS}	= viscous shear force per unit support thickness, N/m
h	= film thickness, m
K	= side leakage factor
K_p	= dimensionless hydrodynamic lubrication coefficient
L	= support thickness, m
ℓ	= span length, m
ℓ_m	= characteristic span length (average of three longest spans), m
m	= total mass per unit length ($m_t + m_h + m_i$), kg/m
N	= number of tube spans
p	= bearing load per unit thickness, N/m
q	= generalized coordinates
P	= tube pitch, m
S	= Stokes number
t	= time, s
t	= tube wall thickness, m
u	= vibration amplitude, m
U	= sliding velocity, m/s
x	= distance along tube, m
x_j	= support location
z	= position along support
ζ	= damping ratio, %
μ	= dynamic viscosity, kg/m-s
ν	= kinematic viscosity, m²/s
ρ	= fluid density, kg/m³
$\delta()$	= delta function
$\phi_1(x)$	= mass normalized mode shape
ω	= angular frequency, $2\pi f$
$\psi[]$	= function of

Subscript

c	= critical
F	= friction
h	= hydrodynamic
i	= mode order
J	= support order
n	= normalized
S	= support
SF	= squeeze-film
T	= total
t	= tube or cylinder
V	= viscous

References

Aikin, J. A., 1977, *Unpublished Report*, Atomic Energy of Canada Ltd.

Aita, S. 1983, *Personal Communication*, April.

Antunes, J., Villard, B. and Axisa, F., 1985, "Cross-Flow Induced Vibration of U-Bend Tubes of Steam Generator," Paper F16/9, *8th International Conference on Structural Mechanics in Reactor Technology*, Brussels, Belgium.

Blevins, R. D., 1975, "Vibration of a Loosely Held Tube," *ASME Journal of Engineering for Industry*, pp. 1301–1304.

Blevins, R. D., 1990, *Flow-Induced Vibration*, Second Edition, Van Nostrand Reinhold Company, New York, NY, USA.

Charreton, C., Béguin, C., Ross, A., Etienne, S. and Pettigrew, M. J., 2015, "Two-Phase Damping for Internal Flow: Physical Mechanism and Effect of Excitation Parameters," *Journal of Fluids and Structures*, 56, pp. 56–74.

Carlucci, L. N. and Brown, J. D., 1983, "Experimental Studies of Damping and Hydrodynamic Mass of a Cylinder in Confined Two-Phase Flow," *ASME Journal of Vibration, Stress and Reliability in Design*, 105, pp. 83–89.

Chen, S. S., 1983, "Design Guide for Calculating Fluid Damping for Circular Cylindrical Structures," *Argonne National Laboratory Report ANL-83-54*.

Chen, S. S., Wambsganss, M. W., and Jendrzejczyk, J. A., 1976, "Added Mass and Damping of a Vibrating Rod in Confined Viscous Fluids," *ASME Journal of Applied Mechanics*, 43, pp 325–329.

Chen, S. S. and Jendrzejczyk, J. A., 1980, "Flow-Velocity-Dependence of Damping in Tube Arrays Subjected to Liquid Cross Flow," *ASME-PVP Conference*, San Francisco, CA, USA, August 12-15, Paper 80-C2/ PVP-129.

Coit, R. S., Peake, C. C. and Lohmeier, A., 1966, "Design and Manufacture of Large Surface Condensers - Problems and Solutions," *Proceedings of the American Power Conference*, Vol. XXXIII, pp. 469–483.

Collinson, A. E., and Warneford, I. P., 1978, "Vibration Tests of Single Heat Exchanger Tubes in Air and Static Water," Paper 3.3, *2nd International Conference on Vibration in Nuclear Plant*, Keswick, UK.

Collinson, A. E. and Taylor, A. F., 1982, "A Development Program for the Minimization of Vibration in a U-Tube Steam Generator Bundle for an LMFBR", Paper 2.1, *3rd International Conference on Vibration in Nuclear Power Plants*, Keswick, UK.

Fisher, N.J., Han, Y., Guérout, F. M. and Janzen, V. P., 2005, "Comparison of Predicted and Observed Fretting-Wear Damage in Nuclear Steam Generators," *Proceedings, ASME-PVP Conference, Vol. 4, Fluid-Structure Interaction*, Denver, Colorado, USA, July 17-21, Paper PVP2005-71390, pp. 535–545.

Gorman, D. J., 1982, *Unpublished Report*, University of Ottawa.

Gorman, D. J., 1983, "Experimental Studies of the Liquid Cross-Flow Induced Vibration of Multispan Heat Exchanger Tube Bundles in Triangular and Square Arrays," *Proceedings of the XX Congress of the I.A.H.R.*, Moscow, Vol. IV, pp. 415–426.

Gorman, D. J., 1988, "Exact Analytical Solutions for the Free Vibration of Steam Generator U-Tubes," *Journal of Pressure Vessel Technology*, 110, pp. 422–429.

Goyder, H. G. D., 1982a, "Measurement of the Natural Frequencies of Damping of Loosely Supported Tubes in Heat Exchangers," Paper 2.7, *3rd International Conference on Vibration in Nuclear Power Plants*, Keswick, UK.

Goyder, H. G. D., 1982b, "The Dynamics of Heat Exchanger Tube Vibration - General Behaviour," *Report AERE 10382, HTFS RS 409*.

Gravelle, A., Ross, A., Pettigrew, M. J. and Mureithi, N. W., 2007, "Damping of Tubes due to Internal Two-Phase Flow," *Journal of Fluids and Structures*, 23, pp. 447–462.

Halle, H., and Wambsganss, M. W., 1980, "Tube Vibration in Industrial Size Heat Exchanger," *Argonne National Laboratory Report ANL-CT-80-18*.

Hartlen, R. T., 1974, "Effect of Support-Point Details upon Natural Frequency and Damping of Heat Exchanger Tubes: Preliminary Laboratory Investigation," *Ontario Hydro Research Report 74-440-K*.

Hartlen, R. T. and Barnstable, S. G., 1971, "Pickering GS Unit 1 Reheaters: Interim Report of Research Findings Regarding Tube Vibration," *Ontario Hydro Research Report 71-424-H*.

Haslinger, K. H., and Thompson J. P., 1980, "Determination of Damping Properties in Lightly Damped Systems," *ASME-WAM Proceeding HTD-Vol. 9*, Chicago, Illinois, pp. 53–63.

Heilker, W. K., 1978, *Unpublished Report*.

Jendrzejczyk, J. A., 1984, "Dynamic Characteristics of Heat Exchanger Tubes Vibrating in a Tube Support Plate Inactive Mode," *Argonne National Laboratory Report ANL-84-39*.

Kim, B. S., Pettigrew, M. J., and Tromp, J. H., 1988, "Vibration Damping of Heat Exchanger Tubes in Liquids: Effects of Support Parameters," *Journal of Fluids and Structures*, 2, pp. 593–614.

Kobate, K., Suzuki, M., Koishikawa, A., and Hashimoto, T., 1980, "Vibration Testing of a Straight Tube Type Steam Generator for FBR," *ASME Publication PVP-41*, pp. 19–31.

Lowery, R. L. and Moretti, P. M., 1975, "Natural Frequencies and Damping of Tubes on Multiple Supports," *ASME AIChE 15th National Heat Transfer Conference*, San Francisco.

Morandin, G. D. and Sauvé, R. G., 2003, "Probabilistic Assessment of Fretting Wear in Steam Generator Tubes under Flow-Induced Vibrations," Paper PVP2003-2081, *Proceedings, ASME-PVP Conference, Vol. 465, Flow-Induced Vibration-2003*, pp. 117–125.

Mulcahy, T. M., 1980, "Fluid Forces on Rods Vibrating in Finite Length Annular Regions," *ASME Journal of Applied Mechanics*, 47, pp. 234–240.

Nagpal, V., Campagna, A. O., and Pettigrew, M.J., 1981, "Tube Vibration in the Entrance Region of a Steam Generator for a 600 MW Nuclear Power Station," *Unpublished Atomic Energy of Canada Report*.

Ouziaux, R., and Perrier, J., 1966, "Mécanique des Fluides Appliquée," *Fluides Incompressibles*, 1, 2nd Edition, Dunod, Paris.

Pettigrew, M. J., Platten, J. L., and Sylvestre, Y., 1973, "Experimental Studies on Flow Induced Vibration to Support Steam Generator Design, Part II: Tube Vibration Induced by Liquid Cross-Flow in the Entrance Region of a Steam Generator," Paper No. 424, *International Symposium on Vibration Problems in Industry*, Keswick, UK. Also, *Atomic Energy of Canada Limited Report AECL-4515*.

Pettigrew, M. J., Tromp, J. H., and Mastorakos, J., 1985a, "Vibration of Tube Bundles Subjected to Two-Phase Cross Flow," *ASME Journal of Pressure Vessel Technology*, 107, pp. 335–343.

Pettigrew, M.J., Carlucci, L.N., Ko, P.L., Holloway, A.G.L. and Campagna, A.O., 1985b, "Computer Techniques to Analyse Shell-and-Tube Heat Exchangers," *Heat Transfer Engineering*, 6, pp. 40–52.

Pettigrew, M. J., Rogers, R.J. and Taylor C.E., 1985c, "Damping of Heat Exchanger Tubes," *Atomic Energy of Canada Limited Report, AECL-8701*.

Pettigrew, M. J., Axisa, F., and Qiao, Z. L., 1985d, "Amortissement des Tubes d'Echangeur de Chaleur en Milieu Liquide," *Report DEMT/85/462, Commissariat a l'Energie Atomique*, CEN Saclay, France.

Pettigrew, M. J., Goyder, H. G. D., Qiao, Z. L., and Axisa, F., 1986a, "Damping of Multispan Heat Exchanger Tubes - Part 1: In Gases," *Symposium on Special Topics of Structural Vibration, ASME Pressure Vessels and Piping Conference*, Chicago, Ill., pp. 81–87.

Pettigrew, M. J., Rogers, R. J., and Axisa, F., 1986b, "Damping of Multispan Heat Exchanger Tubes - Part 2: In Liquids," *Symposium on Special Topics of Structural Vibration, ASME Pressure Vessels and Piping Conference*, Chicago, Ill., pp. 89–98.

Pettigrew, M. J., Taylor, C. E., and Kim, B. S., 1989, "Vibration of Tube Bundles in Two-Phase Cross Flow - Part 1: Hydrodynamic Mass and Damping," *ASME Journal of Pressure Vessel Technology*, 111, pp. 466–477.

Pettigrew, M. J., and Taylor, C. E., 2004, "Damping of Heat Exchanger Tubes in Two-Phase Flow: Review and Design Guidelines," *ASME Journal of Pressure Vessel Technology*, 126, pp. 523–533.

Rao, M.S.M., Laskowski, L.J., Srikantiah, G.S. and Ahluwalia, K.S., 1977, "Prediction of Tube Wear due to Flow-Induced Vibration in PWR Steam Generators," *Proceedings, 4th Int. Symp. on Fluid-Structure Interaction, Aeroelasticity, Flow-Induced Vibration and Noise*, AD-Vol. 53-2, ASME Congress and Exhibition, Dallas, Texas, USA, November 16-21, pp. 257-264.

Rogers, R. J., 1985, *Unpublished Report*, University of New Brunswick.

Rogers, R. J., Taylor, C. E., and Pettigrew, M. J., 1984, "Fluid Effects on Multispan Heat Exchanger Tube Vibration," ASME *PVP Conference Publication* No. H00316, San Antonio, Texas.

Rogers, R. J., and Ahn, K. J., 1986, "Fluid Damping and Hydrodynamic Mass in Finite Length Cylindrical Squeeze Films with Rectilinear Motion," *ASME Flow-Induced Vibration*, PVP-Vol. 104, pp. 99–105.

Scriven, A.H. and Hopley, C.E., 1980, "The Effect of an Internal Two-Phase Flow on the Damping Characteristics of a Vibrating Tube," Paper F2, *European Two-Phase Flow Group Meeting*, Strathclyde University, Glasgow, UK.

Shin, Y. S., Jendrzejczyk, J. A., and Wambsganss, M. W., 1977, "Vibration of a Heat Exchanger Tube with Tube/Support Impact," *ASME Publication 77-JPGC-NE-5*.

Skop, R. A., Ramberg, S. E., and Ferer, K. M., 1976, "Added Mass and Damping Forces on Circular Cylinders," *ASME Paper 76-Pet-3*, Mexico City.

Taylor, C. E., Pettigrew, M. J., Dickinson, T. J., Currie, I. G., and Vidalou, P., 1998, "Vibration Damping in Multispan Heat Exchanger Tubes," *ASME Journal of Pressure Vessel Technology*, 120, No. 3.

Wambsganss, M. W., Halle, H., and Lawrence, W. P., 1981, "Tube Vibration in Industrial Size Test Heat Exchanger (30° Triangular Layout - Six Crosspass Configuration)," *Argonne National Laboratory Report ANL-CT-81-42*.

Welbourne, M.C., 1978, "Flow-Induced Vibration of AGR Heat Exchanger Tubes," Paper 3.2, *2nd International Conference on Vibration in Nuclear Power Plants*, Keswick, UK.

6

Damping of Cylindrical Structures in Two-Phase Flow

Michel J. Pettigrew and Colette E. Taylor

6.1 Introduction

Two-phase cross flow exists in many shell-and-tube heat exchangers such as condensers, evaporators and nuclear steam generators. For example, the U-bend region of nuclear steam generators is particularly vulnerable to vibration due to two-phase flow. In process industries, more than half of the heat exchangers operate in two-phase flow. Two-phase flow also exists in piping systems and in the core of boiling-water nuclear reactors. Nuclear fuel, in particular, may be subjected to two-phase axial flow. Yet, there is still little information available on vibration and damping in two-phase flow.

Excessive vibration may lead to failures due to fatigue or fretting wear (see Chapter 1). Tube failures are very costly from a maintenance and loss of production point of view, and may also have safety implications. Vibration problems may be avoided by thorough vibration analysis, preferably at the design stage. Such analysis requires damping information.

This chapter describes the development of a model to formulate damping in two-phase flow. This model is based on information available in the literature and, particularly, on the results of a comprehensive experimental program undertaken at Chalk River Laboratories. The effects of parameters such as flow velocity, void fraction, confinement, flow regime and fluid properties are discussed. These parameters are taken into consideration in the formulation of a practical design guideline. This chapter outlines the compilation of a database, development of a semi-empirical model and formulation of a design guideline for the calculation of damping due to two-phase flow.

6.2 Sources of Information

So far, most of the work on vibration of cylindrical structures subjected to two-phase flow has been conducted for the nuclear industry. The development of nuclear fuels for boiling-water reactors required information on vibration in axial two-phase flow. Most of this work was done in the late 1960s and 1970s. Unfortunately, very little attention was given to damping except by Carlucci (1980) and Carlucci and Brown (1983). They conducted a systematic study of damping of cylinders in axial two-phase flow simulated by air-water mixtures. A few researchers have reported some damping measurements while investigating turbulence-induced excitation. For example, Gorman

Flow-Induced Vibration Handbook for Nuclear and Process Equipment, First Edition.
Michel J. Pettigrew, Colette E. Taylor, and Nigel J. Fisher.

(1975) has observed much larger damping in two-phase axial flow than in liquid axial flow. While this work is not exactly applicable to cross flow, it is useful to establish trends.

More recently, some fundamental experiments were done with a single cantilevered tube immersed in two-phase mixtures generated by bubbling air through water. Hara and Kohgo (1982) studied the effect of void fraction, confinement and bubble size. Pettigrew and Knowles (1997) investigated the effect of tube frequency and surface tension. The surface tension was varied by the addition of a surfactant in an attempt to simulate the damping characteristics of high-temperature steam-water mixtures. This work is discussed in detail in Section 6.6.

Little work was done on vibration of tube bundles in two-phase cross flow before 1980. In the mid-1970s, we did exploratory vibration experiments on small tube bundles partially exposed to two-phase (air-water) cross flow (Pettigrew and Gorman, 1977). However, we did not measure damping. Heilker and Vincent (1981) have also done some work in air-water cross flow for a limited range of bundle geometries and flow conditions pertaining to their nuclear steam generator design. Their single-span tube bundles were exposed to flow over their entire length. They reported results on fluidelastic instability and turbulence-induced excitation. Unfortunately, they only measured damping at the critical flow velocity for fluide-lastic instability, yielding unrealistically low damping values. Remy (1982) studied vibration excitation mechanisms with a single-span tube bundle of normal-square configuration par-tially exposed to air-water cross flow. Partial exposure made the interpretation of the results, damping in particular, difficult.

Axisa et al (1984) were the first to present results on vibration of a tube bundle subjected to both air-water and steam-water cross flow. They tested a normal-square tube bundle of pitch-over-diameter ratio, $P/D = 1.44$. They found that for fluidelastic instability, the simulation of steam-water flow with air-water mixtures is reasonable. Subsequently, Axisa et al (1985) studied three bundle configurations (i.e., normal-square, normal-triangular, and rotated-triangular tube bun-dles), all of $P/D = 1.44$, in steam-water cross flow. These studies yielded valuable results on damp-ing. Later, Nakamura et al (1986a, 1986b) also reported data on fluidelastic instability in both air-water and steam-water cross flow. They tested a square bundle of cylinders of $P/D = 1.42$, but found it difficult to measure damping in two-phase flow. Thus, no damping data is available.

Taylor et al (1987) took some damping measurements while investigating turbulence-induced excitation in rigid tube rows of $P/D = 1.5$ and 3.0 in air-water cross flow. While this experiment was not specifically designed for damping studies, it provided some useful damping data, particu-larly at higher frequencies (i.e., 100 to 300 Hz). This data was compared against other data in an interesting review paper by Axisa et al (1988). The effect of void fraction between 80% and 100%, and the effect of air-water versus steam-water mixtures, in particular, are discussed.

Most of the research work to date has been in support of nuclear steam generators, which explains the preponderance of tube bundles of square configuration and P/D around 1.5. Generally, researchers had difficulties measuring damping in two-phase cross flow. As a result, reliable damp-ing data is very scarce, although knowledge of damping is as important as that of excitation mechanisms.

In 1983, at Chalk River Laboratories, we undertook a comprehensive experimental program to study tube bundle vibration in two-phase cross flow. Cantilever tube bundles of various configura-tions and pitch-over-diameter ratios were subjected to air-water mixtures to simulate realistic vapor qualities over a practical range of mass fluxes. Eight tube bundles were tested. Normal-triangular, normal-square, rotated-triangular and rotated-square tube bundles, all of $P/D = 1.22$, were tested under HTFS (Heat Transfer and Fluid Flow Service) sponsorship. Normal-triangular, normal-square, and rotated-triangular tube bundles, all of $P/D = 1.47$, and a normal-triangular bundle of $P/D = 1.32$ were tested in support of the CANDU[®1] steam generator technology development program.

The first results from two bundles, namely the normal-square and normal-triangular bundles of $P/D = 1.47$, were published by Pettigrew et al (1985). At the time, these bundles had been subjected to air-water mixtures up to 90% void fraction. Vibration excitation mechanisms and damping were discussed. The tests were done with all tubes in the bundles flexible and free to vibrate. Whereas this arrangement was ideal to study fluidelastic instability, it presented some difficulties for studying damping, mainly due to hydrodynamic coupling between the tubes.

Later, all the tube bundles were tested in two-phase flow up to 99% void fraction. These tests were also conducted with one instrumented flexible tube installed in a bundle of rigid tubes. This yielded more reliable damping results, as well as some valuable insight on the effect of motion of surrounding tubes on vibration response. The results are reported in papers by Pettigrew et al (1989a), Pettigrew et al (1989b) and Taylor et al (1989). These results are used extensively in the formulation of the design guidelines, as discussed in Sections 6.5 and 6.6.

Recently, normal- and rotated-triangular tube bundles of $P/D = 1.5$ were tested in two-phase Freon-22 cross flow. Freon-22 is believed to be a reasonable simulation of steam-water two-phase flow (see Chapter 3). The results showed that for this configuration, damping is similar to that in air-water although the surface tension in Freon-22 is smaller by a factor of nine (Pettigrew et al, 1995 and 2002). This surface tension difference is very useful to understanding damping behavior in two-phase mixtures other than air-water.

A complementary program was carried out at McMaster University on a smaller-scale tube bundle subjected to Freon-11 cross flow (Feenstra et al, 1996). A rotated-triangular tube bundle of $P/D = 1.44$ with tubes of diameter $D = 6.3$ mm was used in this program. These results expand the range of available data.

Fundamental studies were conducted to understand the basic energy absorption mechanisms in two-phase flows as part of the BWC/AECL/ NSERC industrial research chair of Fluid-Structure Interaction held at Polytechnique Montreal. This work is reported in Beguin et al (2009), Gravelle et al (2007), and Charreton et al (2015). A very significant relationship between two-phase damping and the interface surface area between gas and liquid was observed. The effect of viscosity, density and surface tension on two-phase damping was also studied (Beguin, 2010).

6.3   Approach

The study of damping is much more difficult in two-phase than single-phase flow for several reasons. First, damping in two-phase flow depends on void fraction, which is an additional parameter.

Second, damping measurements are difficult to obtain since it is not possible to maintain a stagnant two-phase mixture. Thus, the damping measurements must be taken under some flow. However, at mass fluxes near fluidelastic instability, damping appears to decrease due to coupling between hydrodynamic forces and tube motion, a phenomenon often called "negative damping". Thus, we are caught in the dilemma of wanting damping values in flow regimes corresponding to the mass flux required for instability, but without the effect of negative damping.

Third, damping in two-phase flow is dependent on flow regime. Damping is quite different in continuous-type flow regimes, such as bubbly, froth and wall-type flows, than in intermittent-type flows. Within the category of continuous flow regimes, important differences are expected between bubbly, froth, and wall-type flows. The effect of flow regime is discussed in Sections 6.4.2 and 6.5.8.

In spite of the above difficulties, it is essential to arrive at some design guidelines for damping in two-phase flow. The industry needs the information. We cannot wait for the ultimate results. Therefore, our approach is to formulate the best possible guidelines based on the currently available data. This is done by investigating the effect of each relevant parameter separately. Semi-empirical relationships are developed for each parameter, such as void fraction, confinement, tube mass, two-phase mixture properties, frequency, etc. These relationships are combined into a general damping formulation. The objective is to establish a conservative but realistic design guideline for damping.

The currently available data and the pertinent geometrical and experimental parameters were assembled in the form of a database. Some 600 two-phase damping data points were gathered. A sample page of the database, showing typical data points, is shown in Table 6-1. The relevant parameters in two-phase flow are: void fraction, mass flux, flow velocity, vapor quality, density, viscosity, surface tension, flow direction and flow regime. Two-phase flow parameters are discussed in detail in Section 6.4. Relevant geometrical parameters are: tube diameter, tube length, tube end conditions, confinement or pitch-over-diameter ratio, mass per unit length, tube bundle configuration and orientation. Table 6-2 outlines the geometrical parameters for each reference source of information.

6.4   Two-Phase Flow Conditions

6.4.1   Definition of Two-Phase Flow Parameters

All two-phase fluid properties, such as density, mass flux and void fraction, are taken as homogeneous. We appreciate, however, that the nature of the two-phase flow regime within the tube bundle in cross flow is an important consideration (see Section 6.5.9). For the time being, it is convenient to use homogeneous properties as they are well defined. We also find that they lead to reasonable dynamic parameters in two-phase cross flow (Pettigrew et al, 1989a and 1989b).

The homogeneous void fraction, ε_g, is calculated from the known volume flow rates of gas, $\dot{V}_g$, and liquid, $\dot{V}_\ell$, as:

$$\varepsilon_g = \frac{\dot{V}_g}{\dot{V}_g + \dot{V}_\ell} \tag{6-1}$$

Using the homogeneous void fraction, the homogeneous density, ρ, freestream velocity, U_∞, and freestream mass flux, $\dot{m}_\infty$, are defined as follows:

$$\rho = \rho_\ell \left(1 - \varepsilon_g\right) + \rho_g \varepsilon_g \tag{6-2}$$

$$U_\infty = \frac{\left(\rho_\ell \dot{V}_\ell + \rho_g \dot{V}_g\right)}{\rho A_\infty} = \frac{\dot{V}_\ell + \dot{V}_g}{A_\infty} \tag{6-3}$$

$$\dot{m}_\infty = \rho U_\infty \tag{6-4}$$

where A_∞, is the freestream area. Regardless of tube bundle configuration or orientation, the pitch velocity (sometimes called the reference gap velocity), U_p, is defined as:

Reference	Test No.	Orientation (°)	P/D	D (mm)	D_e/D	m_t (kg/m)	Freq. (Hz)	ε_g (%)	ρ (kg/m³)	ν_{TP} (m²/s)	Mass Flux (kg/m²s)	U_{cr} (m/s)	ζ_T (%)	ζ_S (%)	ζ_V (%)	ζ_{TP} (%)
Feenstra et al (1996)	A	60	1.44	6.35	2.4	0.174	35.7	52	699	4.4×10^{-7}	260		2.91	0.11	0.33	2.47
	B	60	1.44	6.35	2.4	0.174	35.8	58	607	4.8×10^{-7}	220	0.787	2.44	0.11	0.30	2.03
	C	60	1.44	6.35	2.4	0.174	35.4	62	556	5.1×10^{-7}	200	-	2.82	0.11	0.29	2.42
	D	60	1.44	6.35	2.4	0.174	36.1	64	531	5.3×10^{-7}	185	0.714	2.68	0.11	0.27	2.29
Freon-11	E	60	1.44	6.35	2.4	0.174	35.7	71	427	6.0×10^{-7}	160	0.733	2.91	0.11	0.24	2.56
(36°C)	F	60	1.44	6.35	2.4	0.174	35.5	76	355	6.6×10^{-7}	130	0.697	3.15	0.11	0.21	2.83
	G	60	1.44	6.35	2.4	0.174	35.1	80	295	7.2×10^{-7}	105	0.596	3.11	0.11	0.18	2.82
	H	60	1.44	6.35	2.4	0.174	36.7	88	183	8.8×10^{-7}	60	0.715	2.49	0.11	0.13	2.25
Pettigrew et al (2002)	25	30	1.5	12.7	2.56	0.27	23.5	25	893	1.8×17^{-7}	842	0.94	2.21	0.15	0.31	1.75
	40	30	1.5	12.7	2.56	0.27	24.8	40	725	1.9×10^{-7}	841	1.16	4.44	0.15	0.27	4.02
	50	30	1.5	12.7	2.56	0.27	25.6	50	612	2.0×10^{-7}	984	1.61	5.28	0.15	0.24	4.89
	60	30	1.5	12.7	2.56	0.27	26.2	60	500	2.2×10^{-7}	841	1.68	5.19	0.15	0.21	4.82
Freon-22	70	30	1.5	12.7	2.56	0.27	26.7	70	388	2.3×10^{-7}	700	1.80	5.09	0.15	0.18	4.76
(30°C)	80	30	1.5	12.7	2.56	0.27	27.0	80	275	2.4×10^{-7}	464	1.68	4.50	0.15	0.14	4.21
	85	30	1.5	12.7	2.56	0.27	27.3	85	219	2.5×10^{-7}	323	1.47	4.40	0.15	0.11	4.14
	90	30	1.5	12.7	2.56	0.27	27.5	90	163	2.6×10^{-7}	276	1.70	3.94	0.15	0.09	3.70
	95	30	1.5	12.7	2.56	0.27	27.6	95	107	2.7×10^{-7}	183	1.71	3.99	0.15	0.06	3.78
Pettigrew et al (1995)	10	60	1.5	12.7	2.56	0.27	22.8	10	1084	2.0×10^{-7}	-	-	0.54	0.15	0.40	0
	25	60	1.5	12.7	2.56	0.27	23.3	25	911	2.2×10^{-7}	-	-	1.68	0.15	0.37	1.16
	40	60	1.5	12.7	2.56	0.27	23.9	40	737	2.4×10^{-7}	711	0.97	2.79	0.15	0.31	2.33
	50	60	1.5	12.7	2.56	0.27	24.6	50	621	2.5×10^{-7}	692	1.11	3.61	0.15	0.28	3.18
Freon-22	60	60	1.5	12.7	2.56	0.27	25.0	60	505	2.7×10^{-7}	593	1.17	4.31	0.15	0.25	3.91
(23°C)	65	60	1.5	12.7	2.56	0.27	25.3	65	448	2.8×10^{-7}	475	1.06	5.02	0.15	0.23	4.64
	70	60	1.5	12.7	2.56	0.27	25.8	70	390	2.9×10^{-7}	414	1.06	4.25	0.15	0.21	3.89
	75	60	1.5	12.7	2.56	0.27	26.0	75	332	3.0×10^{-7}	318	0.96	4.13	0.15	0.18	3.80
	80	60	1.5	12.7	2.56	0.27	26.3	80	274	3.2×10^{-7}	200	0.73	4.09	0.15	0.16	3.78
	85	60	1.5	12.7	2.56	0.27	26.9	85	216	3.3×10^{-7}	140	0.65	3.37	0.15	0.13	3.09
	90	60	1.5	12.7	2.56	0.27	27.5	90	158	3.5×10^{-7}	94	0.60	2.39	0.15	0.10	2.14

Table 6-2 Experimental Conditions of Test Configurations.

Ref. Authors	Pettigrew et al (1989a, 2001)	Taylor et al (1987, 1988)	Pettigrew et al (1995, 2002)	Feenstra et al (1996)	Axisa et al (1985, 1988)	Nakamura et al (1986a, 1986b)	Carlucci et al (1980, 1983)	Pettigrew and Knowles (1997)
1. Orient	90°, 60°, 45°, 30°	90° (tube row)	60°, 30°	60°	90°, 60°, 30°	90°	Annular	Annular
2. P/D	1.46, 1.32, 1.22	3, 1.5	1.5	1.44	1.44	1.42	--	--
3. D_e/D	2.8, 2.5, 2.4, 1.9	8.3, 2.9	2.6	2.4	2.7	2.7	1.6, 1.2	12, 3.9, 1.9
4. D (mm)	13	30	12.7	6.35	19	19	25.4	13
5. Tube mass (kg/m)	0.33	1.37	0.27	0.17	0.52	0.72	0.9, 1.3, 3.3, 4, 4.3, 4.9	0.33
6. Support	free/fixed	fixed/fixed, free/fixed	free/fixed	free/fixed	fixed/free	supported by wires	fixed/free	free/fixed
7. Fluid	air/water	air/water	Freon-22	Freon-11	air/water, steam/water	air/water, steam/water	air/water	air/water

$$U_p = U_\infty P/(P-D) \tag{6-5}$$

Similarly, the pitch mass flux, $\dot{m}_p$, and tube bundle flow area, A_p, are defined as:

$$\dot{m}_p = \dot{m}_\infty P/(P-D) = \rho U_p \tag{6-6}$$
$$A_p = A_\infty (P-D)/P \tag{6-7}$$

6.4.2 Flow Regime

Some knowledge of flow regime is necessary to understand damping mechanisms in two-phase flow. Flow regime conditions are usually presented in terms of dimensionless parameters in the form of a flow regime map. As shown in Fig. 6-1, the Martinelli parameter, X, and the dimensionless gas velocity, U_g, may be used to get an idea of the flow regime for pertinent flow conditions. These are the conditions corresponding to fluidelastic instability and the conditions prevailing during damping measurements. The Martinelli parameter, X, is defined as:

$$X = \left(\frac{1-\varepsilon_g}{\varepsilon_g}\right)^{0.9} \left(\frac{\rho_\ell}{\rho_g}\right)^{0.4} \left(\frac{\mu_\ell}{\mu_g}\right)^{0.1} \tag{6-8}$$

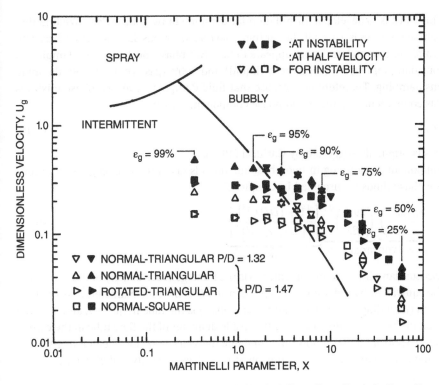

Fig. 6-1 Flow Regime Map for Tube Bundles in Vertical Cross Flow: Symbols Show Flow Conditions for Fluidelastic Instability and Damping Measurements Taken by Pettigrew et al (1989a).

assuming turbulent homogeneous flow and the total mass flux, $\dot{m}$, to be equal to $\dot{m}_\ell + \dot{m}_g$. In Eq. (6-8), ρ is density and μ is dynamic viscosity. The subscripts g and ℓ pertain to gas phase and liquid phase, respectively.

The dimensionless gas velocity for the flow regime map, U_g, is defined as:

$$U_g = \frac{\dot{m}_{P_g}}{\left[d_e g \rho_g \left(\rho_\ell - \rho_g\right)\right]^{0.5}} \tag{6-9}$$

where d_e ($d_e \simeq 2(\text{P-D})$) is the hydraulic diameter and g is acceleration due to gravity.

There is not much information available on flow regime for tube bundles subjected to two-phase cross flow. Grant (1975) has contributed most of the available data. This data has also been reported and discussed by McNaught (1982) and by Collier (1979). We have transposed Grant's flow regime map in terms of the above dimensionless parameters in Fig. 6-1. The test flow conditions for the experiments of Pettigrew et al (1989a) are plotted for comparison.

The assessment of the prevailing flow regime is largely governed by the boundary between continuous (i.e., bubbly, spray, froth) and intermittent flow. As discussed in Section 6.5.8, flow regime is very important in the assessment of damping. Thus, the first calculation to be done in the vibration analysis of a tube bundle subjected to two-phase flow is an assessment of the prevailing flow regime.

6.5 Parametric Dependence Study

The purpose of this work is to develop a formulation for damping in two-phase flow. We find that damping in two-phase mixtures is generally larger than that due to viscous damping alone based on homogeneous two-phase parameters. There appears to be a two-phase mechanism of damping in addition to viscous damping. Of course, at zero (liquid) and 100% (gas) void fraction, damping is equal to viscous damping. Therefore we propose that fluid damping, ζ, in two-phase mixtures is due to a viscous component, ζ_V, and a two-phase component, ζ_{TP}, or

$$\zeta = \zeta_V + \zeta_{TP} \tag{6-10}$$

This approach was originally suggested by Carlucci (1980).

The viscous component of damping in two-phase mixtures is taken to be analogous to viscous damping in single phase fluids as discussed in Chapter 5:

$$\zeta_V = \frac{100\pi}{\sqrt{8}} \underbrace{\left(\frac{\rho_{TP} D^2}{m}\right)}_{\text{MASS RATIO}} \underbrace{\left(\frac{2\nu_{TP}}{\pi f D^2}\right)^{0.5}}_{\text{STOKES NO.}} \underbrace{\left\{\frac{\left[1 + (D/D_e)^3\right]}{\left[1 - (D/D_e)^2\right]^2}\right\}}_{\text{CONFINEMENT}} \tag{6-11}$$

where ζ_V is the viscous damping ratio in percent, ρ_{TP} is the homogeneous density of the two-phase mixture, ν_{TP} is the equivalent two-phase kinematic viscosity, D is the tube diameter, m is the tube mass per unit length (including the hydrodynamic mass, m_h), f is the tube frequency, and D_e is the equivalent diameter of the surrounding tubes or the inside diameter of the flow tube in the case of annular axial flow. This formulation follows the fundamental work done by Chen et al (1976) on this topic. Following MacAdams (1942), the equivalent two-phase viscosity is defined as:

$$\nu_{TP} = \frac{\nu_\ell}{1 + \varepsilon_g \left(\dfrac{\nu_\ell}{\nu_g} - 1\right)} \tag{6-12}$$

where v_ℓ and v_g are the kinematic viscosities of the liquid and gas phase, respectively.

The total damping value, ζ_T, usually includes some structural or support damping, ζ_S. Rogers et al (1984) show that ζ_T, can be formulated in terms of the different damping components by:

$$\zeta_T = \zeta_S \left(\frac{m_t}{m_t + m_h} \right)^{0.5} + \zeta_V + \zeta_{TP} \tag{6-13}$$

where m_t is the tube mass and m_h is the hydrodynamic mass. Since ζ_S is generally small for the data considered here and m_h is relatively small, the above formulation was simplified to:

$$\zeta_T = \zeta_S + \zeta_V + \zeta_{TP} \tag{6-14}$$

To obtain the two-phase component of damping alone, structural and viscous components of damping were subtracted, i.e.,

$$\zeta_{TP} = \zeta_T - \zeta_S - \zeta_V \tag{6-15}$$

We are now ready to study the dependence of relevant parameters on the two-phase component of damping.

6.5.1 Effect of Flow Velocity

Carlucci (1980) and Carlucci and Brown (1983) explored the effect of flow velocity or, more specifically, mass flux on two-phase damping in confined annular flow (axial). They found that mass flux has little effect on two-phase damping, as shown in Figs. 6-2 and 6-3.

Pettigrew et al (1989a) investigated damping of several tube bundles in two-phase cross flow. Damping measurements were taken with increasing mass fluxes up to the critical mass flux for fluidelastic instability, as shown in Fig. 6-4. The results show that damping is not very dependent on mass flux below about two-thirds of the critical mass flux for instability. In Fig. 6-4, damping is taken as the average of that in the lift and drag directions. Figure 6-5 shows the effect of mass flux on damping in the drag and the lift directions separately. Above about one-half of the mass flux for instability, damping generally tends to increase in the drag direction and decrease in the lift direction. Interestingly, fluidelastic instability generally occurs in the lift direction in two-phase cross flow, as discussed by Pettigrew et al (1989b).

In summary, the effect of mass flux on damping is not dominant at mass fluxes well below fluidelastic instability. Near fluidelastic instability, the apparent effect of mass flux is attributed to fluidelastic forces. Therefore, we assume that from a practical point of view, mass flux is not an important parameter in the formulation of damping in two-phase flow.

Figures 6-4 and 6-5 show that above 90% void fraction, damping tends to decrease with mass flux. This discrepancy is attributed to an intermittent flow regime, as discussed in Section 6.5.8.

6.5.2 Effect of Void Fraction

The effect of void fraction was found to be dominant, as shown in Fig. 6-6. Damping increases with void fraction up to a void fraction of about 40%. Damping is maximum between 40 and 70% void fraction. Above 70% void fraction, damping decreases gradually to reach very low values in gas flow.

Axisa et al (1984, 1985, 1986 and 1988) found very similar trends in both air-water and steam-water cross flow. Their results, which are mostly for higher void fractions, are summarized in Fig. 6-7.

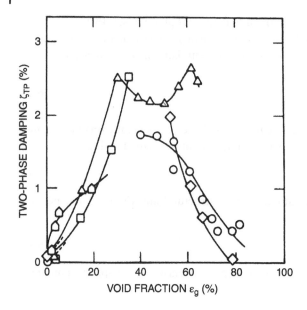

Fig. 6-2 Damping of a Cylinder in Confined Air-Water Axial Flow; Mass Flux: △ = 0, ◇ = 500, □ = 1000, △ = 3000, O = 5000 kg·m^{-2}·s^{-1} (Carlucci, 1980).

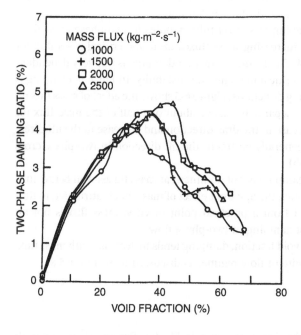

Fig. 6-3 Effect of Mass Flux on Two-Phase Damping Ratio in Annular Flow (Carlucci and Brown, 1983).

Carlucci (1980) and Carlucci and Brown (1983) also found void fraction to be dominant, as shown in Figs. 6-2 and 6-3. Their maximum damping values occur at somewhat lower void fractions (i.e., $\varepsilon_g < 60\%$). However, their void fraction estimates take into account slip between gas and liquid in their axial flow tests. This accounting yields lower void fraction estimates.

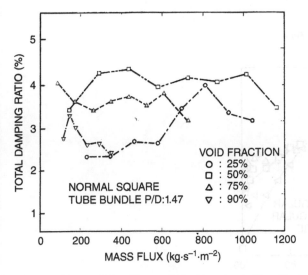

Fig. 6-4 Effect of Mass Flux on Tube Damping in Two-Phase Cross Flow (Pettigrew et al, 1989a).

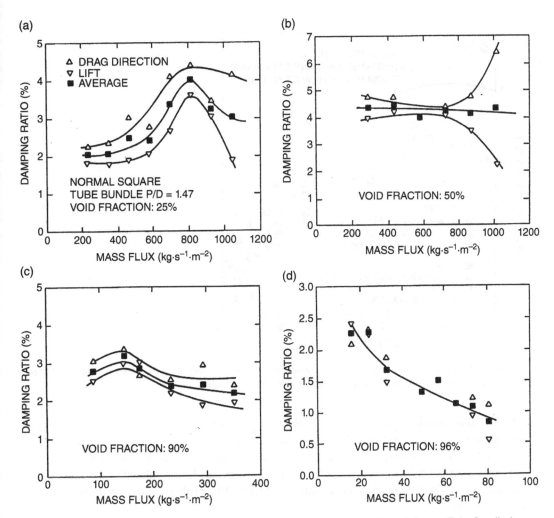

Fig. 6-5 Effect of Mass Flux on Damping in Lift and Drag Directions for a Normal-Square Tube Bundle in Two-Phase Cross Flow (Pettigrew et al, 1989a).

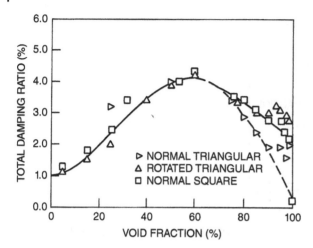

Fig. 6-6 Damping of Tube Bundles of *P/D* = 1.47 in Two-Phase Cross Flow Showing the Effect of Void Fraction (Pettigrew et al, 1989a).

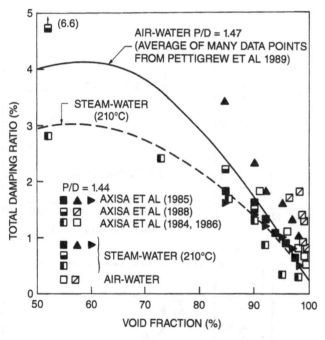

Fig. 6-7 Damping of Tube Bundles in Two-Phase Cross Flow: Comparison of Air-Water and Steam-Water Results: ■ Normal-Square, ▶ Normal-Triangular, ▲ Rotated-Triangular.

We obtained similar results in Freon-22 liquid-vapor mixtures, as shown in Fig. 6-8 (Pettigrew and Knowles, 1997) for rotated-triangular tube bundles of *P/D* = 1.5. Pettigrew et al (1995 and 2001) also observed similar results for normal-triangular tube bundles of *P/D* = 1.5.

From a practical point of view, the effect of void fraction may be formulated by the lower bound curve shown in Fig. 6-9. Figure 6-9 is a summary of Chalk River Laboratories data in air-water two-phase cross flow.

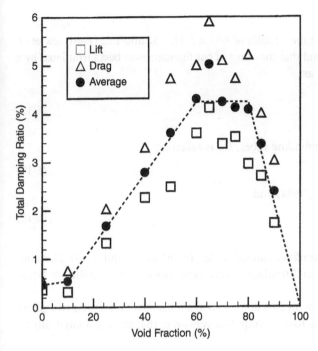

Fig. 6-8 Damping of a Rotated-Triangular Tube Bundle in Freon-22 Two-Phase Cross Flow (Pettigrew and Knowles, 1997).

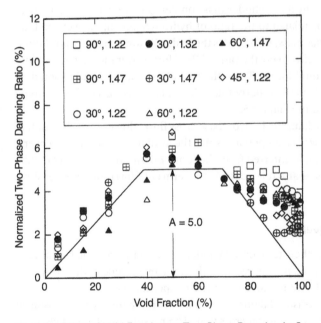

Fig. 6-9 Effect of Void Fraction on Two-Phase Damping in Cross Flow: Proposed Void Fraction Dependence Function, $f(\varepsilon_g)$, Shown. Normalized Two-Phase Damping Ratio,

$$(\zeta_{TP})_n = \zeta_{TP}\left(\frac{\rho_\ell D^2}{m}\right)^{-1}\left\{\frac{\left[1+(D/D_e)^3\right]}{\left[1-(D/D_e)^2\right]^2}\right\}^{-1}$$

6.5.3 Effect of Confinement

Damping measurements were taken for tube bundles of $P/D = 1.47$, 1.32 and 1.22 (Pettigrew et al, 1989a and 2001). From this data, we found that the effect of confinement was best formulated by a confinement function, $f(D_e/D)$, defined as:

$$f(D_e/D) = \left\{ \frac{[1 + (D/D_e)^3]}{[1 - (D/D_e)^2]^2} \right\} \tag{6-16}$$

where the effective diameter of the surrounding tubes, D_e, is taken as:

$$D_e/D = (0.96 + 0.50P/D)P/D \tag{6-17}$$

for triangular tube bundles (Rogers et al, 1984) and

$$D_e/D = (1.07 + 0.56P/D)P/D \tag{6-18}$$

for square tube bundles.

The Fig. 6-9 data was normalized for confinement using Eqs. (6-16), (6-17) and (6-18). This confinement function was used previously to formulate viscous damping for tube bundles in liquid flow, as discussed in Chapter 5.

Carlucci (1980) used a similar confinement function to correlate his data on annular two-phase flow. The confinement function of Eq. (6-16) is a simplified version of that derived analytically by Chen et al (1976).

6.5.4 Effect of Tube Mass

Carlucci and Brown (1983) studied the effect of cylinder mass on damping in confined axial two-phase flow. They found that damping is related to the ratio of hydrodynamic mass over the total cylinder mass, $(\rho D^2/m)$. The total cylinder mass, m, includes the hydrodynamic mass. This relationship makes sense since this parameter is effectively the ratio of fluid forces over inertia forces. Fluid damping should be directly related to fluid forces. Thus, the two-phase damping ratio should be related to $(\rho D^2/m)$. The term $(\rho D^2/m)$ is also similarly found in expressions for viscous damping and squeeze-film damping, as discussed in Chapter 5.

This relationship should also apply to damping in two-phase cross flow. However, verification of this relationship for two-phase cross flow was not possible since the available results do not cover a sufficiently broad range of $(\rho D^2/m)$. It can be inferred from the term $(\rho D^2/m)$ that the effect of tube diameter, D, is small. Since tube mass, m, is also related to D^2, the term $(\rho D^2/m)$ is rather insensitive to diameter.

6.5.5 Effect of Tube Vibration Frequency

Carlucci and Brown (1983) observed in their tests on a cylinder subjected to confined axial two-phase flow that tube vibration frequency did not appreciably affect the magnitude of the two-phase damping ratio, at least in the 20 to 70 Hz range covered in their experiments.

Taylor (1986) and Taylor et al (1987) measured damping with an instrumented cylinder installed in a single row of rigid cylinders subjected to air-water cross flow. Their results for clamped-clamped cylinders of $P/D = 1.5$ and 3.0 (frequency $\approx$300 Hz) and for a cantilevered cylinder of $P/D = 3.0$ (frequency $\approx$110 Hz) are shown in Fig. 6-10. It should be noted that the above experiments were not intended to study damping specifically. They were part of a turbulence-induced excitation investigation (Taylor et al, 1988). Although there was some scatter in the data, the damping values for void fractions greater than 40% are roughly the same as those outlined in Fig. 6-6 in spite of the large

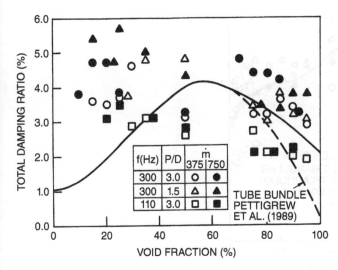

Fig. 6-10 Damping of Tube Rows in Air-Water Cross Flows; ▲ Taylor et al (1987) and ■, ● Taylor (1986); Comparison Against Tube Bundle Data (Pettigrew et al, 1989a).

difference in frequency (i.e., ≈30 versus 300 Hz). This result suggests that frequency is not a dominant factor for damping in two-phase flow.

As discussed in Sections 6.5.1 and 6.5.4, the effects of flow velocity and tube diameter are not significant. Since the effect of frequency is not large, the dimensionless flow velocity, (U/fD), is not a dominant factor in the formulation of two-phase damping.

6.5.6   Effect of Tube Bundle Configuration

Damping results for normal- and rotated-square and for normal- and rotated-triangular tube bundles of $P/D = 1.22$ and $P/D = 1.47$ in air-water are shown in Figs. 6-11 and 6-6, respectively. Both sets of results show that damping is not much affected by tube bundle orientation. The effect of P/D on damping is covered in the confinement function defined in Section 6.5.3.

Axisa et al (1985) reported damping results for normal-square and for both normal- and rotated-triangular tube bundles in steam-water. They found damping to be similar for normal-square and normal-triangular tube bundles. Damping for the rotated-triangular tube bundle was somewhat higher.

From a practical point of view, two-phase damping is not much affected by tube bundle configuration.

6.5.7   Effect of Motion of Surrounding Tubes

Damping measurements can be taken on a bundle of flexible tubes or on a single flexible tube in an otherwise rigid bundle of tubes. Damping results obtained in flexible tube bundles were reported by Pettigrew et al (1985). During these tests, there was hydrodynamic coupling under some conditions. Coupling tends to broaden the frequency response spectrum, making the interpretation of results difficult and often leading to unrealistically large damping values. To resolve this problem, damping measurements were taken with one instrumented tube surrounded by rigid tubes (Pettigrew et al, 1989a). Damping values from the two different types of bundle are compared in Fig. 6-12. As expected, the damping values from the flexible tube bundles are generally higher. The damping data from the rigid bundles is more reliable and probably more realistic. This matter is discussed

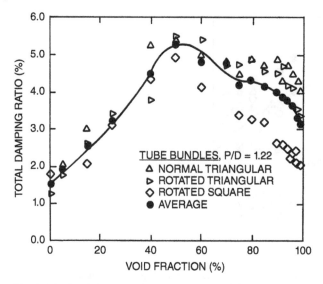

Fig. 6-11 Total Damping for Tube Bundles of *P/D* = 1.22 in Air-Water Cross Flow (Kim et al, 1987).

further in Pettigrew et al (1989a). The important point is the possible effect of hydrodynamic coupling on damping results.

6.5.8 Effect of Flow Regime

The damping results and formulations discussed in this chapter pertain to continuous flow regimes such as bubbly, froth, spray, etc. The effects of flow regime changes or transitions within continuous flows are reflected in the void-fraction function. The situation in intermittent flows is much more

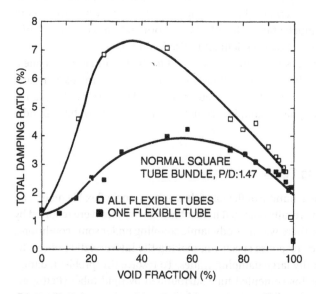

Fig. 6-12 Damping Behavior: Comparison Between All Flexible Tube Bundle and One Flexible Tube Surrounded by Rigid Tubes (Pettigrew et al, 1989a).

complex and is beyond the scope of this analysis. In practice, intermittent flows should be avoided in properly designed components.

6.5.9 Effect of Fluid Properties

The nature of two-phase mixtures (i.e., flow regime, structure of the flow, size of bubbles, etc.) is very much dependent on fluid properties such as surface tension, σ, viscosity and phase density ratio. Thus, two-phase damping should be a function of fluid properties/flow regime, $f(\sigma, v, \rho, T...)$. Unfortunately, there is very little damping data available to cover a broad range of two-phase flow conditions.

The effects of surface tension, void fraction and tube frequency were studied with a single can-tilever cylinder immersed in two-phase mixtures generated by bubbling air through water (Pettigrew and Knowles, 1997). Void fractions up to 30% could be simulated. The surface tension was varied by the addition of a surfactant. Generally, damping was found to increase with surface tension (see Fig. 6-13). This is not entirely surprising. A lower surface tension would result in a much finer two-phase flow structure, which may be less prone to absorb energy, thus explaining the lower damping.

However, the relationship was not always simple. Assuming a relationship of the type:

$$\zeta_{TP} \, \alpha \, \sigma^n \tag{6-19}$$

the surface tension exponent, n, was found to vary between -0.6 and 2.0. However, for the lower tube frequencies and the lower void fractions, the exponent, n, is generally not far from unity, indicating a direct dependence on surface tension. For the higher frequencies and higher void fractions (i.e., $\varepsilon_g > 15\%$), the surface tension dependence is much weaker.

Damping results for higher void fractions (i.e., $\varepsilon_g > 50\%$), in both air-water and steam-water are shown in Fig. 6-7. The results in steam-water are somewhat lower than in air-water. The surface tension of steam-water at 210°C is one-half that of air-water at 20°C. The damping reduction in steam-water could be explained in terms of a reduction in surface tension. However, it should be noted that damping measurements in high-temperature-and-pressure steam-water are extremely difficult to do, resulting in some scatter in the data. This scatter means that some judgment is required to make appropriate use of the limited available results.

Pettigrew et al (1995, 2001) conducted some relevant two-phase damping experiments at Chalk River Laboratories. This work was done on a rotated-triangular tube bundle of $P/D = 1.5$ subjected to two-phase Freon-22 cross flow. Figure 6-14 shows that for this configuration, damping is similar to that in air-water and steam-water, although the surface tension in Freon-22 is smaller than that in air-water by a factor of nine. Similar trends were obtained for a normal-triangular bundle tested as part of the same experimental program. Thus, the effect of surface tension on damping cannot be generalized to all two-phase flow situations. It appears that the nature of the flow regime is very important. In the experiment described earlier, where a single tube was subjected to essentially axial flow at low void fraction, the flow regime was definitely bubbly (Pettigrew and Knowles, 1997). On the other hand, the flow regime in the Freon cross-flow experiment was very different because of the torturous flow path through the tube bundle and the broad range of void fraction and mass flux.

Since the design guidelines are intended primarily for heat exchanger tube bundles in two-phase cross flow, it is deemed appropriate to give more weight to the latter results as they are more relevant in this case. Thus, a rather weak damping dependence on surface tension is now proposed.

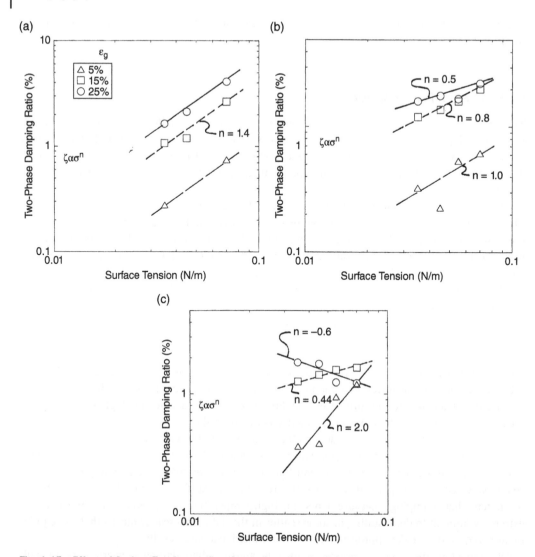

Fig. 6-13 Effect of Surface Tension on Two-Phase Damping for Tube Frequencies of a) 7 Hz, b) 28 Hz, and c) 54 Hz (Pettigrew and Knowles, 1997).

This recommendation is different than what was suggested earlier in Pettigrew et al (1994). However, the conclusion is still tentative until additional supporting results become available.

6.6 Development of Design Guidelines

Having investigated the parameters that govern damping in two-phase flow, design guidelines can now be developed. In summary, the foregoing results show that two-phase damping is strongly dependent on void fraction, fluid properties and flow regimes, directly related to confinement and to the ratio of hydrodynamic mass over cylinder mass, and weakly related to frequency, mass flux or flow velocity, and tube bundle configuration. These dependencies suggest an expression for two-phase damping, ζ_{TP}, of the form:

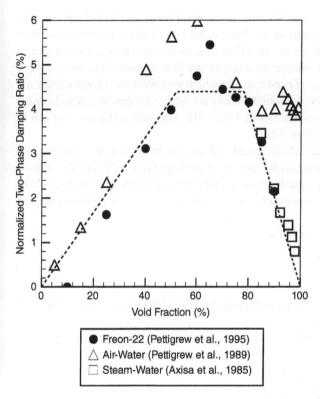

Fig. 6-14 Damping of Rotated Triangular Tube Bundles: Comparison Between Air-Water at 20°C, △ (Pettigrew et al, 1989); Steam-Water at 210°C, □ (Axisa et al, 1985); and Freon-22, ● (Pettigrew et al, 1995).

$$\zeta_{TP} \, \alpha \, \left(\, f\left(\varepsilon_g\right)\right)\left(\, f(\sigma,v,\rho,T...)\right)\left(\frac{\rho D^2}{m}\right)\left(\, f(D_e/D)\right)\left(\, f(U/fD)\right) \tag{6-20}$$

where $f(\varepsilon_g)$ is a function of void fraction, $f(\sigma, v, \rho, T...)$ is a fluid property/flow regime function, $f(D_e/D)$ is a confinement function and $f(U/fD)$ is a vibration frequency/flow velocity function, often called the dimensionless velocity. The effect of motion of surrounding tubes is not considered in the formulation since it contributes hydrodynamic coupling rather than damping, as discussed earlier.

The void-fraction function may be approximated by taking the lower envelope of the data as shown in Fig. 6-9, thus

$$f\left(\varepsilon_g\right) = 1, \text{for } \varepsilon_g = 40 \text{ to } 70\%$$
$$f\left(\varepsilon_g\right) = \varepsilon_g/40, \text{for } \varepsilon_g < 40\% \tag{6-21}$$
$$f\left(\varepsilon_g\right) = 1 - \left(\varepsilon_g - 70\right)/30, \text{for } \varepsilon_g > 70\%$$

The envelope taken is roughly the lower decile curve through the data. This is a conservative but realistic minimum damping level, which is reasonable since vibration problems are usually associated with lower damping values. On the other hand, taking the lower decile is not unduly conservative.

The fluid property/flow regime function is somewhat more difficult to define in the absence of a broad range of damping data on different two-phase mixtures. Flow-regime effects are partly taken care of by the void-fraction function. It should be noted that the current discussions pertain to continuous two-phase flows. We know, nevertheless, that in some cases, particularly at low void fraction, two-phase damping is related to surface tension. On the other hand, for the more relevant cases of tube bundles at higher void fraction, the effect of surface tension is not dominant as discussed earlier. Currently, we cannot completely quantify the fluid property function, $f(\sigma, v, \rho, T...)$. However, we know that for the range of conditions for which we have data, it is not significant. Thus, we assume it is relatively unimportant, which is only acceptable for conditions and configurations close to those for which we have experimental data. This approach is the best that can be done in the absence of more comprehensive data.

The ratio of hydrodynamic mass over cylinder mass, which also represents the ratio of fluid damping forces over inertia forces, is based on the density of the liquid (i.e., $\rho_\ell D^2/m$). The density of the two-phase mixture, ρ_{TP}, could have also been used. Then the void-fraction function would be proportionally different. From a practical point of view, it does not matter which density is used, provided we are consistent.

The confinement function was found to follow:

$$f(D_e/D) = \left\{ \frac{[1 + (D/D_e)^3]}{[1 - (D/D_e)^2]^2} \right\} \tag{6-22}$$

As discussed earlier, the effect of flow velocity and frequency (i.e., U/fD) is small. Thus, it will not be considered further in the formulation.

From the above and, in particular, from Eqs. (6-20) through (6-22), two-phase damping may be approximated by:

$$\zeta_{TP} = A\left(f(\varepsilon_g) \right) \left(\frac{\rho_\ell D^2}{m} \right) \left\{ \frac{[1 + (D/D_e)^3]}{[1 - (D/D_e)^2]^2} \right\} \tag{6-23}$$

where A is an overall coefficient determined experimentally in Fig. 6-15. The above formulation is compared against the available damping data in this figure. The coefficient, A, is equal to 4.0 when ζ_{TP} is expressed in percent.

In Section 6.5, we noted that total damping ratio, ζ_T, of a multi-span heat exchanger tube in two-phase flow comprises support damping, ζ_S, viscous damping, ζ_V, and two-phase damping, ζ_{TP} (see Eq. 6-14). In the same section, the expression for viscous damping in two-phase mixtures was taken to be analogous to viscous damping in single-phase fluids (see Eqs. 6-11 and 6-12).

Depending on the thermalhydraulic conditions (i.e., heat flux, void fraction, flow, etc.), the supports may be dry or wet. If the supports are dry, which is more likely for high heat flux and very high void fraction, only friction damping takes place. Support damping in this case is analogous to damping of heat exchanger tubes in gases (see Chapter 5). Damping due to friction (and impact) in dry supports may be expressed by Eq. (5-3). The above situation is unlikely in well-designed recirculating steam generators. However, it could exist in some areas of once-through steam generators.

When there is liquid between the tube and the support, support damping includes both squeeze-film damping, ζ_{SF}, and friction damping, ζ_F. This situation is analogous to heat exchanger tubes in liquids and the support damping in two-phase mixtures may be evaluated by combining friction damping from Eq. (5-3) and the squeeze-film damping term from Eq. (5-39):

$$\zeta_S = \zeta_{SF} + \zeta_F = \frac{(N-1)}{N} \left[\frac{1460}{f} \left(\frac{\rho_\ell D^2}{m} \right) \left(\frac{L}{\ell_m} \right)^{\frac{1}{2}} + 0.5 \left(\frac{L}{\ell_m} \right)^{\frac{1}{2}} \right] \tag{6-24}$$

Note that in Eq. (6-24), ρ_ℓ is the density of the liquid within the tube support, whereas m is the total mass per unit length of the tube including the hydrodynamic mass calculated with the two-phase homogeneous density.

Fig. 6-15 Comparison Between Proposed Design Guideline and Available Damping Data. Normalized Two-Phase Damping Ratio,

$$(\zeta_{TP})_D = \zeta_{TP}\left(\frac{\rho_\ell D^2}{m}\right)^{-1}\left\{\frac{[1+(D/D_e)^3]}{[1-(D/D_e)^2]^2}\right\}^{-1}$$

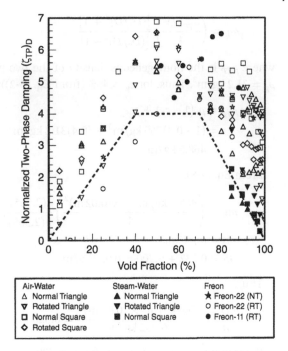

Example 6-1   Calculation of Damping in a Nuclear Steam Generator

The above two-phase damping formulation may be illustrated by an example. Let us assume a typical steam generator such as the one depicted in Fig. 3.6 and having the parameters described and calculated in Examples 3.2, 4.2 and 4.4. These parameters include a tube of diameter, $D = 20$ mm, support thickness, $L = 15$ mm, span length, $\ell_m = 0.59$ m, number of spans, $N = 10$, tube mass per unit length in air including the fluid inside the tube but excluding the hydrodynamic mass, $m_t = 0.732$ kg/m, located within a triangular tube bundle of $P/D = 1.5$ and subjected to a two-phase steam-water mixture at temperature, $T = 275°C$. It will be of interest to estimate damping for void fractions, $\varepsilon_g = 40$ and 85%.

The void-fraction function, $f(\varepsilon_g)$ is determined from Eq. (6-21):

$$f(\varepsilon_g) = 1, \text{for } \varepsilon_g = 40\%$$
$$f(\varepsilon_g) = 1 - (\varepsilon_g - 70)/30, \text{for } \varepsilon_g > 70\%. \text{ Thus,}$$
$$f(\varepsilon_g) = 0.5, \text{for } \varepsilon_g = 85\%$$

The confinement function is based on the equivalent diameter of the surrounding tubes, D_e. For a triangular bundle configuration with $P/D = 1.5$

$$D_e/D = (0.96 + 0.50P/D)P/D = (0.96 + 0.50 \times 1.5)1.5 = 2.565$$

Thus, the confinement function (from Eq. (6-22)) is:

$$\left\{\frac{[1+(D/D_e)^3]}{[1-(D/D_e)^2]^2}\right\} = \left\{\frac{[1+(2.565)^{-3}]}{[1-(2.565)^{-2}]^2}\right\} = 1.47$$

In the mass ratio term, $(\rho_\ell D^2/m)$, the mass, m, includes the hydrodynamic mass, m_h, which is expressed in Eq. (4-8) as follows:

$$m_h = \left(\frac{\rho_{TP}\pi D^2}{4}\right)\frac{\left[(D_e/D)^2 + 1\right]}{\left[(D_e/D)^2 - 1\right]}$$

where ρ_{TP} is the homogeneous density of the two-phase mixture at 275°C, $\rho_\ell = 760\,\text{kg/m}^3$ and $\rho_g = 31.2\,\text{kg/m}^3$. Thus, for $\varepsilon_g = 40\%$ (from Eq. (3-2)):

$$\begin{aligned}
\rho_{TP} &= \rho_\ell(1 - \varepsilon_g) + \rho_g\varepsilon_g \\
&= (1 - 0.4)760\,\text{kg/m}^3 + (0.4)31.2\,\text{kg/m}^3 \\
&= 468.5\,\text{kg/m}^3
\end{aligned}$$

From Eq. (4-8):

$$m_h = \left(\frac{468.5\,\text{kg/m}^3 \times \pi(0.02\,\text{m})^2}{4}\right)\frac{\left[(2.565)^2 + 1\right]}{\left[(2.565)^2 - 1\right]} = 0.200\,\text{kg/m}$$

$$m = m_t + m_h$$

$$m = 0.732 + 0.200 = 0.932\,\text{kg/m}$$

Thus:

$$\frac{\rho_\ell D^2}{m} = \frac{760\,\text{kg/m}^3 \times (0.02\,\text{m})^2}{0.932\,\text{kg/m}} = 0.326$$

Replacing in Eq. (6-23), we get for $\varepsilon_g = 40\%$

$$\zeta_{TP} = 4(1)(0.326)(1.47) = 1.92\%$$

Similarly for $\varepsilon_g = 85\%$,

$$\zeta_{TP} = 4(0.5)(0.384)(1.47) = 1.13\%$$

The total tube damping, ζ_T, may now be calculated using Eq. (6-14) providing that the support damping, ζ_S, (Eq. 6-24) and the viscous damping, ζ_V, (Eq. 6-11) are also evaluated.

The kinematic viscosities of the liquid and gas phases at 275°C are

$$\nu_\ell = 1.26 \times 10^{-7}\text{and } \nu_g = 8.0 \times 10^{-7}\text{m}^2/\text{s}.$$

For $\varepsilon_g = 40\%$,

$$\nu_{TP} = \frac{\nu_\ell}{1 + \varepsilon_g\left(\dfrac{\nu_\ell}{\nu_g} - 1\right)} = \frac{1.26 \times 10^{-7}}{1 + 0.40\left(\dfrac{1.26 \times 10^{-7}}{8.0 \times 10^{-7}} - 1\right)} = 1.9 \times 10^{-7}\text{m}^2/\text{s}$$

$$\zeta_V = \frac{100\pi}{\sqrt{8}}\underbrace{\left(\frac{\rho_{TP}D^2}{m}\right)}_{\text{MASS RATIO}}\underbrace{\left(\frac{2\nu_{TP}}{\pi fD^2}\right)^{0.5}}_{\text{STOKES NO,}}\underbrace{\frac{\left[1 + (D/D_e)^3\right]}{\left[1 - (D/D_e)^2\right]^2}}_{\text{CONFINEMENT}}$$

$$= \frac{100\pi}{\sqrt{8}}\underbrace{\left(\frac{468.5 \times 0.02^2}{0.932}\right)}_{\text{MASS RATIO}}\underbrace{\left(\frac{2 \times 1.9 \times 10^{-7}}{\pi \times 29.5 \times 0.02^2}\right)^{0.5}}_{\text{STOKES NO.}}\underbrace{\left\{\frac{\left[1 + (2.565)^{-3}\right]}{\left[1 - (2.565)^{-2}\right]^2}\right\}}_{\text{CONFINEMENT}}$$

$$= 111.1(0.201)\left(1.03 \times 10^{-5}\right)^{0.5}(1.47) = 0.105\%\ \textit{Ineffective } f = 29.5\,\text{Hz}$$

$$= 111.1(0.201)(2.56 \times 10^{-6})^{0.5}(1.47) = 0.053\% \; \textit{Effective } f = 118 \text{ Hz}$$

$$\zeta_S = \left(\frac{N-1}{N}\right)\left[\frac{1460}{f}\left(\frac{\rho_\ell D^2}{m}\right)\left(\frac{L}{\ell_m}\right)^{0.5} + 0.5\left(\frac{L}{\ell_m}\right)^{0.5}\right]$$

$$= \left(\frac{4}{5}\right)\left[\frac{1460}{29.5}\left(\frac{760 \times 0.02^2}{0.932}\right)\left(\frac{0.015}{1.18}\right)^{0.5} + 0.5\left(\frac{0.015}{1.18}\right)^{0.5}\right]$$

$$= 1.46 + 0.045 = 1.51\% \; \textit{Ineffective}$$

$$= \left(\frac{9}{10}\right)\left[\frac{1460}{118}\left(\frac{760 \times 0.02^2}{0.932}\right)\left(\frac{0.015}{0.59}\right)^{0.5} + 0.5\left(\frac{0.015}{0.59}\right)^{0.5}\right]$$

$$= 0.579 + 0.072 = 0.651\% \; \textit{Effective}$$

Therefore, the total damping at $\varepsilon_g = 40\%$ is as follows:
With ineffective supports:

$$\zeta_T = \zeta_S + \zeta_V + \zeta_{TP} = 1.51 + 0.105 + 1.92 = 3.53\%$$

With effective supports:

$$\zeta_T = \zeta_S + \zeta_V + \zeta_{TP} = 0.651 + 0.053 + 1.92 = 2.62\%$$

Similarly, the total damping at $\varepsilon_g = 85\%$ is as follows:
With ineffective supports:

$$\zeta_T = \zeta_S + \zeta_V + \zeta_{TP} = 1.76 + 0.057 + 1.13 = 2.94\%$$

With effective supports:

$$\zeta_T = \zeta_S + \zeta_V + \zeta_{TP} = 0.752 + 0.028 + 1.13 = 1.91\%$$

6.7 Discussion

6.7.1 Damping Formulation

It could be argued that it is currently premature to formulate an expression for damping in two-phase cross flow. On the other hand, there is a definite need for design guidelines now. The semi-empirical expression proposed above is a first step in that direction. It should be noted that it is based on limited information. In particular, additional data is required on the effect of fluid properties and flow regime. However, the proposed formulation may be used now as a preliminary design guideline for air-water, steam-water and other two-phase mixtures with similar fluid properties. Perhaps this work will stimulate further research in this area.

6.7.2 Two-Phase Damping Mechanisms

The basic energy dissipation mechanism responsible for two-phase damping is not yet understood. Carlucci and Brown (1983) suggested that the larger damping values observed in two-phase flow were due to a detuning effect attributed to the stochastic nature of the hydrodynamic mass in two-phase mixtures. A numerical simulation was done to investigate this mechanism. Preliminary results showed that the hydrodynamic mass variation with time was not sufficient to account for most of the apparent damping.

Another possibility is energy dissipation through increased shear forces in the two-phase mixture beyond those already taken into account by the viscous damping term. With increasing void fraction, the net available volume for the liquid phase decreases. Thus, the shear forces generated by relative tube motion should increase if shear is confined to the liquid phase. This theory, originally proposed by Hara and Kohgo (1986), assumes that no shear occurs in the gas phase. It suggests that the "apparent viscosity" of the two-phase mixture should be larger than that of the single-phase liquid. The difficulty associated with implementing this theory would be to define an equivalent viscosity to model the shear forces relevant to energy dissipation. The situation is probably much more complicated as shear deformation is likely in the gas phase.

As discussed earlier, surface tension appears to play a role in two-phase damping in some cases. This makes sense since the fluid forces that govern flow regime and, thus, the nature of the flow (i.e., bubble size, etc.) are related to surface tension. Fluid forces due to the motion of a tube within a bubbly two-phase mixture should be related to surface tension. Changes in the structure of the two-phase flow (i.e., relative displacement of the gas phase within the liquid phase, formation or breakdown of bubbles, change in gas- or liquid-phase geometry, etc.) would tend to absorb energy, thereby causing damping. As already mentioned, we conducted a simple experiment to explore the effect of surface tension on damping (Pettigrew and Knowles, 1997). Although surface tension was found to play an important role, the relationship appeared more complex. Surface tension variation did not fully explain damping in two-phase flow. The reader should now appreciate that much more research is required to fully understand the nature of damping in two-phase flow.

6.8 Summary Remarks

The available information on damping in two-phase flow was reviewed. The review revealed the following trends:

1) Two-phase damping is very dependent on void-fraction. Damping is maximum between roughly 40 and 70% void fraction.
2) Damping is directly related to the ratio of hydrodynamic mass over tube mass.
3) The effect of tube bundle geometry does not appear significant.
4) Damping is related to confinement (i.e., P/D).
5) Tube frequency and mass flux or flow velocity do not appear to be dominant parameters.

A semi-empirical expression was developed to formulate damping in two-phase cross flow. It is proposed as a preliminary guideline for heat exchanger design.

Nomenclature

A = cross-sectional flow area, m^2
D = tube diameter, m
D_e = equivalent diameter of the flow boundary, m
d_e = hydraulic diameter (used twice the minimum gap), m

f	= tube vibration frequency, Hz
g	= acceleration due to gravity, 9.81 m/s^2
L	= support thickness, m
ℓ_m	= characteristic span length, m
m	= total mass per unit length ($m_t + m_h$), kg/m
$\dot{m}$	= mass flux, kg/(m^2-s)
N	= number of tube spans
n	= exponent
P	= tube pitch, m
T	= temperature, °C
U	= fluid velocity, m/s
U_g	= dimensionless gas velocity, as defined in Eq. (6-9)
$\dot{V}$	= volume rate of flow, m^3/s
X	= Martinelli parameter
ε_g	= void fraction, %
ζ	= damping ratio
μ	= dynamic fluid viscosity, kg/m^2-s
ν	= kinematic fluid viscosity, m^2/s
ρ	= fluid density, kg/m^3
σ	= surface tension, N/m

Subscripts

F	= friction
g	= gas or vapor
h	= hydrodynamic
ℓ	= liquid
P	= pitch
S	= support
SF	= squeeze-film
T	= total
TP	= two-phase
t	= tube or cylinder
V	= viscous
∞	= free stream

Note

1 CANDU = Canada Deuterium Uranium is a registered trademark of Atomic Energy of Canada Limited.

References

Axisa, F., Villard, B., Gibert, R. J., Hetsroni, G. and Sundheimer, P., 1984, "Vibration of Tube Bundles Subjected to Air-Water and Steam-Water Cross Flow: Preliminary Results on Fluidelastic Instability," *Proceedings of ASME Symposium on Flow-Induced Vibrations*, Vol. 2, New Orleans, pp. 269–284.

Axisa, F., Boheas, M.A. and Villard, B., 1985, "Vibration of Tube Bundles Subjected to Steam-Water Cross Flow: A Comparative Study of Square and Triangular Arrays," Paper No. B1/2, *8th International Conference on Structural Mechanics in Reactor Technology*, Brussels.

Axisa, F., Villard, B., and Sundheimer, P., 1986, "Flow-Induced Vibration of Steam Generator Tubes," *Electric Power Research Institute Report EPRI-NP4559*.

Axisa, F., Wullschleger, M., Villard, B. and Taylor, C., 1988, "Two-Phase Cross Flow Damping in Tube Arrays," *Proceedings of ASME Pressure Vessel and Piping Conference*, Pittsburg, USA, PVP Vol. 133, Damping 1988, pp. 9-15.

Béguin, C., 2010, "Modélisation des Écoulements Diphasiques: Amortissement, Forces Interfaciales et Turbulence Diphasique," Doctoral Thesis, École Polytechnique de Montréal, Montreal, Canada.

Béguin, C., Anscutter, F., Ross, A., Pettigrew, M. J. and Mureithi, N. W., 2009, "Two-Phase Damping and Interface Surface Area in Tubes with Vertical Internal Flow," *Journal of Fluids and Structures*, 25(1), pp. 178–204.

Carlucci, L. N., 1980, "Damping and Hydrodynamic Mass of a Cylinder in Simulated Two-Phase Flow," *ASME Journal of Mechanical Design*, 102, pp. 597–602. *AECL Report, AECL-7315*.

Carlucci, L. N. and Brown, J. D., 1983, "Experimental Studies of Damping and Hydrodynamic Mass of a Cylinder in Confined Two-Phase Flow," *ASME Journal of Vibration, Stress and Reliability in Design*, 105, pp. 83–89.

Charreton, C., Béguin, C., Ross, A., Étienne, S. and Pettigrew, M. J., 2015, "Two-Phase Damping for Internal Flow: Physical Mechanism and Effect of Excitation Parameters," *Journal of Fluids and Structures*, 56, pp. 56-74.

Chen, S. S., Wambsganss, M. W., Jendrzyczyk, J. A., 1976, "Added Mass and Damping of a Vibrating Rod in Confined Viscous Fluids, " *ASME Journal of Applied Mechanics*, pp. 325–329.

Collier, J. G., 1979, "Two-Phase Gas-Liquid Flows within Rod Bundles," *Turbulent Forced Convection in Channels and Bundles: Theory and Applications to Heat Exchangers and Nuclear Reactors*, Vol. 2, Hemisphere Publishing Corporation, Washington, DC, pp. 125–131.

Feenstra, P. A., Weaver, D. S. and Judd, R. L., 1996, "Damping and Fluidelastic Instability of a Tube Array in Two-Phase R-11 Cross Flow", *Proceedings of Symposium on Flow-Induced Vibration-1996, ASME Pressure Vessels and Piping Conference*, Montreal, ASME PVP-Vol. 328, pp. 89–102.

Gorman, D. J., 1975, "Experimental and Analytical Study of Liquid and Two-Phase Flow-Induced Vibration in Reactor Fuel Bundles," *2nd National Congress on Pressure Vessel and Piping*, San Francisco, USA, *ASME Paper 75-PVP-52*.

Grant, I. D. R., 1975, "Flow and Pressure Drop with Single Phase and Two-Phase Flow in the Shell-Side of Segmentally Baffled Shell-and-Tube Heat Exchangers," *NEL Report No. 590*, National Engineering Laboratory, Glasgow, pp. 1–22.

Gravelle, A., Ross, A., Pettigrew, M. J. and Mureithi, N. W., 2007, "Damping of Tubes due to Internal Two-Phase Flow," *Journal of Fluids and Structures*, 23, pp. 447–462.

Hara, F. and Kohgo, O., 1982, "Added Mass and Damping of a Vibrating Rod in a Two-Phase Air-Water Mixed Fluid," ASME Publication PVP, Vol. 63, Orlando, Florida, pp. 1–8.

Hara, F., and Kohgo, O., 1986, "Numerical Approach to Added Mass and Damping of a Vibrating Circular Cylinder in a Two-Phase Bubble Fluid," *Proceedings of Int. Conf. on Computational Mechanics*, Tokyo, VII, pp. 255–260.

Heilker, W. J. and Vincent, R. Q., 1981, "Vibration in Nuclear Heat Exchangers Due to Liquid and Two-Phase Flow," *ASME Journal of Engineering for Power*, 103, pp. 358–365.

Kim, B. S., Pettigrew, M. J., Taylor, C. E. and Tromp, J. H., 1987, "Flow-Induced Vibration of Heat Exchanger Tubes in Two-Phase Cross Flow," *Unpublished Chalk River Laboratory Data*.

McAdams, W. H., Woods, W. K. and Heroman L. C., 1942, "Vaporization Inside Horizontal Tubes - II - Benzene - Oil Mixtures," *Trans. ASME*, 64, pp. 193–200.

McNaught, J. M., 1982, "Two-Phase Forced Convection Heat Transfer During Condensation on Horizontal Tube Bundles," *7th International Heat Transfer Conference*, Vol. 5, Munich, pp. 125–131.

Nakamura, T., Fujita, K., Kawanishi, K. and Saito, I., 1986a, "A Study on the Flow Induced Vibration of a Tube Array by a Two-Phase Flow (2nd Report: Large Amplitude Vibration of Steam-Water Flow)," *Trans. of Japanese Society of Mechanical Engineers*, 52, C.

Nakamura, T., Yamaguchi, N., Tsuge, A., Fujita, K., Sakata, K. and Saito, I., 1986b, "Study on Flow-Induced Vibration of a Tube Array by a Two-Phase Flow (1st Report: Large Amplitude Vibration by Air-Water Flow)," *Trans. of Japanese Society of Mechanical Engineers*, 52(473), C.

Pettigrew, M. J. and Gorman, D. J., 1977, "Experimental Studies on Flow-Induced Vibration to Support Steam Generator Design, Part III: Vibration of Small Tube Bundles in Liquid and Two-Phase Cross Flow," Paper No. 424, *International Symposium on Vibration Problems in Industry*, Keswick, UK, *AECL Report, AECL-5804*.

Pettigrew, M. J., Tromp, J. H. and Mastorakos, J., 1985, "Vibration of Tube Bundles Subjected to Two-Phase Cross Flow," *ASME Journal of Pressure Vessel Technology*, 107, pp. 335–343.

Pettigrew, M. J., Goyder, H. G. D., Qiao, Z. L. and Axisa, F., 1986a, "Damping of Multispan Heat Exchanger Tubes - Part 1: In Gases," *Symposium on Special Topics of Structural Vibration, ASME Pressure Vessels and Piping Conference*, Chicago, Vol. 104, pp. 81–88.

Pettigrew, M. J., Rogers, R. J. and Axisa, F., 1986b, "Damping of Multispan Heat Exchanger Tubes - Part 2: In Liquids," *Symposium on Special Topics of Structural Vibration, ASME Pressure Vessels and Piping Conference*, Chicago, Vol. 104, pp. 89–98.

Pettigrew, M. J., Taylor, C. E. and Kim, B. S., 1989a, "Vibration of Tube Bundles in Two-Phase Cross Flow - Part 1: Hydrodynamic Mass and Damping," *ASME Journal of Pressure Vessel Technology*, 111, pp. 466–477.

Pettigrew, M. J., Kim, B. S., Taylor, C. E. and Tromp, J. H., 1989b, "Vibration of Tube Bundles in Two-Phase Cross Flow - Part 2: Fluidelastic Instability," *ASME Journal of Pressure Vessel Technology*, 111, pp. 478–487.

Pettigrew, M. J., Taylor, C. E. and Yasuo, A., 1994, "Vibration Damping of Heat Exchanger Tube Bundles in Two-Phase Flow," *Pressure Vessel Research Council Bulletin WRC 389*, pp. 1–41.

Pettigrew, M. J., Taylor, C. E., Jong, J. H. and Currie, I. G., 1995, "Vibration of a Tube Bundle in Two-Phase Freon Cross Flow," *ASME Journal of Pressure Vessel Technology*, 117, pp. 321–329.

Pettigrew, M. J., and Knowles, G. D., 1997, "Some Aspects of Heat Exchanger Tube Damping in Two-Phase Mixtures," *Journal of Fluids and Structures*, 11(8), pp. 929–945.

Pettigrew, M. J., Taylor, C. E., and Kim, B. S., 2001, "The Effects of Tube Bundle Geometry on Vibration in Two-Phase Cross Flow," *ASME Journal of Pressure Vessel Technology*, 123(4), pp. 414–420.

Pettigrew, M. J., Taylor, C. E., Janzen, V. P., and Whan, T., 2002, "Vibration Behavior of Rotated Triangular Tube Bundles in Two-Phase Cross Flows", *ASME Journal of Pressure Vessel Technology*, 124 (2), pp. 144–153.

Remy, F. N., 1982, "Flow Induced Vibration of Tube Bundles in Two-Phase Cross Flow," Paper No. 1.9, *Proceedings of 3rd International Conference on Vibration in Nuclear Plants*, Vol. 1, Keswick, UK, pp. 135–160.

Rogers, R. J., Taylor, C. E. and Pettigrew, M. J., 1984, "Fluid Effects on Multi-Span Heat Exchanger Tube Vibration," *Proceedings of ASME PVP Conference*, San Antonio, Texas, June, ASME Publication H00316: Topics in Fluid Structure Interaction, pp. 17–26.

Taylor, C. E., Pettigrew, M. J., Axisa, F. and Villard, B., 1987, "Damping Measurements in Two-Phase Cross Flow," *Proceedings of 11th CANCAM Conference*, Edmonton, Canada, pp. B148–B149.

Taylor, C. E., Pettigrew, M. J., Axisa, F. and Villard, B., 1988, "Experimental Determination of Single and Two-Phase Cross Flow-Induced Forces on Tube Rows," *ASME Journal of Pressure Vessel Technology*, 110, pp. 22–28.

Taylor, C.E., Pettigrew, M.J., Currie, I.G. and Kim, B.S., 1989, "Vibration of Tube Bundles in Two-Phase Cross Flow - Part 3: Turbulence Induced Excitation," *ASME Journal of Pressure Vessel Technology*, 111, pp. 488–500.

7

Fluidelastic Instability of Tube Bundles in Single-Phase Flow

Michel J. Pettigrew and Colette E. Taylor

7.1　Introduction

There are several flow-induced vibration excitation mechanisms that could cause excessive vibration in both process and nuclear system components. These are: fluidelastic instability, periodic-wake-shedding resonance, turbulence-induced excitation and acoustic resonance. Of these, fluidelastic instability is by far the most important. In our experience, most vibration problems in components such as heat exchangers and steam generators are related to fluidelastic instability.

Excessive vibration due to fluidelastic instability may cause tube failures either by fatigue or fretting wear. To avoid such problems, it is necessary to perform a thorough flow-induced vibration analysis at the design stage. Such analysis requires sound information on fluidelastic instability.

This chapter addresses the question of fluidelastic instability of heat exchanger tube bundles subjected to single-phase cross flow. Here, single-phase flow means either liquid or gas flow.

7.2　Nature of Fluidelastic Instability

Generally, in a tube bundle, the fluid forces on one tube are affected by its motion and the motion of neighboring tubes. This situation creates an interaction between fluid forces and tube motion. Fluidelastic instability is possible when the interaction between the motions of individual tubes is such that it results in fluid force components that are both proportional to tube displacements and in-phase with tube velocities. Instability occurs when, during one vibration cycle, the energy absorbed by a tube from the fluid forces exceeds the energy dissipated by damping. Then, the vibration amplitude of the tube increases rapidly and, in theory, would become extremely large. In practice, the vibration amplitude is limited by the presence of surrounding tubes or by other non-linearities such as support plates.

Figure 7-1 shows a typical vibration amplitude versus flow velocity relationship for a tube bundle in cross flow. Beyond a given threshold velocity, the vibration amplitude becomes very large and the tube bundle becomes unstable. The threshold velocity is usually called the critical velocity for fluidelastic instability, U_{pc}.

Flow-Induced Vibration Handbook for Nuclear and Process Equipment, First Edition.
Michel J. Pettigrew, Colette E. Taylor, and Nigel J. Fisher.

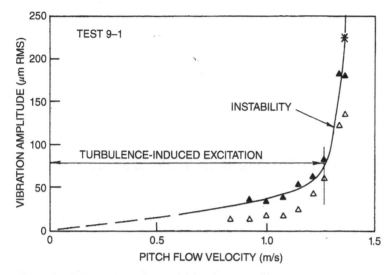

Fig. 7-1 Typical Vibration Response versus Flow Velocity Relationship for Tube Bundles Showing Fluidelastic Instability: Normal Triangle *P/D* = 1.54, *D* = 13 mm, Closed Symbols: Lift Direction, Open Symbols: Drag Direction (Pettigrew and Gorman, 1978).

Fluidelastic instability is very distinct from periodic-wake-shedding resonance. Periodic-wake-shedding resonance occurs when the flow velocity is such that the frequency of vortex shedding coincides with the natural frequency of the tube. Such a resonance would disappear if the flow velocity is increased significantly, as shown in Fig. 7-2. This behavior does not happen in the case of fluidelastic instability. A significant increase in flow velocity would simply result in more violent vibration.

During instability, the motion of one tube is somewhat related to the motion of neighboring tubes. There is usually an organized pattern of tube motion within the tube bundle. This pattern is sometimes called bundle vibration mode. There are many possible modes of vibration for a given bundle, depending on the relative motion and phase of the tubes. Thus, fluidelastic instability may be possible for several bundle modes of vibration. In some cases, the tube motion is mostly in the lift direction, in others, mostly in the drag direction and, sometimes, the motion is orbital. From a practical point of view, the specific bundle mode in which instability is occurring is not important as long as instability is avoided.

The presence of fluidelastic instability is usually characterized by the following features. As already discussed, a sudden large increase in vibration amplitude with increasing flow velocity indicates instability. The motion between adjacent tubes is usually well correlated, suggesting strong hydrodynamic coupling. A sudden change in vibration pattern within the tube bundle is also indicative of instability. Finally, the tube vibration response spectra are significantly different during instability. The response peak becomes much narrower, giving the appearance of much reduced damping. This behavior indicates coupling between fluid forces and tube motion, which is characteristic of fluidelastic instability phenomena. Not all the above characteristics are necessarily present in practical cases, due to a variety of reasons, such as non-linearities. From a practical point of view, we define fluidelastic instability as severe vibration that is not due to either periodic-wake-shedding resonance or turbulence-induced excitation.

Gorman (1978).

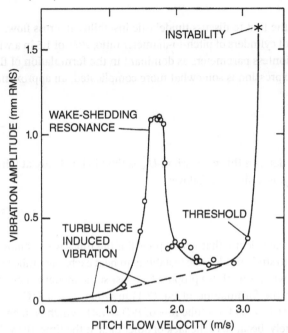

Fig. 7-2 Vibration Response of a Normal-Triangular Tube Bundle of *P/D* = 1.33 Showing Periodic-Wake-Shedding Resonance and Fluidelastic Instability (Gorman, 1976).

7.3 Fluidelastic Instability: Analytical Modelling

Since the mid-1960s, several models and theories have been proposed to formulate fluidelastic instability in tube bundles subjected to cross flow. Ideally, it should be a time-domain model to consider non-linearities due to tube-to-support clearances and due to fluid-force-tube-motion inter-action. However, the tube probably does not vibrate back and forth across the available clearance between tube and support. If it did, it would get damaged relatively quickly. Most likely, the tube stays in contact with the support and is essentially pinned from a practical point of view. This allows for simple linear analyses until more sophisticated non-linear time-domain analyses become avail-able to the practical engineer. A non-linear model should also be able to predict the post-instability behavior of the tube bundle during a loss of coolant scenario.

In the simplest model, fluidelastic instability is expressed in terms of a dimensionless velocity, U_{pc}/fD, and a dimensionless mass-damping term, $2\pi\zeta m/\rho D^2$, as follows

$$\frac{U_{pc}}{fD} = K\left(\frac{2\pi\zeta m}{\rho D^2}\right)^b \tag{7-1}$$

where U_{pc} is the critical velocity for fluidelastic instability, f the tube natural frequency, D the tube diameter, ζ the damping ratio, ρ the density of the fluid, and m the tube mass per unit length, which includes the hydrodynamic mass and the mass of the fluid inside the tube. Note that ζ is expressed as the damping ratio, not as a percentage. In the simplest model, the exponent b is 0.5. The flui-delastic instability constant K is determined experimentally as explained below.

7.4 Fluidelastic Instability: Semi-Empirical Models

To our knowledge, Roberts (1966) was the first to discuss fluidelastic instability in cross flow. He studied the aerolastic behavior of a row of cylinders of pitch-to-diameter ratio, P/D, of 1.5 in a wind tunnel. He identified the above dimensionless parameters as dominant in the formulation of fluidelastic instability. Although Roberts' expression is somewhat more complicated, an approximate simplified expression

$$\frac{U_{pc}}{fD} = 9.8 \left(\frac{2\pi\zeta m}{\rho D^2} \right)^{0.5} \tag{7-2}$$

may be deduced from his experimental data. In this expression, U_{pc} is the pitch velocity at which instability occurs. The pitch velocity, U_p, is defined as follows:

$$U_p = \frac{U_\infty P}{(P-D)} \tag{7-3}$$

where U_∞ is the free stream velocity or the velocity that would prevail if the tubes were removed.

Connors (1970) was probably the first to study fluidelastic instability in heat exchanger tube bundles. He did his experimental work in a wind tunnel. He proposed a quasi-static model to describe fluidelastic instability. This work led to the formulation of Eq. (7-1) in which $b = 0.5$. Connors reported a fluidelastic instability constant $K = 9.9$ for a tube row of $P/D = 1.41$, which is in agreement with Roberts' findings. Unfortunately, because of the lack of other data at the time, the value of $K = 9.9$ was used for tube bundles by some designers.

In the early 1970s, a comprehensive program was undertaken at Chalk River Laboratories to study flow-induced vibration of nuclear heat exchangers. This work was done in collaboration with Canadian industries and universities. Realistic tube bundles of both triangular and square configurations of P/D between 1.22 and 1.57 were tested in liquid flow. This work led to the recommendation of

$$\frac{U_{pc}}{fD} = 3.3 \left(\frac{2\pi\zeta m}{\rho D^2} \right)^{0.5} \tag{7-4}$$

as a design guideline to avoid fluidelastic instability in heat exchangers. Details may be found in Gorman (1976), Pettigrew et al (1978) and Pettigrew and Gorman (1978).

Since circa 1975, a number of researchers have studied fluidelastic instability in tube bundles. Several new models have been proposed and a number of design guidelines have been recommended. It is not the purpose of this chapter to review all this work in detail. However, we shall discuss some of the more interesting models and, in particular, those that are directly relevant to the formulation of practical design guidelines. For additional information, the reader is referred to some excellent reviews by Price (1995), Païdoussis (1983), Chen (1984) and Weaver and Fitzpatrick (1988).

Generally, the research work on fluidelastic instability falls into two categories. The first is aimed at obtaining a basic understanding of fluidelastic instability. This work is usually done on well-defined and well-instrumented arrays of cylinders, often in wind tunnels because of simplicity, often with only one flexible cylinder in an otherwise rigid array of cylinders, often with externally added damping, sometimes with a single row of cylinders and most of the time with cylinders with well-defined support conditions.

The second category, on the other hand, is aimed at producing design information. This type of work is usually done on realistic heat exchanger tube bundles, often in liquid flow, in a few cases in two-phase flow, sometimes with replicated heat exchanger tube-support conditions, sometimes on complete, or on sections of, real heat exchangers and mostly in large-scale test facilities. While the latter category is much less elegant, it often provides information that is much closer to real heat exchanger designs.

At this time, it is important to review some typical models and design guidelines that have been proposed for fluidelastic instability. These are outlined in Table 7-1. As already discussed, the simplest model is of the form

$$\frac{U_{pc}}{fD} = K\left(\frac{2\pi\zeta m}{\rho D^2}\right)^{0.5} \tag{7-5}$$

Based on this model, Pettigrew and Gorman (1978) recommended $K = 3.3$ as a design guideline for all tube bundle configurations; a few years later, Chen (1984) recommended $K = 2.35$ for normal-square (90°) tube bundles, of $(2\pi\zeta m/\rho D^2) > 0.7$ and $K = 2.8$ for rotated-triangular (60°) bundles of $(2\pi\zeta m/\rho D^2) > 1.0$. A slight variation of this model uses Eq. (7-1)

$$\frac{U_{pc}}{fD} = K\left(\frac{2\pi\zeta m}{\rho D^2}\right)^{b} \tag{7-6}$$

where Weaver and Fitzpatrick (1988) recommend: $K = 2.5$ and $b = 0.48$ for normal-square (90°) tube bundles, $K = 4$ and $b = 0.48$ for rotated-square (45°) tube bundles, $K = 3.2$ and $b = 0.40$ for normal-triangular (30°) tube bundles and $K = 4.8$ and $b = 0.30$ for rotated-triangular (60°) tube bundles, in all cases for $(2\pi\zeta m/\rho D^2) > 0.3$; and, $b = 0$ and $K = 1.4, 2.2, 2.0$ and 1.0 for the 90°, 45°, 30° and 60° tube bundles, respectively, for $(2\pi\zeta m/\rho D^2) \leq 0.3$. Weaver and Fitzpatrick's (1988) recommendations are shown in Fig. 7-3. Chen (1984) used the same formulation to make the recommendation $K = 2.1$ and $b = 0.15$ for square bundles with $(2\pi\zeta m/\rho D^2) < 0.7$ and $K = 2.8$ and $b = 0.17$ for rotated-triangular bundles with $(2\pi\zeta m/\rho D^2) < 0.1$. Chen's recommendations are shown in Fig. 7-4. The above recommendations lead to somewhat discontinuous curves, as shown in Figs. 7-3 and 7-4. Fig. 7-3d, in particular, shows a sudden step at $(2\pi\zeta m/\rho D^2) = 0.3$ for the rotated-triangular bundle. Interestingly, Fig. 7-4c shows a similar step at $(2\pi\zeta m/\rho D^2) = 2.0$, however, for a different configuration, the normal-triangular bundle. These results illustrate the difficulties in interpreting experimental data on fluidelastic instability.

A somewhat more sophisticated model has been proposed by several researchers (Chen and Jendrzejczyk, 1983; Gibert et al, 1977, and Tanaka and Takahara, 1981):

$$\frac{U_{pc}}{fD} = K(2\pi\zeta)^a \left(\frac{m}{\rho D^2}\right)^{b} \tag{7-7}$$

In this model, the mass and damping terms are separated. For example, Tanaka and Takahara (1981) suggested $a = 0.20$ and $b = 0.33$ for heavy fluids (liquid) based on testing a square tube bundle of $P/D = 1.33$.

The addition of a dimensionless parameter including the term P/D has been proposed by Païdoussis (1981) and Chen (1984), thus:

$$\frac{U_{pc}}{fD} = K(2\pi\zeta)^a \left(\frac{m}{\rho D^2}\right)^{b} \left(\frac{P}{D} - C\right)^{c} \tag{7-8}$$

Table 7-1 Typical Fluidelastic Instability Models and Suggested Design Guidelines (Pettigrew and Taylor, 1991)

Fluidelastic Instability Model	Author	Suggested Design Guidelines					Applicability		Comments
		K	a	b	c	C	$\left(\frac{2\pi\zeta m}{\rho D^2}\right)$	Bundle type	
$\dfrac{U_{pc}}{fD} = K\left(\dfrac{2\pi\zeta m}{\rho D^2}\right)^b$	Pettigrew and Gorman (1978)	3.3		0.5			All	All	
	Connors (1978)	2.9		0.5			All	90°	P/D=1.42
	Chen (1978)	2.35		0.5			>0.7	90°	
	"	2.80		0.5			>1.0	60°	
	"	2.10		0.15			<0.7	90°	
	"	2.80		0.17			<1.0	60°	
	Weaver and	2.5		0.48			>0.3	90°	
	Fitzpatrick	4.0		0.48			>0.3	45°	
	(1988)	3.2		0.40			>0.3	30°	
	"	4.8		0.30			>0.3	60°	
	"	1.4		0			≤0.3	90°	
	"	2.2		0			≤0.3	45°	
	"	2.0		0			≤0.3	30°	
	"	1.0		0			≤0.3	60°	
$\dfrac{U_{pc}}{fD} = K(2\pi\zeta)^a\left(\dfrac{m}{\rho D^2}\right)^b$	Tanaka et al (1981)			0.5			>~2.0	90°	P/D=1.33
	"		0.2	0.33			<~2.0	90°	P/D=1.33
$\dfrac{U_{pc}}{fD} = K(2\pi\zeta)^a$ $\left(\dfrac{m}{\rho D^2}\right)^b\left(\dfrac{P}{D}-C\right)^c$	Païdoussis (1981)	5.8	0.4	0.4	0.5	1	All	All	
	Chen (1984)	3.54	0.5	0.5	1.0	0.5	All	45°	
	"	3.58	0.1	0.1	1.0	0.9	<2.0	30°	
	"	6.53	0.5	0.5	1.0	0.9	>2.0	30°	
	Connors (1978)	1.76	0.5	0.5	1.0	-0.21		90°	P/D>1.41
$\dfrac{U_{pc}}{fD} = F(...)\left(\dfrac{2\pi\zeta m}{\rho D^2}\right)^b$	Lever and Weaver (1986)			0.5					F(...) is derived analytically
	Teh and Goyder (1988)			0.5					F(...) is derived empirically

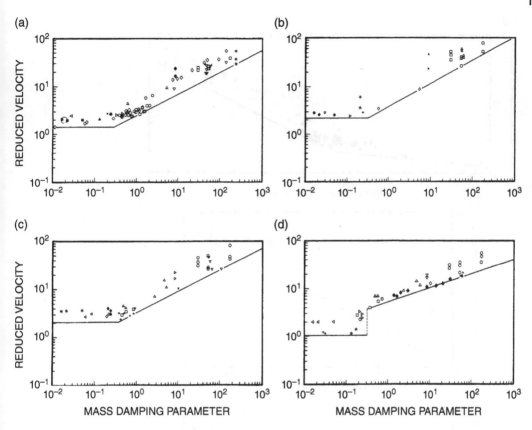

Fig. 7-3 Fluidelastic Instability Data and Design Recommendations for Various Tube Configurations: a) Normal Square, b) Rotated Square, c) Normal Triangular, d) Rotated Triangular (Weaver and Fitzpatrick, 1988).

Based on Eq. (7-8), Païdoussis (1981) suggested a design guideline where $K = 5.8$, $a = 0.4$, $b = 0.4$, $c = 0.5$ and $C = 1$. Similarly, Chen (1984) proposed $K = 3.54$, $a = 0.5$, $b = 0.5$, $c = 1.0$ and $C = 0.5$ for rotated-square tube bundles.

Teh and Goyder (1988) and Lever and Weaver (1986) proposed a slightly different version of Eq. (7-8):

$$\frac{U_{pc}}{fD} = F\left(\frac{2\pi\zeta m}{\rho D^2}\right)^b \tag{7-9}$$

where F is a function of several parameters such as Reynolds Number, P/D, tube bundle configuration, turbulence level, etc. The function, F, was obtained experimentally by Teh and Goyder (1988) for square, rotated-triangular and normal-triangular tube bundles. The function, F, is derived analytically in the case of the Lever and Weaver (1986) model.

Not surprisingly, the designer may be confused by the abundance of models and design guidelines. It is the purpose of this chapter to review this work and arrive at simple and practical design guidelines.

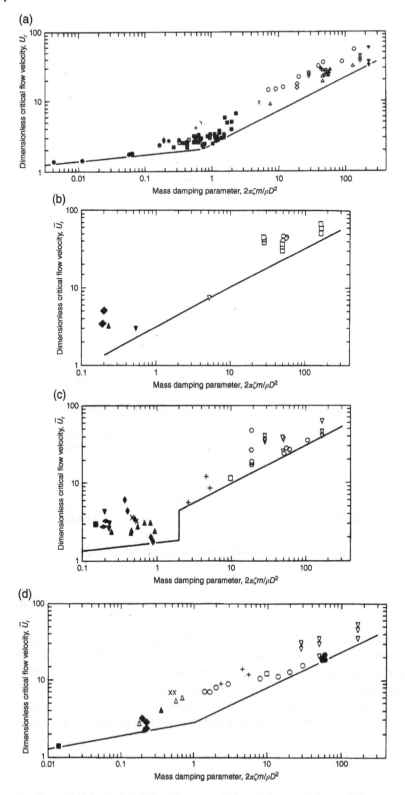

Fig. 7-4 Fluidelastic Instability Diagrams: a) Normal Square, b) Rotated Square, c) Normal Triangular, d) Rotated Triangular (Chen, 1984).

7.5 Approach

Since our aim is to formulate practical design guidelines, our approach has to be very pragmatic. We have reviewed most of the available data from the literature and data available from CEN Saclay in France and from Chalk River Laboratories of Atomic Energy of Canada Ltd. Nearly 300 data points were gathered in a comprehensive database. The experimental parameters extracted from the references are outlined in Table 7-2. Typical data points from the database are given in Table 7-3.

The data was compared against the available models to select the most appropriate. Then, conservative but realistic design guidelines were derived from the best fit of the available data to the chosen model. More weight was given to data originating from realistic heat exchanger configurations. Obviously, practical design guidelines must not contradict experience on real heat exchangers.

7.6 Important Definitions

Inconsistent definition of terms is a significant difficulty in reviewing data from other researchers. Flow velocity, damping, frequency, threshold velocity for instability, etc. are often defined differently. Sometimes important information is missing. Some assumptions were necessary. In other cases, missing parameters could be deduced by calculation. Such assumptions or calculations are noted in Table 7-3. Uniform definitions of important parameters were adopted in this document, as discussed below. Often the data of a given researcher had to be normalized or corrected to be compatible with the uniform definitions.

7.6.1 Tube Bundle Configurations

Most heat exchanger tube bundles are either of triangular or square configuration. For each configuration, there are two principal tube bundle orientations: normal (30°) and rotated (60°) for triangular bundles, and normal (90°) and rotated (45°) for square bundles. The angle pertains to the flow direction relative to the tube orientation, as shown in Fig. 7-5. In practice, it is very difficult to predict the exact flow direction everywhere in a shell-and-tube heat exchanger. In fact, all flow directions are usually possible in a given heat exchanger. Practical design guidelines must cover the worst tube bundle orientation. Thus, in this document, we consider the two principal bundle orientations together. This leads to only one set of design guidelines without concern for bundle orientation.

7.6.2 Flow Velocity Definition

Throughout our research program, we have been using a so-called "reference gap velocity," U_r, to present our results. Many other researchers (e.g., Chen, 1984; Weaver and Fitzpatrick, 1988) are now using the same velocity definition, although they do not necessarily call it reference gap velocity. Pitch velocity has been used extensively. Since it is shorter and equally meaningful, we propose to use it in this document. Thus, regardless of tube bundle configuration or orientation, the reference gap velocity, U_r, or pitch velocity, U_p, is defined as previously shown in Eq. (7-3)

$$U_r = U_p = U_\infty P/(P - D)$$

where P is the pitch of the tube bundle and U_∞ the free-stream flow velocity. The so defined pitch velocity is the true gap velocity for both normal-triangular and normal-square configurations. It is

Table 7-2 Summary of References and Tube Bundle Geometries (Pettigrew and Taylor, 1991)

Test Series	References	Orientation (deg)	P/D	L (L_e) (m)	D (mm)	w (mm)	Material	m_t (kg/m)	End Condition
Remy82	Remy (1982) Remy & Bai (1982)	90	1.44	1.0 (0.144)	10.0	1.0	304L	0.220	1
Nak86a	Nakamura et al (1986a)	90	1.42		19.05	1.0	SS	0.720	7
Nak86b	Nakamura et al (1986b)	90	1.42		19.05	1.0	SS	0.720	7
Sop83	Soper (1983)	30, 60 45, 90	1.25, 1.27, 1.35, 1.52, 1.78	0.456	25.4, 25.0	2.0	acrylic	0.278, 0.177	5
Axisa84	Axisa et al (1984)	90	1.44		19.05	1.09	I600	0.492	1
H&V81	Heilker & Vincent (1981)	30, 60, 45, 90	1.33, 1.36, 1.41, 1.50	0.914	22.0, 19.05, 22.2	1.2, 1.22, 1.27	I800,1600	0.631, 0.577, 0.706	4
P&G77	Pettigrew & Gorman (1977)	30, 60, 90	1.5, 1.47	1.22 (0.051), 0.914(0.051)	12.7, 12.9	0.76, 1.12	SS, I600	0.319, 0.439	4
Pett87	Pettigrew et al (1987)	30, 60, 45, 90	1.22, 1.35, 1.47	0.6	13.0	1.07	SS	0.330	5
Chen81	Chen & Jendrzejczyk (1981)	90, 60	1.35, 1.52, 1.5, 1.6	0.914 (0.305)	15.90	3.18	Brass		4
Con78	Connors (1978)	30, 60, 90	1.41, 1.69, 2.12		18, 24				7
Con80	Connors (1980)	30, 60	1.32, 1.25		24				7
J&S84	Johnson & Schneider (1984)	60	1.5		15.9				4
Lub86	Lubin et al (1986)	90	1.55	0.325	17.0	3.5	SS		1
W&K82	Weaver & Koroyannakis (1982)	60	1.375	0.305	25.4	1.02	Al	0.367	2
W&K83	Weaver & Koroyannakis (1983)	60	1.375	0.305	25.4		acrylic	0.607	2
W&AR85	Weaver & AbdRabbo (1985)	90	1.5	0.305	25.4		acrylic	0.607	2
AR&W86	AbdRabbo & Weaver (1986)	45	1.41	0.305	25.4		acrylic	0.607	5
Scot87	Scott (1987)	45, 90, 30, 60	1.7, 1.33, 1.5, 1.375, 1.73	0.305	25.4		acrylic	0.607	2

W&G78	Weaver & Grover (1978)	60	1.375	0.305	25.4			0.115	2
W&EK 81	Weaver & ElKashlan (1981)	60	1.375	0.305	25.4		AL, SS, Brass	0.157, 0.325, 0.889, 1.18, 0.151, 1.78	2
W&Y84	Weaver & Yeung (1984)	90, 45, 30, 60	1.5	0.316	12.7		acrylic, AL, Brass		2
M&O87	Minkami & Ohtomi (1987)	30, 60	1.3	0.335 (0.20)	12.0		SS	0.920	5
Gor76	Gorman (1976)	30, 60, 90, 45	1.33, 1.54, 1.36, 1.3	0.94	19.05, 13.0	1.17, 1.68	SS		6
P&G78	Pettigrew & Gorman (1978)	30, 60, 90, 45	1.57, 1.23, 1.47, 1.50	0.94	19.05, 13.0	1.17, 1.68	SS		4
Hal86	Halle et al (1986)	30, 60, 90, 45	1.25		19.05	1.2			9
Hal88	Halle et al (1988)	30, 60, 90, 45	1.42		19.05	1.2			9
Eis80	Eisenger (1980)		1.25, 1.3						9
God88	Bai (1982), Godon (1984) Godon & Lebert (1988)	30	1.37, 1.42, 1.31, 1.41, 1.33		19.0, 26.0, 17.0, 18.0	0.4 to 1.0	Ti, Fe/St, Ad/Br	0.3 to 1.122	9
Har74	Hartlen (1974)	30, 60, 45, 90	1.25, 1.375, 1.5, 1.28, 1.4, 1.53	0.762 (0.610)	12.7, 12.5, 12.2			0.092, 0.083, 0.0565,	1
Tan81	Tanaka & Takahara (1981)	90	1.33	0.3	30.0			1.62, 1.98	8

* End Conditions:

1) Clamped-Clamped (C-C)
2) One tube plus neighbors are Clamped-Free and all other tubes are C-C
3) One tube is flexible in lift direction and all other tubes are C-C
4) Pinned-Pinned
5) Clamped-Free

6) Clamped-Pinned
7) Clamped-Clamped with piano wire (translation)
8) Unknown
9) Multi-span Heat Exchanger Tubes

Table 7-3 Database on Fluidelastic Instability of Tube Bundles in Single-Phase Cross Flow (Pettigrew and Taylor, 1991)

Test Series	Orient	P/D	Tube Dia mm	Tube Mass Freq kg/m	(air) Hz	Damp Rat (air) %	Density kg/m³	Freq Hz	Damp Rat (tot) %	Tot Mass kg/m	Crit Vel m/s	Mass Damping	Reduced Crit Vel	FEI Constant	Comments	Footnotes
H&V81	45	1.500	19.05	0.577	59.8	1.600	1000.00	46.8	2.070	0.942	2.81	3.38E-01	3.15E+00	5.42		
H&V81	60	1.330	19.05	0.577	59.8	1.530	1000.00	45.0	' 1.970	1.017	1.61	3.47E-01	1.88E+00	3.19		
H&V81	30	1.330	19.05	0.577	59.8	1.530	1000.00	45.0	1.970	1.017	1.89	3.47E-01	2.20E+00	3.74		
H&V81	90	1.410	22.20	0.706	70.1	1.080	1000.00	53.4	1.390	1.218	2.85	2.16E-01	2.40E+00	5.17		
H&V81	45	1.410	22.20	0.706	70.1	1.080	1000.00	53.4	1.390	1.218	3.46	2.16E-01	2.92E+00	6.28		
P&G77	30	1.500	12.700	0.319		2.670	1000.00	17.0	2.670	0.452	0.99	4.70E-01	4.59E+00	6.69		
P&G77	60	1.500	12.700	0.319		2.670	1000.00	17.0	2.670	0.452	1.52	4.70E-01	7.04E+00	10.27		
P&G77	60	1.470	12.900	0.439		2.480	1000.00	30.0	2.480	0.576	2.67	5.39E-01	6.90E+00	9.39		
P&G77	90	1.470	12.900	0.439		2.480	1000.00	30.0	2.480	0.576	1.63	5.39E-01	4.21 E+00	5.74		
Pett87	30	1.220	13.000	0.330		0.200	1000.00	25.4	1.020	0.560	0.68	2.12E-01	2.06E+00	4.47	not	
Pett87	60	1.220	13.000	0.330		0.200	1000.00	25.4	1.020	0.560	0.63	2.12E-01	1.91 E+00	4.14	published	
Pett87	45	1.220	13.000	0.330		0.200	1000.00	25.9	0.930	0.540	0.83	1.87E-01	2.47E+00	5.70		
Pett87	90	1.220	13.000	0.330		0.200	1000.00	25.9	0.930	0.540	0.81	1.87E-01	2.41 E+00	5.57		
Pett87	30	1.350	13.000	0.330		0.200	1000.00	25.9	0.940	0.540	0.78	1.89E-01	2.32E+00	5.33		
Pett87	30	1.470	13.000	0.330		0.200	1000.00	26.2	0.890	0.530	1.10	1.75E-01	3.23E+00	7.71		
Pett87	60	1.470	13.000	0.330		0.200	1000.00	26.2	0.890	0.530	0.63	1.75E-01	1.85E+00	4.42	Streamwise	
Pett87	90	1.470	13.000	0.330		0.200	1000.00	26.5	0.840	0.510	0.87	1.59E-01	2.53E+00	6.33	P/D	
Chen81	90	1.500	15.90				1000.00	23.9	2.300	1.280	0.81	7.32E-01	2.13E+00	2.49		8
Chen81	90	1.500	15.90				1000.00	24.3	3.300	1.280	0.95	1.05E+00	2.46E+00	2.40	1.50	
Chen81	90	1.500	15.90				1000.00	24.3	3.900	1.280	1.08	1.24E+00	2.80E+00	2.51	1.50	
Chen81	90	1.500	15.90				1000.00	24.7	4.800	1.280	1.28	1.53E+00	3.26E+00	2.64	1.50	
Chen81	90	1.500	15.90				1000.00	25.1	6.400	1.280	2.04	2.04E+00	5.11 E+00	3.58	1.50	
Chen81	90	1.350	15.90				1000.00	22.5	2.400	1.280	0.86	7.63E-01	2.42E+00	2.76	1.42	
Chen81	90	1.350	15.90				1000.00	22.6	2.900	1.280	1.05	9.23E-01	2.92E+00	3.04	1.42	

Chen81	90	1.350	15.90	22.7	4.100	1.280	1.30	1.30E+00	3.60E+00	3.15	1.42
Chen81	90	1.350	15.90	22.7	5.000	1.280	2.10	1.59E+00	5.82E+00	4.61	1.42
Chen81	90	1.520	15.90	24.0	2.300	1.280	0.92	7.32E-01	2.41E+00	2.82	1.60
Chen81	90	1.520	15.90	24.0	3.700	1.280	1.03	1.18E+00	2.70E+00	2.49	1.60
Chen81	90	1.520	15.90	24.6	4.500	1.280	1.18	1.43E+00	3.02E+00	2.52	1.60
Chen81	90	1.520	15.90	25.0	6.300	1.280	1.61	2.00E+00	4.05E+00	2.86	1.60
Chen81	90	1.520	15.90	25.4	7.300	1.280	2.66	2.32E+00	6.59E+00	4.32	1.60
Chen81	90	1.500	15.90	29.3	1.400	1.240	1.21	4.31 E-01	2.60E+00	3.95	1.42
Chen81	90	1.500	15.90	29.8	2.200	1.240	1.32	6.78E-01	2.79E+00	3.38	1.42
Chen81	90	1.500	15.90	29.9	3.000	1.240	1.40	9.25E-01	2.94E+00	3.06	1.42
Chen81	90	1.500	15.90	30.2	4.100	1.240	1.76	1.26E+00	3.67E+00	3.26	1.42
Chen81	90	1.500	15.90	30.5	5.400	1.240	3.27	1.66E+00	6.74E+00	5.23	1.42
Chen81	90	1.500	15.90	22.4	1.800	0.800	0.67	3.58E-01	1.88E+00	3.15	1.42
Chen81	90	1.500	15.90	22.7	2.500	0.800	0.77	4.97E-01	2.13E+00	3.02	1.42
Chen81	90	1.500	15.90	22.9	3.200	0.800	0.86	6.36E-01	2.36E+00	2.96	1.42
Chen81	90	1.500	15.90	23.1	4.300	0.800	1.53	8.55E-01	4.17E+00	4.51	1.42
Chen81	90	1.500	15.90	23.7	2.900	0.510	0.95	3.68E-01	2.52E+00	4.16	1.42
Chen81	90	1.500	15.90	24.0	4.200	0.510	1.19	5.32E-01	3.12E+00	4.27	1.42
Chen81	90	1.500	15.90	24.2	5.000	0.510	1.26	6.34E-01	3.27E+00	4.11	1.42
Chen81	90	1.600	15.90	24.0	2.200	1.280	1.09	7.00E-01	2.86E+00	3.41	1.52
Chen81	90	1.600	15.90	24.2	2.700	1.280	1.19	8.59E-01	3.09E+00	3.34	1.52

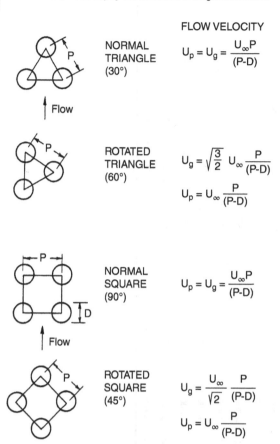

FLOW VELOCITY

Fig. 7-5 Principal Tube Bundle Configurations.

NORMAL TRIANGLE (30°)

$$U_p = U_g = \frac{U_\infty P}{(P-D)}$$

ROTATED TRIANGLE (60°)

$$U_g = \sqrt{\frac{3}{2}}\, U_\infty \frac{P}{(P-D)}$$

$$U_p = U_\infty \frac{P}{(P-D)}$$

NORMAL SQUARE (90°)

$$U_p = U_g = \frac{U_\infty P}{(P-D)}$$

ROTATED SQUARE (45°)

$$U_g = \frac{U_\infty}{\sqrt{2}} \frac{P}{(P-D)}$$

$$U_p = U_\infty \frac{P}{(P-D)}$$

not for the rotated-triangular and rotated-square configurations. For the latter, the pitch velocity is only an indication of the flow velocity between the tubes.

This definition of velocity has two advantages. Firstly, it enables designers to directly compare the two principal orientations of a tube bundle from a vibration viewpoint. As already discussed, flow in all directions is usually possible in one or another region of actual heat exchangers. Secondly, vibration excitation forces appear related to the flow velocity between the tubes, and thus, to the pitch velocity. Some designers use a gap velocity, U_g, in heat transfer and pressure drop calculations. Gap velocity definitions are shown in Fig. 7-5. The pitch velocity can easily be deduced from the gap velocity.

7.6.3   Critical Velocity for Fluidelastic Instability

Different researchers often use different criteria to determine the critical velocity for fluidelastic instability. For example, Heilker and Vincent (1981) used the velocity at which tube rattling occurred as a criterion, while Soper (1983) used the point at which a tangent to the post-critical response intersected the velocity axis, whereas Teh and Goyder (1988) took the velocity at which damping appears to be nil. These differences are a great source of disparity in the data. Here again we take a very practical approach. When the instability threshold is well defined, we simply take the flow velocity at which it occurs as the critical instability velocity (see Fig. 7-6). When it is not well defined, we take the velocity at which excessive vibration amplitude occurs (see Fig. 7-7). This

Fig. 7-6 Well-Defined Fluidelastic Threshold

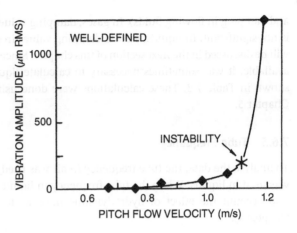

Fig. 7-7 Less Well-Defined Fluidelastic Threshold

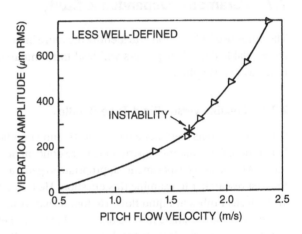

amplitude is normally between 250 and 750 μm root-mean-square (rms), depending on tube frequency and tube size (i.e., span length and flexural rigidity). Such vibration levels are not normally acceptable in heat exchangers. In our database development, we have attempted to normalize the results of other researchers to our definition of fluidelastic instability. This was not always easy or even possible because of lack of information. In cases where the tubes are partially subjected to cross flow, the flow velocities were normalized to obtain the equivalent uniformly distributed velocity, as outlined in Pettigrew et al (1978).

7.6.4 Damping

Damping is a great source of controversy for two reasons. Firstly, damping is difficult to measure accurately, particularly in tube bundles. Secondly, different damping definitions have been used by different researchers to analyze their own results (i.e., damping in air, damping in still liquid and damping in flowing fluids). In gases, damping values in air are acceptable since fluid damping is not significant. In liquids, we use damping values in still liquids. The reasons for this approach will be discussed in the next section of this chapter. Since damping values in liquids were not always available, it was sometimes necessary to calculate liquid damping to complete the database, as shown in Table 7-3. These calculations were done using the damping formulation discussed in Chapter 5.

7.6.5 Tube Frequency

To analyse the data, the tube frequency in air was used in gases; the tube frequency in still liquid was used in liquids. When the tube frequency in liquid was not given, it was calculated by taking into account the effect of hydrodynamic mass as formulated by Pettigrew et al (1989a) (see Chapter 4).

7.7 Parametric Dependence Study

The purpose of this section is to identify the more important parameters by correlating them against the available data. This process will lead to formulation of the most suitable practical model for fluidelastic instability.

7.7.1 Flexible versus Rigid Tube Bundles

Some experiments were done with realistic tube bundles where all the tubes were free to vibrate. Other more fundamental experiments were done on one flexible tube in an otherwise rigid array of tubes. Are the latter experiments appropriate to generate design data? Gorman (1977) was probably the first researcher to examine this question. He tested brass tubes, which are more flexible, in an array of steel tubes in liquid flow. He found that instability occurred at a flow velocity effectively 20% greater for the single flexible tube. We have completed a series of experiments on both flexible tube bundles and single flexible tubes in rigid arrays (Pettigrew et al, 1989b and Kim et al, 1987). We found instability took place at a slightly higher flow velocity (i.e., ≈10% higher) for the single flexible tube experiments. Teh and Goyder (1988) have conducted elegant experiments on a single flexible cylinder constrained to vibrate in the lift direction only. The single flexible cylinder was installed in several otherwise rigid arrays of different configurations subjected to air flow in a wind tunnel. They found reasonable agreement with the results of Soper (1983) in similar arrays of flexible cylinders in the same wind tunnel. Thus, it can be said that in some cases, the difference is not large. On the other hand, some researchers have reported considerable differences in behavior between single and multiple flexible cylinder arrays. For example, Fig. 7-8 compares the vibration response of one flexible tube to that of seven flexible tubes for a rotated-triangular array (Lever and Rzentkowski, 1988). The critical velocity for fluidelastic instability is significantly different. Andjelic and Popp (1989) arrived at similar conclusions for a normal-triangular array.

In Fig. 7-9, we present all the fluidelastic instability data for bundles with all flexible tubes in terms of the dimensionless velocity, U_p/fD, and the mass-damping parameter, $2\pi\zeta m/\rho D^2$ A line at $K = 3.0$ and $b = 0.5$ has been drawn to represent the overall trend in the data for further comparison. Figure 7-10 presents Teh and Goyder's (1988) data for a single flexible cylinder in normal-triangular tube arrays. Païdoussis' (1987) results for a single flexible cylinder in a normal-square rigid array of $P/D = 1.5$ are shown in Fig. 7-11. Comparison of these figures clearly indicates that

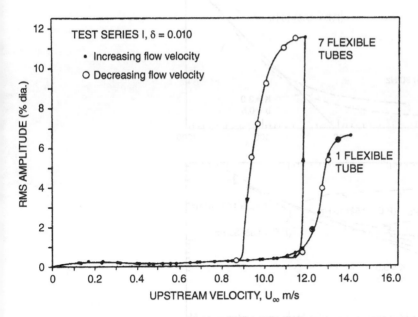

Fig. 7-8 Vibration Response of One Flexible Tube versus Seven Flexible Tubes in 60° Arrays from Lever and Rzentkowski (1988).

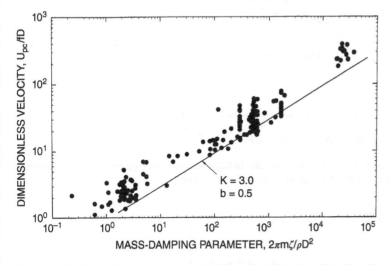

Fig. 7-9 Outline of Fluidelastic Instability Data for All-Flexible Tube Bundles.

the fluidelastic instability behavior is generally different for flexible tube bundles than for a single flexible tube in an otherwise rigid array of tubes. For the latter tube arrays, the relationship between dimensionless flow velocity and dimensionless mass-damping is far from simple and is somewhat discontinuous between liquid flow and gas flow regions. Thus, for the purpose of formulating design guidelines, only data from tests with all flexible tube bundles is considered.

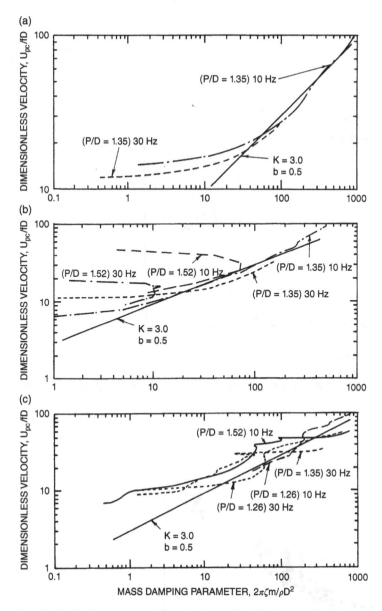

Fig. 7-10 Fluidelastic Instability of a Single Flexible Tube in Rigid Arrays Subjected to Gas Flow: a) Normal Triangular, b) Rotated Triangular, c) Normal Square from Teh and Goyder (1988).

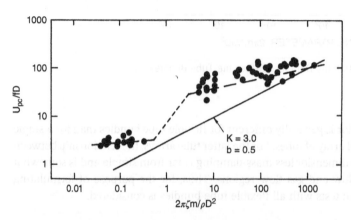

Fig. 7-11 Fluidelastic Instability of a Single Flexible Tube in a Normal-Square Rigid Array of *P/D* = 1.5 Subjected to Both Liquid and Gas Flow, with Data from Païdoussis (1987).

7.7.2 Damping

In gases, structural damping (mostly friction and impact damping at the support) is the dominant energy dissipation mechanism, as discussed in Chapter 5. Fluid damping is not significant except perhaps for very dense gases. Thus, we generally use damping values in air for data analysis.

Fluid damping is much more important in liquids. Some researchers do not consider fluid damping in their analysis. A possible reason is that it is very difficult to measure in tube bundles because of hydrodynamic coupling, etc. Another reason is that, at fluidelastic instability, damping appears very small, which is characteristic of fluidelastic instability phenomena. However, this is not necessarily a valid reason for ignoring fluid damping. Looking at the problem a different way, it can be said that, below critical flow velocities for instability, the tube bundle is stable largely because of fluid damping. Thus, the latter may very well be a governing parameter in fluidelastic instability phenomena.

Some researchers have reported that the exponent, b, in Eq. (7-5) becomes smaller for lower values of the mass-damping parameter. See, for example, Fig. 7-4c. In this figure, the data points for $(2\pi\zeta m/\rho D^2) < 0.2$ came from tests performed by Tanaka and Takahara (1981) on a square tube array in water flow. Only structural damping was used in the analysis of this data. We have calculated the stagnant fluid damping for this case using the formulation of Pettigrew et al (1986). The fluidelastic instability results using fluid damping are compared to the original analysis in Fig. 7-12. Interestingly, the results now fit along a straight line of exponent $b = 0.5$.

In Fig. 7-9, the data for flexible tube bundles is presented in terms of the mass-damping parameter, $2\pi\zeta m/\rho D^2$, in which the damping ratio, ζ, includes stagnant fluid damping. The same data is compared to that using only structural damping in Fig. 7-13. The data collapse along a straight line when fluid damping is included. Thus, it is desirable to consider fluid damping in the formulation of fluidelastic instability. We also arrived at the same conclusion while analyzing fluidelastic instability data for two-phase flow (Pettigrew et al, 1989b)

Flow-dependent damping is another aspect of the damping question. It is not normally a dominant damping mechanism under heat exchanger conditions. Chen (1981) has shown that it is directly related to the dimensionless flow velocity, U_p/fD, and that it is practically insignificant for the tube bundles he tested for $U_p/fD < 3.0$, which is the range of interest for heat exchangers in liquids. Furthermore, it is very difficult to differentiate energy dissipating fluid forces from fluidelastic excitation forces, particularly near instability. Thus, we assume that flow-dependent damping is part of the fluidelastic instability phenomenon, since both are related to flow-dependent fluid dynamic forces.

It should be noted that in fundamental experiments in wind tunnels, the mass-damping parameter is often changed by changing damping. In real heat exchangers, fluid density governs the mass-damping parameter. For example, in high-pressure steam, the density may be 100 kg/m^3, whereas in power condensers it may be 0.01 kg/m^3. This is a factor of 10^4, which is simulated in wind tunnel studies by similarly changing damping. Thus, laboratory conditions may be very different than realistic heat exchanger operating conditions (e.g., fluid density, viscosity, Reynolds Number, etc.). The difference should be carefully considered in applying laboratory data to heat exchanger design. For example, data from tests in wind tunnels may not necessarily apply to condensers.

7.7.3 Pitch-to-Diameter Ratio, *P/D*

Some researchers found the pitch-to-diameter ratio, *P/D,* to have a significant effect on fluidelastic instability for some tube bundle configurations. For instance, both Hartlen (1974) and Soper (1983) observed the instability factor, K, to increase with *P/D* for normal-triangular and rotated-square tube bundles. On the other hand, they found no significant effect of *P/D* for normal-square and rotated-triangular tube bundles. Unfortunately, the designer cannot take advantage of the effect of *P/D,*

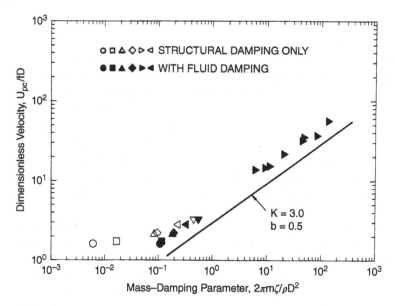

Fig. 7-12 Comparison of Fluidelastic Instability Results with Only Structural Damping versus with Both Structural and Fluid Damping Using Tanaka and Takahara (1981) Data.

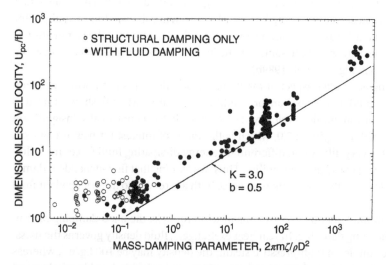

Fig. 7-13 Fluidelastic Instability Data for Flexible Tube Bundles: Comparison of Data with and without Using Fluid Damping.

since it does not apply to all tube bundle orientations. In practice, all orientations are possible in a given heat exchanger tube bundle.

The fluidelastic instability factor is plotted against P/D for all the available data in Figs. 7-14a, 7-14b, 7-14c and 7-14d for the normal-triangular, rotated-triangular, normal-square and rotated-square tube bundles, respectively. The results show that, although there is a trend with P/D in some cases, the trend is not general and, thus, cannot be used for practical design guidelines. Thus, P/D is not taken into account in the recommended fluidelastic instability formulation.

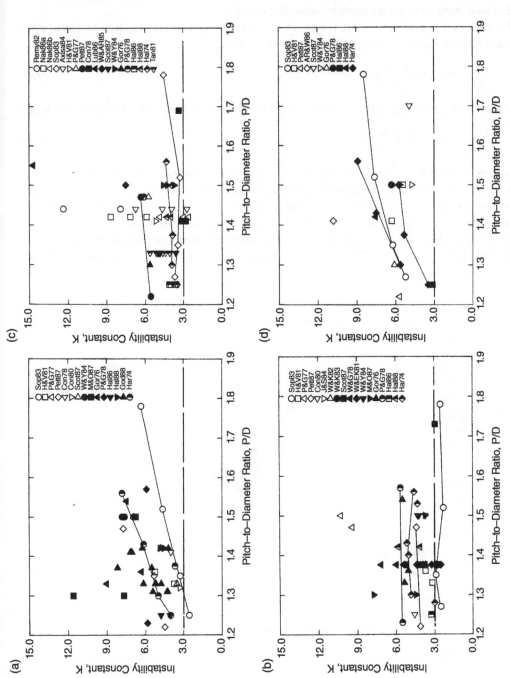

Fig. 7-14 Effect of Pitch-to-Diameter Ratio on Fluidelastic Instability Constant: a) Normal Triangular, b) Rotated Triangular, c) Normal Square, d) Rotated Square.

7.7.4 Fluidelastic Instability Formulation

Figure 7-9 would suggest a fluidelastic instability formulation of the form

$$\frac{U_{pc}}{fD} = K\left(\frac{2\pi\zeta m}{\rho D^2}\right)^{0.5} \tag{7-10}$$

The exponent $b = 0.5$ seems very appropriate as it follows the data reasonably well over values of the mass-damping parameter covering some five decades.

An attempt was made to explore the suitability of more sophisticated models. As discussed in Section 7.3, several authors suggested a model of the form

$$\frac{U_{pc}}{fD} = K(2\pi\zeta)^a\left(\frac{m}{\rho D^2}\right)^b \tag{7-11}$$

Values of the damping exponent, a, between 0.2 and 0.3 have been suggested. Thus, the model

$$\frac{U_{pc}}{fD} = K(2\pi\zeta)^{b/2}\left(\frac{m}{\rho D^2}\right)^b \tag{7-12}$$

was tried on the data, as shown in Fig. 7-15. A line with the slope of $b = 0.5$ has been drawn as a lower envelope of the data. This corresponds to a damping exponent $a = 0.25$. Generally, the agreement with the data is no better than for the simpler model of Fig. 7-9. The more sophisticated model may fit some of the data better, as suggested by several researchers. However, this is not generally true. Thus, we have no reason at this time to recommend more sophisticated models. Also, as shown in Figs. 7-9, 7-16 and 7-17, a mass-damping parameter exponent, $b = 0.5$, generally fits well

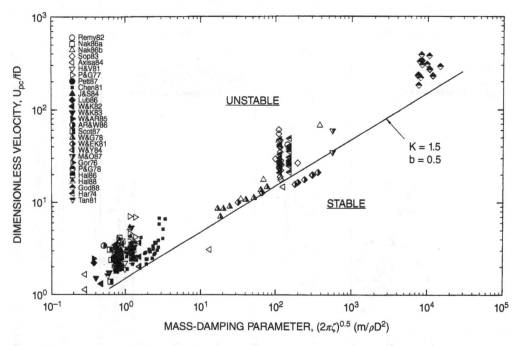

Fig. 7-15 Fluidelastic Instability Data Presented in Terms of Modified Mass-Damping Parameter, $(2\pi\zeta)^{0.5}(m/\rho D^2)$.

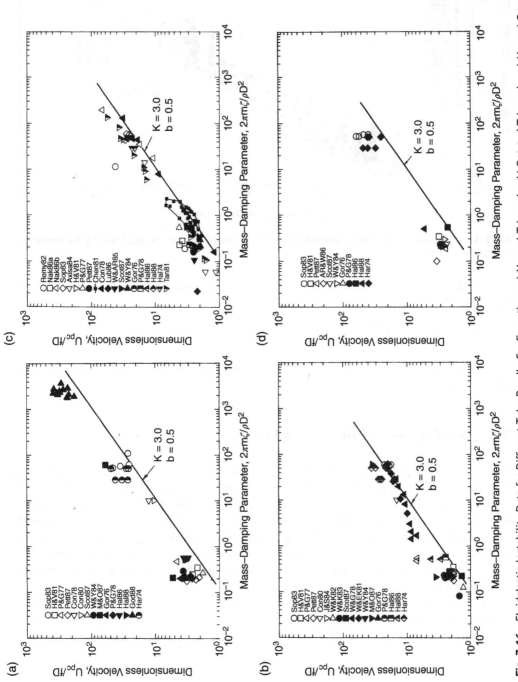

Fig. 7-16 Fluidelastic Instability Data for Different Tube Bundle Configurations: a) Normal Triangular, b) Rotated Triangular, c) Normal Square, d) Rotated Square.

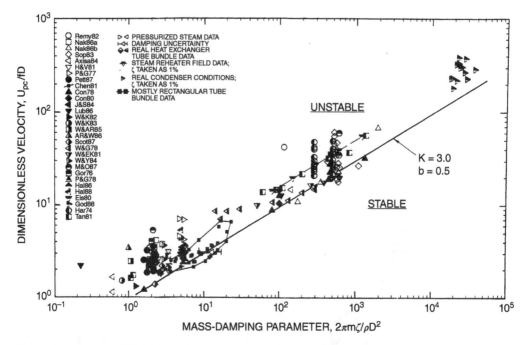

Fig. 7-17 Summary of Fluidelastic Instability Data for All-Flexible Tube Bundles Showing Recommended Design Guideline of *K*=3.0.

with the data. Thus, we recommend the simple fluidelastic instability formulation given earlier in Eq. (7-5)

$$\frac{U_{pc}}{fD} = K\left(\frac{2\pi\zeta m}{\rho D^2}\right)^{0.5}$$

as a practical design guideline. An appropriate value for the instability constant, *K*, needs to be specified as discussed next.

7.8   Development of Design Guidelines

The data of Table 7-3 are presented in terms of dimensionless flow velocity and dimensionless mass-damping in Fig. 7-16 for the different tube bundle configurations. We now have to choose an appropriate instability constant, *K*, as a design guideline. To take a lower envelope of all data would be unrealistic and overly conservative. The scatter in the data is often due to unrepresentative laboratory conditions (very high or low damping, unrealistic gas densities, etc.) or to non-uniform definitions of terms as already discussed. Thus, to avoid unduly penalizing the designer with extreme data points, we propose to take roughly a "lower decile" line through the data. We believe this is a conservative and realistic approach.

This approach leads to an instability constant *K* = 3.0, as shown in Fig. 7-16, for both triangular tube bundles and square tube bundles. It may be deduced from Figs. 7-16a and 7-16b that twelve data points out of 108 are below the *K* = 3.0 line for triangular tube bundles. Thus roughly 90% of the data is above $K = 3.0$, which corresponds to the lower decile line. Similarly for the square tube bundles, five data points out of 93 are below the $K = 3.0$ line (see Figs. 7-16c and 7-16d). Thus, more than 90% of the data is above $K = 3.0$.

The data of Chen and Jendrzejczyk (1981), shown in Fig. 7-16c, was not considered in the above analysis, since most of this data is for rectangular (not square) arrays. However, had it been considered, it would not have changed the above guideline. Some results from Hartlen (1974) were also not considered, since they were qualified by the author as being preliminary and tentative. These results were obtained with plastic tubes with very high damping. However, all other results from Hartlen (1974) were considered.

All the data points for all tube bundles are presented in Fig. 7-17. The "lower decile," $K = 3.0$, design guideline is also shown. Only 17 out of 200 points are below that line. All of these seventeen data points may be questioned for a variety of reasons, such as insufficient information, unrealistic damping and ill-defined critical velocity for instability. However, they are not necessarily in error. On the other hand, all data points from realistic situations, such as Halle et al's (1986 and 1988) data on real multi-span heat exchanger tube bundles, Godon and Lebert's (1988) data on real condenser configurations in steam flow, Gorman's (1976) and Pettigrew and Gorman's (1978) data on realistic tube bundles in liquid flow and Eisinger's (1980) data on actual steam reheaters, are above $K = 3.0$.

Interestingly, Fig. 7-17 covers a range of mass-damping parameters from 5×10^{-2} to 5×10^3, which is a factor of 10^5! It includes data from highly damped light cylinders in liquid flow, heat exchanger tubes in liquid flow, tube bundles in high-pressure high-density steam flow, tube bundles in wind tunnels, and condenser tube bundles in very-low-pressure low-density steam flow.

Since the mid-1970s, hundreds of heat exchangers have been designed following a similar guideline (i.e., $K = 3.3$) for the Canadian nuclear industry. These guidelines (i.e., $K = 3.0$ and $K = 3.3$) are also used for non-nuclear heat exchangers by most Canadian manufacturers and many manufacturers in the USA and elsewhere. To our knowledge, none of the heat exchangers that have been designed according to the guidelines have failed, and all those that did fail did not satisfy these guidelines.

The performance of ten operating heat exchangers was reviewed and compared against the design guidelines (Pettigrew and Campagna, 1979). Both failure histories and satisfactory performances were considered, as shown in Fig. 7-18. This comparison confirms that the above design guidelines are a reasonable design criterion. Many more such comparisons have been done since that time with similar results.

Example 7-1 Fluidelastic Instability Calculation in a Single-Phase Heat Exchanger

As an example, we take a typical multi-span heat exchanger tube of diameter, $D = 20$ mm, mass including hydrodynamic mass, $m = 1$ kg/m, natural frequency, $f = 50$ Hz, tube span length, $\ell = 1.0$ m, damping ratio including squeeze-film damping and viscous damping in stagnant fluid, $\zeta = 1.5\%$ and shell-side density, $\rho = 1000$ kg/m^3. This tube, which is illustrated in Fig. 7-19, is subjected to a pitch flow velocity, $U_p = 1.0$ m/s.

We need to calculate the critical velocity for fluidelastic instability U_{pc} to make sure the actual velocity is comfortably lower. Thus,

$$\frac{U_{pc}}{fD} = K \left(\frac{2\pi\zeta m}{\rho D^2} \right)^{0.5}$$

CASE NO	COMPONENT TYPE	REGION	P/D	D (mm)	MODE ORDER	f (Hz)	$\dfrac{2\pi\zeta m}{\rho D^2}$	$\dfrac{U_p}{U_{pc}}$
NSPG-1	Steam Generator	Inlet	1.50	12.7	1st	42.5	0.44	1.36
NSPG-2	" "	Outlet	1.50	12.7	1st	48.9	11.75	0.85
DPSG-1	" "	U-tube	1.38	12.7	1st	11.4	16.2	2.27
DPSG-2	" "	U-tube	1.38	12.7	1st	30.2	20.4	0.11
PASG	" "	U-tube	1.50	12.7	1st	17.0	10.4	1.30
BASG	" "	U-tube	1.57	13.0	1st	35.5	16.2	0.49
DPBC-1	Cooler	Outlet	1.5	12.7	1st	103.0	0.156	0.77
DPBC-2	"	Inlet	1.5	12.7	1st	87.1	0.181	0.75
DPBC-3	"	Interior	1.5	12.7	1st	104.0	0.152	0.76
PASC-1	"	Outlet	1.26	12.5	1st	62.1	0.256	1.73
PASC-2	"	Outlet	1.26	12.5	3rd	209.	0.076	1.74
PAMX-1	Heat Exchanger	Interior	1.50	12.7	1st	37.1	0.43	2.04
PAMX-2	" "	Interior	1.50	12.7	1st	134.0	0.12	1.07
GBHX-1	" "	Inlet	1.25	19.1	1st	17.7	0.751	0.70
GBHX-2	" "	Inlet	1.25	19.1	8th	42.7	0.311	1.37
GBHX-3	" "	Inlet	1.25	19.1	8th	69.2	0.192	0.35
GBCX-1	Cooler	Inlet	1.25	19.1	1st	18.15	0.581	1.71
PHRC-1	"	Inlet	1.33	19.1	15th	177.44	0.050	1.55

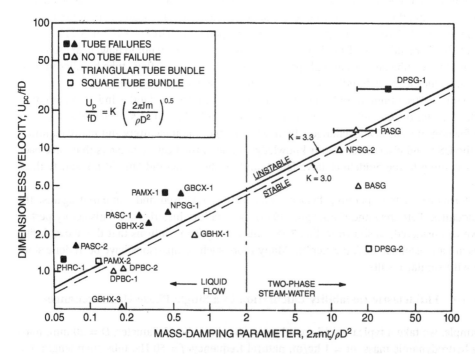

Fig. 7-18 Fluidelastic Instability Analysis of Real Heat Exchangers: Comparison of Operating Experience versus Fluidelastic Instability Constants $K = 3.0$ and $K = 3.3$ (Pettigrew and Campagna, 1979).

Taking $K = 3.0$, we get

$$\frac{U_{pc}}{50\,\text{Hz} \times 0.02\,\text{m}} = 3.0\left[\frac{2\pi \times 0.015 \times 1.0\,\text{kg/m}}{1000\,\text{kg/m}^3 \times (0.02\,\text{m})^2}\right]^{0.5}$$

$$U_{pc} = 1.456\,\text{m/s}$$

Since the flow velocity in the heat exchanger is considerably lower than the critical velocity for fluidelastic instability, i.e.,

$$\frac{U_p}{U_{pc}} = \frac{1.0\,\text{m/s}}{1.456\,\text{m/s}} = 0.687 < 1$$

this heat exchanger is well designed and should not experience severe vibration due to fluidelastic instability.

Example 7.2   Fluidelastic Instability Calculation in a Single-Phase Heat Exchanger with Window Region

Following the heat exchanger with window region case (see Fig. 3-5) introduced in Examples 3.1, 4.1 and 5.2, the following fluidelastic parameters are known: $D = 0.02$ m, $m = 1.16$ kg/m, $f = 135.6$ Hz, $f_{window} = 33.9$ Hz, $\rho = 1000$ kg/m^3, $\zeta = 0.85\%$ (single-phase), $\zeta_{window} = 1.96\%$ (single-phase), $U_p = 1.5$ m/s.

The critical velocity in the heat exchanger can be calculated as follows:

$$\frac{U_{pc}}{fD} = K\left(\frac{2\pi\zeta m}{\rho D^2}\right)^{0.5} = 3.0\left(\frac{2\pi \times 0.0085 \times 1.16}{1000 \times 0.02^2}\right)^{0.5} = 1.18$$

$$U_{pc} = 1.18fD = 1.18 \times 135.6 \times 0.02 = 3.2\,\text{m/s}$$

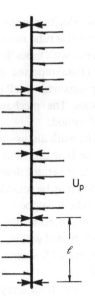

Fig. 7-19 Multi-Span Heat Exchanger Tube Schematic with Cross-Flow Pattern.

This critical velocity is well above the actual pitch velocity in the main section of the heat exchanger. Therefore, as shown below, this heat exchanger should not experience fluidelastic instability in the main section.

$$\frac{U_p}{U_{pc}} = \frac{1.5}{3.2} = 0.47 < 1.0\ \text{OK}$$

In the window region of this heat exchanger, the tube span length is twice as long leading to a frequency that is four times lower than in the main section of the heat exchanger. As a result, the critical velocity will be lower in this region of the heat exchanger.

$$\frac{U_{pc}}{fD} = K\left(\frac{2\pi\zeta m}{\rho D^2}\right)^{0.5} = 3.0\left(\frac{2\pi \times 0.0196 \times 1.16}{1000 \times 0.02^2}\right)^{0.5} = 1.79$$

$$U_{pc} = 1.79fD = 1.79 \times 33.9 \times 0.02 = 1.22\,\text{m/s}$$

$$\frac{U_p}{U_{pc}} = \frac{1.5}{1.22} = 1.23 > 1.0\ \text{unstable behavior is possible}$$

Since the pitch velocity is larger than the calculated critical velocity, it is possible that fluidelastic instability will occur in the window region.

7.9   In-Plane Fluidelastic Instability

In heat exchangers with U-tubes, the tubes can vibrate in out-of-plane (OOP) and in-plane (IP) directions, as described in Chapter 4. Contrary to OOP instability, the possibility of IP instability has raised less concerns because the IP modes have higher frequencies and considerably higher flow velocity at instability. Also, many of the reported cases of fluidelastic instability involve purely transverse tube motion.

Tests with tubes flexible only in the transverse direction have been reported. Weaver and Schneider (1983) reported on the effect of Flat-bar supports (FURs) on tube stability in the U-bend region. Supports with small clearances were observed to be effective in stabilizing both IP and OOP tube modes. The mechanism by which the IP modes were stabilized was, however, not clearly understood.

The work discussed here pertains to the question of fluidelastic instability in arrays solely flexible in the flow direction. This work is motivated by the need to quantify the risk of IP instability in steam generator designs. Flat-Bar supports, sometimes termed anti-vibration bars, in the U-bend region may be considered effective in the OOP even in the presence of small gaps, because nonlinear tube-to-support interaction (impacting) introduces additional damping, while tube-to-support contact increases the effective stiffness. However, small tube-to-support gaps may render the tube unsupported within the plane, allowing for the possibility of significant IP vibration due to reduced effective IP frequency.

Simple tests in a wind tunnel were conducted to determine if fluidelastic instability could be achieved on an array of tubes preferentially flexible in the flow direction (Mureithi et al, 2005). Experiments were conducted in a wind tunnel with a 305 mm x 305 mm test section. The tube bundle, shown in Fig. 7-20, consisted of an array of light-weight tubes of 40.4 mm diameter in a rotated-triangular configuration of pitch-to-diameter ratio, $P/D = 1.37$. Each tube was attached with a mild-steel thin plate to one end of the test section. The plate had a 1.5 mm x 15 mm rectangular section, so that vibration was constrained to only one direction. All tubes were originally designed as cantilever tubes. For the purposes of this study, each tube could be independently restrained from vibrating by means of a stopper inserted at the end of the tube, hence simulating a rigidly fixed tube. Five series of tests of different tube bundle configurations were performed (Mureithi et al, 2005).

Figure 7-21 shows typical vibration spectra for low flow velocities, with minimal excitation due to turbulence. At pitch velocity, $U_p = 6.44$ m/s, a strong instability developed. The rms vibration response versus flow velocity is shown in Fig. 7-22. Increasing the flow velocity to 7.0 m/s resulted in peak amplitudes of the order of the inter-tube gap. Coincidentally with the large increase in vibration amplitude, a coalescence of the individual tube frequencies to a single modal frequency was observed (see Fig. 7-23), confirming again the onset of fluidelastic instability. The results demonstrate that fluidelastic instability can occur in a tube bundle preferentially flexible in the flow direction subjected to single-phase cross flow.

A summary of the test results is presented in Table 7-4. The results show that a single flexible tube within a rotated-triangular array cannot undergo in-plane fluidelastic instability. However, the same single tube (flexible in the cross-flow direction) was able to go unstable in cross flow. The results also show that in-flow instability can occur with two or more adjacent flexible tubes. With just two flexible tubes, the in-flow instability velocity is double that when all tubes are flexible.

The tests in air flow discussed here were exploratory tests to investigate instability in the flow direction. To arrive at final conclusions regarding stability behavior in actual steam generators, tests in two-phase flow are required.

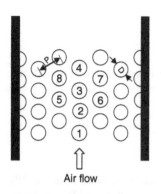

Air flow

Fig. 7-20 Wind-Tunnel, Rotated-Triangular Test Array.

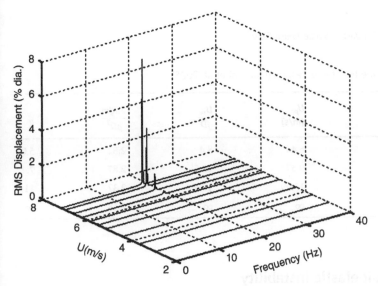

Fig. 7-21 Response Spectra Variation with Flow Velocity for Tube 2 in a Fully Flexible Bundle within a Wind Tunnel.

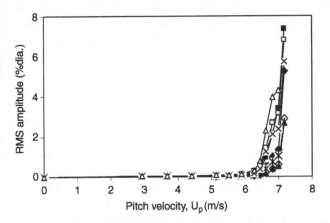

Fig. 7-22 Vibration Response for the Flexible Bundle within Wind Tunnel: ◆, Tube 1; ■, Tube 2; *, Tube 3; ▲, Tube 4; ◇, Tube 5; □, Tube 6; X, Tube 7; △, Tube 8.

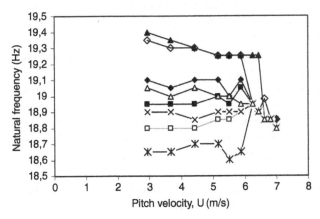

Fig. 7-23 Response Frequency versus Flow Velocity for the Flexible Bundle within Wind Tunnel ◆, Tube 1; ■, Tube 2; *, Tube 3; ▲, Tube 4; ◇, Tube 5; □, Tube 6; X, Tube 7; △, Tube 8.

Table 7-4 Summary of In-Plane Instability Results (Mureithi et al, 2005)

	U_{pc} (m/s)	$\dfrac{U_{pc}}{fD}$	$\dfrac{2\pi\zeta m}{\rho D^2}$	K
all tubes flexible	6.44	8.4	2.82	5.0
one column flexible	9.57	12.5	2.87	7.4
2 tubes flexible	13.99	18.3	2.37	11.9
1 tube flexible	∞	∞	2.18	∞

7.10 Axial Flow Fluidelastic Instability

For cylinders in axial flow, the pertinent fluid forces are the fricitional forces, the fluid acceleration forces and, in some cases, the drag forces (e.g. cylinders with one free end). Instabilities appear in the form of either buckling or flutter-like oscillations. Fluidelastic instabilites are possible with both internal and external flow. In a very approximate way, the critical velocity for fluidelastic instability on a cylinder of length, L, flexural rigidity, EI, and hydrodynamic mass, m_h, is governed by the dimensionless velocity, $\tilde{U}$:

$$\tilde{U} = VL\sqrt{m_h/EI} \tag{7-13}$$

Fig. 7-24 Clamped-Free Cylinder Experiencing 4th-Mode Buckling In Confined Liquid Axial Flow (Païdoussis and Pettigrew, 1979).

where V is the axial flow velocity in the flow channel. For a comprehensive analytical treatment of the fluidelastic behavior of cylindrical structures in axial flow, the reader is referred to the early work of Pettigrew and Païdoussis (1975) and Païdoussis and Pettigrew (1979). A summary paper by Païdoussis (1987) provides further references, including some work on predicting the stability of a cluster of cylinders that showed good agreement to experimental results.

As expected from theory, the mode order of instabilities increased with flow velocity and flexural rigidity and decreased with increased hydrodynamic mass due to annular confinement. Figure 7-24 shows a flexible cylinder experiencing fourth-mode buckling while being subjected to confined liquid flow. Typical results in liquid axial flow are shown in Fig. 7-25a and 7-25b. Figure 7-25a shows a typical flutter spectrum, while Fig. 7-25b shows a spectrum of combined buckling and free-end vibration.

This phenomenon has been found in industry to the extent that some fretting wear has resulted, but no failures have been noted. Fortunately the flow velocities in most industrial components are much lower than those required for fluidelastic instability in single-phase axial flow conditions. For this reason, research in this area is limited and a design guideline to avoid this phenomenon has not been developed, but suggestions to prevent occurrence are provided in texts such as Kaneko et al (2014).

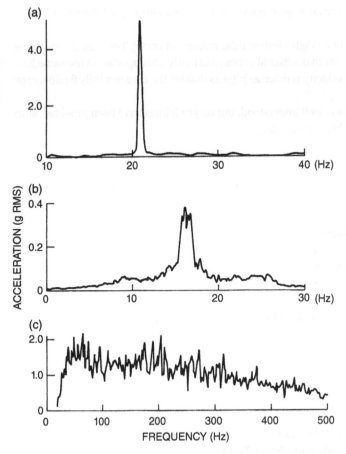

Fig. 7-25 Selected Frequency Spectra for Fluidelastic Instability of Clamped-Free Flexible Cylinder in Axial Flow (Pettigrew and Païdoussis, 1975): a) Oscillations in Liquid Flow, U = 4.26 m/s, b) Buckling and Free-End Vibration in Liquid Flow, U = 3.05 m/s, and c) Broadband Vibration Response in Two-Phase Air-Water Flow, Mass Flux = 2700 kg/(m²s), ε_g = 15%.

7.11 Concluding Remarks

Most of the available data on fluidelastic instability in tube bundles subjected to single-phase cross-flow has been reviewed. This data was compared against the existing expressions to formulate fluidelastic instability. From a practical design point of view, it is suggested that:

- Fluidelastic instability may be expressed in terms of dimensionless flow velocity, U_p/fD, and dimensionless mass-damping, $2\pi\zeta m/\rho D^2$.
- A simple expression of the form $\dfrac{U_{pc}}{fD} = K\left(\dfrac{2\pi\zeta m}{\rho D^2}\right)^{0.5}$ is appropriate for design purposes.
- A fluidelastic instability constant $K = 3.0$ is a reasonable design criterion for both square and triangular tube bundles.
- Although the effect of pitch-to-diameter ratio, P/D, is significant in some cases, the trend is not general. Thus, the term P/D could not be included in the fluidelastic instability expression.
- Fluid damping should be included in the mass-damping parameter.

- At the present time, there is no advantage to considering more sophisticated fluidelastic instability models.
- In-plane fluidelastic instability of a single flexible tube would not occur. Thus, an in-flow instability would require having at least two adjacent tubes essentially unsupported at the same location. In this case, the instability velocity is twice as high as that for the case of a fully flexible array of tubes.
- Axial-flow fluidelastic instability is well understood, but no guidelines have been provided, since it is generally not a significant issue in industry.

Nomenclature

a	=	damping term exponent
b	=	mass damping term exponent
c	=	pitch-to-diameter term exponent
C	=	pitch-to-diameter term constant
C_m	=	confinement coefficient
D	=	tube diameter, m
D_e	=	equivalent diameter of surrounding tubes, m
EI	=	flexural rigidity, Pa·m^4
f	=	frequency, Hz
F	=	fluidelastic instability function
K	=	fluidelastic instability constant
ℓ	=	tube span length, m
L	=	tube length, m
L_e	=	tube length subjected to cross flow, m
m	=	total mass per unit length $(m_t + m_h + m_i)$, kg/m
m_h	=	hydrodynamic mass per unit length, kg/m
P	=	tube pitch, m
U	=	flow velocity, m/s
$\tilde{U}$	=	dimensionless velocity
V	=	axial velocity, m/s
ζ	=	damping ratio, %
ρ	=	fluid density, kg/m^3

Subscript

a	=	air (damping)
c	=	critical
g	=	gap
h	=	hydrodynamic (mass)
p	=	pitch
r	=	reference gap
s	=	structural (damping)

t = tube (mass)
T = total (damping)
V = viscous
∞ = free stream

References

Abd-Rabbo, A., and Weaver, D.S., 1986, "A Flow Visualization Study of Flow Development in a Staggered Tube Array," *Journal of Sound and Vibration*, 106, pp. 241–256.

Andjelic, M., and Popp, K., 1989, "Stability Effects in a Normal Triangular Cylinder Array," *Journal of Fluids and Structures*, 3, pp. 165–185.

Axisa, F., Villard, B., Gilbert, R.J., Hetsroni, G., and Sundheimer, P., 1984, "Vibration of Tube Bundles Subjected to Air-Water and Steam-Water Cross-Flow: Preliminary Results on Fluidelastic Instability," *Proceedings of ASME Symposium on Flow-Induced Vibrations*, New Orleans, LA., 2, pp. 269–284.

Bai, D., 1982, "Flow-Induced Vibrations of Multi-Span Tube Bundles of Large Condensers Experimental Studies on Full-Scale Models in Steam Cross-Flow," *Proceedings of the Third Keswick International Conference on Vibration in Nuclear Plant*, Keswick, UK, Vol. 1. pp. 217–230.

Chen, S.S., 1981, "Fluid Damping for Circular Cylindrical Structures," *Journal of Nuclear Engineering and Design*, 75, pp. 351–373.

Chen, S.S., 1984, "Guidelines for the Instability Flow Velocity of Tube Arrays in Cross Flow," *Journal of Sound and Vibration*, 93, pp. 439–455.

Chen, S.S., and Jendrzejczyk, J.A., 1981, "Experiments on Fluidelastic Instability in Tube Banks Subjected to Liquid Cross-Flow," *Journal of Sound and Vibration*, 78, pp. 355–381.

Chen, S.S., and Jendrzejczyk, J.A., 1983, "Stability of Tube Arrays in Cross-Flow," *Journal of Nuclear Engineering and Design*, 75, pp. 352–373.

Connors, H.J., 1970, "Fluidelastic Vibration of Tube Arrays Excited by Cross-Flow," *Flow-Induced Vibration in Heat Exchangers*, ASME-WAM, New York, pp. 42–56.

Connors, H.J., 1978, "Vibration of Heat Exchanger Tube Arrays," *Journal of Mechanical Design*, 100, pp. 347–353.

Connors, H.J., 1980, "Fluidelastic Vibration of Tube Arrays Excited by Non-Uniform Cross-Flow," *Flow-Induced Vibration of Power Plant Components*, ASME Publication PVP-41, pp. 93–107.

Eisinger, F.K., 1980, "Prevention and Cure of Flow-Induced Vibration Problems in Tubular Heat Exchangers," *Journal of Pressure Vessel Technology*, 102, pp. 138–145.

Gibert, R.G., Chabrerie, J., and Sagner, M., 1977, "Vibrations of Tube Arrays in Transversal Flow," Paper F 6/3, *Transactions of the 4th International Conference on Structural Mechanics in Reactor Technology*, San Francisco, CA, August 15-19.

Godon, J.L., 1984, "Flows and Flow-Induced Vibrations in Large Condensers," *Vibration in Heat Exchangers*, ASME Winter Annual Meeting, New Orleans, LA, 3, pp. 1–16.

Godon, J.L., and Lebert, J., 1988, "Influence of the Tube Support Plate Clearance on Flow-Induced Vibration in Heat Transfer Equipment," *ASME Winter Annual Meeting*, Chicago, Ill., 5, pp. 177–186.

Gorman, D.J., 1976, "Experimental Development of Design Criteria to Limit Liquid Cross-Flow-Induced Vibration in Nuclear Reactor Heat Exchange Equipment," *Journal of Nuclear Science and Engineering*, 61, pp. 324–336.

Gorman, D.J., 1977, "Experimental Study of Peripheral Problems Related to Liquid Flow-Induced Vibration in Heat Exchangers and Steam Generators," Paper F6/2, *Proceedings, 4th International Conference on Structural Mechanics in Reactor Technology*, San Francisco, CA.

Halle, H., Chenoweth, J.M., and Wambsganss, M.W., 1986, "Shellside Flow-Induced Tube Vibration in Typical Heat Exchangers Configurations; Overview of a Research Program," *Flow-Induced Vibration*, Proceedings of ASME-PVP Conference, PVP-Vol. 104, Chicago, Ill., pp. 161–169.

Halle, H., Chenoweth, J.M., and Wambsganss, M.W., 1988, "Shellside Waterflow-Induced Tube Vibration in Heat Exchanger Configurations with Tube Pitch-to-Diameter Ratio of 1.42." *Flow-Induced Heat Transfer Equipment*, Vol. 5, ASME Annual Meeting, Chicago, Ill., pp. 1-16.

Hartlen R.T., 1974, "Wind-Tunnel Determination of Fluid-Elastic Vibration Thresholds for Typical Heat-Exchanger Tube Patterns," *Ontario Hydro Research Report 74-30-K*, Aug.

Heilker, W.J., and Vincent, R.Q., 1981, "Vibration in Nuclear Heat Exchangers Due to Liquid and Two-Phase Flow," *ASME Journal of Engineering for Power*, 103, pp. 358–365.

Johnson, D.K., and Schneider, W.G., 1984, "Flow-Induced Vibration of a Tube Array with an Open Lane," *Proceedings of ASME Symposium on Flow-Induced Vibrations*, New Orleans, LA, Dec., Vol. 3, Vibration in Heat Exchangers, pp. 63–72.

Kaneko, S., Nakamura, T., Inada, F., Kato, M., Ishikara, K., Nishihara, T., Mureithi, N.W. and Langthjem, M.A., 2014, *Flow-Induced Vibrations: Classifications and Lessons from Practical Experiences, 2nd Edition*, Elsevier, Academic Press, London.

Kim, B.S., Pettigrew, M.J. and Tromp, J.H., 1988, "Vibration Damping of Heat Exchanger Tubes in Liquids: Effects of Support Parameters", *Proceedings of the 6th International Conference on Pressure Vessel Technology*, Beijing, China, September.

Kim, B.S., Pettigrew, M.J., Taylor, C.E., and Tromp, J.H., 1987, "Flow-Induced Vibration of Heat Exchanger Tubes in Two-Phase Cross-Flow," *Unpublished Chalk River Laboratory Data*.

Lever, J.H., and Weaver, D.S., 1986, "On the Stability of Heat Exchanger Tube Bundles, Part I: Modified Theoretical Model," *Journal of Sound and Vibration*, 107, pp. 375–392.

Lever, J.H. and Rzentkowski, G., 1988, "An Investigation into the Post-Stable Behaviour of a Tube Array in Cross-Flow," *Proceedings of the International Symposium on Flow-Induced Vibration and Noise*, ASME-WA Meeting, Chicago, Ill., Vol. 3, pp. 95–110.

Lubin, B.T., Letendre, R.P., Quinn, J.W., and Kenny, R.A., 1986, "Comparison of the Response of a Scale Model and Prototype Design Tube Bank Structure to Cross-Flow-Induced Fluid Excitation," Flow-Induced Vibration, *Proceedings of ASME PVP Conference*, Chicago, Ill., PVP-Vol. 104, pp. 171–177.

Minkami, K., and Ohtomi, K., 1987, "Flow Direction and Fluid Density Effects on the Fluidelastic Vibrations of a Triangular Array of Tubes," Paper B2, *International Conference on Flow-Induced Vibrations*, Bowness-on-Windermere, UK, pp. 65–75.

Mureithi, N.W., Zhang, C., Ruël, M. and Pettigrew, M.J., 2005, "Fluidelastic Instability Tests on an Array of Tubes Preferentially Flexible in the Flow Direction," *Journal of Fluids and Structures*, 21, pp. 75–87.

Nakamura, T., Yamaguchi, N., Tsuge, A., Fujita, K., Sakata, K., and Saito, I., 1986a, "Study on Flow-Induced Vibration of a Tube Array by a Two-Phase Flow (1st Report: Mechanical Engineers), *Trans. of Japanese Society of Mechanical Engineers*, 52, No. 473, C.

Nakamura T., Fujita, K., Kawanishi, K., and Saito, I., 1986b, "A Study on the Flow-Induced Vibration of a Tube Array by a Two-Phase Flow (2nd Report: Large Amplitude Vibration by Steam-Water Flow)," *Trans. of Japanese Society of Mechanical Engineers*, 52, C.

Païdoussis, M.P., 1981, "Fluidelastic Vibration of Cylinder Arrays in Axial and Cross-Flow: State of the Art," *Journal of Sound and Vibration*, 76, pp. 31–60.

Païdoussis, M.P., 1983, "A Review of Flow-Induced Vibrations in Reactors and Reactor Components," *Journal of Nuclear Engineering and Design*, 74, pp. 31–60.

Païdoussis, M.P., 1987, "Flow-Induced Instabilities of Cylindrical Structures," *Applied Mechanics Review*, 40(2), pp. 163–175.

Païdoussis, M. P. and Pettigrew, M. J., 1979, "Dynamics of Flexible Cylinders in Axisymmetrically Confined Flow," *Journal of Applied Mechanics*, 46, pp. 37–44.

Pettigrew, M.J. and Païdoussis, M.P., 1975, "Dynamics and Stability of Flexible Cylinders Subjected to Liquid and Two-Phase Axial Flow in Confined Annuli," *Transactions of the 3rd International Conference on Structural Mechanics in Reactor Technology*, Paper No D2/6, London, UK, (also *Atomic Energy of Canada Report, AECL-5502*).

Pettigrew, M.J., and Gorman, D.J., 1977, "Experimental Studies on Flow-Induced Vibration to Support Steam Generator Design, Part III: Vibration of Small Tube Bundles in Liquid and Two-Phase Cross-Flow," Paper No. 424, *International Symposium on Vibration Problems in Industry*, Keswick, UK.

Pettigrew, M.J., and Gorman, D.J., 1978, "Vibration of Heat Exchange Components in Liquid and Two-Phase Cross-Flow," Paper 2:3, *Proceedings of the B.N.E.S. Conference on Vibration in Nuclear Plant*, Keswick, UK.

Pettigrew, M.J., and Campagna, A.O., 1979, "Heat Exchanger Tube Vibration: Comparison Between Operating Experiences and Vibration Analyses," *Proceedings of IAHR-IUTAM Symposium on Practical Experiences with Flow-Induced Vibrations*, Karlsruhe, Germany, Sep. 3-6.

Pettigrew, M. J. and Taylor, C. E., 1991, "Fluidelastic Instability of Heat Exchanger Tube Bundles: Review and Design Recommendations," *ASME Journal of Pressure Vessel Technology*, 113(2), pp. 242–256.

Pettigrew, M.J., Sylvestre, Y., and Campagna, A.O., 1978, "Vibration Analysis of Heat Exchanger and Steam Generator Designs," *Nuclear Engineering and Design*, 48, pp. 97–115.

Pettigrew, M.J., Rogers, R.J., and Axisa, F., 1986, "Damping of Multispan, Heat Exchanger Tubes, Part 2: In Liquids," *Symposium on Special Topics of Structural Vibration, ASME Pressure Vessels and Piping Conference*, Chicago, Ill., July, 104, pp. 89–98.

Pettigrew, M.J., Kim, B.S., Taylor, C.E., and Tromp, J.H., 1987, *Unpublished Chalk River Laboratory Data*.

Pettigrew, M.J., Taylor, C.E., and Kim, B.S., 1989a, "Vibration of Tube Bundles in Two-Phase Cross-Flow; Part 1: Hydrodynamic Mass and Damping," *ASME Journal of Pressure Vessel Technology*, 111, pp. 466–477.

Pettigrew, M.J., Taylor, C.E., Tromp, J.H., and Kim, B.S., 1989b, "Vibration of Tube Bundles in Two-Phase Cross-Flow; Part 2: Fluidelastic Instability," *ASME Journal of Pressure Vessel Technology*, 111, pp. 478–487.

Price, S.J., 1995, "A Review of Theoretical Models for Fluidelastic Instability of Cylinder Arrays in Cross-Flow", *Journal of Fluids and Structures*, 9, pp. 463–518.

Price, S.J., and Païdoussis, M.P., 1989, "The Flow-Induced Response of a Single Flexible Cylinder in an In-line Array of Rigid Cylinders," *Journal of Fluids and Structures*, 3, pp. 61–82.

Remy, F.N., 1982, "Flow-Induced Vibration of Tube Bundles in Two-Phase Cross-Flow," *BNES International Conference on Vibration in Nuclear Plant*, Keswick, UK, Log No. 68.

Remy, F.N., and Bai, D., 1982, "Comparative Analysis of Cross-Flow Induced Vibrations of Tube Bundles," Paper F3, *International Conference on Flow-Induced Vibrations in Fluid Engineering*, Reading, UK, September 14-16.

Roberts, B.W., 1966, "Low Frequency, Aeroelastic Vibrations in a Cascade of Circular Cylinders," *Mechanical Engineering Science*, Monograph, No. 4, Sep., pp. 1–29.

Scott, P.M., 1987, "Flow Visualization of Cross-Flow Induced Vibration in Tube Arrays," *Masters of Eng. Thesis*, McMaster University, Hamilton, Ontario, April.

Soper, B.M.H., 1983, "The Effect of Tube Layout on the Fluid-Elastic Instability of Tube Bundles in Cross-Flow," *ASME Journal of Heat Transfer*, 105, pp. 744–750.

Tanaka, H., and Takahara, S., 1981, "Fluidelastic Vibration of a Tube Array in Cross-Flow," *Journal of Sound and Vibration*, 77, pp. 19–37.

Teh, C.E., and Goyder, H.G.D., 1988, "Data for the Fluidelastic Instability of Heat Exchanger Tube Bundles," *Flow-Induced Vibration and Noise in Cylinder Arrays*, ASME-WAM, Chicago, Ill., Vol. 3, pp. 77–94.

Weaver, D.S., and Grover, L.K., 1978, "Cross-Flow Induced Vibrations in a Tube Bank - Turbulent Buffeting and Fluidelastic Instability," *Journal of Sound and Vibration*, 59(2), pp. 277–294.

Weaver, D.S., and El-Kashlan, M., 1981, "The Effect of Damping and Mass on Flow-Induced Response of Various Tube Arrays in Water," *Journal of Sound and Vibration*, 76, pp. 283–293.

Weaver, D.S., and Koroyannakis, D., 1982, "The Cross-Flow Response of a Tube Array in Water - A Comparison with the Same Array in Air," *ASME Journal of Pressure Vessel Technology*, 104, pp. 139–104.

Weaver, D.S., and Koroyannakis, D., 1983, "Flow-Induced Vibrations of Heat Exchanger U-Tubes, A Simulation to Study the Effects of Asymmetric Stiffness," *ASME Journal of Vibration, Stress and Reliability in Design*, 105, pp. 67–75.

Weaver, D. S. and Schneider, W., 1983, "The Effect of Flat Bar Supports on the Cross-Flow Response of Heat Exchanger U-Tubes," *ASME Journal of Eng. Power*, 105, pp. 775–781.

Weaver, D.S., and Yeung, H.C., 1984, "The Effect of Tube Mass on the Flow-Induced Response of Various Tube Arrays in Water," *Journal of Sound and Vibration*, 93(3), pp. 409–425.

Weaver, D.S., and Abd-Rabbo, A., 1985, "A Flow Visualization Study of a Square Array of Tubes in Water Cross-Flow," *ASME Journal of Fluids Engineering*, 107, pp. 354–363.

Weaver, D.S., and Fitzpatrick, J.A., 1988, "A Review of Flow-Induced Vibrations in Heat Exchanger Tube Arrays," *Journal of Fluids and Structures*, 2, pp. 73–93.

8

Fluidelastic Instability of Tube Bundles in Two-Phase Flow
Michel J. Pettigrew and Colette E. Taylor

8.1 Introduction

Many industrial components operate in two-phase flow. For example, two-phase cross flow exists in many shell-and-tube heat exchangers, such as condensers, evaporators, reboilers, and nuclear steam generators. In fact, more than half of process heat exchangers operate in two-phase flow according to industry experts.

Excessive vibration due to fluidelastic instability may lead to component failures due to fatigue and fretting wear. The nature of fluidelastic instability forces and existing models are discussed in detail in Chapter 7 (Sections 7.2 to 7.5). Such vibration problems may be avoided by thorough vibration analysis, preferably at the design stage. This analysis requires an understanding of vibration excitation mechanisms in two-phase flows.

The topic of two-phase flow-induced vibration has been reviewed by Pettigrew and Taylor (1993). Thus, a complete review is not necessary here. Suffice it to say that a comprehensive study on vibration of tube bundles subjected to two-phase air-water cross flow was conducted in the 1980s (Pettigrew et al, 1989a, 1989b; Taylor et al, 1989). Eight tube bundles of different configurations and pitch-to-diameter ratios were tested. This work provided useful results on fluidelastic instability, random turbulence excitation and two-phase damping. However, air-water mixtures are quite different from the vapor-liquid mixtures encountered in most process equipment. The effect of dynamic characteristics of two-phase mixtures on flow-induced vibration is still largely unknown. Studies in Freon two-phase flow shed some light on this question (Pettigrew et al, 1995, 2002 and 2009). These studies are complementary to the earlier work in air-water inasmuch as similar test sections and techniques are used so that results can be easily compared. Table 8-1 outlines the characteristics of air-water, steam-water, Freon-22 and Freon-134a vapor-liquid mixtures.

8.2 Previous Research

In spite of the preponderance of components subjected to two-phase flows, relatively little information exists on fluidelastic instability in two-phase flow. This is not entirely surprising because the understanding of single-phase flow-induced vibration, a simpler topic, is still very much under development. Vibration in two-phase flow is much more complex because it depends on two-phase flow regime (i.e., the dynamic characteristics of the two-phase mixture), and it involves a new

Flow-Induced Vibration Handbook for Nuclear and Process Equipment, First Edition.
Michel J. Pettigrew, Colette E. Taylor, and Nigel J. Fisher.

Table 8-1   Physical Properties of Freon-22, Freon-134a, Air-Water and Steam-Water.

Property	Phase	Unit	Freon-22		Freon-134a		Air-Water	Steam-Water
Temperature		°C	23	30	26	39	20	257
Pressure		MPa	1	1.2	0.69	1.0	0.1	4.5
Density	Gas	kg/m^3	42.2	50.7	33.3	48.7	1.2	22.6
	Liquid	kg/m^3	1200	1174.0	1202	1150	998.5	789.0
	Gas/Liquid	ratio	0.035	0.043	0.028	0.042	0. 0012	0.029
Dynamic Viscosity	Gas	Pa·s	1.3×10^{-5}	1.4×10^{-5}	1.27×10^{-5}	1.34×10^{-5}	1.8×10^{-5}	1.8×10^{-5}
	Liquid	Pa·s	2.0×10^{-4}	1.95×10^{-4}	2.06×10^{-4}	1.75×10^{-4}	1.0×10^{-3}	1.0×10^{-4}
	Gas Liquid	ratio	0.065	0.072	0.062	0.076	0.018	0.174
Kinematic Viscosity	Gas	m^2/s	3.1×10^{-7}	2.8×10^{-7}	3.8×10^{-7}	2.7×10^{-7}	1.5×10^{-5}	7.9×10^{-7}
	Liquid	m^2/s	1.7×10^{-7}	1.6×10^{-7}	1.7×10^{-7}	1.5×10^{-7}	1.0×10^{-6}	1.3×10^{-7}
	Gas/Liquid	ratio	1.848	1.740	2.224	1.808	15.00	6.10
Surface Tension	Liquid	N/m	0.0082	0.0072	0.0079	0.0063	0.0727	0.0246

dimension, which is void fraction or vapor quality. Two-phase flow experiments are also much more expensive and difficult to conduct as they require pressurized loops with the ability to produce two-phase mixtures. Nevertheless, some progress was made in the last three decades. So far, most of this work was done for the nuclear industry.

8.2.1   Flow-Induced Vibration in Two-Phase Axial Flow

The development of nuclear fuels for boiling-type reactors required information on vibration in axial two-phase flow. Most of this work was done in the late 1960s and in the 1970s. For example, Gorman (1971) studied the vibration behavior of a single cylinder in two-phase air-water annular flow. He investigated, in particular, random turbulence excitation. Dallavalle et al (1973), in Italy, explored different two-phase mixtures, such as water-nitrogen and acetone-nitrogen to simulate steam-water mixtures. Pettigrew and Gorman (1973a) conducted experiments on an electrically heated cylinder to investigate the effect of nucleate boiling in steam-water annular flow. Païdoussis and Pettigrew (1979) studied fluidelastic instability of a flexible cylinder subjected to confined two-phase axial flow. Pettigrew (1993) studied the vibration behavior of nuclear fuel bundles under actual nuclear reactor core conditions in steam-water axial flow. Fundamental studies on axial flow-induced vibration were done in Japan on small test sections in air-water flow (Hara, 1978 and 1980).

8.2.2 Flow-Induced Vibration in Two-Phase Cross Flow

Little work was done on vibration of tube bundles in two-phase cross flow before 1980. Years ago, we did exploratory vibration experiments on small tube bundles partially exposed to two-phase (air-water) cross flow (Pettigrew and Gorman, 1973b). However, we did not measure damping. Heilker and Vincent (1981) have also done some work in air-water cross flow for a limited range of bundle geometries and flow conditions pertaining to their nuclear steam generator design. Their single-span tube bundles were exposed to flow over their entire length. They reported results on fluidelastic instability and turbulence-induced excitation. Remy (1982) studied vibration excitation mechanisms with a single-span tube bundle of normal-square configuration partially exposed to air-water cross flow. This partial exposure made the interpretation of results, damping in particular, difficult.

Axisa et al (1984), in France, were the first to present results on vibration of a tube bundle subjected to both air-water and steam-water cross flow. They tested a normal-square tube bundle of pitch-to-diameter ratio, $P/D = 1.44$. They found that the simulation of steam-water flow with an air-water mixture is reasonable for fluidelastic instability. Subsequently, they studied three bundle configurations, i.e., normal-square, normal-triangular, and rotated-triangular tube bundles, all of $P/D = 1.44$, in steam-water cross flow (Axisa et al, 1985). They reported valuable results on fluidelastic instability and damping. Later, Nakamura et al (1986a, 1986b and 1991) also reported data on fluidelastic instability in both air-water and steam-water cross flow. They tested a square bundle of cylinders of $P/D = 1.42$. They found it difficult to measure damping in two-phase flow.

8.2.3 Damping Studies

Unfortunately, very little attention was given to damping except by Carlucci (1980) and Carlucci and Brown (1983). They conducted a systematic study of damping of cylinders in axial two-phase flow simulated by air-water mixtures. About the same time, some fundamental experiments were done with a single cantilevered tube immersed in two-phase mixtures generated by bubbling air through water. Hara (1982) studied, in particular, the effect of void fraction, confinement, and bubble size. Pettigrew and Knowles (1997) also investigated the effect of tube frequency and surface tension. The surface tension was varied by the addition of a surfactant in an attempt to simulate the damping characteristics of high-temperature steam-water mixtures. However, this technique was limited to 30% void fraction.

Taylor et al (1988) investigated turbulence-induced excitation in rigid tube rows of $P/D = 1.5$ and 3.0 in air-water cross flow. While this experiment was not specifically designed for damping studies, it provided some useful damping data, particularly at higher frequencies (i.e., 100 to 300 Hz). This damping data is compared against other data in an interesting review paper by Axisa et al (1988). The effect of void fraction between 80 and 100%, and the effect of air-water versus steam-water mixtures in particular were discussed. Most of the research work to date on tube bundles in cross flow has been in support of nuclear steam generators. This emphasis explains the preponderance of tube bundles of square configuration and of P/D around 1.5. See Chapter 6 for detailed information on damping in two-phase flow.

8.3 Fluidelastic Instability Mechanisms in Two-Phase Cross Flow

Generally, three basic flow-induced vibration mechanisms are considered in two-phase cross flow. They are fluidelastic instability, periodic wake shedding and random excitation due to flow turbulence. Practical situations may involve either isolated cylinders or bundles of cylinders. Tube bundles are prevalent in heat exchanger components. In cross flow, fluidelastic instabilities are known to occur in bundles of cylinders because interaction with surrounding cylinders is required for this phenomenon. The emphasis in the following discussion is on bundles of cylinders.

Fluidelastic instability is by far the most important vibration excitation mechanism for tube bundles subjected to cross flow. In our experience, most heat exchanger tube vibration problems are related to fluidelastic instability. The topic of fluidelastic instability of tube bundles in single-phase cross flow was reviewed by Pettigrew and Taylor (1991) (see Chapter 7). Experimental results suggested that fluidelastic instability is equally important in two-phase cross flow.

Generally, in a tube bundle, the fluid forces on one tube are affected by its motion and the motion of neighboring tubes. This situation creates an interaction between fluid forces and tube motion. Fluidelastic instability is possible when the interaction between the motions of individual tubes is such that it results in fluid force components that are both proportional to tube displacement and in-phase with tube velocities. Instability occurs when, during one vibration cycle, the energy absorbed by a tube from the fluid forces exceeds the energy dissipated by damping. Then, the vibration amplitude of the tube increases rapidly and, in theory, would become extremely large. In practice, the vibration amplitude is limited by the presence of surrounding tubes or by other nonlinearities, such as clearances at the supports.

Apart from large vibration amplitudes, several other features usually characterize fluidelastic instabilities: significant correlation between the motions of adjacent tubes, suggesting strong hydro-dynamic coupling; much narrower vibration response spectrum, giving the appearance of much reduced damping; and sudden change in vibration pattern. The foregoing characteristics have been observed while testing tube bundles in two-phase cross flow, thereby demonstrating the existence of fluidelastic instability (Pettigrew et al, 1989b). For example, Fig. 8-1 shows typical tube response

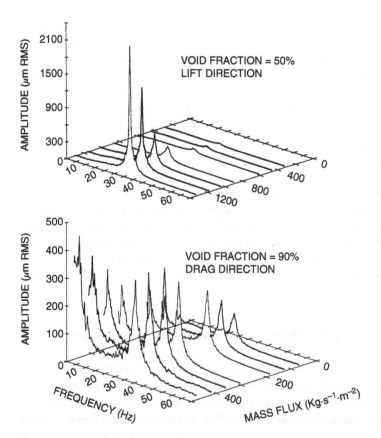

Fig. 8-1 Typical Tube Response Spectra for Increasing Mass Fluxes at Constant Void Fraction.

spectra for increasing mass fluxes at constant void fraction. The response spectrum at the highest mass flux, where instability occurs, is much narrower, clearly indicating fluidelastic behavior.

Typical results presented in the form of vibration response versus mass flux curves are shown in Figs. 8-2 and 8-3. These figures show the difference in vibration response between a tube surrounded by flexible tubes and a tube surrounded by rigid tubes. Fluidelastic instability may occur at much lower mass fluxes for bundles with flexible tubes. This greater susceptibility to instability illustrates the effect of hydrodynamic coupling between the motions of surrounding tubes. This

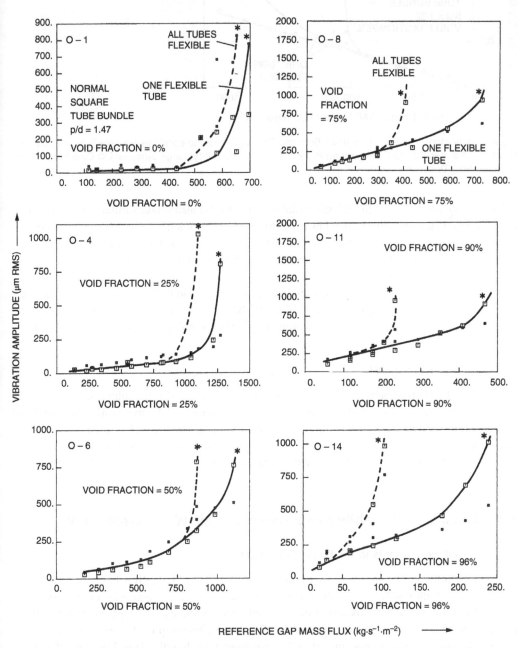

Fig. 8-2 Typical Vibration Response: Comparison Between Flexible Tube Bundle and One Flexible Tube in Rigid Tube Bundle for Normal-Square Configuration P/D = 1.47 at Various Void Fractions. Stars: Instability; Large Symbols: Lift Direction; Small Symbols: Drag Direction.

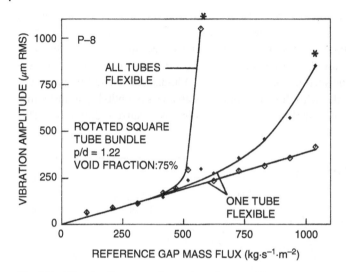

Fig. 8-3 Vibration Response: Comparison Between Flexible Tube Bundle and One Flexible Tube in Rigid Tube Bundle for Rotated-Square Configuration *P/D* = 1.22. Stars: Instability; Large Symbols: Lift Direction; Small Symbols: Drag Direction.

behavior is characteristic of fluidelastic instability. Figure 8-3 also shows similar tube response in both the lift and the drag direction at low mass fluxes where turbulence-induced excitation prevails. At higher mass fluxes, near the threshold for fluidelastic instability, the vibration response generally increases dramatically and usually becomes much larger in the lift direction. These observations clearly indicate fluidelastic behavior in two-phase cross flow.

Fluidelastic instability may be formulated in terms of a dimensionless flow velocity, U_p/fD, and a dimensionless mass-damping parameter, $2\pi\zeta m/\rho D^2$, (Connors, 1970). For the simple case of a tube bundle subjected to uniform flow over its entire length, the foregoing parameters are simply related by the constant K

$$U_p/fD = K\left(2\pi\zeta m/\rho D^2\right)^n \tag{8-1}$$

where U_p is the pitch velocity, f is the tube frequency in the fluid, D is the tube diameter, m is the mass per unit length including the hydrodynamic mass, ζ is the total damping ratio (structural and fluid damping), ρ is the two-phase fluid density, and n is an exponent which is often around 0.5, as discussed in Chapter 7. Note that ζ is expressed as a ratio, not as a percentage. The instability constant, K, is obtained from experimental data using the same methods described in Chapter 7.

8.4 Fluidelastic Instability Experiments in Air-Water Cross Flow

8.4.1 Initial Experiments in Air-Water Cross Flow

In 1983, we undertook a comprehensive experimental program at Chalk River Laboratories to study tube bundle vibration in two-phase cross flow. The experiments were conducted in two-phase test loops. Generally, the loops consist of a liquid supply system and an air supply system which are brought together at the test section, as shown in Fig. 8-4. A photograph of the test section and part of the air-water loop is shown in Fig. 8-5. Cantilever tube bundles (see Fig. 8-6) of various

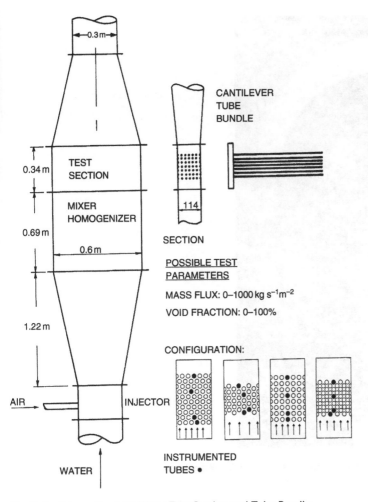

Fig. 8-4 Schematic of Air-Water Test Section and Tube Bundle.

configurations and pitch-to-diameter ratios, P/D, were subjected to air-water mixtures. Realistic vapor qualities were simulated over a practical range of mass fluxes by air-water.

The results from the above experimental program are presented in Sections 8.4 and 8.5.

We tested eight tube bundles in air-water. A normal-triangular (30°), normal-square (90°), rotated-triangular (60°), and rotated-square (45°) tube bundle, all of $P/D = 1.22$, were tested. A normal-triangular, normal-square, and a rotated-triangular tube bundle, all of $P/D = 1.47$, and a normal-triangular bundle of $P/D = 1.32$ were tested in support of the CANDU[®1] steam generator technology development program.

The results from two bundles, namely the normal-square and normal-triangular bundle of $P/D = 1.47$, were published by Pettigrew et al (1985). At the time, these bundles were subjected to air-water mixtures simulating up to 93% void fraction. Damping and vibration excitation mechanisms were investigated. The tests were done with all the tubes in the bundles flexible and free to vibrate. Whereas this is ideal to study fluidelastic instability, it presents some difficulties for studying damping, mainly due to hydrodynamic coupling between the tubes, as explained in Chapter 6. The results are outlined in Table 8-2.

Fig. 8-5   Photograph of Air-Water Test Section and Part of Air-Water Loop.

Fig. 8-6   Photograph of Cantilever Tube Bundle with Normal-Triangular Configuration and *P/D* = 1.47.

Table 8-2 Summary of Test Conditions and Results for Early Air-Water Experiments (Pettigrew et al, 1985).

Test no.	Geom. (°)	P/D	e_g (%)	ρ (kg/m³)	m (kg/m)	$\dot{m}_{pc}$ (kg/m²s)	U_{pc} (m/s)	f (Hz)	ζ (%)	$\frac{U_{pc}}{fD}$	$\frac{2\pi\zeta m}{\rho D^2}$	K
B13-B16	90	1.47	25	750	0.466	1448	1.93	27.7	7.3	5.36	1.69	4.1
B1-B4	90	1.47	50	500	0.422	1128	2.26	29.1	8.0	5.96	2.51	3.8
B17-B20	90	1.47	75	250	0.379	808.0	3.23	30.7	6.8	8.10	3.84	4.1
B21-B24	90	1.47	80	200	0.363	728.0	3.64	31.4	5.0	8.92	3.37	4.9
B5-B8	90	1.47	85	150	0.370	488.0	3.25	31.1	4.5	8.05	4.12	4.0
B9-B12	90	1.47	90	100	0.351	325.0	3.25	31.9	2.7	7.84	3.53	4.2
B25-B28	90	1.47	93	70	0.354	245.0	3.50	31.8	2.4	8.47	4.51	4.0
B29-B32	90	1.47	96	40	0.349	151.0	3.78	32.0	1.5	9.08	4.87	4.1
B33-B36	90	1.47	98	20	0.349	80.0	4.00	32.0	0.9	9.62	5.84	4.0
C41-C44	30	1.47	5	950	0.512	879.0	0.93	26.1	2.3	2.73	0.461	4.0
C45-C48	30	1.47	15	850	0.512	964.0	1.13	26.1	3.2	3.34	0.717	4.0
C9-C12	30	1.47	25	750	0.516	1286	1.71	26.0	6.4	5.07	1.64	4.0
C13-C16	30	1.47	50	500	0.452	1029	2.06	27.8	6.8	5.70	2.28	3.8
C17-C20	30	1.47	75	250	0.404	727.0	2.91	29.4	5.5	7.61	3.30	4.2
C21-C24	30	1.47	80	200	0.385	517.0	2.59	30.1	4.0	6.61	2.86	3.9
C25-C28	30	1.47	85	150	0.383	454.0	3.03	30.2	4.4	7.71	4.17	3.8
C29-C32	30	1.47	90	100	0.361	274.0	2.74	31.1	2.2	6.78	2.95	3.9
C33-C36	30	1.47	93	70	0.361	207.0	2.96	31.1	2.0	7.31	3.83	3.7

Later, all the tube bundles were tested in two-phase flow up to 99% void fraction. The tests were also done with one instrumented flexible tube installed in a bundle of rigid tubes. This yielded more reliable damping results, as well as some valuable insight on the effect of the motion of surrounding tubes on vibration response. This program was completed and the results are reported in Pettigrew et al (1989a, 1989b) and Taylor et al (1989), as shown in Table 8-3.

8.4.2 Behavior in Intermittent Flow

Values of the dimensionless velocity, U_p/fD, and dimensionless mass-damping parameter, $2\pi\zeta m/\rho D^2$, were deduced from the experimental data where damping information was available. Figure 8-7 shows the results for several tube bundles subjected to air-water cross flow (Pettigrew et al, 1989b). There appears to be two regions of instability. A region where the exponent n of the mass-damping parameter is roughly 0.5, as expected from previous work in single-phase flow is observed at void fractions below roughly 80 to 90%. The fluidelastic behavior above 80 to 90% void fraction appears very different. In this region, the exponent n is much lower, approximately 0.1. This behavior is explained in terms of flow-regime changes. As discussed in Chapter 3, the transition between the two fluidelastic instability regions corresponds to the transition between continuous flow and intermittent flow, as may be seen in Fig. 8-8. In the continuous (bubbly or froth) flow region below 80 to 90% void fraction, the formulation of the fluidelastic mechanism is as expected, with

Table 8-3 Summary of Test Conditions and Results for Damping in Rigid Bundles and Fluidelastic Instability in Flexible Bundles for Air-Water Experiments (Pettigrew et al, 1989b).

Test no.	Geom. (°)	P/D	ε_g (%)	ρ (kg/m³)	m (kg/m)	$\dot{m}_{pc}$ (kg/m²s)	U_{pc} (m/s)	f (Hz)	ζ (%)	$\frac{U_{pc}}{fD}$	$\frac{2\pi\zeta m}{\rho D^2}$	K
J-7	30	1.32	5	950	0.52	856	0.90	26.3	2.0	2.64	0.41	4.1
J-11	30	1.32	15	850	0.52	856	1.01	26.4	2.8	2.94	2.94	3.7
J-15	30	1.32	25	750	0.49	1012	1.35	27.2	3.2	3.82	0.77	4.4
J-19	30	1.32	40	600	0.46	857	1.43	27.9	4.7	3.94	1.35	3.4
J-23	30	1.32	50	501	0.43	779	1.56	28.8	4.7	4.16	1.51	3.4
J-27	30	1.32	60	401	0.42	780	1.95	29.3	4,5	5.11	1.75	3.9
J-31	30	1.32	70	301	0.41	703	2.34	29.8	4.2	6.03	2.10	4.2
J-35	30	1.32	75	251	0.39	625	2.49	30.2	3.9	6.35	2.28	4.2
J-39	30	1.32	80	201	0.39	548	2.73	30.3	3.9	6.93	2.83	4.1
J-43	30	1.32	85	151	038	471	3.12	30.6	3.5	7.84	3.31	4.3
J-47	30	1.32	90	101	0.38	315	3.12	30.7	3.4	7.82	4.77	3.6
J-51	30	1.32	93	71	0.38	238	3.35	30.9	3.2	8.34	6.30	3.3
Q-4	30	1.47	25	750	0.45	1200	1.60	28.2	3.2	4.36	0.71	5.2
Q-6	30	1.47	50	501	0.42	1000	2.00	29.4	4.0	5.23	1.23	4.7
Q-8	30	1.47	75	251	0.38	700	2.79	30.8	3.4	6.97	1.91	5.1
Q-9	30	1.47	80	201	0.37	630	3.14	31.1	2.9	7.76	1.96	5.5
Q-10	30	1.47	85	151	0.36	500	3.31	31.4	2.4	8.11	2.11	5.6
Q-12	30	1.47	90	101	0.36	350	3.46	31.6	1.9	8.43	2.57	5.3
Q-14	30	1.47	93	71	0.36	250	3.52	31.7	1.9	8.53	3.48	4.6
Q-16	30	1.47	95	51	0.36	175	3.42	31.6	1.9	8.33	5.03	3.7
Q-18	30	1.47	96	41	0.36	150	3.65	31.7	2.0	8.84	6.44	3.5
Q-20	30	1.47	97	31	0.36	110	3.53	31.7	1.8	8.57	7.82	3.1
Q-22	30	1.47	98	21	0.36	80	3.78	31.7	2.0	9.17	12.7	2.6
Q-24	30	1.47	99	11	0.35	45	4.02	31.9	1.8	9.70	21.3	2.1
R-1	60	1.47	5	950	0.50	688	0.72	26.7	1.1	2.09	0.22	4.5
R-2	60	1.47	15	850	0.50	688	0.81	26.8	1.5	2.32	0.33	4.1
R-3	60	1.47	25	750	0.49	750	1.00	27.0	2.0	2.85	0.49	4.1
R-4	60	1.47	40	600	0.47	775	1.29	27.7	3.4	3.58	0.99	3.6
R-5	60	1.47	50	501	0.45	700	1.40	28.3	3.9	3.80	1.30	3.3
R-6	60	1.47	60	401	0.43	689	1.72	28.9	4.2	4.58	1.66	3.6
R-7	60	1.47	75	251	0.41	500	1.99	29.5	3.4	5.20	2.08	3.6
R-9	60	1.47	85	151	0.40	320	2.12	30.0	3.0	5.43	2.95	3.2
R-11	60	1.47	90	101	0.39	220	2.18	30.2	3.0	5.54	4.35	2.7
R-13	60	1.47	93	71	0.39	160	2.25	30.2	3.2	5.73	6.60	2.2
R-15	60	1.47	95	51	0.39	120	2.35	30.4	3.1	5.94	8.77	2.0
R-17	60	1.47	96	41	0.39	100	2.43	30.5	3.0	6.13	10.5	1.9
R-19	60	1.47	97	31	0.39	75	2.41	30.4	2.9	6.09	13.5	1.7

Table 8-3 (Continued)

Test no.	Geom. (°)	P/D	ε_g (%)	ρ (kg/m³)	m (kg/m)	$\dot{m}_{pc}$ (kg/m²s)	U_{pc} (m/s)	f (Hz)	ζ (%)	$\dfrac{U_{pc}}{fD}$	$\dfrac{2\pi\zeta m}{\rho D^2}$	K
R-21	60	1.47	98	21	0.39	55	2.60	30.4	2.8	6.57	19.1	1.5
R-23	60	1.47	99	11	0.39	28	2.50	30.5	2.9	6.31	37.3	1.0
O-2	90	1.47	5	950	0.50	753	0.79	26.9	1.3	2.27	0.25	4.5
O-3	90	1.47	15	850	0.49	826	0.97	27.0	1.8	2.76	0.39	4.4
O-4	90	1.47	25	750	0.47	984	1.31	27.7	2.5	3.64	0.57	4.8
O-5	90	1.47	32	680	0.46	1013	1.49	27.9	3.4	4.11	0.87	4.4
O-6	90	1.47	50	501	0.43	898	1.79	28.9	4.0	4.78	1.28	4.2
O-7	90	1.47	60	401	0.43	840	2.10	28.9	4.3	5.58	1.70	4.3
O-8	90	1.47	75	251	0.40	581	2.32	30.1	3.5	5.93	2.07	4.1
O-9	90	1.47	80	201	0.40	466	2.32	30.2	3.4	5.91	2.47	3.8
O-10	90	1.47	85	151	0.39	389	2.58	30.4	3.1	6.53	2.98	3.8
O-11	90	1.47	90	101	0.38	234	2.32	30.6	2.8	5.83	3.93	2.9
O-12	90	1.47	93	71	0.39	176	2.48	30.6	2.8	6.25	5.56	2.7
O-13	90	1.47	94	61	0.38	147	2.41	30.9	2.7	6.00	6.11	2.4
O-14	90	1.47	96	41	0.38	104	2.53	30.7	2.8	6.34	9.56	2.1
O-15	90	1.47	97	31	0.37	75	2.41	31.0	2.4	5.98	10.8	1.8
O-16	90	1.47	98	21	0.37	55	2.60	31.3	2.1	6.39	13.6	1.7
O-17	90	1.47	99	11	0.37	29	2.60	31.4	2.2	6.38	26.8	1.2

a mass-damping exponent, n, around 0.5. Not surprisingly, the behavior in the intermittent flow regime is quite different and could lead to severe vibration problems. From a practical point of view, intermittent flow regimes should be avoided in industrial components.

8.4.3 Effect of Bundle Geometry

Figure 8-7b shows the effect of pitch-to-diameter ratio, P/D, on fluidelastic instability for normal-triangular tube bundles. The fluidelastic instability constant, K, is approximately 4.9 for $P/D = 1.47$ and approximately 3.5 for the smaller $P/D = 1.35$. Additional tests on tube bundles of $P/D = 1.22$ confirmed this trend (Pettigrew et al, 2001). The results for all tube bundles in this series of tests in continuous flow are summarized in Fig. 8-9. The instability constant, K, appears to be closely related to the minimum dimensionless flow path width, thus:

$$K \propto (P - D)/D \tag{8-2}$$

This trend was somewhat unexpected, being quite different than in single-phase flow, as shown in Fig. 7-14b of Chapter 7, where for both normal-square (90°) and rotated-triangular (60°) bundles the effect of P/D is small (Pettigrew and Taylor, 1991). A possible explanation is that hydrodynamic

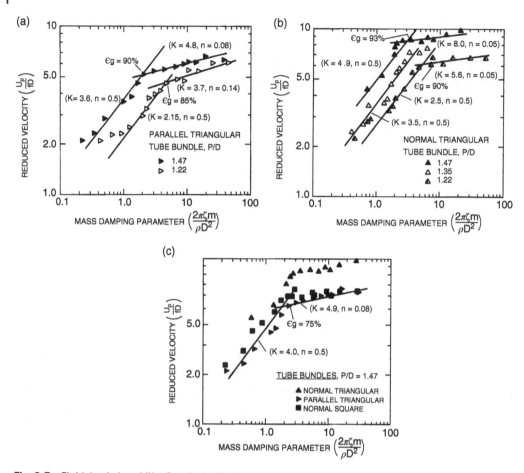

Fig. 8-7 Fluidelastic Instability Results in Air-Water Cross Flow.

coupling between tubes, which is required for fluidelastic instability, is much enhanced by confinement in two-phase flow.

The fluidelastic instability results for three different bundles of $P/D = 1.47$ may be compared in Fig. 8-7c. Fluidelastic instability occurs at somewhat lower flow velocity for the rotated-triangular (or parallel-triangular) bundle, $K = 3.6$, (see Fig. 8-7a) than for the normal-triangular bundle, $K = 4.9$, (see Fig. 8-7b). There is no obvious reason for this behavior. The flow in rotated-triangular bundles tends to stream through because of the clear flow path between the tubes, whereas this is not the case for normal-triangular tube bundles (compare Fig. 8-10 and Fig. 8-11). This characteristic may play a part in the observed behavior. However, this difference has little practical significance because all bundle orientations are possible in a given heat exchanger.

8.4.4 Flexible versus Rigid Tube Bundle Behavior

Figure 8-3 shows the difference in vibration response between a tube surrounded by flexible tubes and a tube surrounded by rigid tubes. Generally, the vibration levels are the same at low mass fluxes. This means that turbulence-induced excitation is not affected by the motion of surrounding tubes. On the other hand, fluidelastic instability occurs at much higher mass fluxes for a flexible

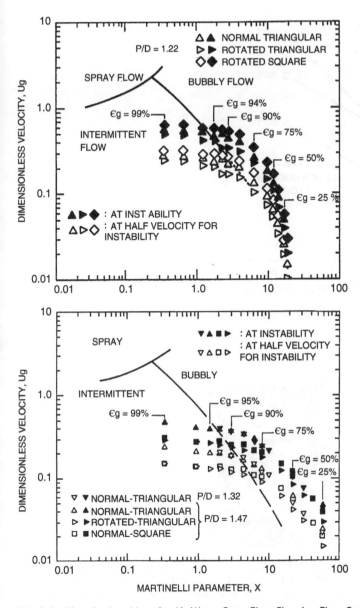

Fig. 8-8 Flow Regime Maps for Air-Water Cross Flow Showing Flow Conditions at Fluidelastic Instability.

tube surrounded by rigid tubes than for all-flexible tube bundles (i.e., by a factor of up to two in some cases). Thus, bundles of flexible tubes are necessary to study fluidelastic instability in two-phase cross flow.

The above behavior is a somewhat controversial issue. Some researchers, Gorman (1977) and Lever and Weaver (1982), for example, have found that fluidelastic instability is not affected by the motion of surrounding tubes in single-phase flow. Similarly, we found in this study that the critical mass flux for instability is only slightly higher for rigid tube bundles subjected to liquid flow. On the other hand, some authors have found a considerable difference in instability behavior

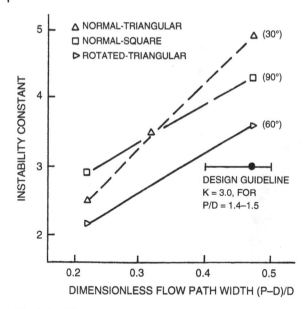

Fig. 8-9 Effect of *P/D* on Fluidelastic Instability Constant in Air-Water Cross Flow.

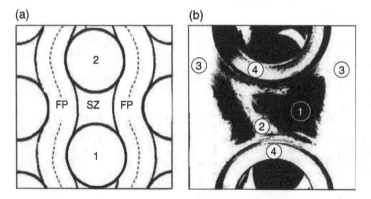

Fig. 8-10 Two-Phase Flow Structure in a Rotated-Triangular Tube Bundle: a) Simplified Sketch (FP - Flow Path, SZ - Stagnation Zone); b) Black and White Picture Taken During Tests Illustrating the Important Components of the Flow (1) Mixture with Low Void Fraction in Stagnation Zone, (2) Oscillating Filament of Air-Water Mixture, (3) Flow Path of Air-Water Mixture, (4) Tubes.

between rigid and flexible tube bundles. For example, Price and Païdoussis (1986) reported that the effect of surrounding tube motion is very significant. Thus, we cannot say at this time whether the foregoing behavior is limited to two-phase flows or whether it is generic to fluidelastic instability mechanisms.

8.4.5 Hydrodynamic Coupling

Hydrodynamic coupling between tubes was investigated by measuring the coherence between the vibration responses of adjacent tubes. This work was done on a closely packed normal-triangular

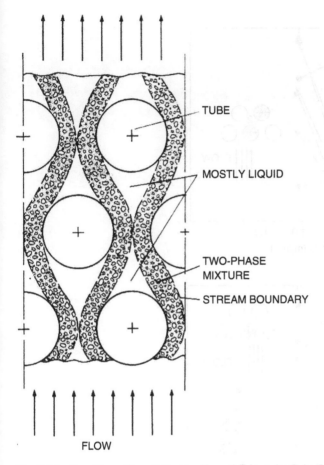

Fig. 8-11 Two-Phase Flow Paths in a Normal-Triangular Tube Bundle.

tube bundle (i.e., $P/D = 1.22$) to emphasize hydrodynamic coupling. The coherence function was measured in both the lift and drag direction and for tubes located side-by-side and one-behind-the-other.

Typical results of coherence as a function of the ratio of mass flux divided by the critical mass flux for instability are shown in Figs. 8-12a and 8-12b. The coherence function $\gamma_{12}(f)$ is defined as

$$\gamma_{12} = |S_{12}(f)|^2/|S_{11}(f)S_{22}(f)| \tag{8-3}$$

where $S_{12}(f)$ is the cross-power spectral density function and $S_{11}(f)$ and $S_{22}(f)$ are power spectral density functions of the vibration response of Tube 1 and adjacent Tube 2, respectively. The coherence results outlined here are for the value of the coherence function at the main response frequency, f, of the tubes.

The coherence is generally low in two-phase cross flow for mass flux ratios below 0.8. In liquid flow, the coherence is relatively high for all mass flux ratios. This result shows that there is much more hydrodynamic coupling in liquid flow than in two-phase flow below instability. On the other hand, hydrodynamic coupling is very significant (i.e., coherence above 0.5) at instability for both liquid and two-phase flows up to a void fraction of roughly 80%. This result helps to explain fluidelastic instability in two-phase cross flow.

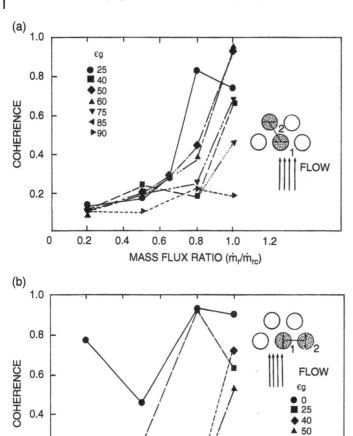

Fig. 8-12 Hydrodynamic Coupling Coherence as a Function of Mass Flux Ratio.

However, it is not clear why hydrodynamic coupling decreases above 80% void fraction. Perhaps hydrodynamic coupling is inhibited by intermittent flow or perhaps the structure of the two-phase mixture in the stream flow paths is such that it reduces hydrodynamic coupling at higher void fractions (i.e., froth flow).

8.5 Analysis of the Fluidelastic Instability Results

8.5.1 Defining Critical Mass Flux and Instability Constant

As mentioned earlier, fluidelastic instability may be formulated in terms of a dimensionless flow velocity, U_p/fD, and a dimensionless mass-damping parameter, $2\pi\zeta m/\rho D^2$. For the simple case of a tube bundle subjected to uniform flow over the entire length, the foregoing parameters are simply related by

$$U_{pc}/fD = K\left(2\pi\zeta m/\rho D^2\right)^n \tag{8-4}$$

where the critical reference gap velocity (critical pitch velocity), U_{pc}, is taken as $\dot{m}_p/\rho$ in two-phase flow ($\dot{m}_p$ is the pitch mass flux), f is the tube natural frequency in the fluid, D is the tube diameter, m is the mass per unit length including the hydrodynamic mass, m_h, ρ is the homogeneous density of the two-phase mixture, ζ is the total damping ratio (structural and fluid damping), and n is an exponent which is often around 0.5, as discussed in Chapter 7. Using this relationship, we can now deduce the instability constant, K, and the exponent, n, knowing the critical mass flux for fluidelastic instability derived from the vibration response results, such as Figs. 8-2 and 8-3.

All necessary parameters have now been defined or evaluated except for the critical pitch mass flux for fluidelastic instability, $\dot{m}_{pc}$. Theoretically, the vibration amplitude should be very large when instability occurs. In practice, the critical mass flux at which instability occurs is not always well defined in the vibration response curves. This is attributed to nonlinearities and possibly to the nature of the two-phase flow. When instability is well defined (see, for example, Test O-11, Fig. 8-2), the velocity or mass flux at which it occurs is used for the calculations. When instability is not well defined (e.g., one-tube flexible Test O-14, Fig. 8-2), following earlier work in liquid flow (Pettigrew and Gorman, 1978) and in two-phase flow (Pettigrew et al, 1985), we used the mass flux corresponding to a vibration response of roughly 750 μm rms. The value of 750 μm rms is a convenient and well-defined criterion. It corresponds to a vibration level that could cause tube damage. Figure 8-3 shows some of the instability data points used for the calculations. Another approach is to limit the work-rate to a specified value (see Chapter 12).

The calculated values of the dimensionless parameters, U_{pc}/fD and $2\pi\zeta m/\rho D^2$ and all the necessary flow parameters are listed in Table 8-3. The values of the instability constant, K, in Table 8-3, are based on the exponent n being equal to 0.5, although this assumed value is open to question, as discussed later. The damping values used in the calculations are those measured in the rigid tube bundles at roughly half the mass flux for instability. The damping measurements are described in detail in Chapter 6. The damping values are listed in Table 8-3.

8.5.2 Comparison with Results of Other Researchers

All the available air-water and steam-water data on fluidelastic instability in two-phase cross flow from other researchers as of 1990 is compiled in Table 8-4. The data was obtained from a variety of test sections in different laboratories: Pettigrew and Gorman (1973b), Heilker and Vincent (1981), Remy (1982), Remy and Bai (1982), Axisa et al (1984 and 1985), and Nakamura et al (1986a and 1986b).

The interpretation of the data was sometimes difficult because of inconsistent definitions of parameters such as damping, critical velocity for instability, void fraction, etc. Wherever possible the parameters were normalized or corrected to be consistent with the definitions given in this chapter.

Two-phase flow conditions are based on homogeneous void fraction, density, velocity, etc. Most researchers either had great difficulties or did not measure damping at all in their two-phase cross-flow experiments. Two researchers reported reasonable damping measurements, Axisa et al (1984 and 1985), and Pettigrew et al (1989b). In two experiments (Pettigrew and Gorman, 1973b), and Remy (1982), the tube bundles were subjected to flow over only a portion of their span length (i.e., partial admission). In these experiments, the results were normalized to take into account the effect of the non-uniform flow velocity distribution. Axisa et al (1984 and 1985) and Nakamura et al (1986b) contributed some much needed experimental results on fluidelastic instability in steam-water cross

Table 8-4 Summary of Test Conditions and FEI Results from Other Researchers (Pettigrew et al, 1989b).

Reference	Press. (MPa)	Geom. (°)	P/D	D (mm)	e_g (%)	ρ (kg/m³)	m_t (kg/m)	m (kg/m)	U_{pc} (m/s)	f (Hz)	ζ_s (%)	ζ (%)
Remy (1982)		90	1.44	10.00	89	112	0.220	0.300	6.6	43.6	0.10	1.00
& Remy & Bai		90	1.44	10.00	95.5	46	0.220	0.300	10.9	43.6	0.10	1.00
(1982)		90	1.44	10.00	98.5	16	0.220	0.300	16.5	43.6	0.10	1.00
air-water		90	1.44	10.00	99.7	4	0.220	0.300	23	43.6	0.10	1.00
Nakamura et al		90	1.42	19.05	7.7	923	0.720	1.070	0.78	12.5	0.64	
(1986a)		90	1.42	19.05	13.8	862	0.720	1.047	0.87	12.6	0.64	
air-water		90	1.42	19.05	34.9	652	0.720	0.967	0.86	13.1	0.64	
		90	1.42	19.05	7.5	925	0.720	1.071	1.59	20.5	0.61	
		90	1.42	19.05	16.9	831	0.720	1.035	1.77	20.9	0.61	
		90	1.42	19.05	57.4	426	0.720	0.882	2.35	22.6	0.61	
		90	1.42	19.05	75.3	248	0.720	0.814	2.43	23.5	0.61	
		90	1.42	19.05	88.2	119	0.720	0.765	2.55	24.3	0.61	
		90	1.42	19.05	97.8	23	0.720	0.729	4.55	24.8	0.61	
		90	1.42	19.05	60.8	393	0.720	0.869	2.55	27.3	0.74	
		90	1.42	19.05	92.7	74	0.720	0.748	4.13	29.4	0.74	
		90	1.42	19.05	98.3	18	0.720	0.727	5.9	29.9	0.74	
Nakamura et al	~0.5	90	1.42	19.05	46.5	491	0.720	0.907	2.15	14.3	5.00	
(1986b)	~0.5	90	1.42	19.05	75	230	0.720	0.808	2.0	15.1	5.00	
steam-water	~0.5	90	1.42	19.05	82	166	0.720	0.783	2.25	15.3	5.00	
	~0.5	90	1.42	19.05	87	122	0.720	0.766	2.3	15.5	5.00	
	~3.0	90	1.42	19.05	7.4	764	0.720	1.010	1.35	13.2	8.00	
	~3.0	90	1.42	19.05	76.9	201	0.720	0.796	1.3	14.8	8.00	
	~5.8	90	1.42	19.05	7.4	710	0.720	0.990	1.35	13.0	11.00	
	~5.8	90	1.42	19.05	21.4	607	0.720	0.950	1.4	13.2	11.00	
	~5.8	90	1.42	19.05	64.3	292	0.720	0.831	1.4	14.1	11.00	
	~5.8	90	1.42	19.05	83.3	152	0.720	0.778	1.2	14.6	11.00	
	~5.8	90	1.42	19.05	95.2	64	0.720	0.744	2.1	15.0	11.00	
Heilker &		60	1.36	22.00	72	281	0.631	0.792	3.51	60.6	1.46	
Vincent (1981)		60	1.36	22.00	84	161	0.631	0.741	5.98	62.6	1.51	
air-water		60	1.36	22.00	87	131	0.631	0.706	6.67	64.2	1.54	
		60	1.36	22.00	65	351	0.631	0.831	2.92	59.1	1.43	
		45	1.5	19.05	50	501	0.577	0.760	4.53	52.1	1.83	
Axisa et al	2.5	30	1.44	19.05	98	29.1	0.520	0.526	12	73.2	0.2	0.50
(1984)	2.5	30	1.44	19.05	97	39.2	0.520	0.529	12.5	72.5	0.2	0.80
Axisa et al	2.5	30	1.44	19.05	95	51.8	0.520	0.532	11.2	72.2	0.2	1.00

Table 8-4 (Continued)

Reference	Press. (MPa)	Geom. (°)	P/D	D (mm)	ε_g (%)	ρ (kg/m³)	m_t (kg/m)	m (kg/m)	U_{pc} (m/s)	f (Hz)	ζ_s (%)	ζ (%)
(1985)	2.5	30	1.44	19.05	92	79.7	0.520	0.540	12	71.8	0.2	1.20
steam-	2.5	30	1.44	19.05	90	101	0.520	0.546	11	71.8	0.2	1.40
water												
	2.5	30	1.44	19.05	85	145	0.520	0.559	11	71.2	0.2	1.60
	2.5	60	1.44	19.05	98	29.1	0.520	0.526	10.5	71.8	0.2	1.00
	2.5	60	1.44	19.05	97	39.2	0.520	0.529	10	71.0	0.2	1.30
	2.5	60	1.44	19.05	95	51.8	0.520	0.532	10	72.1	0.2	1.60
	2.5	60	1.44	19.05	92	79.7	0.520	0.540	8.9	70.0	0.2	1.80
	2.5	60	1.44	19.05	90	101	0.520	0.546	11	69.8	0.2	2.30
	2.5	60	1.44	19.05	85	145	0.520	0.559	9.2	69.0	0.2	3.40
	2.5	90	1.44	19.05	98	29.1	0.520	0.526	8.4	76.0	0.2	0.50
	2.5	90	1.44	19.05	97	39.2	0.520	0.529	8.2	75.4	0.2	0.80
	2.5	90	1.44	19.05	95	5.8	0.520	0.532	8.1	74.5	0.2	1.00
	2.5	90	1.44	19.05	92	79.7	0.520	0.540	8	74.0	0.2	1.30
	2.5	90	1.44	19.05	90	101	0.520	0.546	8.2	73.9	0.2	1.60
	2.5	90	1.44	19.05	85	145	0.520	0.559	7.8	73.7	0.2	1.80
Pettigrew &		30	1.5	12.7	90	103	0.319	0.456	3.3	17.0	2.67	2.67
Gorman		60	1.5	12.7	90	103	0.319	0.456	3.11	17.0	2.67	2.67
(1973b)												
air-water		60	1.5	12.7	80	202	0.319	0.456	1.84	17.0	2.67	2.67

flow. It should be noted that most of the foregoing work was done in support of nuclear steam generators; thus, for tube bundles of P/D between 1.4 and 1.5 and, in particular, for square-bundle configurations.

To avoid the issue of damping, several researchers, in particular Remy and Bai (1982), have recommended grouping together the fluidelastic instability constant, K, and damping. This approach leads to the following equation:

$$U_{pc}/fD = \left(K\sqrt{2\pi\zeta}\right)\left(m/\rho D^2\right)^{1/2} \tag{8-5}$$

Accordingly, we have plotted the data of Table 8-3 for continuous flow and all the data of Table 8-4 in terms of the dimensionless mass parameter $(m/\rho D^2)$ in Fig. 8-13. This approach leads to a reasonable collapse of the data, except for the results of partial flow admission tests (Pettigrew and Gorman, 1973b, and Remy and Bai, 1982). Partial admission tests yield high values of U_{po}/fD. This result is not surprising since the flow area is not well controlled in a partial admission test. The effective flow area increases away from the inlet resulting in a lower flow velocity U_{pc} than that calculated from the inlet flow area. Interestingly, the common data points for both partial admission tests are in close agreement.

Remy and Bai (1982) recommend a combined instability-damping parameter, $K\sqrt{2\pi\zeta} = 2.24$, to avoid instability in two-phase cross flow. Although this value fits their data well, it is not conservative, as shown in Fig. 8-13. A value of $K\sqrt{2\pi\zeta} = 1$ would be a reasonable design criterion based on the data of Fig. 8-13. However, it would be too conservative in many cases.

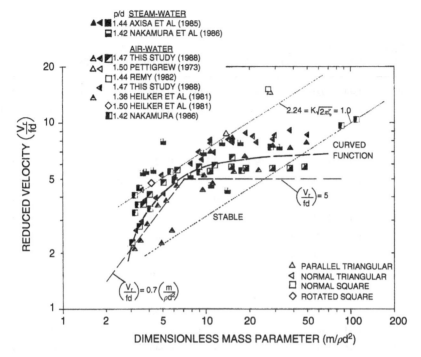

Fig. 8-13 Air-Water and Steam-Water Fluidelastic Instability Results Plotted Using the Dimensionless Mass Parameter.

The data of Fig. 8-13 appears to follow a curved function of $(m/\rho D^2)$. In fact, ε_g is directly related to the inverse of the density, $1/\rho$. Since the variation in m/D^2 is small, ε_g is directly related to $(m/\rho D^2)$. Thus, the fluidelastic instability data appears to follow a curved function of void fraction. This curved function of void fraction is similar in shape to that found for two-phase damping, as discussed in Chapter 6. This result suggests the strong influence of two-phase damping on fluidelastic instability in two-phase cross flow.

The data for which two-phase damping values were available was plotted in Fig. 8-14 as a function of the dimensionless mass-damping parameter, $2\pi\zeta m/\rho D^2$, in which ζ is the total damping (i.e., structural and two-phase damping). The agreement between the expression $U_{pc}/fD = K\sqrt{2\pi\zeta m/\rho D^2}$ and the experimental data is reasonable. The data for partial admission tests does not fit as well for reasons already discussed. The agreement between the data of Axisa et al (1985) in steam-water and the data presented in this chapter is remarkably good, particularly for the normal-square bundle data. This agreement suggests that dimensionless velocity, U_{pc}/fD, and dimensionless mass-damping, $2\pi\zeta m/\rho D^2$, are reasonable parameters to formulate fluidelastic instability in two-phase cross flow. It appears that this formulation is equally valid for air-water, steam-water and probably other liquid-vapor mixtures. However, knowledge of damping is required.

8.5.3 Summary of Air-Water Tests

Eight tube bundles of triangular and square configurations and P/D of 1.47, 1.32, and 1.22 were subjected to a broad range of two-phase cross-flow conditions. The flow conditions encompassed two main flow regimes: continuous and intermittent. The tests were done on bundles with all flexible tubes and on bundles with one flexible tube surrounded by rigid tubes. Fluidelastic instabilities

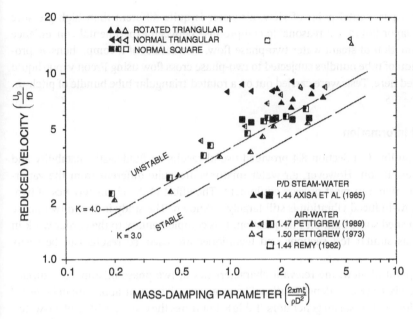

Fig. 8-14 Air-Water and Steam-Water Fluidelastic Instability Results Plotted Using the Mass-Damping Parameter.

have been observed for all tube bundles and all flow conditions. The critical flow velocity for fluidelastic instability is significantly lower for all-flexible tube bundles.

The fluidelastic instability behavior is different for intermittent flow than for continuous flow regimes such as bubbly or froth flows. For continuous flows, the observed instabilities satisfy the relationship $U_{pc}/fD = K\sqrt{2\pi\zeta m/\rho D^2}$ in which the minimum instability constant K was found to be approximately 4.0 for bundles of $P/D = 1.47$ and significantly less for smaller P/D. The lowest critical velocities for fluidelastic instability were observed with rotated-triangular tube bundles.

The results are in reasonable agreement with those of other researchers in air-water and steam-water. The latter agreement is particularly relevant to practical design guidelines. For intermittent flow, the observed instabilities do not follow the foregoing formulation. Significantly lower flow velocities resulted in instability.

In summary, fluidelastic instabilities are possible in two-phase cross flow. The parameters governing the fluidelastic behavior were reasonably well understood from test results in air-water and steam-water. However, many questions remained such that further work was undertaken, as outlined below. In particular, experiments in Freon were conducted to complement the air-water tests.

8.6 Tube Bundle Vibration in Two-Phase Freon Cross Flow

8.6.1 Introductory Remarks

Two-phase cross flow exists in the U-bend region of recirculating-type nuclear steam generators. In most heat exchangers, two-phase means vapor-liquid such as steam-water. So far, relatively little work has been done to study flow-induced vibration of tube bundles subjected to steam-water cross flow. Most of the work has been done in two-phase flow simulated by air-water mixtures. Although convenient from an experimental point of view, air-water is quite different than high-pressure steam-water. Freon vapor-liquid is a reasonable compromise between experimental convenience and appropriate simulation of steam-water two-phase flow. The results of a comprehensive program to study vibration of tube bundles subjected to two-phase cross flow using Freon vapor-liquid mixtures are presented here. Tests were carried out on a rotated-triangular tube bundle of pitch-to-diameter ratio, $P/D = 1.5$.

8.6.2 Background Information

The air-water work outlined in Section 8.4 provided useful results on fluidelastic instability and random turbulence excitation. However, air-water mixtures are quite different from the vapor-liquid mixtures encountered in most process equipment. The effect of the characteristics of two-phase mixtures on flow-induced vibration is still largely unknown. The study in Freon two-phase flow was intended to shed some light on this question. It is complementary to the earlier work in air-water inasmuch as similar test sections and techniques are used, so results can be easily compared.

From a vibration point of view, the relevant characteristics of two-phase mixtures are surface tension, vapor-liquid viscosity and density ratios. Steam-water mixtures at approximately 265°C are of prime interest for nuclear steam generators. Table 8-1 outlines the characteristics of air-water, steam-water and Freon vapor-liquid mixtures. The characteristics of Freon mixtures can be varied to some extent by changing the temperature. Freon characteristics are generally much closer to those of steam-water than air-water. Testing in Freon is much simpler than in steam-water since equivalent pressures and temperatures are much lower. It is also much cheaper in energy requirements because the latent heat of evaporation is much lower.

8.6.3 Experiments in Freon Cross Flow

The general approach is to subject a realistic tube bundle to two-phase Freon cross flow. Vibration excitation mechanisms and damping are deduced from the vibration response. This approach requires a test section connected to a Freon flow loop.

Test Section. The test section consists of a flow channel built inside a pressure vessel (see Figs. 8-15 and 8-16). A cantilevered triangular tube bundle is inserted in the flow channel through a bolted flange (see Fig. 8-17). The flow channel is approximately 100 x 610 mm. Half-tubes are installed on each side of the flow channel to minimize wall effects. A mixer-homogenizer is provided upstream of the test section to ensure uniform and homogeneous two-phase cross flow. Vibration of the tube free-end could be observed through a flanged transparent window.

Stainless steel Type 304 tubes, 609 mm long, 12.7 mm (0.5 in.) diameter and 0.9 mm wall thickness were used for the test section (see Fig. 8-17). The tubes had a natural frequency in air of 28.5 Hz, which is realistic for steam generators. The tubes were plugged at the free-end. The tube pitch was 19.05 mm (0.75 in.), which gives a pitch-to-diameter ratio of $P/D = 1.5$.

Tests were also done with one flexible tube surrounded by rigid tubes to avoid the effect of hydrodynamic coupling between the tubes. This configuration was useful to study damping and random turbulence excitation. Rigidity was achieved by supporting the free-end of the tubes with a special insert fixed to the free-end of the tubes, which increased the frequency of the tubes to 150 Hz. Thus, they were essentially rigid compared to the flexible tube. The single flexible tube was in the interior of the tube bundle.

The tube vibration response was measured with strain gages installed inside the tubes near the cantilevered end. Upstream, interior and downstream tubes were instrumented. Two pairs of

Fig. 8-15 Photograph of Freon Pressure Vessel Test Section Showing Instrumented End of Tube Bundle.

Fig. 8-16 Schematic of Freon Test Section.

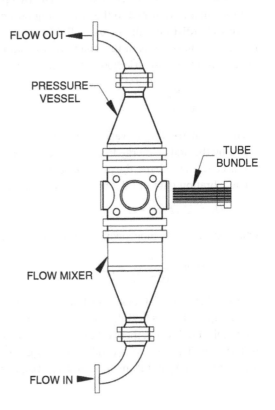

FLOW OUT

PRESSURE
VESSEL

TUBE
BUNDLE

FLOW MIXER

FLOW IN

Fig. 8-17 Photograph of Rotated-Triangular Tube Bundle Instrumented with Strain Gages.

diametrically opposite strain gages were installed at 90° from each other to measure vibration in the flow direction (drag) and in the direction normal to the flow (lift).

Flow Loop. The test section was installed in an existing Freon loop. The working fluid was Freon-22 at 1.0 MPa pressure and 23.3°C saturation temperature at the test section. Because of environmental concerns, Freon-134a was used in later tests (Pettigrew et al, 2002). The Freon vapor was provided by 1.2 MW electrical heaters. The two-phase mixture was condensed downstream of the test section and returned to the circulating pump.

Flow Conditions. All two-phase fluid properties such as density, mass flux and void fraction are taken as homogeneous in this chapter. For example, the homogeneous void fraction, ε_g, is calculated from the volumetric flow rate of vapor, $\dot{V}_g$, and liquid, $\dot{V}_\ell$ as

$$\varepsilon_g = \frac{\dot{V}_g}{\dot{V}_g + \dot{V}_\ell} \tag{8-6}$$

The volumetric flow of vapor is obtained from heat balance calculations, which are largely governed by the added energy from the electrical heaters.

It is convenient to use the pitch velocity, U_p, (sometimes called the reference gap velocity) to correlate vibration data in tube bundles. It is defined as

$$U_p = U_\infty P/(P-D) \tag{8-7}$$

where U_∞ is the homogeneous free-stream velocity. Similarly, the pitch mass flux,

$$\dot{m}_p = \dot{m}_\infty \frac{P}{(P-D)} = \rho U_p \tag{8-8}$$

In Test Series A, all tubes in the bundle were flexible and free to vibrate; whereas, in Series B only one interior tube was flexible, while the remaining tubes were fixed at the free-end and, thus, essentially rigid. For brevity, these tube bundles are henceforth called "flexible" and "rigid" tube bundles, respectively. The tests were conducted in liquid Freon and at several void fractions from 40 to 90% for Test Series A, and from 10 to 90% for Test Series B.

The tube bundles were subjected to increasing mass flux until the vibration amplitude was sufficiently high to indicate the onset of fluidelastic instability. During a given test, the void fraction was kept constant. The mass fluxes, void fractions and corresponding calculated results are summarized in Table 8-5. Table 8-5 also has the flow conditions and results for the Freon tests described in Pettigrew et al (2002) and Pettigrew and Taylor (2009)

Table 8-5 Summary of Freon Test Conditions and Results.

Fluid	Geom. (°)	P/D	ε_g (%)	ρ (kg/m^3)	m_h (kg/m)	m (kg/m)	$\dot{m}_{pc}$ (kg/m^{2}s)	U_{pc} (m/s)	f (Hz)	ζ (%)	$\frac{U_{pc}}{fD}$	$\frac{m}{\rho D^2}$	$\frac{2\pi\zeta m}{\rho D^2}$	K
Freon-22	60	1.5	0	1200	0.207	0.473	734	0.61	22.6	0.47	2.13	2.45	0.07	7.9
1 MPa	60	1.5	40	737	0.127	0.394	732	0.99	23.9	2.79	3.28	3.31	0.58	4.3
23°C	60	1.5	50	621	0.107	0.374	713	1.15	24.6	3.61	3.67	3.73	0.85	4.0
Pettigrew	60	1.5	55	563	0.097	0.364	638	1.13	—	—	—	4.00	—	—
et al	60	1.5	60	505	0.087	0.354	611	1.21	25.0	4.31	3.80	4.34	1.18	3.5
(1995)	60	1.5	65	448	0.077	0.344	491	1.10	25.3	5.02	3.41	4.76	1.50	2.8
	60	1.5	70	390	0.067	0.334	428	1.10	25.8	4.25	3.36	5.31	1.42	2.8
	60	1.5	75	332	0.057	0.324	330	0.99	26.0	4.13	3.01	6.06	1.57	2.4
	60	1.5	80	274	0.047	0.314	208	0.76	26.3	4.09	2.27	7.11	1.83	1.7
	60	1.5	85	216	0.037	0.304	146	0.68	26.9	3.37	1.98	8.73	1.85	1.5
	60	1.5	90	158	0.027	0.294	99	0.63	27.5	2.39	1.79	11.54	1.73	1.4
Freon-	60	1.5	0	1175	0.202	0.469	825	0.70	38.8	0.52	1.43	2.47	0.08	5.0
134a	60	1.5	25	891	0.153	0.420	1100	1.23	40.3	2.38	2.41	2.92	0.44	3.6
0.74 MPa	60	1.5	40	721	0.124	0.391	1100	1.52	41.9	3.39	2.87	3.36	0.72	3.4
34°C	60	1.5	50	608	0.104	0.371	1100	1.81	42.9	3.83	3.32	3.79	0.91	3.5
Pettigrew	60	1.5	60	495	0.085	0.352	1050	2.12	44.1	4.11	3.79	4.41	1.14	3.6
et al	60	1.5	70	381	0.065	0.332	820	2.15	45.1	3.38	3.76	5.41	1.15	3.5
(2002)	60	1.5	80	268	0.46	0.313	550	2.05	45.8	2.66	3.53	7.25	1.21	3.2
	60	1.5	85	211	0.036	0.303	425	2.01	46.0	2.66	3.45	8.91	1.49	2.8
	60	1.5	90	154	0.026	0.293	325	2.11	46.3	2.38	3.58	11.79	1.77	2.7
	60	1.5	95	98	0.17	0.284	225	2.31	46.9	1.60	3.87	18.02	1.81	2.9
	60	1.5	98	64	0.011	0.278	125	1.97	47.4	1.52	3.26	27.10	2.60	2.0
Freon-22	30	1.5	25	911	0.157	0.424	841	0.92	19.7	2.21	3.69	2.88	0.40	5.8
1.2 MPa	30	1.5	40	737	0.127	0.394	841	1.14	24.8	4.44	3.63	3.31	0.93	3.8
30°C	30	1.5	50	621	0.107	0.374	983	1.58	25.6	5.28	4.87	3.73	1.24	4.4
Pettigrew	30	1.5	60	505	0.087	0.354	841	1.67	26.2	5.19	5.00	4.34	1.42	4.2
&	30	1.5	70	390	0.067	0.334	699	1.80	26.7	5.09	5.29	5.31	1.70	4.1
Taylor	30	1.5	80	274	0.047	0.314	463	1.69	27.0	4.50	4.94	7.11	2.01	3.5
(2009)	30	1.5	85	216	0.037	0.304	323	1.50	27.3	4.40	4.32	8.73	2.41	2.8
	30	1.5	90	158	0.027	0.294	276	1.75	27.5	3.94	5.01	11.54	2.86	3.0
	30	1.5	95	100	0.017	0.284	183	1.83	27.6	3.99	5.21	17.60	4.41	2.5
2.0 MPa	30	1.5	60	485	0.083	0.350	746	1.54	26.0	4.87	4.67	4.48	1.37	4.0
	30	1.5	80	287	0.049	0.316	463	1.61	26.7	3.77	4.76	6.83	1.62	3.8

8.7 Freon Test Results and Discussion

8.7.1 Results and Analysis

Vibration excitation mechanisms and damping were deduced from the tube bundle vibration response. Typical vibration response spectra are shown in Fig. 8-18. As expected, the tubes were vibrating at roughly their natural frequency. Typical vibration response results are presented as rms (root-mean-square) vibration response at the tube free-end versus mass flux curves in Fig. 8-19. Generally, the threshold velocity for fluidelastic instability is much better defined in Freon (Fig. 8-19) than in air-water (Fig. 8-2). Also, the vibration response to random turbulence excitation below instability is much lower than in air-water, as expected (see Fig. 8-20). The "turbulence" level in two-phase Freon flow should be much lower, because the surface tension and the density and viscosity ratios between the two phases are much lower. These parameters govern the characteristics of two-phase mixtures resulting in a much smoother, thus hydraulically less noisy, two-phase flow. The much-better-defined fluidelastic instability thresholds in

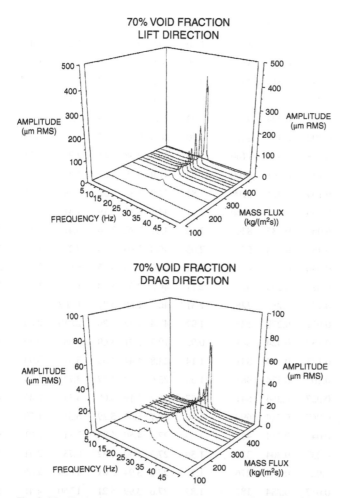

Fig. 8-18 Typical Vibration Response Spectra for Two-Phase Freon Cross Flow.

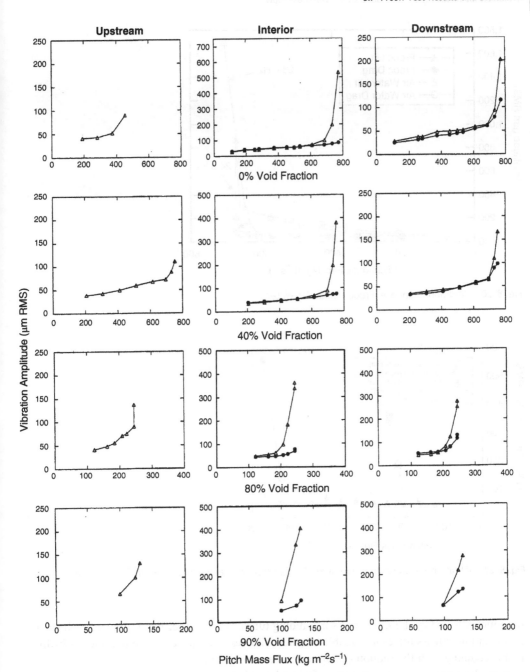

Fig. 8-19 Typical Vibration Response Curves for Two-Phase Freon Cross Flow: Flexible Tube Bundle; △ Lift Direction; • Drag Direction.

Freon are also explained in terms of lower turbulence level. High turbulence levels are known to disturb fluidelastic instability phenomena.

The difference in vibration response between the flexible bundles and the rigid bundles is small at mass fluxes below instability (see Fig. 8-21). This result means that random turbulence excitation is

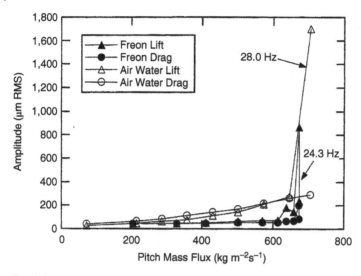

Fig. 8-20   Vibration Response: Freon versus Air-Water.

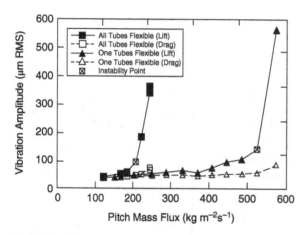

Fig. 8-21   Vibration Response In Freon at 80% Void Fraction: Flexible versus Rigid Tube Bundle.

not much affected by the motion of surrounding tubes. On the other hand, fluidelastic instability occurs at lower mass fluxes for flexible bundles. This result indicates the importance of hydrodynamic coupling with the motion of adjacent tubes at instability.

Hydrodynamic mass and damping are fluid-related dynamic parameters that are required to analyze flow-induced vibration. Hydrodynamic mass is defined as the equivalent external mass of fluid vibrating with the tube. Hydrodynamic mass and damping are discussed in detail in Chapter 4, and Chapters 5 and 6, respectively.

There appears to be another instability region at void fractions above ~80% for air-water and above ~65% for Freon, where the slope of the instability curve is considerably lower (see Fig. 8-22). In fact, the change in behavior is dramatic for Freon two-phase flow. At 90% void fraction, the instability constant is $K = 1.36$, which is less than half the recommended design guideline of $K = 3.0$!

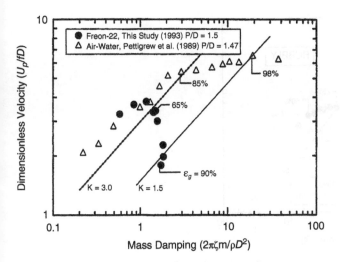

Fig. 8-22 Fluidelastic Instability Results in Two-Phase Cross Flow: Comparison Between Freon-22 and Air-Water.

8.7.2 Proposed Explanations

The foregoing behavior was explained in terms of flow regime change for the air-water results. As shown in Fig. 8-23, the transition between the two fluidelastic instability regions corresponds to the transition between continuous flow (spray, bubbly) and intermittent or churn flow. Figure 8-23 is a Grant flow regime map for two-phase cross flow (the axes for this map are defined in Chapter 3). In the air-water tests, there were several other indications of intermittent flow regime above 80% void fraction, such as low-frequency vibration response in the drag direction (see Fig. 8-1) and visual observation of intermittent flow regime in the test section. Also some evidence of a different flow regime could be seen in the hydrodynamic mass and damping results.

In Freon two-phase flow, the transition between the two instability regions also corresponds to the transition between continuous and intermittent flow. However, there was no other evidence of an intermittent flow regime. There is no indication of low-frequency vibration response in the drag direction (see Fig. 8-18). This surprising behavior is still in search of a satisfactory explanation. It is most likely related to flow regime changes, possibly transitions to intermittent flow or churn flow. Obviously, a lot more work needs to be done to understand vibration and flow regime in tube bundles subjected to two-phase cross flow.

8.7.3 Concluding Remarks

While these experiments may have provided a better understanding of tube bundle vibration in two-phase cross flow, they also raised a number of questions that are yet to be answered. These questions are generally related to the effect of two-phase flow regime on vibration excitation mechanisms and damping.

Damping values in Freon-22 and air-water mixtures are very similar, as shown in Fig. 6-14 in Chapter 6. This is puzzling since the physical properties of air-water and Freon vapor-liquid mixtures are very different. There are no significant trends leading to the identification of energy dissipation mechanisms for two-phase damping.

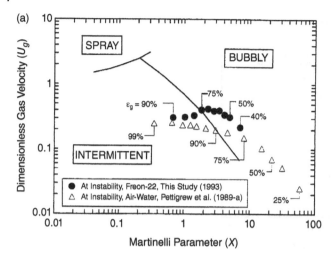

Fig. 8-23a Flow Regime Map for Freon and Air-Water Two-Phase Cross Flow Showing Conditions at Fluidelastic Instability.

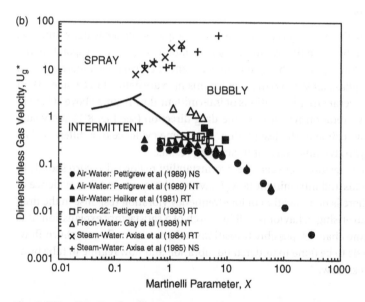

Fig. 8-23b Flow Regime Map for Steam-Water, Freon and Air-Water Two-Phase Cross Flow Showing Conditions at Fluidelastic Instability for Several Experimental Programs (Jong et al, 1994).

The very large decrease in critical velocity for fluidelastic instability above approximately 65% void fraction was unexpected and remains largely unexplained (see Fig. 8-22). The presence of intermittent flow is a possible explanation. However, it does not explain why the instability behavior is so different in Freon than in air-water. Perhaps the answer lies in the detailed characteristics of flow regime between the tubes. From a practical point of view, however, it is of concern that some nuclear components, such as steam generators, may be subjected to the fluidelastic instability conditions encountered here.

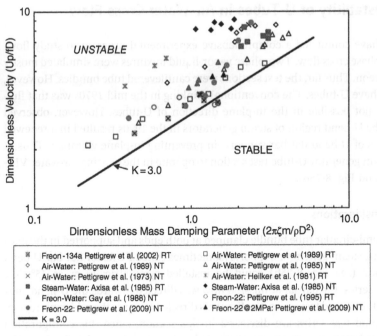

Fig. 8-24 Freon Results Included in Two-Phase Fluidelastic Instability Results Plotted Using the Mass-Damping Parameter.

Turbulence-induced excitation is much lower in Freon than in air-water, as expected. Furthermore, the effect of void fraction and mass flux above roughly 50% void fraction is small, which is quite different than in air-water. This is also attributed to flow regime effects.

Figure 8-24 adds Freon results from Pettigrew et al (1995), Pettigrew et al (2002), and Pettigrew and Taylor (2009) (listed as Pettigrew et al (2009) in the legend) to the two-phase instability results plotted versus mass-damping parameter in Fig. 8-14. The addition of the Freon results (including results from Gay et al (1988)) have not changed the lower boundary line of $K = 3.0$.

8.7.4 Summary Findings

A rotated-triangular tube bundle of $P/D = 1.5$ was subjected to two-phase cross flow over a range of void fractions from zero to 90% in a Freon test loop. The results revealed:

1) Well-defined fluidelastic instabilities have been observed in vapor-liquid Freon flow. Below roughly 65 to 70% void fraction, in continuous flow regimes, these instabilities may be simply formulated in terms of a dimensionless flow velocity and a dimensionless mass-damping term.
2) Above roughly 65% void fraction, the fluidelastic behavior was completely different. Critical flow velocities for instability were much lower than expected. This behavior is explained in terms of flow regime effects, probably intermittent flow regimes.
3) Hydrodynamic mass is roughly related to the homogeneous mixture density.
4) Two-phase damping is very dependent on void fraction. Damping is maximum between 50% and 80% void fraction, where the two-phase damping ratio reaches approximately 4%.
5) Random turbulence excitation forces are much lower in vapor-liquid Freon than in air-water. The effect of void fraction and mass flux is relatively small above 50% void fraction.

8.8 Fluidelastic Instability of U-Tubes in Air-Water Cross Flow

As discussed earlier, we have conducted a comprehensive experimental program to study flow-induced vibration in two-phase cross flow. Two-phase vapor-liquid mixtures were simulated mostly by air-water and also by Freon. Thus far, the test sections were cantilevered tube bundles. However, nuclear steam generators have U-tubes. The conventional thinking in the mid-1970s was that fluidelastic instabilities were not possible in the in-plane direction of U-tubes. However, observed fretting-wear damage in the U-bend region of steam generators in the 1990s resulted in a renewed interest in the effectiveness of U-bend flat-bar supports in preventing in-plane vibration. Thus, it was decided to build a steam generator U-tube test section to operate in the existing air-water VIB-FLO loop (see Fig. 8-25a and Fig. 8-25b).

8.8.1 Experimental Considerations

The test section is a 180° semi-circular tube bundle clamped at both ends and supported in the middle with typical flat-bar-type steam generator supports (sometimes called FURs for flat-bar U-bend restraints). The test section (see Fig. 8-25b) could be installed symmetrically or at 45° (see Fig. 8-25a). In the first test series, the partial-admission flow was directed towards the U-bend apex at the support location. In the second test series, the flow was directed towards the mid-span region. The results from this second test series are discussed in this section, since this configuration offers a more realistic simulation of a steam generator U-bend region. More details about the test section may be found in Janzen et al (2005).

(a)

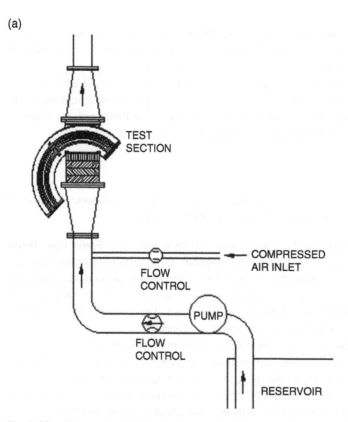

Fig. 8-25a   Schematic of the Air-Water U-Bend Test Section and Flow Loop.

(b)

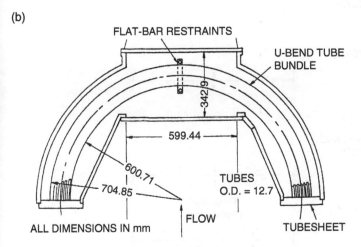

Fig. 8-25b   Schematic of Air-Water U-Bend Test Section.

8.8.2   U-Tube Dynamics

For tubes tightly supported at the apex, the lowest frequency mode becomes the OOP (out-Of-plane) anti-symmetric mode at 33 Hz (see Figs. 8-26 and 8-27). The results showed that for the lowest modes of vibration, in liquid flow and at low void fraction (25%), fluidelastic instability in the out-of-plane direction led to amplified tube response. Unfortunately, testing at higher void fraction was not possible due to test loop limitation.

8.8.3   Vibration Response

We expected severe rattling of the tubes due to OOP fluidelastic instability within the available clearance. However, this was not the case. For tubes with clearance supports, tube-to-support interaction is insignificant for low level vibration and the supports are "inactive". When the OOP vibration amplitudes approach the available clearance, tubes come into contact with the supports, which then become "active" or effective. The switch from inactive to active support may be accompanied by a so called mode switch, as revealed by a change in the OOP vibration frequency from 10 Hz to 33 Hz, as shown in Fig. 8-28. An accompanying increase in tube-to-support interaction, leading to higher apparent damping ratios and enhanced work-rate, would also be expected. In situations where this process is driven by fluidelastic instability, Taylor et al (1991) and Janzen et al (2005) referred to it as Amplitude Limited Fluidelastic Instability (ALFI). ALFI was observed in the first test series (where the flow was directed towards the U-bend apex), but only at zero and 25% void fraction, suggesting that random turbulence is likely the dominant excitation mechanism in steam generator U-bends, which typically operate at two-phase void fraction of 60% to 90%.

8.8.4   Out-of-Plane Vibration

Out-of-plane fluidelastic instability was observed in the second test series at zero void fraction (Fig. 8-29) and at 25% void fraction, corresponding to the steep rise in vibration amplitude for zero void fraction. As expected and observed in the first test series, motion in the OOP is limited by the available clearance when loose FURs are present (see Fig. 8-29). When ALFI occurs, at zero and 25% void fraction, the maximum rms response in the first out-of-plane mode is 0.50 mm with a

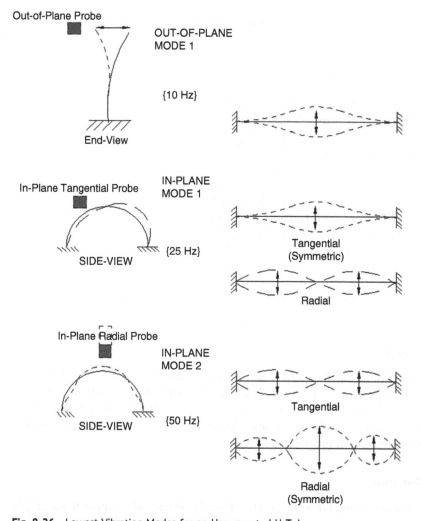

Fig. 8-26 Lowest Vibration Modes for an Unsupported U-Tube.

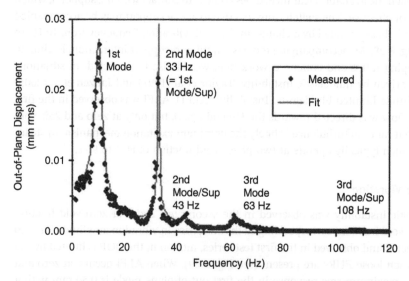

Fig. 8-27 Out-of-Plane U-Tube Frequency-Response Spectrum for 90% Void Fraction, at a Pitch Velocity of 8.7 m/s, with a 1.5 mm Tube-to-Support Clearance. Peaks Corresponding to the Lowest Vibration Modes are Labeled (Sup = Tube Supported at Apex).

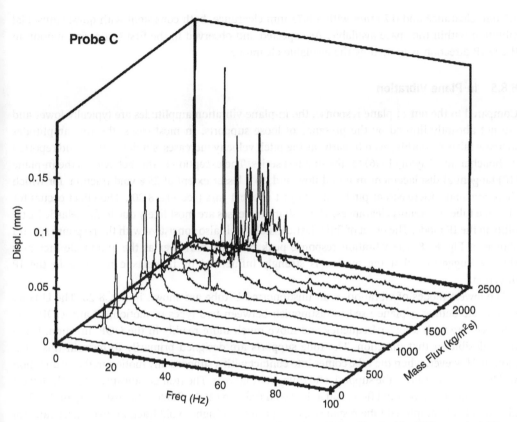

Fig. 8-28 Out-of-Plane Vibration Response Spectra as a Function of Mass Flux for Liquid Flow and a 0.75 mm Tube-to-Support Clearance. Note the Mode Switch at ~1750 kg/m²s Mass Flux ($U_P \approx 1.7$ m/s).

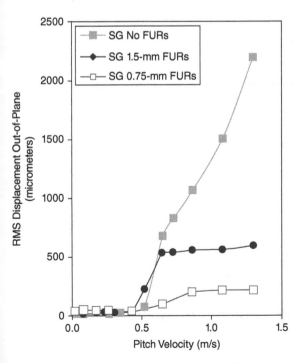

Fig. 8-29 Out-of-Plane Vibration Amplitudes Measured by Strain Gages, in Liquid Flow

1.5 mm clearance and 0.20 mm with a 0.75 mm clearance, both consistent with quasi-sinusoidal vibration within the space available. As expected and observed in the first test series, motion in the OOP direction is limited by the available clearance.

8.8.5 In-Plane Vibration

Compared to the out-of-plane response, the in-plane vibration amplitudes are typically lower and are not abruptly limited by the presence of loose supports. In most cases, the rms amplitudes increase fairly smoothly, even linearly, as the pitch velocity increases, similar to the trend reported by Boucher and Taylor (1996) for the first test series. The exception to this behavior is the in-plane (IP) tangential displacement in liquid flow and, to a lesser extent at 25% void fraction, for which there are sharp increases at pitch velocity just above 1 m/s (see Fig. 8-30). The effect occurs for all three tube-to-support clearances. These sharp increases are most likely due to fluidelastic instability in the IP mode. The onset of fluidelastic instability is also consistent with the response spectra shown in Fig. 8-28; the vibration response peak becomes narrower as the amplitude increases sharply, suggesting that the motion is strongly influenced by fluidelastic response in the IP direction.

The lowest modes of vibration of an unsupported U-tube are shown in Fig. 8-26. The U-bend semi-circle defines a plane used to distinguish between two types of motion, IP and OOP. The tubes vibrate with a lower frequency and move much more readily in the OOP direction, being more flexible in that direction. The tube supports are designed primarily to restrain this OOP motion. However, when the tube comes into contact with a support, any motion in the IP direction will lead to sliding along the support, and thus, fretting wear. The results indicate that for the lowest mode of vibration, in liquid flow, and at low void fraction (25%), fluidelastic instability in the OOP direction led to amplified tube responses, as expected. At higher void fraction, two-phase random

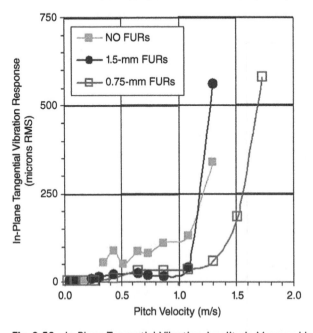

Fig. 8-30 In-Plane Tangential Vibration Amplitude Measured by Strain Gages, in Liquid Flow.

excitation was dominant. The FUR severely limited the OOP motion, while IP motion was damped but not completely restricted in liquid flow (see Fig. 8-29).

In summary, IP fluidelastic instability has been observed in two-phase cross flow at void fractions below 25%. It could occur in the range of 70 to 95% void fraction, which is of relevance to nuclear steam generators. However, this supposition requires experimental verification.

Some progress in understanding two-phase IP vibration is described in the following section using straight tubes restricted to vibrate in the in-flow direction.

8.9 In-Plane (In-Flow) Fluidelastic Instability

8.9.1 In-Flow Experiments in a Wind Tunnel

Contrary to OOP instability, the possibility of IP instability has raised less concerns. The IP modes have higher frequencies and considerably higher flow velocity at instability. Also, many of the reported cases of fluidelastic instability involve purely transverse tube motion. However, as discussed in Chapter 7, Section 7.9, IP instability has been shown to occur in a bundle of tubes preferentially flexible in the flow direction subjected to air flow in a wind tunnel. These air-flow tests were exploratory to investigate instability in the flow direction. It is clear that, to arrive at final conclusions regarding stability behavior in actual steam generators, tests in two-phase flow are required.

8.9.2 In-Flow Experiments in Two-Phase Cross Flow

Janzen et al (2005) observed fluidelastic instability for an IP mode in water flow and in low-void fraction (25%) air-water flow. The logical next step is to verify that fluidelastic instability can occur for an array of cylinders flexible only in the streamwise direction subjected to high void fraction (70 to 95%) two-phase cross flow. This range of void fraction corresponds to conditions in the U-bend region of nuclear steam generators.

Air-water tests by Violette et al (2006) have provided some insight into the possibility of IP flow-induced vibration. The test section used by Violette et al (2006) is connected to an air-water test loop. The reader is referred to a paper by Pettigrew et al (2005) for a description of the two-phase flow test loop. The test section, shown in Fig. 8-31, has a flow area of 0.038 m^2 (0.2 m x 0.19 m). It

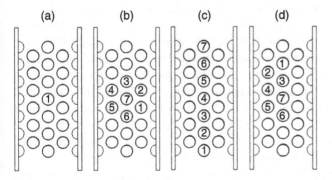

Fig. 8-31 Flexible Tube Assemblies a) Single Tube, b) Central Cluster, c) Single Column and d) Two Columns.

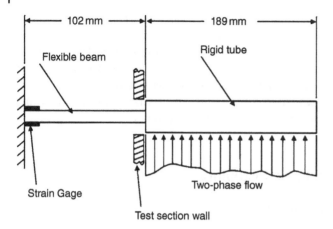

Fig. 8-32 Configurations of Flexible Tubes Tested Within the Test Section.

includes 12 rigid tubes and seven flexible tubes in a rotated-triangular configuration. Each tube has a diameter of 38.0 mm and the pitch-to-diameter ratio is 1.5. The test section was designed so that it is possible to switch the positions of flexible and rigid tubes. This design offers the possibility of experimenting with different configurations of flexible and rigid cylinders.

A typical instrumented tube assembly is shown in Fig. 8-32. Each assembly consists of a rigid cylinder attached to a flexible cantilever beam. Two sets of flexible beams were used. The first set has a rectangular cross-section, which makes the assembly much more flexible in one direction. The second set has a circular cross-section of 11.0 mm diameter, which makes the assembly axisymmetrically flexible. For the latter, the natural frequency in air for both directions is 30 Hz. Two different sets of rectangular beams were tested: one of 0.00415 m x 0.025 m cross-section and one of 0.00675 m x 0.025 m. The flexible beam was oriented so the tube assembly would be more flexible in the flow direction, resulting in tube streamwise frequencies in air of 14 Hz (81 Hz for the transverse direction) and 28 Hz (103 Hz in the transverse direction) for the thinner and thicker beam, respectively.

8.9.3 Single-Tube Fluidelastic Instability Results

Tube rms vibration amplitudes were evaluated from the averaged spectra. Figure 8-33 shows the rms response of the tube flexible only in the flow direction for all void fractions tested (i.e., 65%, 80%, 90% and 95%). Clearly, no instability was observed up to the maximum velocity tested. Moreover, it can be seen for the three lowest void fractions that the amplitude curves start decreasing beyond a certain velocity. Examining the behavior of the tube response spectra with flow pitch velocity (Fig. 8-34), it is apparent that the response spectra become broader beyond that velocity. This broadening is due to impacting between the extremity of the flexible tube and the test section wall. The very high flow velocities cause a large deflection of the flexible tube. From this position, small vibration amplitudes make the tube impact with the test section wall and limit its movement. These results indicate that fluidelastic instability does not occur for a single tube flexible in the flow direction up to the maximum flow velocity tested. Instability would have normally been expected for a single axisymetrically flexible tube at lower flow velocities.

Results obtained for the axisymetrically flexible tube are shown in Fig. 8-35. As expected, the response of the tube in the flow direction (drag direction) is much less important than that

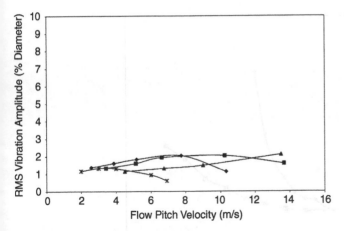

Fig. 8-33 Response versus Flow-Pitch Velocity for a Single Flexible Tube for Various Void Fractions: Void Fractions: X, 65%; ◆, 80%; ■, 90%; ▲, 95%.

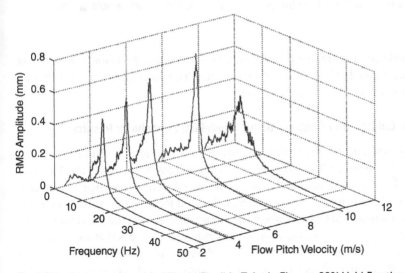

Fig. 8-34 Response Spectra of Single Flexible Tube in Flow at 80% Void Fraction.

in the lift direction at instability. In fact, no instability was observed whatsoever in the drag direction. Fluidelastic instability in the lift direction occurs for all void fractions tested, i.e., zero, 20%, 40%, 50%, 60%, 80% and 90%, respectively, at 2, 2.4, 3.75, 4.5, 5.5, 9.5 and 12 m/s. The critical flow velocity appears to be less well defined for the high void fractions (80% and 90%).

Figures 8-33 and 8-35 show that the response to two-phase flow turbulence at flow velocities below instability does not increase proportionally with flow velocity. This somewhat unexpected trend has been observed before in two-phase flows. It is explained in terms of changes in the structure of the two-phase flow. Sometimes low flow velocities result in larger characteristic flow structures and somewhat intermittent flows. These flows cause larger vibration excitation forces. With increasing flow velocity, the turbulence scale diminishes, resulting in lower excitation forces in

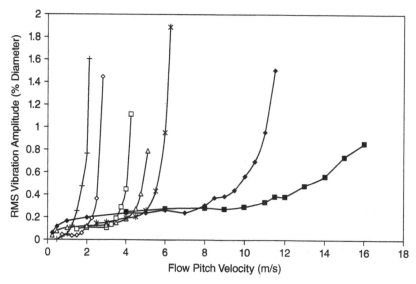

Fig. 8-35 Response in Lift Direction versus Flow Pitch Velocity for the Single Tube Flexible in All Directions for the Following Void Fractions: for Tube 1, + zero, ◇ 20%, □ 40%, △ 50%, x 60%; ◆ 80%, ■ 90%.

spite of the larger flow velocity. The smaller turbulence scale also reduces the spatial correlation of the excitation forces, decreasing further the vibration response. The effect of intermittent flow on random excitation forces is discussed in detail in Chapter 10.

8.9.4 Single Flexible Column and Central Cluster Fluidelastic Instability Results

The results with an axisymetrically flexible single column are shown in Fig. 8-36. This figure shows the rms response in the lift direction of Tube 4 for all void fractions below 80% and of Tube 7 for void fractions of 80% and above. The response of both tubes in the drag direction is negligible compared to the lift direction. This difference in response was also observed for the other tubes within the column. As discussed in Violette et al (2006), there is significant difference in the instability results between the central-cluster and single-flexible-column configurations for tubes that are axisymmetrically flexible. However, the tendency of the instability to develop only for downstream tubes at high void fraction (80%, 90%, and 95%) that was observed for the central cluster is emphasized for the single flexible column. Table 8-6 shows that, for all void fractions tested, the critical flow pitch velocities are similar to those obtained for the central-cluster configuration.

When the single column was tested with the tubes flexible in only the flow direction, instability did not occur (similar to the single-flexible-tube results). Since instability did not occur, these cases are not shown in Table 8-6.

8.9.5 Two Partially Flexible Columns.

Tests were done with the flexible tubes placed over two adjacent columns to get a configuration that is situated between a single column and a cluster of tubes. The results obtained for this configuration are shown in Fig. 8-37, which shows the rms response of Tube 7 with pitch flow velocity. Figure 8-37 and Table 8-6 show that instability occurred at slightly greater velocities than for the central-cluster configuration, i.e., 8 m/s for 80%, 10.5 m/s for 90% and 14 m/s for 95% void

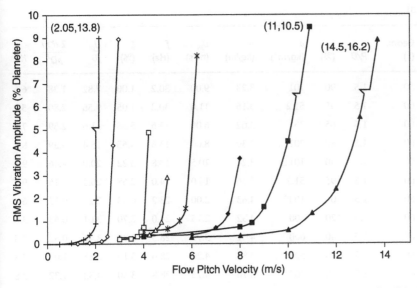

Fig. 8-36 Response in Lift Direction versus Flow Pitch Velocity for the Single Flexible Column Configuration, Tube Flexible in All Directions for the Following Void Fractions: for Tube 4, + zero, ◊ 20%, ☐ 40%, △ 50%, * 60%; for Tube 7 ♦ 80%, ■ 90%, ▲95%.

Table 8-6 Summary of Air-Water Test Conditions and Results (Violette, 2005).

Tube Flexibility	Geom. (°)	P/D	ε_g (%)	ρ (kg/m³)	m (kg/m)	U_{pc} (m/s)	f (Hz)	ζ (%)	$\frac{U_{pc}}{fD}$	$\frac{2\pi\zeta m}{\rho D^2}$	K
Single tube flexible in all directions	60	1.5	0	1000	4.63	2.00	24.8	0.63	2.12	0.13	6.0
	60	1.5	20	800	4.32	2.40	25.6	2.70	2.46	0.63	3.1
	60	1.5	40	600	4.01	3.75	27.2	2.80	3.62	0.81	4.0
	60	1.5	50	500	3.85	4.50	28.2	3.10	4.20	1.03	4.1
	60	1.5	60	400	3.70	5.50	28.6	3.30	5.05	1.32	4.4
	60	1.5	80	200	3.39	9.50	29.9	2.00	8.34	1.47	6.9
	60	1.5	90	101	3.23	12.0	30.8	1.00	10.2	1.38	8.7
One column of tubes flexible in all directions	60	1.5	0	1000	4.63	2.60	22.9	0.63	2.29	0.13	6.5
	60	1.5	20	800	4.32	2.50	23.7	2.70	2.77	0.63	3.5
	60	1.5	40	600	4.01	4.00	28.2	2.80	3.72	0.81	4.1
	60	1.5	50	500	3.85	4.50	29.3	3.10	4.03	1.03	4.0
	60	1.5	60	400	3.70	5.75	29.0	3.30	5.20	1.32	4.5
	60	1.5	80	200	3.39	7.50	30.2	2.00	6.52	1.47	5.4

(Continued)

Table 8-6 (Continued)

Tube Flexibility	Geom. (°)	P/D	ε_g (%)	ρ (kg/m³)	m (kg/m)	U_{pc} (m/s)	f (Hz)	ζ (%)	$\dfrac{U_{pc}}{fD}$	$\dfrac{2\pi\zeta m}{\rho D^2}$	K
	60	1.5	90	101	3.23	9.00	30.2	1.00	7.82	1.38	6.6
	60	1.5	95	51.2	3.16	11.0	30.2	1.05	9.56	2.80	5.7
Two columns of tubes flexible in the flow direction	60	1.5	65	351	3.62	6.00	13.6	5.58	11.6	2.59	7.2
	60	1.5	80	201	3.39	8.00	13.6	4.65	15.4	3.39	8.4
	60	1.5	90	1011	3.24	10.5	13.8	3.22	20.0	4.46	9.5
	60	1.5	95	51.2	3.16	14.0	14.0	2.68	26.3	7.15	9.8
Two columns of tubes flexible in all directions	60	1.5	0	1000	4.63	2.00	23.2	0.63	2.26	0.13	6.4
	60	1.5	20	800	4.32	2.75	24.0	2.70	3.01	0.63	3.8
	60	1.5	40	600	4.01	3.75	28.5	2.80	3.46	0.81	3.8
	60	1.5	50	500	3.85	4.20	28.6	3.10	3.85	1.03	3.8
	60	1.5	60	400	3.70	4.75	28.8	3.30	4.33	1.32	3.8
	60	1.5	80	200	3.39	7.00	30.1	2.00	6.10	1.47	5.0
	60	1.5	90	101	3.23	8.25	30.4	1.00	7.12	1.38	6.1
	60	1.5	95	51.2	3.16	10.5	30.4	1.05	9.07	2.80	5.4
Cluster of tubes flexible in the flow direction	60	1.5	65	351	3.62	6.00	13.3	5.58	11.8	2.59	7.3
	60	1.5	80	201	3.39	7.75	13.8	4.65	14.8	3.39	8.0
	60	1.5	90	101	3.24	9.50	13.9	3.22	17.9	4.46	8.5
	60	1.5	95	51.2	3.16	13.5	14.0	2.68	25.3	7.15	9.5
	60	1.5	85	151	3.31	13.0	27.8	3.40	12.3	3.23	6.8
	60	1.5	90	101	3.23	14.5	27.7	3.15	13.8	4.36	6.6
Cluster of tubes flexible in all directions	60	1.5	0	1000	4.63	2.00	23.0	0.63	2.28	0.13	6.4
	60	1.5	20	800	4.32	2.75	23.8	2.70	3.03	0.63	3.8
	60	1.5	40	600	4.01	4.00	28.6	2.80	3.67	0.81	4.1
	60	1.5	50	500	3.85	4.50	28.8	3.10	4.10	1.03	4.0
	60	1.5	60	400	3.70	5.50	28.8	3.30	5.01	1.32	4.4
	60	1.5	80	200	3.39	7.75	29.9	2.00	6.80	1.47	5.6
	60	1.5	90	101	3.23	9.50	30.4	1.00	8.20	1.38	7.0
	60	1.5	95	51.2	3.16	10.5	30.4	1.05	9.07	2.80	5.4

fraction. For 65% void fraction, clear instability was not obtained before the operational flow limit of the experimental set-up was reached. This limit is approximately 7 m/s. It is reasonable to assume that the fluidelastic instability would have been reached at a flow velocity near this limit. The unstable vibration modes are practically identical to those obtained for the central-cluster configuration.

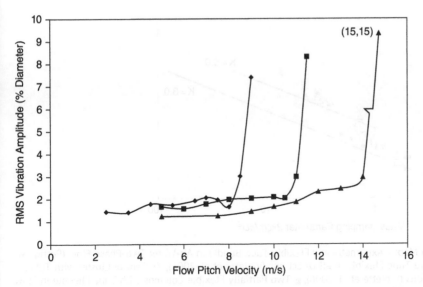

Fig. 8-37 Response of Tube 7 versus Flow Pitch Velocity for the Partially Flexible Columns Case, Tubes Flexible in Flow Direction for the Following Void Fractions: ◆ 80%, ■ 90%, ▲95%.

8.9.6 In-Flow Fluidelastic Instability Results and Discussion.

The tests confirmed that fluidelastic instability can occur in a rotated-triangular tube bundle flexible only in the flow direction when subjected to high void fraction two-phase cross flow. The fluidelastic test results reported in the previous section are compared to existing data for the rotated-triangular configuration (Pettigrew et al, 1989) on the instability diagram of Fig. 8-38. On this diagram, the abscissa is the mass-damping parameter and the ordinate is the reduced velocity. The constant K is the proportionality constant defined by Eq. (8-5), as follows: $U_{pc}/fD = K\sqrt{2\pi\zeta m/\rho D^2}$.

Results obtained for the axisymetrically flexible tube with a mass-damping parameter value less than approximately 1.5 are in good agreement with those reported by Pettigrew et al, 1989b. These results collapse very well around the $K = 3$ line. However, above 1.5, there is a jump in the reduced-velocity value.

From the results obtained for the tube flexible only in the flow direction, it can be concluded that several flexible tubes are required for fluidelastic instability to occur (multiple degrees of freedom system). Since the damping-controlled mechanism needs only one degree of freedom to cause instability, it can be deduced that the stiffness-controlled mechanism produced the instability for the tubes flexible only in the flow direction.

8.10 Design Recommendations

8.10.1 Design Guidelines

Figure 8-14 leads to a design guideline for fluidelastic instability of the form introduced in Eq. (8-5)

$$U_{pc}/fD = K\sqrt{2\pi\zeta m/\rho D^2}$$

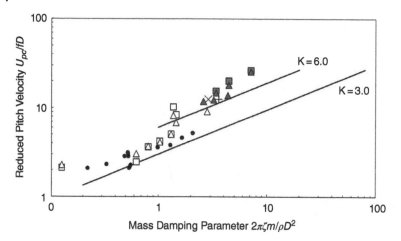

Fig. 8-38 Instability Map: • Axisymetrically Flexible Tube Bundle in Air-Water Two-Phase Flow (Pettigrew et al, 1989b), □ Single Tube Flexible in all Directions (Violette et al, 2006), △ Central Cluster with Tubes Flexible in all Directions (Violette et al, 2006), ■ Two Partially Flexible Columns with Tube Flexible in Flow Direction (Violette et al, 2006), ▲ Central Cluster with Tubes Flexible in Flow Direction (Violette et al, 2006), + Wind Tunnel Result for the Central Cluster with Tubes Flexible in Flow Direction (Mureithi et al, 2005), and × Wind Tunnel Results for a Single-Flexible-Column Configuration with Tubes Flexible in Flow Direction (Mureithi et al, 2005).

where a minimum value of $K = 3$ may be used as a general criterion and $K = 4$ may be used for square bundles. This applies to tube bundles of P/D between 1.4 and 1.5 in continuous two-phase cross flow (bubbly or froth liquid-vapor mixtures). This formulation requires damping values for two-phase flow conditions. The damping information given in Chapter 6 may be used for this purpose.

The foregoing recommendation would be conservative for bundles of P/D ratios larger than 1.5. On the other hand, lower values of K must be used for P/D ratios smaller than 1.4.

To avoid the question of two-phase damping, the data and curve of Fig. 8-13 may be used as a design guideline, although this approach is clearly not as desirable. This criterion may be formulated quite simply by

$$U_{pc}/fD = 5, \quad \text{for} \quad m/\rho D^2 \geq 7 \tag{8-9a}$$

and

$$U_{pc}/fD = 0.7m/\rho D^2, \quad \text{for} \quad m/\rho D^2 < 7 \tag{8-9b}$$

as shown in Fig. 8-13. The latter guideline should not be used much beyond the range of parameters (i.e., tube mass, diameter, flow conditions, etc.) for which the experimental data has been obtained. In practice, U_{pc}/fD, should not exceed the values calculated with Eq. (8-9) in a well-designed heat exchanger or nuclear steam generator. The foregoing formulation, although convenient, has no physical significance. In Eq. (8-9a), the fact that U_{pc}/fD is independent of $m/\rho D^2$ simply means that, at higher void fractions, damping decreases as $m/\rho D^2$ increases, thus giving the erroneous impression that $m/\rho D^2$ has a relatively weak influence on fluidelastic instability.

8.10.2 Fluidelastic Instability with Intermittent Flow

The fluidelastic instability behavior of tube bundles subjected to intermittent two-phase flow is not well understood. Intermittent flow should be avoided in heat exchanger design. Aside from being undesirable from a thermal performance point of view, it leads to fluidelastic instability at significantly lower mean flow velocity. The values of K and n given in Fig. 8-7 for intermittent flow should not be used for design purposes. They are given merely to emphasize the very significant difference between continuous and intermittent flow. Intermittent flow is characterized by a cycle comprising reflooding, transient, burst of two-phase mixture and transient back to reflooding. One part of this cycle is more prone to fluidelastic instability. If intermittent flow cannot be avoided in a given heat exchanger design, the fluidelastic instability analysis should be based on the most severe part of this cycle. This recommendation also applies to any components where flow oscillations occur, which is not unusual in two-phase flow.

Example 8-1   Fluidelastic Instability in a Steam Generator U-Bend

Fluidelastic Instability Parameters:

From previous calculations, temperature of 275°C for $\varepsilon_g = 85.6\%$.

From Example 3-2: $D = 0.02$ m, $P/D=1.5$, $\rho_{TP} = 137$ kg/m^3, $U_p = 5.09$ m/s.

From Example 4-2: $m = 0.791$ kg/m.

From Example 4-4: $f = 118$ Hz, $f_{ineffective} = 29.5$ Hz.

From Example 6-1: $\zeta = 1.91\%$ (U-bend $\zeta_{ineffective} = 2.94\%$).

Fluidelastic Instability Calculation:

Using Eq. (8-5) and K = 3 (from Fig. 8-14 or Fig. 8-24), we can calculate the critical pitch velocity,

$$\frac{U_{pc}}{fD} = K\left(\frac{2\pi\zeta_T m}{\rho D^2}\right)^{\frac{1}{2}} = 3.0\left(\frac{2\pi \times 0.0191 \times 0.791}{137 \times 0.02^2}\right)^{0.5} = 3.95$$

$$U_{pc} = 3.95\, fD = 3.95 \times 118 \times 0.02 = 9.32 \text{ m/s}$$

To determine if the U-bend tubes will experience instability, the critical pitch velocity is compared to the actual pitch velocity,

$\dfrac{U_p}{U_{pc}} = \dfrac{5.09}{9.32} = 0.55 < 1.0$. Fluidelastic instability is unlikely.

However, if every 2nd support is ineffective:

$$\frac{U_{pc}}{fD} = K\left(\frac{2\pi\zeta_T m}{\rho D^2}\right)^{\frac{1}{2}} = 3.0\left(\frac{2\pi \times 0.0294 \times 0.791}{137 \times 0.02^2}\right)^{0.5} = 4.90$$

$$U_{pc} = 4.90\, fD = 4.90 \times 29.5 \times 0.02 = 2.89 \text{ m/s}$$

$\dfrac{U_p}{U_{pc}} = \dfrac{5.09}{2.89} = 1.76 > 1.0$. Fluidelastic instability is likely to occur, since the pitch velocity is significantly larger than the critical pitch velocity.

8.11 Fluidelastic Instability in Two-Phase Axial Flow

A short introduction to fluidelastic instability in axial flow is provided in Section 7.10. Some early work on fluidelastic instability in axial two-phase flow is presented in Pettigrew and Païdoussis (1975) and Païdoussis and Pettigrew (1979) and remains a substantial portion of the work available in this area. In this work, flexible cylinders were subjected to both liquid and two-phase axial flow. The effects of several parameters such as flow velocity, void fraction and flexural rigidity were explored. The cylinders behaved as expected in liquid flow (see Section 7.10). The dynamic behavior of the cylinders was completely different in two-phase axial flow.

In spite of some experimental effort, it was not possible to obtain fluidelastic instabilities. The vibration response of all cylinders was severe and broadband in nature. Typical results in liquid and two-phase axial flow are shown in Fig. 8-39. The first two parts of the figure show liquid flow results. Figure 8-39a shows a typical flutter spectrum, while Fig. 8-39b shows a spectrum of combined buckling and free-end vibration. In Fig. 8-39c, the two-phase flow results show no sign of fluidelastic buckling or oscillations at discrete frequencies and no well-defined mode of vibration.

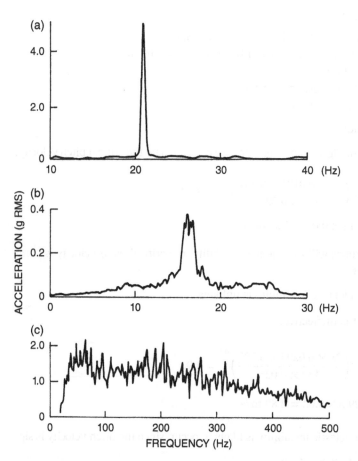

Fig. 8-39 Selected Frequency Spectra for Fluidelastic Instability of Clamped-Free Flexible Cylinder in Axial Flow (Pettigrew and Païdoussis, 1975): a) Oscillations in Liquid Flow, $U = 4.26$ m/s, b) Buckling and Free-End Vibration in Liquid Flow, $U = 3.05$ m/s, and c) Broadband Vibration Response in Two-Phase Air-Water Flow, $\dot{m} = 2700$ kg m^{-2}s^{-1}, $\varepsilon_g = 15\%$.

The current theory does not predict the observed behavior. A possible explanation is that the hydrodynamic mass is small in two-phase flow and that hydrodynamic coupling may be minimal because of an annular-type of flow regime. The relationship shown in Eq. (7-13) suggests stability if the hydrodynamic mass is effectively very small. Other possible reasons are the high fluid damping in two-phase mixtures and the high turbulence level, which may inhibit the development of instability. Obviously, this area requires further study and no guideline has been developed. Fortunately, this work suggests that fluidelastic behavior is less likely in two-phase flow than in single-phase flow and we know that the flow velocities in most industrial components are much lower than those required for fluidelastic instability in single-phase axial flow conditions.

8.12 Concluding Remarks

The available data on fluidelastic instability in tube bundles subjected to two-phase cross flow has been reviewed. This data was compared against the existing expressions to formulate fluidelastic instability. From a practical design point of view, it is suggested that:

- Two-phase fluidelastic instability may be expressed in terms of dimensionless flow velocity, U_p/fD, and dimensionless mass-damping, $2\pi\zeta m/\rho D^2$.
- A simple expression of the form $\dfrac{U_{pc}}{fD} = K\left(\dfrac{2\pi\zeta m}{\rho D^2}\right)^{0.5}$ is appropriate for design purposes in two-phase flow.
- A two-phase fluidelastic instability constant $K = 3.0$ is a reasonable design criterion for both square and triangular tube bundles.
- Two-phase damping should be included in the mass-damping parameter.
- At the present time, there is no advantage to considering more sophisticated fluidelastic instability models.
- In-flow fluidelastic instability for a single flexible tube would not occur. Thus, an in-flow instability would require having at least two adjacent tubes essentially unsupported at the same location. In this case, the instability velocity is twice as high as that for the case of a fully flexible array of tubes. The current design guideline is also suitable for in-flow fluidelastic instability.
- Two-phase axial-flow fluidelastic instability has not been seen in experiments

Nomenclature

D = tube diameter, m

f = frequency, Hz

K = fluidelastic instability constant

m = total mass per unit length ($m_t + m_h$), kg/m

$\dot{m}$ = mass flux, kg/(m^2·s)

n = mass-damping exponent

P = tube pitch, m

$S_{12}(f)$ = cross-power spectral density function, (N/m)^2·s

$S_{11}(f)$ = power spectral density function of the vibration response of Tube 1, (N/m)^2·s

$S_{22}(f)$ = power spectral density function of the vibration response of Tube 2, (N/m)^2·s

U = fluid velocity, m/s

U_g^* = dimensionless gas velocity, m/s

$\dot{V}$ = volumetric flow rate, m^3/s

X = Martinelli parameter

γ_{12} = coherence function between Tubes 1 and 2

ε_g = void fraction

ζ = damping ratio

ρ = fluid density, kg/m^3

Subscripts

c = critical

g = gas or vapor

ℓ = liquid

p = pitch

r = reference gap (equal to pitch)

∞ = free stream

Note

1 CANDU = CANada Deuterium Uranium is a registered trademark of AECL.

References

Axisa, F., Villard, B., Gibert, R. J., Hetsroni, G., and Sundheimer, P., 1984, "Vibration of Tube Bundle Subjected to Air-Water and Steam-Water Cross Flow: Preliminary Results on Fluidelastic Instability," *Proceedings of ASME Symposium of Flow-Induced Vibration*, Vol. 2, New Orleans, LA, Dec., pp. 269–284.

Axisa, F., Boheas, M.A. and Villard, B., 1985, "Vibration of Tube Bundles Subjected to Steam-Water Cross-Flow: A Comparative Study of Square and Triangular Arrays," *8th International Conference on Structural Mechanics in Reactor Technology*, Brussels, Belgium, Aug., Paper No. B 1/2.

Axisa, F., Wullschleger, M., Villard, B. and Taylor, C., 1988, "Two-Phase Cross Flow Damping in Tube Arrays," *Proceedings of ASME Pressure Vessel and Piping Conference*, Pittsburg, USA, PVP Vol. 133, Damping 1988, pp. 9–15.

Boucher, K. M. and Taylor, C. E., 1996, "Tube Support Effectiveness and Wear Damage Assessment in the U-Bend Region of Nuclear Steam Generators," *ASME PVP-Vol. 328, Flow-Induced Vibration*, pp. 285–296.

Carlucci, L. N., 1980, "Damping and Hydrodynamic Mass of a Cylinder in Simulated Two-Phase Flow," *Journal of Mechanical Design*, 102, pp. 597–602.

Carlucci, L. N. and Brown, J. D., 1983, "Experimental Studies of Damping and Hydrodynamic Mass of a Cylinder in Confined Two-Phase Flow," *ASME Journal of Vibration, Stress and Reliability in Design*, 105, pp. 83–89.

Connors, H. J., 1970, "Fluidelastic Vibration of Tube Arrays Excited by Cross-Flow," *Proceedings of the Symposium on Flow-Induced Vibration in Heat Exchangers*, ASME Winter Annual Meeting, New York, Dec., pp. 42–56.

Dallavalle, F., Rossini, T. and Vanoli, E., 1973, "Vibrations Induced by a Two-Phase Parallel Flow to a Rod Bundle: Preliminary Experiments on the Surface Tension and Gas Density Effects," *Proceedings from UKAEA/NPL International Symposium on Vibration Problems in Industry*, Paper 525, Keswick.

Gay, N., Decembre, P. and Launay, J., 1988, "Comparison of Air-Water to Water-Freon Two-Phase Cross-Flow Effects on the Vibration Behaviour of a Tube Bundle," *ASME-WAM 1988 International Symposium on Flow-Induced Vibration and Noise*, (M. P. Païdoussis, principal editor), Chicago, Vol. 3, pp. 139–158.

Gorman, D. J., 1971, "An Analytical and Experimental Investigation of the Vibration of Cylindrical Reactor Fuel Elements in Two-Phase Parallel Flow," *Nuclear Science and Engineering*, 44, pp. 277–290.

Gorman, D. J., 1977, "Experimental Study of Peripheral Problems Related to Liquid Flow Induced Vibration of Heat Exchangers and Steam Generators," *Trans. of 4th International Conference on Structural Mechanics in Reactor Technology*, Paper F6/2, San Francisco, Calif., Aug.

Hara, F., 1980, "Two-Phase Flow-Induced Parametric Vibrations in Structural Systems: Pipes and Nuclear Fuel Pins," *Report of the Institute of Industrial Science*, 28(4), University of Tokyo.

Hara, F., 1982, "Two-Phase Cross Flow-Induced Forces Acting on a Circular Cylinder," *ASME Symposium on Flow-Induced Vibration of Circular Cylindrical Structures*, PVP-Vol. 63 (S. S. Chen, M. P. Païdoussis and M. K. Au-Yang editors), Orlando, pp. 9–17.

Hara, F. and Yamashita, T., 1978, "Parallel Two-Phase Flow-Induced Vibrations in Fuel Pin Model," *Journal of Nuclear Science Technology*, 15, pp. 346–354.

Heilker, W., and Vincent, R. Q., 1981, "Vibration in Nuclear Heat Exchangers Due to Liquid and Two-Phase Flow," *ASME Journal of Engineering for Power*, 103, pp. 358–365.

Janzen, V. P., Hagberg, E. G., Pettigrew, M. J. and Taylor, C. E., 2005, "Fluidelastic Instability and Work-Rate Measurements of Steam Generator U-Tubes in Air-Water Cross-Flow," *Journal of Pressure Vessel Technology*, 127, pp. 84–91.

Jong, J. H., Currie, I. G., Pettigrew, M. J. and Taylor, C. E., 1994, "Fluidelastic Instability of a Tube Bundle in Two-Phase Freon-22 Cross Flow," *Proceedings of the 12th Symposium on Engineering Applications of Mechanics, Interactions of Fluids, Structures and Mechanics*, June 27-29, McGill University, Montreal, pp. 425–434.

Lever, J. H., and Weaver, D. S., 1982, "A Theoretical Model for Fluidelastic Instability in Heat Exchanger Tube Bundles," *ASME Journal of Pressure Vessel Technology*, 104, pp. 147–158.

Mureithi, N. W., Zhang, C., Ruël, M. and Pettigrew, M. J., 2005, "Fluidelastic Instability Tests on an Array of Tubes Preferentially Flexible in the Flow Direction," *Journal of Fluids and Structures*, 21(1), pp. 75–87.

Nakamura, T., Yamaguchi, N., Fujita, K., Sakata, K., and Saito, I., 1986a, "Study on Flow Induced Vibration of a Tube Array by a Two-Phase Flow (1st Report: Large Amplitude Vibration by Air-Water Flow)," *Trans. of Japanese Society of Mechanical Engineers*, 52, (473), C.

Nakamura, T., Fujita, K., Kawanishi, K., and Saito, 1986b, "A Study on the Flow-Induced Vibration of a Tube Array by a Two-Phase Flow (2nd Report: Large Amplitude Vibration by Steam-Water Flow)," *Trans. of Japanese Society of Mechanical Engineers*, 52, C.

Nakamura, T., Fujita, K., Kawanishi, K., Yamaguchi, N. and Tsuge, A., 1991, "Study on the Vibrational Characteristics of a Tube Array Caused by Two-Phase Flow—Part 2 - Fluidelastic Vibration," *ASME Symposium on Flow-Induced Vibration and Wear 1991*, PVP-Vol. 206 (M. K. Au-Yang and F. Hara editors), San Diego, pp. 25–30.

Païdoussis, M. P. and Pettigrew, M. J., 1979, "Dynamics of Flexible Cylinders in Axisymmetrically Confined Flow," *Journal of Applied Mechanics*, 46, pp. 37–44.

Pettigrew, M. J., 1993, "The Vibration Behaviour of Nuclear Fuel under Reactor Conditions," *Journal of Nuclear Science and Engineering*, 114, pp. 179–189.

Pettigrew, M. J. and Gorman, D. J., 1973a, "Experimental Studies on Flow-Induced Vibration to Support Steam Generator Design, Part I: Vibration of a Heated Cylinder in Two-Phase Axial Flow," *Proceedings of the UKAEA/NPL International Symposium on Vibration Problems in Industry*, Keswick, UK, Paper 424.

Pettigrew, M. J., and Gorman, D. J., 1973b, "Experimental Studies on Flow-Induced Vibration to Support Steam Generator Design, Part II: Vibration of Small Tube Bundles in Liquid and Two-Phase Cross Flow," *International Symposium on Vibration Problems in Industry*, Paper No. 426, Keswick, UK, also, *Atomic Energy of Canada Limited Report, AECL-5804*.

Pettigrew, M. J. and Païdoussis, M. P., 1975, "Dynamics and Stability of Flexible Cylinders Subjected to Liquid and Two-Phase Axial Flow in Confined Annuli," *Transactions of the 3rd International Conference on Structural Mechanics in Reactor Technology*, Paper No D2/6, London, UK.

Pettigrew, M. J., and Gorman, D. J., 1978, "Vibration of Heat Exchange Components in Liquid and Two-Phase Cross-Flow," *International Conference on Vibration in Nuclear Plant*, Paper 2.3, Keswick, UK.

Pettigrew, M. J. and Taylor, C. E., 1991, "Fluidelastic Instability of Heat Exchanger Tube Bundles: Review and Design Recommendations," *ASME Journal of Pressure Vessel Technology*, 113(2), pp. 242–256.

Pettigrew, M. J. and Taylor, C. E., 1994, "Two-Phase Flow-Induced Vibration: An Overview," *ASME Journal of Pressure Vessel Technology*, 116, pp. 233–253.

Pettigrew, M. J. and Knowles, G. D., 1997, "Some Aspects of Heat Exchanger Tube Damping in Two-Phase Mixtures," *Journal of Fluids and Structures*, 11(8), pp. 929–945.

Pettigrew, M. J. and Taylor, C.E., 2009 "Vibration of a Normal Triangular Tube Bundle Subjected to Two-Phase Freon Cross Flow," *Journal of Pressure Vessel Technology*, 131, 051302 (1-7).

Pettigrew, M. J., Tromp, J. H., and Mastorakos, J., 1985, "Vibration of Tube Bundles Subjected to Two-Phase Cross-Flow," *ASME Journal of Pressure Vessel Technology*, 107, pp. 335–343.

Pettigrew, M. J., Taylor, C. E. and Kim, B. S., 1989a, "Vibration of Tube Bundles in Two-Phase Cross-Flow - Part 1: Hydrodynamic Mass and Damping," *ASME Journal of Pressure Vessel Technology*, 111, pp. 466–477.

Pettigrew, M. J., Kim, B. S., Taylor, C. E. and Tromp, J. H., 1989b, "Vibration of Tube Bundles in Two-Phase Cross-Flow - Part 2: Fluidelastic Instability," *ASME Journal of Pressure Vessel Technology*, 111, pp. 478–487.

Pettigrew, M. J., Taylor, C. E., Jong, J. H., and Currie, I. G., 1995, "Vibration of a Triangular Tube Bundle in Two-Phase Freon Cross Flow," *ASME Journal of Pressure Vessel Technology*, 117, pp. 321–329.

Pettigrew, M. J., Taylor, C. E. and Kim, B. S., 2001, "The Effects of Bundle Geometry on Heat Exchanger Tube Vibration in Two-Phase Cross Flow," *Journal of Pressure Vessel Technology*, 123, pp. 414–420.

Pettigrew, M. J., Taylor, C. E., Janzen, V. P. and Whan, T., 2002, "Vibration Behavior of Rotated Triangular Tube Bundles in Two-Phase Cross Flows," *ASME Journal of Pressure Vessel Technology*, 124, pp. 144–153.

Pettigrew, M. J., Zhang, C., Mureithi, N. W. and Pamfil, D., 2005, "Detailed Flow Measurements in a Rotated Triangular Tube Bundle Subjected to Two-Phase Cross Flow," *Journal of Fluids and Structures*, 20(4), pp. 567–575.

Price, S. K., and Païdoussis, M. P., 1986, "A Constrained-Mode Analysis of the Fluidelastic Instability of Double Row of Flexible Circular Cylinders Subject to Cross-Flow: A Theoretical Investigation of System Parameters," *Journal of Sound and Vibration*, 105, pp. 121–142.

Remy, R. M., 1982, "Flow-Induced Vibration of Tube Bundles in Two-Phase Cross-Flow," Paper 1.9, *Proceedings of 3ʳᵈ Int'l Conference on Vibration in Nuclear Plants*, Vol. 1, Keswick, UK, pp. 135–160.

Remy, F. N., and Bai, D., 1982, "Comparative Analysis of Cross-Flow-Induced Vibrations of Tube Bundles," *International Conference on Flow Induced Vibrations in Fluid Engineering*, Reading, England, September 14-16.

Taylor, C. E., Pettigrew, M. J., Axisa, F. and Villard, B., 1988, "Experimental Determination of Single and Two-Phase Cross Flow Induced Forces on Tube Rows," *ASME Journal of Pressure Vessel Technology*, 110, pp. 22–28.

Taylor, C. E., Pettigrew, M. J., Currie, I. G. and Kim, B. S., 1989, "Vibration of Tube Bundles in Two-Phase Cross-Flow - Part 3: Turbulence Induced Excitation," *ASME Journal of Pressure Vessel Technology*, 111, pp. 488–500.

Taylor, C. E., Pettigrew, M. J. and Tromp, J. H., 1991, "Vibration of Steam Generator U-Bend Tubes: Effectiveness of Flat-Bar Restraints," *Proceedings of 1991 Pressure Vessels and Piping Conference*, (M. K. Au-Yang and F. Hara editors), San Diego, California, USA, June, PVP-Vol. 206, pp. 1–8.

Violette, R., 2005, "Instability Fluidélastique d'un Faiseau, de Tubes d'Échangeurs de Chaleur Soumis a un Écoulement Diphasique Transverse," Master's Thesis, École Polytechnique de Montréal, Montreal, Canada.

Violette, R., Pettigrew, M. J. and Mureithi, N. W., 2006, "Fluidelastic Instability of an Array of Tubes Preferentially Flexible in the Flow Direction Subjected to Two-Phase Cross Flow," *ASME Journal of Pressure Vessel Technology*, 128, pp. 148–159.

Weaver, D. S. and Schneider, W., 1983, "The Effect of Flat Bar Supports on the Cross-Flow Response of Heat Exchanger U-Tubes," *ASME Journal of Eng. Power*, 105, pp. 775–781.

Weaver, D. S., and Fitzpatrick, J. A., 1987, "A Review of Flow-Induced Vibrations in Heat Exchangers," Paper A1, *Proceedings of the International Conference on Flow-Induced Vibrations*, Bowness-on-Windermere, England, May 12–14, pp. 1–17.

9

Random Turbulence Excitation in Single-Phase Flow

Colette E. Taylor and Michel J. Pettigrew

9.1 Introduction

Many process components such as heat exchangers operate with a single-phase fluid. As well, tubes near the inlet region and preheater of two-phase heat exchangers, such as a nuclear steam generator, are exposed to single-phase cross flow. In single-phase cross flow, three major mechanisms can lead to tube failure in a shell-and-tube heat exchanger: fluidelastic instability, periodic wake shedding, and random excitation forces. Both fluidelastic instability and periodic wake shedding can cause large amplitude vibrations in a tube bundle and quickly lead to catastrophic tube failure. However, random excitation forces cause comparatively small-amplitude vibrations that will not lead to short-term failure. These vibrations do, however, lead to continuous rubbing of a tube against its supports, thereby resulting in progressive damage to the tube as a result of fretting wear. As reactors age, it is increasingly important to be able to assess the effects of random excitation vibration to determine the maximum safe operating lifetime for each component. As well, the production of more reliable and longer lasting steam generators requires an accurate assessment of tube response to random excitation during the design process. To this end, it is necessary to have accurate design guidelines.

Although a great deal of experimental and theoretical work on flow-induced vibration is available, it is widely agreed that design-oriented information focusing on random excitation vibration is insufficient. In preparation for this chapter, a database containing most of the available single-phase random turbulence excitation data for tube bundles was compiled. Using the available data, this chapter presents a guideline for determining the random excitation forces in tube bundles subjected to single-phase cross flow.

9.2 Theoretical Background

It is now generally accepted that all flow-induced excitation mechanisms fall into three mathematically distinct classifications: 1) forced; 2) self-excited; and, 3) self-controlled. When referring to a vibrating cylinder, the mechanisms that fall into these categories are commonly known as:

1) Turbulent buffeting (also referred to as random turbulence, turbulence-induced excitation and random fluctuating forces);
2) Fluidelastic instability; and,
3) Vortex shedding (also referred to as periodic wake shedding) and acoustic resonance.

Flow-Induced Vibration Handbook for Nuclear and Process Equipment, First Edition.
Michel J. Pettigrew, Colette E. Taylor, and Nigel J. Fisher.
© 2022 John Wiley & Sons, Inc. This Work is a co-publication between ASME Press and John Wiley & Sons, Inc.

Forced vibration is an equilibrium problem. In a generalized coordinate system, the forcing function, $F(t)$, is a function of time and not a function of the structural motion. This functional relationship implies that systems subjected to forced vibration can be modelled by rigid bodies. The use of a rigid body can be a significant simplification in experimental work.

Self-excited vibration is an eigenvalue problem. In this case, the forcing function depends upon the structural displacement. This mechanism will not occur if the structure is rigid. This is a stability problem that is best avoided because the amplitudes become very large and damage is certain (see Chapters 7 and 8).

Self-controlled vibration occurs if there is periodicity in the flow. If this periodicity is not at the tube natural frequency, the amplitudes will generally be small, and one is faced with a basic linear equilibrium problem. When the frequency of the flow periodicity coincides with a tube natural frequency, the tube deflection will increase, causing the equilibrium problem to become nonlinear and the forcing function to become amplitude dependent (see Chapter 11).

This chapter is concerned with low-amplitude forced vibrations. The derivation and simplification of the forced vibration equations for randomly fluctuating fluid forces are given in Sections 9.2.1 through 9.2.4. This derivation was developed using background information from Blevins (1990), Thomson (1981) and especially from Meirovitch (1967).

In the development of design guidelines for random vibration, it is not possible to directly measure the fluid forces. As a result, the forces are deduced from the effect they have on tubes subjected to fluid flow. In some experimental work, the tubes were allowed to vibrate and vibration amplitudes were measured using strain gages. In other work, the tubes were kept rigid and reaction forces on the tube were measured. In either case, it is necessary to understand random vibration theory and the simplifications that can be assumed due to the nature of flow in the experimental test sections.

9.2.1 Equation of Motion

A heat exchanger tube subjected to homogeneous cross flow is modelled as a continuous system (beam) subjected to random excitation. The equation of motion of a uniform beam in bending (no tension term) with forced excitation can be expressed as follows for a tube subjected to cross flow:

$$m\frac{\partial^2 y(x,t)}{\partial t^2} + c\frac{\partial y(x,t)}{\partial t} + EI\frac{\partial^4(x,t)}{\partial x^4} = F(x,t) \tag{9-1}$$

where, $0 \leq x \leq L$

m is the total mass (tube + fluid added mass) per unit length, c is the total damping coefficient, E is Young's modulus of elasticity for the tube material,

I is the tube area moment of inertia,

$y(x,t)$ is the tube displacement normal to the x-direction, and $F(x,t)$ is the flow excitation force per unit length. Defining $y(x,t)$ as the sum of the displacements over all modes of vibration and introducing generalized coordinates, one obtains:

$$y(x,t) = \sum \phi_i(x)q_i(t) \tag{9-2}$$

where, $\phi_i(x)$ is the normalized mode shape defined as $\int_0^L \phi_i^2(x)dx = L$, and $q_i(t)$ is the generalized coordinate.

The resulting equation of motion is

$$\ddot{q}_i(t) + 2\zeta_i\omega_i\dot{q}(t) + \omega_i^2 q_i(t) = F_i(t) \tag{9-3}$$

where,

$$F_i(t) = \frac{1}{M_i L}\int_0^L \phi_i^2(x)F(x,t)dx,$$

$$M_i = \frac{1}{L}\int_0^L \phi_i^2(x)m(x)dx$$

$$\zeta_i = \frac{1}{2m\omega_i L}\int_0^L \phi_i^2(x)c(x)dx$$

$F_i(t)$ is the generalized force per unit length for the ith mode,
M_i is the generalized mass per unit length for the ith mode,
ζ_i is the generalized damping per unit length for the ith mode,
ω_i is the radial frequency for the ith mode, $\omega_i = (\lambda_i/L)^2(EI/M_i)^{0.5}$,
λ_i is the eigenvalue for the ith mode, and
L is the tube length.

If the length of the tube subjected to flow, L_e, is not equal to L, the integration limits must reflect the excited length for the terms which are fluid dependent. Also note that $M_i = m(x)$ and $2m\omega_i\zeta_i = c(x)$ if m and c are not functions of x. These assumptions are suitable for the tubes and flows used in test sections.

9.2.2 Derivation of the Mean-Square Response

The cross-correlation function of the tube displacement at positions x and x' along the tube can be defined as

$$\overline{y(x,t)y(x',t+\tau)} = \lim_{T\to\infty}\frac{1}{T}\int_{-\frac{T}{2}}^{\frac{T}{2}} y(x,t)y(x',t+\tau)dt \tag{9-4}$$

Combining Eqs. (9-2) and (9-4) and taking subscripts r and s to represent different modes, we obtain

$$\overline{y(x,t)y(x',t+\tau)} = \lim_{T\to\infty}\frac{1}{T}\int_{-\frac{T}{2}}^{\frac{T}{2}}\left[\sum_{r=1}^{\infty} q_r(t)\phi_r(x)\right]\left[\sum_{r=1}^{\infty} q_s(t+\tau)\phi_s(x')\right]dt$$

$$= \sum_{r=1}^{\infty}\sum_{s=1}^{\infty}\phi_r(x)\phi_s(x')\lim_{T\to\infty}\frac{1}{T}\int_{-\frac{T}{2}}^{\frac{T}{2}} q_r(t)q_s(t+\tau)dt \tag{9-5}$$

Using Parseval's Theorem to introduce the cross-spectral density, $S_{q_r q_s}(f)$, letting $x = x'$ and setting $\tau = 0$, the mean-square response becomes

$$\overline{y^2(x)} = \sum_{r=1}^{\infty}\sum_{s=1}^{\infty}\phi_r(x)\phi_s(x)\int_{-\infty}^{\infty} S_{q_r q_s}(f)df \tag{9-6}$$

A transfer function, $H_i(f)$, can be defined in the frequency domain such that the input-force cross-spectral density, S_{FF}, is related to the cross-spectral density of the displacements $q_r(t)$ and $q_s(t)$ as follows:

$$S_{q_r q_s}(f) = \frac{1}{\omega_r^2 \omega_s^2} H_r(f) H_s^*(f) S_{F_r F_s}(f) \tag{9-7}$$

Where, $H_i(f) = \dfrac{1}{1 - (f/f_i)^2 + i2\zeta_i(f/f_i)}$, and f_i is the tube fequency for the ith mode.

Therefore,

$$\overline{y^2(x)} = \sum_{r=1}^{\infty} \sum_{s=1}^{\infty} \phi_r(x)\phi_s(x) \int_{-\infty}^{\infty} \frac{1}{\omega_r^2 \omega_s^2} H_r(f) H_s^*(f) S_{F_r F_s}(f) df \tag{9-8}$$

Using the definition for the cross-spectral density function, one obtains

$$S_{F_r F_s}(f) = \int_{-\infty}^{\infty} \left[\lim_{T \to \infty} \frac{1}{T} \int_{-\frac{T}{2}}^{\frac{T}{2}} F_r(t) F_s(t+\tau) dt \right] e^{-j2\pi f \tau} d\tau \tag{9-9}$$

$$= \int_{-\infty}^{\infty} \left[\lim_{T \to \infty} \frac{1}{T} \int_{-\frac{T}{2}}^{\frac{T}{2}} \frac{1}{M_r M_s L^2} \left[\int_0^L \phi_s(x') F(x', t+\tau) dx' \left[\int_0^L \phi_r(x) F(x, t) dx \right] \right] dt \, e^{-j2\pi f \tau} d\tau \right.$$

$$= \frac{1}{M_r M_s L^2} \int_0^L \phi_s(x') \int_0^L \phi_r(x) \left[\int_{-\infty}^{\infty} \lim_{T \to \infty} \frac{1}{T} \int_{-\frac{T}{2}}^{\frac{T}{2}} F(x,t) F(x', t+\tau) dt \, e^{-j2\pi f \tau} d\tau \right] dx dx'$$

If one applies, once more, the definition for the cross-spectral density, the expression within the square brackets in the final line of Eq. (9-9) is simply the cross-spectral density of the input forces, $S_{F_x F_{x'}}$. Combining Eq. (9-8) with the last line of Eq. (9-9), one obtains the equation for the mean-square tube deflection in terms of the cross-spectral density of the input forces.

$$\overline{y^2(x)} = \sum_{r=1}^{\infty} \sum_{s=1}^{\infty} \frac{1}{\omega_r^2 \omega_s^2 M_r M_s L^2} \phi_r(x) \phi_s(x)$$

$$\int_{-\infty}^{\infty} H_r(f) H_s^*(f) \left[\int_0^L \phi_s(x') \int_0^L \phi_r(x) S_{F_x F_{x'}}(f) dx dx' \right] df \tag{9-10}$$

The presence of a second position along the tube has been reintroduced to the equation. This is only significant in that $S_{F_x F_{x'}}$ is a cross-spectral density. The other x and x' variables should be considered as domains of integration.

9.2.3 Simplification of Tube Vibration Response

Some simplification of Eq. (9-10) is useful to facilitate interpretation of experimental data. Several assumptions concerning the nature of the input and response force spectra can be made:

1) The randomly fluctuating forces are homogeneous along the tube so that one-sided power spectral densities at any point along the tube are equal $S_{F_x}(f) = S_{F_{x'}}(f) = S_F(f)$. This assumption permits one to define the cross-spectral density as $2S_{F_x F_{x'}} = S_F(f)\gamma(x,x',f)$ where, $\gamma(x, x',f)$ is the correlation factor;
2) $S_F(f)$ is constant or at least broadband over the frequency range close to f_i;
3) $\gamma(x,x',f)$ is independent of frequency over the frequency bandwidth of the resonant peak;
4) The resonant peak dominates the response spectrum; and,

5) A single mode dominates the tube displacement, and the coupling between modes is negligible. This assumption is valid if the tube damping ratio is small and the modes are reasonably separated.

When these assumptions are applied to Eq. (9-10), one obtains:

$$\overline{y^2(x)}_i = \frac{\phi_i^2(x)S_F(f)J_i^2}{16\pi^4 f_i^4 M_i^2} \int_{-\infty}^{\infty} H_i(f)H_i^*(f)df \qquad (9\text{-}11)$$

where,

$$J_i^2 = \frac{1}{L^2}\int_0^L \phi_i(x')\int_0^L \phi_i(x)\gamma(x,x')dxdx'$$

and J_i^2 is the joint acceptance for the ith mode. The joint acceptance represents the distribution and effectiveness of the excitation forces relative to the ith mode.

9.2.4 Integration of the Transfer Function

The evaluation of the definite integral of the transfer function for the ith mode is made using the assumption that the transfer peak is only lightly damped ($\lambda \ll 1$).

$$\int_{-\infty}^{\infty} H_i(f)H_i^*(f)df = 2\int_0^{\infty} |H_i(f)|^2 df = 2\int_0^{\infty} \frac{df}{\left[1-\left(\frac{f}{f_i}\right)^2\right]^2 + 4\zeta_i^2\left(\frac{f}{f_i}\right)^2} \qquad (9\text{-}12)$$

Let $X = f/f_i$, then $dX = 1/f_i\, df$, and the integral becomes

$$\int_{-\infty}^{\infty} H_i(f)H_i^*(f)df = 2f_i\int_0^{\infty} \frac{dX}{(1-X^2)^2 + 4\zeta_i^2 X^2}$$

Let $X = 1 + \delta x$, then $dX = d(\delta x)$, and the terms of the integrand become

$$(1-X^2)^2 \Rightarrow [1-(1+\delta x)]^2 = [1-(1+\delta x)]^2[1+(1+\delta x)]^2$$
$$= 4(\delta X)^2 + 4(\delta X)^3 + (\delta X)^4$$
$$4\zeta_i^2 X^2 \Rightarrow 4\zeta_i^2(1+\delta X)^2 = 4\zeta_i^2 + 6\zeta_i^2\delta X + 4\zeta_i^2(\delta X)^2$$

If one keeps only the terms up to the second order, the integral becomes

$$2f_i\int_0^{\infty} \frac{d(\delta X)}{4(\delta X)^2 + 4\zeta_i^2} = \frac{f_i}{2\zeta_i^2}\int_0^{\infty} \frac{d(\delta X)}{\left(\frac{\delta X}{\zeta_i}\right)^2 + 1}$$

Let $\beta = \delta X/\zeta_i$, then $d\beta = 1/\zeta_i\, d(\delta X)$ and the value of the integral is

$$\frac{f_i}{2\zeta_i^2}\int_0^{\infty} \frac{\zeta_i d\beta}{\beta^2 + 1} = \frac{\pi f_i}{4\zeta_i} \qquad (9\text{-}13)$$

9.2.5 Use of the Simplified Expression in Developing Design Guidelines

Using the result given in Eq. (9-13), the mean-square tube response for the dominant mode, i, can be expressed as

$$\overline{y^2(x)}_i = \frac{\phi_i^2(x)S_F(f)J_i^2}{64\pi^3 f_i^3 M_i^2 \zeta_i} \qquad (9\text{-}14)$$

The first mode of vibration is assumed to dominate the other modes. This is a reasonable assumption for single-span tubes such as those used in simple test sections, but it could prove to be inaccurate in a multi-span tube (i.e., a tube with several supports along its length). Henceforth, all mode dependent variables will refer to the first mode unless otherwise noted.

The joint acceptance can be calculated if the correlation factor is known. Values of the spanwise correlation factor for a tube array (Blevins et al, 1981) have been reported, but the available data is for tubes subjected to air cross flow. Due to the general lack of information, several approaches have been used to deal with the correlation factor.

One approach taken by early guideline developers such as (Pettigrew et al, 1991 and Taylor et al, 1989) has been to set the correlation factor equal to unity. Real heat exchangers and the experimental rigs that were used do not have perfectly correlated force fields. In reality, the correlation factor is always less than unity. Consequently, if the correlation factor is assumed to be equal to unity, the experimentally obtained magnitude of the excitation force power spectral density will be less than the true magnitude. However, one can assume that a test rig will provide a more correlated force field than an actual heat exchanger. This assumption means that the excitation force power spectral density magnitudes measured in a test rig will predict conservative tube amplitudes when applied to an operating heat exchanger.

A second approach has been to set the correlation factor to unity and then add an additional factor to account for the effect of the excited tube length (Pettigrew and Taylor, 1994 and Taylor et al, 1996). The difference in span lengths and/or tube lengths between a test rig and an operating heat exchanger could be important when comparing the power spectral density magnitudes. Although the actual correlation length is not known, one can account for the effect of differing excited tube lengths. Antunes (1986) has shown that for values of $\lambda_c/L \ll 1$, the joint acceptance is proportional to the correlation length divided by the excited tube length, λ_c/L_e, if the correlation factor is represented by a function that is exponentially decreasing with separation distance along the tube, such as $\gamma(x, x') = \exp(-|x - x'|/\lambda_c)$. The rigorous proof carried out by Antunes was graphically introduced in the field of jet noise by Prof. A. Powell in the mid-1950s (Crandall, 1958). Although the actual value of the correlation factor is not known, the excitation force power spectral density can be scaled properly when differing excited tube lengths are used.

A third approach, used by several authors such as Pettigrew and Taylor (2003) and de Langre and Villard (1998), has been creating an expression for the joint acceptance using a proportionality coefficient, a_i, as $J_i^2 = a_i\lambda_c/L_e$. Then, the expression defining the first-mode mean-square of tube response, $\overline{y^2(x)}_1$, in terms of the power spectral density of the excitation force, $S_F(f)$, can be written as follows:

$$\overline{y^2(x)}_1 = \frac{\phi_1^2(x)S_F(f)a_1\lambda_c/L_e}{64\pi^3 f^3 m^2 \zeta_1} \tag{9-15}$$

This approach is fully developed in Section 9.4. Table 9-1 gives values of $(\phi_{1\ max})^2$ and a_1 for a variety of end conditions.

Table 9-1 Modal Factor (a_1) and Mode Shape ($\phi_{1\ max}$)^2 Constants for Mode 1.

	Rigid Tube	Clamped-Clamped	Clamped-Pinned	Clamped-Free	Pinned-Pinned
Modal Factor	2	0.8	0.9	0.5	1.1
Mode Shape	Translation	2.522	2.278	4.0	2.0

9.3 Literature Search

A great deal of research has been conducted on flow-induced vibration in the past 50 years. However, much of this work has focused on fluidelastic instability and vortex shedding because these mechanisms can quickly lead to catastrophic failures. As well, most random excitation research has been conducted on single tubes and tube rows, rather than on tube bundles.

One of the earliest sources of data on single-phase random excitation in tube bundles was presented by Pettigrew and Gorman (1978). This work covered a variety of bundle orientations, though only in a narrow reduced-frequency band. Chen and Jendrzejczyk (1987) gave the results of tests on a normal-square bundle over a wide range of reduced frequencies. Axisa et al (1990) presented new tests on normal-triangular, square and rectangular bundles in air and water. Their results were compared with row data from Taylor et al (1988), and an upper boundary was proposed. Oengören and Ziada (1992) proposed a boundary based on tests on a normal-square bundle in air cross flow, as well as bundle data from Chen and Jendrzejczyk (1987) and row data from Taylor et al (1988). Blevins (1990) presented another bounding spectrum based on data from Axisa et al (1990), Taylor et al (1988), and Chen and Jendrzejczyk (1987). Tests on a normal-square bundle in water are reported in Taylor et al (1996). Results from a normal-triangular bundle in air are provided in Oengören and Ziada (1992). Wolgemuth (1994), while focused primarily on the effect of tube-support clearances, provided valuable data on the effect of highly turbulent flow on random excitation forces in a tube bundle. Table 9-2 summarizes the geometric characteristics for the tube bundles used in each of these studies.

9.4 Approach Taken

In this chapter, the effect of many different parameters such as bundle geometry, upstream turbulence, and fluid properties will be considered as bounding spectra are developed. These upper bounds can be used as guidelines by heat exchanger designers, troubleshooters and analysts.

The experimental studies examined in this chapter collected random excitation data using one of two methods:

1) Random excitation determined from reaction forces. These tests used force transducers to directly measure the reaction forces.
2) Random excitation determined from tube response. In these tests, strain gages were used to measure tube displacement. In this case, tube displacement was used to calculate excitation force power spectral densities, using Eq. (9-15).

To be able to compare data and find an upper bound, the results must be presented as a normalized excitation force spectrum. Researchers in this field have used various methods of normalizing their results. Therefore, it was necessary to select one means of normalization and apply it to all the data. The method adopted was the "equivalent power spectral density" method, described by Axisa et al (1990).

The power spectral density can be rendered dimensionless, $\tilde{S}_F(f_R)$, using the dynamic pressure head, as follows:

$$\tilde{S}_F(f_R) = \frac{S_F(f)}{\left(\frac{1}{2}\rho U_P^2 D\right)^2} \frac{U_p}{D} \tag{9-16}$$

Table 9-2 Summary of Bundle Geometries.

Reference	Bundle Orientation	Tube Diameter (mm)	P/D (drag)	Excited Length (m)	Total Mass (kg/m)	Natural Freq. (Hz)	End Conditions	Notes
Taylor (1996)	90	30	1.5	0.3	-	268	Clamped-Clamped	Water flow
Pettigrew & Gorman (1978)	30	19	1.33	0.051	1.19	40	Clamped-Free	Water flow
	30	13	1.36, 1.54	0.051	0.52	30	Clamped-Free	"
	30	19	1.57	0.051	1.19	40	Clamped-Free	"
	45	13	1.3	0.051	0.52	30	Clamped-Free	"
	60	19	1.33, 1.57	0.051	1.19	40	Clamped-Free	"
	60	13	1.36	0.051	0.52	30	Clamped-Free	"
Oengoren & Ziada (1992)	90	20	1.26, 1.5, 1.95, 3.0	0.2	-	900	Clamped-Clamped	Air flow
Oengoren & Ziada (1995)	30	31	1.61	0.2	-	>1000	Clamped-Clamped	Air flow
	30	18	2.08	0.2	-	>1000	Clamped-Clamped	
	30	22	3.41	0.2	-	>1000	Clamped-Clamped	
Chen & Jendrezjczyk (1987)	90	25.4	1.75	0.3	-	-	Clamped-Clamped	Water flow
Axisa et al (1990)	90	24	1.25 (2.16), 1.25 (1.44)	0.25	0.624	78.8	Clamped-Free	#1 Air flow
	90	20	1.5 (2.6), 1.5 (1.73)	0.25	0.433	68.6	Clamped-Free	"
	90	20	1.5	0.25	0.244	57.8	Clamped-Free	"
	30	24	1.25 (1.08)	0.25	0.624	78.8	Clamped-Free	"
	30	20	1.5 (1.3)	0.25	0.433	68.6	Clamped-Free	"
	30	15	2.0 (1.73), 2.0 (1.15)	0.25	0.244	60.4	Clamped-Free	"
	90	25	1.5	0.48	0.3	33	Clamped-Free	#2 Air Flow
	60	38	1.18	0.43	1.71	109	Clamped-Free	#3 Water Flow
	90	19.05	1.44 (80 U-tubes)	0.3 (average)	-	-	Clamped-Free	#4 Water Flow
Wolgemuth (1994)	60	16	1.38	0.87	0.6093	36	Clamped-Pinned	Water flow

where, f_R is the reduced frequency, defined as fD/U_p, ρ is the fluid density, U_p is the pitch velocity, and D is the tube diameter. A difficulty arises in the calculation of $S_F(f)$ because the correlation length, λ_c, is rarely known. Axisa et al (1990) presented an equivalent power spectral density, $S_F(f_R)_e$, defined as

$$S_F(f_R)_e = \frac{\lambda_c}{L_e} S_F(f_R) \quad \text{or} \quad \tilde{S}_F(f_R)_e = \frac{\lambda_c}{L_e} \tilde{S}_F(f_R) \tag{9-17}$$

to overcome this difficulty.

Substituting Eq. (9-15) and Eq. (9-16) into Eq. (9-17), one obtains

$$\tilde{S}_F(f_R)_e = \frac{\overline{y^2(x)}_1 \, 64\pi^3 f_1^3 m^2 \zeta_1}{\phi^2 a_1} \frac{1}{\left(\frac{1}{2}\rho U_p^2 D\right)^2} \frac{U_p}{D} \tag{9-18}$$

Using Eq. (9-18), the mean-square of tube displacement can be found without knowledge of the correlation length. Instead, a small correlation length has been assumed, as discussed in the previous section on random vibration theory (Section 9.2). To correctly compare spectra obtained from experimental rigs with varying geometries, it is necessary to reference a single excited tube length, L_0. In this work, a reference length of 1 m is applied, as follows:

$$\tilde{S}_F(f_R)_{e \, @L_0=1m} = \tilde{S}_F(f_R)_e \times \frac{L_e}{1} \tag{9-19}$$

9.5 Discussion of Parameters

9.5.1 Directional Dependence (Lift versus Drag)

Figure 9-1 gives three examples of lift and drag results, from different papers in the literature. Based on this figure, it can be said that the excitation forces in the lift direction are either greater than or equal to the forces in the drag direction. Using this fact, the remainder of this chapter deals only with lift direction results.

9.5.2 Bundle Orientation

The standard tube bundle orientations are normal square, rotated square, normal triangular and rotated triangular. In a real heat exchanger, both normal and rotated flow orientations are possible within a given tube bundle. Therefore, it is necessary to consider the bundle orientation with the highest excitation force spectrum for design purposes. The bounds for normal-square and normal-triangular bundles of similar P/D are shown in Fig. 9-2. The differences in excitation force magnitude between the bundles are not significant for $P/D < 3.0$. Similarly, Pettigrew and Gorman (1978) ($1.23 < P/D < 1.57$) did not find any significant differences between various bundle orientations.

9.5.3 Pitch-to-Diameter Ratio (P/D)

Oengören and Ziada (1995) tested normal-triangular tube bundles with a large range of pitch-to-diameter ratios. Oengören and Ziada (1992) tested normal-square bundles over a smaller range of P/D. The results of these tests are shown in Fig. 9-3a and Fig. 9-3b, respectively. The data does not reveal a clear trend. From $P/D = 1.26$ to 1.95, there is an increase in dimensionless equivalent power

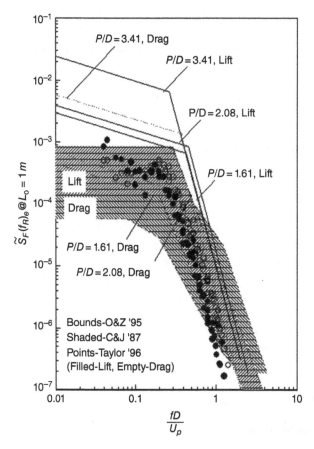

Fig. 9-1 Directional Dependence (Lift versus Drag).

spectral density magnitude with pitch-to-diameter ratio. However, at $P/D = 2.08$ and greater, there is no such trend.

Most heat exchangers are designed with pitch-to-diameter ratios between 1.2 and 1.6. The proposed guideline in this chapter will be based on data from P/D of 1.5 to 1.75. The designer of tube bundles with smaller P/D may wish to use the trends shown in Fig. 9-3 to appropriately reduce the proposed upper bound that is presented in the summary of this chapter.

9.5.4 Upstream Turbulence

Highly turbulent flow, such as that found in the inlet region of a steam generator, is observed to have a significant effect on tube response. Figure 9-4 (Wolgemuth, 1994) shows tube responses for the first two rows and two interior rows when a bundle was exposed to highly turbulent flow.

High upstream turbulence can cause a significant increase in tube response. However, this effect is observed only in the first few rows of the bundle. Wolgemuth (1994) also showed that by the fifth row, tube response is similar to that caused by uniform flow.

Because upstream turbulence can cause a significant increase in tube response, a separate bound will be proposed for use as a guideline when designing tube bundles that will be exposed to highly turbulent upstream flow.

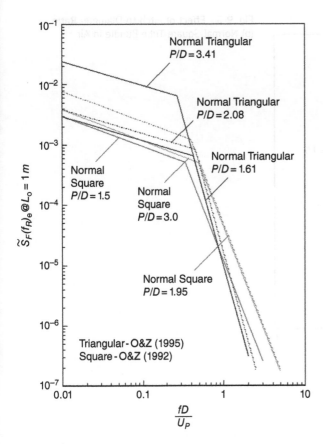

Fig. 9-2 Effect of Tube Bundle Orientation.

Fig. 9-3 Effect of Pitch-to-Diameter Ratio
(a) Normal-Triangular Tube Bundle in Air.

(a)

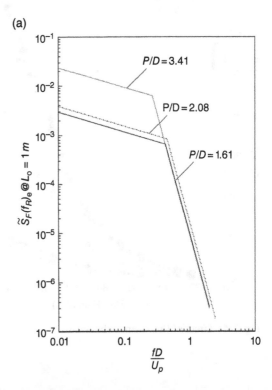

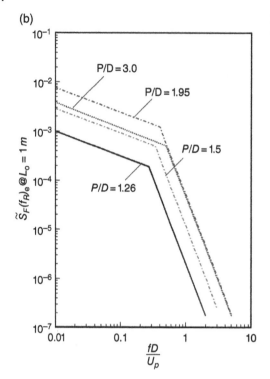

(b)

Fig. 9-3 Effect of Pitch-to-Diameter Ratio (b) Normal-Square Tube Bundle in Air.

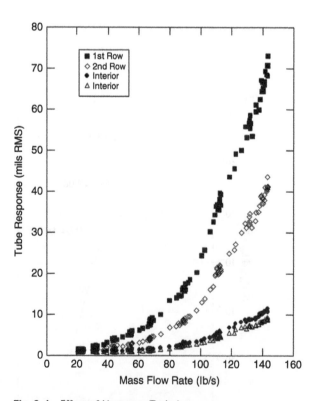

Fig. 9-4 Effect of Upstream Turbulence.

9.5.5 Fluid Density (Gas versus Liquid)

Figure 9-5 shows dimensionless power spectral density results for air and water data. Most heat exchangers using gas as a secondary fluid will have very high flow rates and consequently, low reduced frequencies. Examination of Fig. 9-5 shows that most air data fall below $f_R = 0.2$. Very few water heat exchangers have flow rates that result in reduced frequencies below 0.2. Thus, for $f_R < 0.2$, the proposed guidelines will place a greater emphasis on air results.

9.5.6 Summary

The proposed boundary spectra of dimensionless equivalent power spectral density of the excitation force per unit length resulting from single-phase cross flow for bundles with P/D between 1.5 and 1.75 are shown in Fig. 9-6. The general collapse of the data is surprisingly good. An upper bound is drawn over most of the data and then a second boundary is provided for tubes that are subjected to highly turbulent inlet flows. The boundaries are defined as follows:

$$\text{Interior}: \begin{array}{ll} 4 \times 10^{-4}\left(fD/U_p\right)^{-0.5}, & 0.01 < fD/U_p < 0.5 \\ 5 \times 10^{-5}\left(fD/U_p\right)^{-3.5}, & 0.5 < fD/U_p \end{array} \tag{9-20}$$

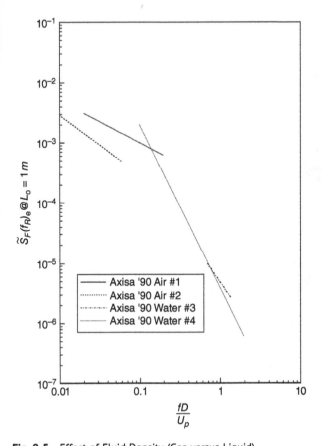

Fig. 9-5 Effect of Fluid Density (Gas versus Liquid).

Inlet: $\quad 1 \times 10^{-2}\left(fD/U_p\right)^{-0.5}, \quad 0.01 < fD/U_p < 0.5$

$\qquad\quad 1.25 \times 10^{-3}\left(fD/U_p\right)^{-3.5}, \quad 0.5 < fD/U_p$

$$(9\text{-}21)$$

A comparison with previously proposed guidelines is presented in Fig. 9-7. The guideline proposed in this chapter is based on the largest amount of data from the literature and, therefore, provides an upper bound. The bound from Blevins (1990) includes the effects of vortex shedding.

9.6 Design Guidelines

This section provides the designer with a methodology to estimate random excitation forces for specified flow conditions. These guidelines are based on a limited database and should not be applied outside the limitations given.

The lower bound equation, shown in Fig. 9-6, should be used when upstream turbulence is less than or equal to the turbulence within the tube bundle. The upper bound, shown in Fig. 9-6 should be used if upstream turbulence exceeds the turbulence inside the tube bundle.

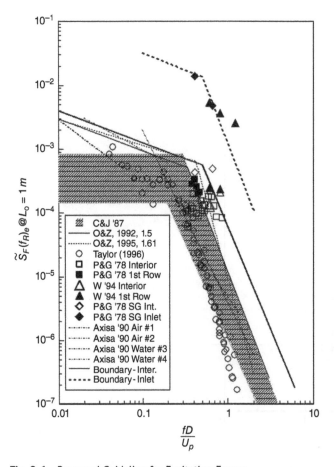

Fig. 9-6 Proposed Guideline for Excitation Forces.

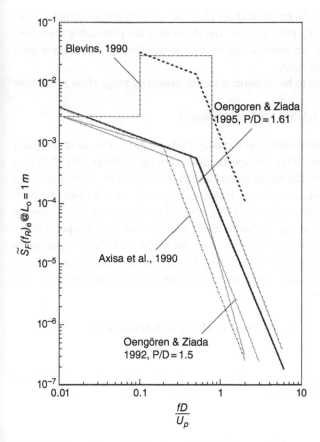

Fig. 9-7 Comparison with Previous Guidelines.

Combining Eq. (9-16) and Eq. (9-17), the product of random excitation power spectral density and correlation length, $S_F(f)\lambda_c$, can be calculated from the equivalent power spectral density.

$$S_F(f)\lambda_c = \tilde{S}_F(f_R)_e \times \frac{\left(\frac{1}{2}\rho U_p^2 D\right)^2 D}{U_p} \times L_e \tag{9-22}$$

For a single-span tube, the maximum mean-square vibration response, $\overline{y^2}$, can be calculated from the power spectral density of the excitation force from Eq. (9-15) as follows:

$$\overline{y^2(x)}_1 = \frac{[S_F(f)\lambda_c]\phi_1^2(x)a_1}{64\pi^3 f_1^3 m^2 \zeta_1 L_e}$$

where, $\phi_1(x)$ is the normalized mode shape for the 1st mode, a_1 is the modal factor for the 1st mode, f_1 is the tube natural frequency for the 1st mode, m is the total tube mass (tube mass + hydrodynamic mass) per unit length, ζ_1 is the damping ratio for the 1st mode, and L_e is the excited tube length (the length of the tube that is subjected to flow). Table 9-1 contains values of $(\phi_{1\,max})^2$ and a_1 for various end conditions.

Several assumptions have been made in the derivation of Eq. (9-15) for a single-span tube:

- A single mode dominates the tube displacement and coupling between modes is negligible. This assumption is valid if the tube damping ratio is small and the modes are reasonably separated.
- The random fluctuating forces are homogeneous along the tube so that the one-sided power spectral densities are equal within a given span.
- The power spectral density is assumed to be constant over the frequency range close to the tube natural frequency.
- The correlation length is very small relative to the span length.

Calculation of the tube response for multi-span heat exchanger tubes requires the use of computer codes (Au-Yang, 2001 and Pettigrew et al, 1991). The spectra defined by Eq. (9-20) and Eq. (9-21) are incorporated into the computer code and the modal factors and mode shapes are calculated within the code. For multi-span tubes, the random excitation forces and response are calculated for individual velocity regions. The number of velocity regions along a tube span depends on the complexity of the velocity and density distributions along the tube and the limitations of the computer code.

Future research in the area of single-phase random excitation forces should include experiments to provide additional data regarding the upstream turbulence found in inlet regions of actual heat exchangers. These experiments should examine various inlet configurations to determine the full range of inlet (upstream) excitation forces.

Example 9-1 Random Excitation in Process Heat Exchanger Interior Flow

Random Excitation Parameters:
From previous calculations in Examples 3.1, 4.1, 4.3 and 5.2: $D = 0.02$ m, $m = 1.16$ kg/m, $f = 135.6$ Hz (interior tube), $\rho = 1000$ kg/m^3, $U_p = 1.5$ m/s, $\zeta = 0.0085$, $\ell = L_e = 0.5$ m, pinned-pinned tube (see Table 9-1).
Random Excitation Calculation:
Using the interior tube design guidelines in Eq. (9-20)

$$f_R = \frac{fD}{U_p} = \frac{135.6 \times 0.02}{1.5} = 1.808$$

$$\tilde{S}_F(f_R)_{e @ L_o = 1m} = 5 \times 10^{-5}(fD/U_p)^{-3.5} = 5 \times 10^{-5} \cdot (1.808)^{-3.5} = 6.29 \times 10^{-6}$$

Using Eqs. (9.22) and (9.19)

$$S_F(f)\lambda_c = \tilde{S}_F(f_R)_{e @ L_o = 1m}\left(\frac{1}{L_e}\right)\frac{\left(\frac{1}{2}\rho U_p^2 D\right)^2 DL_e}{U_p}$$

$$= 6.29 \times 10^{-6}\frac{0.02(0.5 \times 1000 \times 1.5^2 \times 0.02)^2}{1.5}$$

$$S_F(f)\lambda_c = 4.25 \times 10^{-5}(N/m)^2 s \cdot m$$

Finally, calculating the root-mean-square (rms) tube displacement for pinned end conditions using Eq. (9-15)

$$\overline{y^2}(L/2) = \frac{a_1\phi_1^2(S_F(f)\lambda_c)}{64\pi^3 f_1^3 m^2 \zeta_1 L_e} = \frac{1.1 \times 2 \times 4.25 \times 10^{-5}}{64\pi^3 \times 135.6^3 \times 1.16^2 \times 0.0085 \times 0.5}$$

$$= 3.3 \times 10^{-12}m^2$$

Therefore, the rms tube displacement is as follows:

$$[\overline{y^2}(L/2)]^{\frac{1}{2}} = 1.8 \times 10^{-6}m = 1.8 \ \mu m \ rms$$

9.7 Random Turbulence Excitation in Axial Flow

Random turbulence excitation due to single-phase axial flow can also exist in process heat exchangers and nuclear components. Random excitation forces result from boundary layer pressure fluctuations generated by fluid flow. In single-phase flow, fluid-borne pressure fluctuations may be divided into two groups: near field and far field. Near-field pressure fluctuations are generated locally as the fluid flows around the structure of interest. They are generally broadband in nature. Far-field pressure fluctuations are caused by upstream components such as pumps, valves or other piping elements. They are transmitted by the fluid as travelling pressure waves or are convected as turbulence. It should be noted that near field pressure fluctuations generally result in insignificant tube motion in most heat exchangers, but some very flexible structures could be susceptible. In contrast, far field pressure pulsations have been known to cause significant damage to tubes in heat exchangers and to fuel bundles or strings in nuclear reactors.

The analytical formulation for random excitation in axial flow is identical to that for cross flow. In fact, considerable effort has gone into the prediction of random excitation forces due to axial flow. This work will not be repeated here but the reader is encouraged to review these efforts in texts such as Chen (1987), Blevins (1990), Au-Yang (2001) and Kaneko et al (2014). In addition, some of these texts provide axial flow power spectra for various axial flow situations such as confined annular flow, boundary-layer-type turbulence and cavitating annular flow, as well as examples from industry to illustrate application of both prediction and problem-solving techniques

Nomenclature

a_i	= modal factor for the ith mode
c	= total damping coefficient
D	= tube diameter, m
E	= Young's modulus of elasticity, Pa
$F(x.t)$	= flow excitation force per unit length
$F_i(t)$	= generalized force per unit length for the ith mode
f_R	= reduced frequency, fD/U_p
f_i	= tube natural frequency for the ith mode, Hz
$H_i(f)$	= transfer function for the ith mode
I	= tube area moment of inertia, m^4
J_i^2	= joint acceptance for the ith mode
L	= tube length, m
L_e	= excited tube length, m
L_o	= reference excited tube length, m
M_i	= generalized mass per unit length for the ith mode, kg/m
m	= total mass per unit length ($m_h + m_t$), kg/m
m_h	= hydrodynamic mass, kg/m
m_t	= tube mass, kg/m
P	= pitch between tubes, m
$q_i(t)$	= generalized coordinate
$S_F(f)$	= auto-power spectral density of the input force per unit length, (N/m)^2·s
$S_{F_r F_s}(f)$	= input force cross-spectral density, (N/m)^2·s

$S_{q_r q_s}(f)$ = displacement cross-spectral density, $m^2 \cdot s$

$\tilde{S}_F(f_R)$ = dimensionless power spectral density

$S_F(f_R)_e$ = equivalent power spectral density, $(N/m)^2 \cdot s$

$q_i(t)$ = displacement for the ith mode, m

t = time, s

U_p = pitch velocity, m/s

U_∞ = free stream (or upstream) velocity, m/s

x = distance along the tube, m

$y(x,t)$ = tube displacement normal to the x-direction, m

$\overline{y^2(x)}_i$ = i^{th} mode mean-square of tube response, m^2

$\gamma(x,x',f)$ = correlation factor

ζ_i = generalized damping ratio for the ith mode

λ_i = eigenvalue for the ith mode

λ_c = correlation length, m

ρ = fluid density, kg/m^3

τ = time, s

ϕ_i = normalized mode shape for the ith mode

ω_i = radial frequency for the ith mode, $\omega_i = (\lambda_i/L)^2 (EI/M_i)^{0.5}$

References

Antunes, J., 1986, *Contribution à l'Étude des Vibrations de Faisceaux de Tubes en Écoulement Transversal*, Doctoral Thesis, Université de Paris VI, Centre des Études Nucléaires de Saclay, France.

Au Yang, M. K., 2001, *Flow-Induced Vibration of Power and Process Plant Components*, Professional Engineering Publishing Ltd., New York, NY.

Axisa, F. Antunes, J. and Villard, B., 1990, "Random Excitation of Heat Exchanger Tubes by Cross Flows," *Journal of Fluids and Structures*, 4, pp. 321–341.

Blevins, R. D., 1990, *Flow-Induced Vibration*, 2nd Edition, Van Nostrand Reinhold Company, New York.

Blevins, R. D., Gibert, R. J. and Villard, B., 1981, "Experiment on Vibration of Heat Exchanger Tube Arrays in Cross-Flow," Paper B6/9, *Proceedings of the 6th International SMiRT Conference*, Paris.

Chen, S-S, 1987, *Flow-Induced Vibration of Circular Cylindrical Structures*, Hemisphere Publishing Corporation, Washington, DC.

Chen, S. and Jendrzejczyk, J., 1987, "Fluid Excitation Forces Acting on a Square Tube Array," *Journal of Fluids Engineering*, 109, pp. 415–423.

Crandall, S. H., 1958, *Random Vibration*, Technology Press of the Massachusetts Institute of Technology and John Wiley & Sons, N.Y.

De Langre, E. and Villard, B., 1998, "An Upper Bound on Random Buffeting Forces Caused by Two-Phase Flows Across Tubes," *Journal of Fluids and Structures*, 12, pp. 1005–1023.

Kaneko, S., Nakamura, T., Inada, F., Kato, M., Ishikara, K., Nishihara, T., Mureithi, N.W. and Langthjem, M.A., 2014, *Flow-Induced Vibrations: Classifications and Lessons from Practical Experiences*, 2nd Edition, Elsevier, Academic Press, London.

Meirovitch, L., 1967, *Analytical Methods in Vibration*, MacMillan Company, New York, N.Y.

Oengören, A. and Ziada, S., 1992, "Unsteady Fluid Forces Acting on a Square Tube Bundle in Air Cross Flow," *1992 International Symposium on Flow-Induced Vibration and Noise, Vol. 1: FSI/FIV in Cylinder*

Arrays in Cross Flow, (M.P. Païdoussis, W.J. Bryan, J.R. Stenner, D.A. Steininger - editors), Anaheim, pp. 55–74.

Oengören, A. and Ziada, S., 1995, "Vortex-Shedding, Acoustic Resonance and Turbulent Buffeting in Normal Triangular Tube Arrays," *Proceedings of the 6th International Conference on Flow-Induced Vibration* (P.W. Bearman - editor), London, UK, pp. 295–313.

Pettigrew, M. J. and Gorman, D. J., 1973, "Experimental Studies on Flow-Induced Vibration to Support Steam Generator Design, Part 3: Vibration of Small Tube Bundles in Liquid and Two-Phase Cross flow," *Proceedings of the UKAEA/NPL International Symposium in Vibration Problems in Industry*, Keswick, UK, Paper 426 (also *Atomic Energy of Canada Limited Report, AECL-5804*).

Pettigrew, M. J. and Gorman, D. J., 1978, "Vibration of Heat Exchanger Components in Liquid and Two-Phase Cross Flow," *Proceedings of the BENS International Conference on Vibration in Nuclear Power Plants*, Keswick, UK, Paper 2.3 (also *Atomic Energy of Canada Limited Report, AECL-6184*).

Pettigrew, M. J. and Taylor, C. E., 1994, "Two-Phase Flow-Induced Vibration: An Overview," *Journal of Pressure Vessel Technology*, 116, pp. 233–253.

Pettigrew, M. J. and Taylor, C. E., 2003, "Vibration Analysis of Shell-and-Tube Heat Exchangers: An Overview-Part 2: Vibration Response, Fretting-Wear, Guidelines," *Journal of Fluids and Structures*, 18, pp. 485–500.

Pettigrew, M. J., Carlucci, L. N., Taylor, C. E. and Fisher, N. J., 1991, "Flow-Induced Vibration and Related Technologies in Nuclear Components," *Nuclear Engineering and Design*, 131, pp. 81–100.

Taylor, C. E., Currie, I. G., Pettigrew, M. J., and Kim, B. S., 1989. "Vibration of Tube Bundles in Two-Phase Cross Flow - Part 3: Turbulence-Induced Excitation," *ASME Journal of Pressure Vessel Technology*, 111, pp. 488–500.

Taylor, C. E., Pettigrew, M. J. and Currie, I. G., 1996, "Random Excitation Forces in Tube Bundles Subjected to Two-Phase Cross Flow," *ASME Journal of Pressure Vessel Technology*, 118, pp. 265–277.

Taylor, C. E., Pettigrew, M. J., Axisa, F. and Villard, B., 1988, "Experimental Determination of Single- and Two-Phase Cross Flow-Induced Forces on Tube Rows," *ASME Journal of Pressure Vessel Technology*, 110, pp. 22–28.

Thomson, W. T., 1981, *Theory of Vibration with Applications*, 2nd Edition, Prentice-Hall, Englewood Cliffs, N.J.

Wolgemuth, G. A., 1994, *Unpublished Chalk River Laboratories Data*.

10

Random Turbulence Excitation Forces Due to Two-Phase Flow

Colette E. Taylor and Michel J. Pettigrew

10.1 Introduction

A majority of process heat exchangers operate with a two-phase mixture on the shell-side. Today's economics require that thermal efficiency is continually improved in these units. Such improvement invariably leads to higher flow rates and more flexible tubes. To prevent tube failures in these heat exchangers, an improved understanding of random excitation mechanisms in two-phase cross flow is required. Random turbulence excitation forces can cause low-amplitude tube motion that will result in long-term fretting wear or fatigue damage.

This chapter reviews previous guidelines for two-phase cross flow forces and describes the development of a new guideline. Data from many sources is used to ensure that the guideline is comprehensive, but most of the development is done with data from two extensive air-water and Freon experimental programs conducted on large-scale flow loops in France (Centre d'Études Nucléaires de Saclay (CENS)) and Canada (Chalk River Laboratories (CRL), Chalk River, ON). A summary of the tube-bundle geometries used in these test programs is provided in Table 10-1. Strain gages (CRL) and force transducers (CENS) were used to measure the vibration response of a centrally located tube as the tube array was subjected to a wide range of void fractions and flow rates. Details regarding the air-water CENS and CRL experiments are given in Taylor et al (1988 and 1989, respectively). Raw data and further experimental detail are available in Taylor (1994). The Freon test program details are found in Pettigrew et al (1995 and 2002) and Pettigrew and Taylor (2009). Chapter 8 also provides some details concerning the air-water and Freon experimental facilities and test programs at CRL.

As this chapter develops, scaling techniques for two-phase excitation forces, steam-water, Freon and air-water data from a wide variety of test facilities and test conditions are compared. An emphasis is placed on the need to understand the effect of the flow regimes that are present in two-phase flow in tube arrays.

10.2 Background

The study of forced vibration due to two-phase cross flow began in 1970, with the first two-phase research reported by Pettigrew and Gorman (1973b). Their very early design guideline is shown in Fig. 10-1. They excited a tube bundle, with a 1 m tube length, by introducing air-water cross flow

Flow-Induced Vibration Handbook for Nuclear and Process Equipment, First Edition.
Michel J. Pettigrew, Colette E. Taylor, and Nigel J. Fisher.

Table 10-1   Tube Bundle Geometries from CENS and CRL.

	CENS	CRL	
	Air-Water Tests	**Air-Water Tests**	**Freon Tests**
Tube Diameter	30 mm	13 mm	12.7 mm
Tube Length (Excited Length)	0.7 m (0.3 m)	0.6 m (0.6 m)	0.609 m (0.609 m)
Wall thickness	2 mm	1.07 mm	0.9 mm
Tube Material	Stainless Steel	Stainless Steel	Stainless Steel
End Conditions	Clamped-Clamped	Clamped-Free, Clamped-Pinned	Clamped-Free
Natural Frequency in air	325 Hz	33 Hz, 160 Hz	28.5 Hz
Bundle configurations	Normal square, 90°	Normal square, 90° Rotated Triangular, 60° (also known as Parallel Triangular) Normal Triangular, 30°	Rotated Triangular, 60° (Freon-22 & Freon-134a) Normal Triangular, 30° (Freon-22)

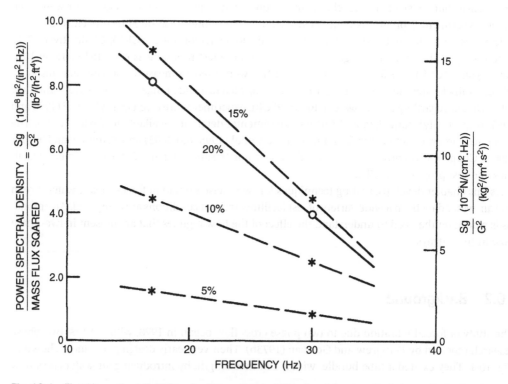

Fig. 10-1   First Normalized Guideline for Power Spectral Density of Random Turbulence Excitation in Two-Phase Flow at Several Steam Qualities (Pettigrew and Gorman, 1973b).

over a central span of 0.0508 m. Referenced in a second two-phase cross-flow paper by Pettigrew and Gorman (1978), this research remained the only source of two-phase random vibration data until the 1980s. While excellent reviews by Païdoussis (1982), Pettigrew (1981) and Blevins (1979) provided a complete overview of single-phase flow-induced vibration in nuclear power-plant components, only Pettigrew (1981) presented two-phase results.

In the early 1980s, Heilker and Vincent (1981), Nakamura et al (1982) and Remy (1982) reported experimental results for tube bundles subjected to air-water cross flow. Hara (1982) presented analytical and experimental results on a single circular cylinder in air-water cross flow.

Further experiments in two-phase cross flow were reported in the mid-1980s. The results of tests that covered a wide void-fraction range of 5% to 93% were reported by Pettigrew et al (1985). At about the same time, Axisa et al presented high void-fraction steam-water data (1985) and air-water data (1986). Both of these test programs emphasized fluidelastic instability, not random vibration. Two years later, some of these two-phase results were discussed in a review paper by Weaver and Fitzpatrick (1987), but no data was presented.

In the late 1980s and early 1990s, the number of papers in the area of two-phase flow-induced vibration increased dramatically. Researchers in England (Goyder, 1987 and 1988), France (Gay et al, 1988 and Axisa et al, 1990), Japan (Hara, 1987a and 1987b; Hara and Iijima, 1988; and Nakamura et al, 1991) and Canada (Lian et al, 1992 and 1997) were adding to the database with air-water, steam-water and Freon test data. As well, the CRL and CENS results used extensively in this chapter were being reported (Taylor et al, 1988, 1989, and 1996; Axisa et al, 1990; and Pettigrew and Taylor, 1993) and used in developing two-phase design guidelines for random excitation of heat-exchanger tubes (Taylor et al, 1989; Pettigrew et al, 1991; and Axisa and Villard, 1992). The Pettigrew et al (1991) guideline (shown in Fig. 10-2) has been widely used in steam generator and process component design.

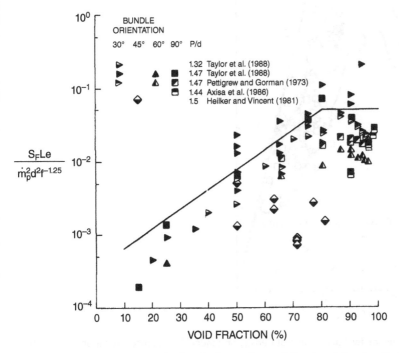

Fig. 10-2 Early Normalized Guideline for Power Spectral Density of Random Turbulence Excitation in Two-Phase Flow (Pettigrew et al, 1991).

In the late 1990s and early 2000s, a significant number of experimental tests in Freon were conducted and published (Pettigrew et al, 1995 and 2002; Pettigrew and Taylor, 2009; and Feenstra, 2000) and the next generation of design guidelines were being developed. In his book on flow-induced vibration, Au-Yang (2001) placed an upper boundary over the data presented in Taylor et al (1996) to provide a two-phase random excitation guideline (see Fig. 10-3). De Langre and Villard (1998) developed the first dimensionless upper bound for two-phase buffeting forces (see Fig. 10-4). Five years later, Pettigrew et al (2003) adopted and published the de Langre and Villard (1998) design guideline with minor changes.

The data used in this chapter is almost entirely from experimental work carried out prior to 2000. There was relatively little two-phase work done in the following decade However, some significant work in understanding random turbulence excitation due to two-phase cross flow across tube bundles has more recently been carried out at institutions such as École Polytechnique in Montreal Canada (e.g., Zhang et al, 2005; and Sim and Mureithi, 2013), the University of California Los Angeles (e.g., Mitra et al, 2009), Mitsubishi Industries in Japan (Nishida et al, 2019), the University of Sao Paulo in Brazil (Ribatski et al, 2019), the University of Blumenau, Blumenau, Brazil (da Silva et al, 2019) and the Escuela Politécnica Nacional, Quito, Ecuador (Álvarez-Briceño et al, 2018; and Álvarez-Briceño and Oliveira, 2020). Most of this work was published after this chapter was originally written, but a brief review indicates that the experimental data is similar to the results presented in this chapter. The most recent papers in this list reach into the physical underpinnings of two-phase flow to develop more theoretical predictions of tube behavior and numerical simulations of two-phase fluid-structure interaction.

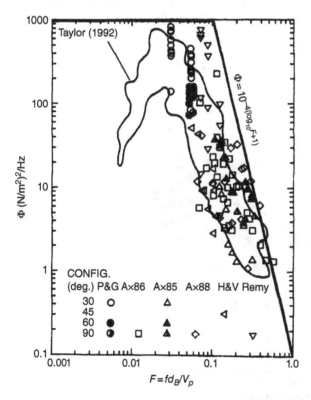

Fig. 10-3 Normalized Guideline for Power Spectral Density of Random Turbulence Excitation in Two-Phase Flow (Au-Yang, 2001).

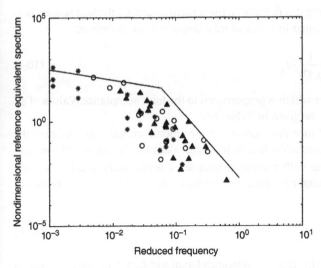

Fig. 10-4 Early Dimensionless Guideline for Power Spectral Density of Random Turbulence Excitation in Two-Phase Flow (de Langre and Villard, 1998).

10.3 Approach Taken to Data Reduction

From random vibration theory (see Chapter 9), we know that the mean-square tube response for the dominant mode, i, can be expressed as

$$\overline{y^2(x)}_i = \frac{\phi_i^2(x)S_F(f)J_i^2}{64\pi^3 f_i^3 M_i^2 \zeta_i} \tag{10-1}$$

where, for the ith mode, $\phi_i(x)$ is the normalized mode shape, $S_F(f)$ is the force power spectral density per unit length, J_i^2 is the joint acceptance, f_i is the tube natural frequency, M_i is the generalized mass per unit length, and ζ_i is the generalized damping ratio. The first mode of vibration is assumed to dominate the other modes. This is a reasonable assumption for single-span tubes such as those used in simple test sections. The joint acceptance can be calculated as $J_i^2 = a_i \lambda_c / L_e$ where a_i is the modal correlation factor which is proportional to the joint acceptance, λ_c is the correlation length and L_e is the excited tube length, when $\lambda_c / L_e \ll 1$.

The power spectral density, $S_F(f)$, can be rendered dimensionless using a pressure scaling factor, p_o, and a frequency scaling factor, f_o, as follows:

$$\widetilde{S}_F(f_R) = \frac{S_F(f)}{(p_o D)^2} f_o \tag{10-2}$$

where, f_R is the reduced frequency, defined as f/f_o, and D is the tube diameter.

A difficulty arises in the estimation of $S_F(f)$ because the correlation length, λ_c, is rarely known. Axisa et al (1990) present an equivalent power spectral density, $S_F(f_R)_e$, defined as follows:

$$S_F(f_R)_e = \frac{\lambda_c}{L_e} S_F(f_R) \tag{10-3}$$

where, L_e is the excited tube length. Using this definition, the dimensionless equivalent power spectral density for the 1st mode can be defined in terms of tube displacement as follows:

$$\tilde{S}_F(f_R)_e = \frac{\overline{y^2(x)}_1 64\pi^3 f_1^3 m^2 \zeta_1}{\phi_1^2 a_1} \frac{1}{(p_o D)^2} f_o \tag{10-4}$$

where, a_1 is the modal correlation factor which is proportional to the joint acceptance. Values of ϕ_1^2 and a_1 for a variety of end conditions are given in Table 9-1.

Using Eq. (10-4), the mean-square of tube displacement can be found without knowledge of the correlation length. Instead, a small correlation length has been assumed. To correctly compare spectra obtained using experimental rigs with varying geometries, it is necessary to define a reference power spectral density, $S_F(f)_e^o$, based on a reference excited tube length, L_o, and a reference tube diameter, D_o, as

$$S_F(f)_e^o = S_F(f)_e \times \frac{L_e}{L_o} \times \frac{D_o}{D} \tag{10-5}$$

where, L_e is the excited tube length. In this chapter, reference lengths of $L_o = 1$ m and $D_o = 0.02$ m are applied.

When the experimental test program measures the forces directly, the reference equivalent spectrum is calculated as

$$S_F(f)_e^o = S_F(f) \times \frac{1}{a_1} \times \frac{L_e}{L_o} \times \frac{D_o}{D} \tag{10-6}$$

where a_1 is calculated assuming the tubes are rigid.

In single-phase flow, it is possible to collapse the excitation force power spectral density (PSD) curves, $S_F(f)$, using a dynamic pressure head, $p_o = \frac{1}{2}\rho U_p^2$, to reduce the excitation forces and a pitch-velocity-to-diameter ratio, $f_o = \rho U_p/D$, to reduce the frequency component. These scaling factors are defined using the fluid density, ρ, the pitch velocity, U_p and the tube outer diameter, D.

The first approach in two-phase research was to use these scaling factors to collapse two-phase cross-flow force data. Several researchers including Taylor (1994) and de Langre and Villard (1998) have shown that neither scaling factor is suitable.

A second approach was an empirical approach looking at the dependence of the key variables such as tube bundle geometry, mass flux, frequency and void fraction. The original design guidelines (Fig. 10-1 through Fig. 10-3) were developed with this type of parametric approach, but they did not manage to present the data in a dimensionless format. In this chapter, this parametric dependence approach will guide the development of dimensionless upper bounds for two-phase random excitation forces.

The next two sections in this chapter outline the relationship between two-phase random excitation forces and the various flow and geometric parameters. Dimensionless forms for frequency (abscissa) and power spectral density (ordinate) are developed in Sections 10.4 and 10.5, respectively.

10.4 Scaling Factor for Frequency

Power spectral densities are commonly plotted with frequency as the independent variable. The independent variable is generally in the form of a dimensionless frequency that has been reduced by suitable length and velocity scales. For single-phase flow, and void fractions of 10% or less

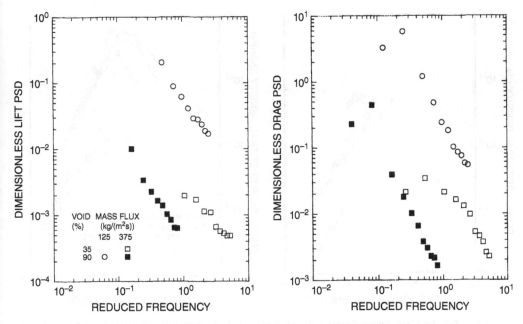

Fig. 10-5 CENS Air-Water Power Spectral Densities for a Normal-Square Tube Bundle with $P/D = 1.5$ Collapsed Using Single-Phase-Type Dimensionless Expressions (Eq. (10-1) and Eq. (10-2)).

(Taylor et al, 1988), the reduced frequency, f_R, defined in Eq. (10-2) precisely collapses each spectrum along the abscissa so that cut-off frequencies are lined up at $f_R = 0.2$ (see Fig. 9-1). Vortex shedding, and, therefore, the Strouhal frequency, $f_s = f_R$, plays a large role in the forced-vibration forces for these very low void fractions (see Chapter 11 for more detail regarding vortex shedding forces). Although 10% void fraction is not a sudden transition point, it is recognized (Hara, 1987b) as a significant point of change from single-phase-like to two-phase characteristics.

For higher void fraction two-phase flow, a different reduced frequency had to be developed. At higher void fractions, broad frequency peaks are found in the two-phase power spectra (see Fig. 10-5). These frequency peaks often dominate the power spectrum and significantly affect the root-mean-square (rms) excitation forces. Therefore, an understanding of the dimensionless frequency required to align these two-phase frequency peaks is essential. The development of a suitable dimensionless frequency was a major hurdle in the data reduction process. The problem was approached in two steps. First, an appropriate velocity scale was determined, a relatively simple task; second, a length scale was defined.

10.4.1 Definition of a Velocity Scale

To isolate the effect of velocity, power spectra for a range of mass fluxes at a single void fraction are plotted in Fig. 10-6 versus the single-phase-type reduced frequency. Since the data is all from the same tube array, the tube-diameter length scale is not affecting the effectiveness of the reduced frequency. Considering only the horizontal alignment of the drag spectra, the broad peaks are lined up at approximately $f_R = 0.4$. The homogeneous pitch velocity, U_p, appears to be a reasonable choice as a velocity scale. Drag direction excitation forces are given more emphasis in determining a suitable

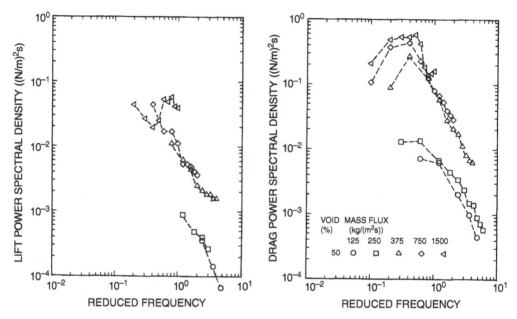

Fig. 10-6 CENS Air-Water Power Spectral Densities at 50% Void Fraction for a Normal-Square Tube Bundle with *P/D* = 1.5 Collapsed Using Single-Phase-Type Reduced Frequency.

frequency scale, because the drag direction has generally higher excitation forces than the lift direction in two-phase flow.

10.4.2 Definition of a Length Scale

The need for a different reduced frequency becomes clear when the effect of void fraction is isolated in Fig. 10-7. The single-phase-type reduced frequency is not able to collapse the broad spectral peaks.

To understand the nature of the broad frequency peaks, it is necessary to determine the flow mechanism that causes them to exist. Taylor (1994) examined the possibility that broad frequency peaks were related to acoustic pressure pulses. Taylor notes that acoustic pressure pulses, like vortex shedding, are a recognized frequency-related vibration phenomenon, but these pressure pulsations did not explain the frequency peaks found in two-phase data.

In an effort to understand the nature of the broad frequency peaks, a peak frequency associated with each measured drag power spectrum from the CENS bundle data was plotted against void fraction (see Fig. 10-8). The error bars shown in Fig. 10-8 indicate a large degree of uncertainty in defining the peak frequencies. Dashed lines are added to highlight the general trend between peak frequencies and void fraction. In general, the peak frequency decreases with increasing void fraction.

An obvious difference between single- and two-phase flows is the presence of a bubble-like structure in a two-phase mixture. The size of bubbles could be used as a length scale parameter. A bubble size and pitch velocity can be related to the frequency at which bubbles are passing a tube. To be an appropriate length scale, the increase in bubble size with void fraction would have to be proportional to the decrease in peak frequency with void fraction.

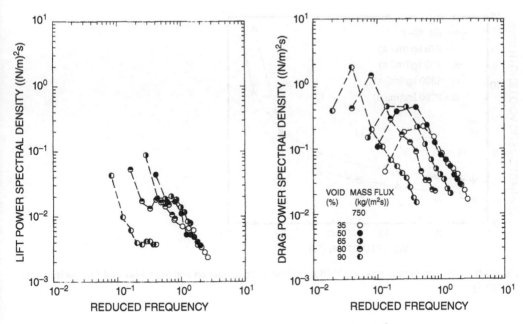

Fig. 10-7 CENS Air-Water Power Spectral Densities at a Mass Flux of 750 kg/(m²s) for a Normal-Square Tube Bundle with *P/D* = 1.5 Collapsed Using Single-Phase-Type Reduced Frequency.

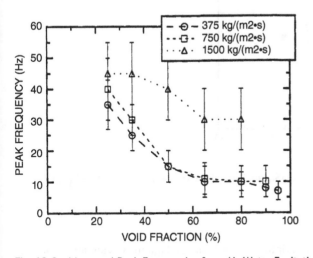

Fig. 10-8 Measured Peak Frequencies from Air-Water Excitation Force Drag Power Spectral Densities.

A limited number of "bubble size" measurements were conducted during the CENS test series. The average "bubble diameters" are plotted in Fig. 10-9. A trend to increasing "bubble diameter" with void fraction is seen at all mass fluxes. In Fig. 10-10, these CENS measurements are compared against "bubble diameter" measurements made by Van der Welle (1985). Van der Welle not only observed similar trends but also measured very similar "bubble diameters". It should be noted that

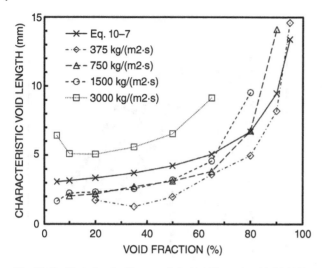

Fig. 10-9 Variation in Characteristic Void Length with Void Fraction and Mass Flux and Comparison with Eq. (10-7).

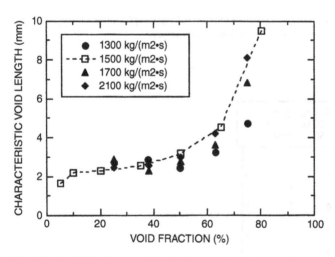

Fig. 10-10 CENS Characteristic Void Length Data (Square Symbols at 1500 kg/m²s)) Compared with Van der Welle (1985) Data (Filled Symbols).

"bubble diameter" is not an appropriate descriptor for this measured quantity. Van der Welle uses the term "bubble diameter," but he notes that theory dictates that bubble diameter should decrease with void fraction. In practice, the theory, which is based on a model for drop diameters (involving surface tension and a Weber number), is only successful for void fractions up to 20%. In Fig. 10-9, a decrease in bubble diameter with void fraction is observed in the 3000 kg/m²s data but not at the other mass fluxes. Van der Welle states that the significant deviation at higher void fractions is highly plausible because the flow will be annular rather than bubbly, which means the model loses its physical basis. In the remainder of this chapter, the measured "bubble diameter" shall be referred to as the characteristic void length, D_B.

An expression for the characteristic void length in terms of known flow parameters was required. A multiple linear regression exercise was used to define a characteristic-void-length expression that is based on suitable flow parameters. The most influential flow parameters were determined to be the two-phase velocity, U_p, and the liquid fraction, $(1-\varepsilon_g)$. It is somewhat surprising to find that the characteristic void length depends on velocity. In both Van der Welle's data (Fig. 10-10) and the data from this study (Fig. 10-9), the higher-void-fraction data at a constant void fraction shows an increase in D_B with mass flux. However, the reason why the void characteristic lengths are larger at a mass flux of 3000 kg/m²s than at the other mass fluxes is not clear. Clearly, the effect of velocity and flow regime on D_B are not well understood.

Taylor (1994) developed a characteristic-void-length expression that included both velocity and void fraction dependence, but this expression did not lead to a dimensionless reduced frequency. A few years later, de Langre and Villard (1998) introduced a characteristic void length that generally followed Taylor's approach but did not include the relatively small velocity dependence. This characteristic void length, D_B, is empirically defined as follows:

$$D_B = 0.1D/\sqrt{1-\varepsilon_g} \qquad (10\text{-}7)$$

A plot of characteristic void length comparing measured values against the calculated fit based on Eq. (10-7) is given in Fig. 10-9. The plotted equation follows the mid-velocity curves rather well and is acceptable as a suitable length scale.

Further characteristic-void-length measurements are necessary to determine the effects of tube-array geometry and flow regime on the characteristic void length. For example, void measurements and flow regime measurements should be taken in a gap between tubes rather than just below a gap, as they were in Taylor (1994). A more sophisticated model for void characteristic length might have different expressions for each flow regime.

10.4.3 Dimensionless Reduced Frequency

A dimensionless frequency defined using D_B and U_p is calculated as follows:

$$\text{Reduced Frequency, } f_R = f D_B/U_p \qquad (10\text{-}8)$$

The effectiveness of this reduced frequency is illustrated in Fig. 10-11, which plots the spectra from Fig. 10-7 against the reduced frequency defined in Eq. (10-8). The alignment of the peak frequencies in Fig. 10-11 shows a marked improvement compared to the alignment in Fig. 10-7.

10.4.4 Effect of Frequency

Although two-phase frequency spectra vary in shape, there is always a cut-off frequency after which the excitation forces begin to decrease as the dimensionless frequency increases. In single-phase flow, a reduced frequency dependence of $f_R^{-3.5}$ is found (see Chapter 9). Two-phase data has an average high-frequency slope that is less than that found in single-phase flow (see Fig. 10-11).

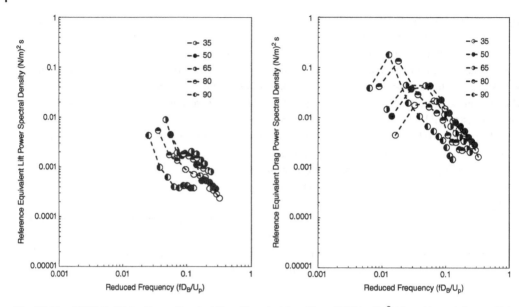

Fig. 10-11 CENS Air-Water Power Spectral Densities at a Mass Flux of 750 kg/(m²s) for a Normal-Square Tube Bundle with $P/D = 1.5$ are Collapsed Using fD_B/U_P.

10.5 Scaling Factor for Power Spectral Density

In single-phase flow, it is possible to collapse the PSD curves using the dynamic pressure head to reduce the excitation forces and a pitch-velocity-to-diameter ratio to reduce the frequency component of the PSD (see Eq. (10-1) and Eq. (10-2)). The inability of this single-phase method to collapse two-phase PSDs was noted in Section 10.3. The single-phase normalization does not account for the effects of flow regime and void fraction.

The need to understand two-phase flow patterns in tube bundles is illustrated in Fig. 10-12, where the PSD is plotted against a void-fraction ratio, $\varepsilon_g/(1 - \varepsilon_g)$. At low void fractions, the PSD increases rapidly with void fraction and mass flux; in contrast, at high void fractions, there is a distinct change in the dependence on void fraction and mass flux. In Fig. 10-12, the PSD at high void fractions decreases somewhat as void fraction and mass flux increase; however, this decreasing trend is not clear in all of the results. What is clear is that the PSD no longer increases with either void fraction or mass flux. This change in behavior can only be explained after examining relevant flow pattern maps. The following section outlines the flow regimes that are found in two-phase cross flow and suggests an explanation for the behavior noted in Fig. 10-12.

It should be noted that the reduced frequency used in Figs. 10-12 to 10-15 is based on the characteristic void length, d_B, defined in Taylor (1994) and Taylor et al (1996), rather than the definition for D_B used in this chapter. This small change in definition does not have a significant effect on the arguments or conclusions in this chapter.

10.5.1 Effect of Flow Regime

Flow regime maps can be examined to assist in understanding the possible flow patterns within a tube bundle. Significant development in the field of two-phase flow pattern maps has occurred in the past few years, as described in Chapter 3, Section 3.2.5. Recently, flow regime maps for vertical

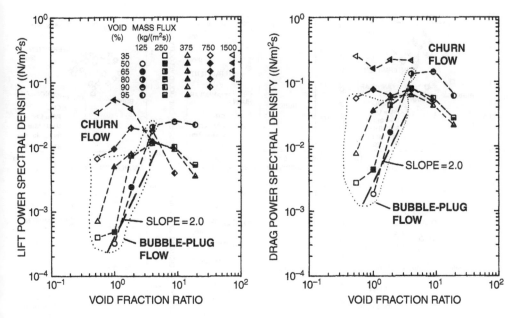

Fig. 10-12 CENS Air-Water Power Spectral Densities at fd_B/U_P = 0.1 for a Normal-Square Tube Bundle with P/D = 1.5.

cross flow in tube bundles have been developed by Kanizawa and Ribatski (2016). They have developed flow maps with flow-regime transitions that are defined using transition expressions that include fluid property and geometric considerations. The development of each transition equation, as well as a physical explanation of the dominant forces in each flow regime, is given in Kanizawa and Ribatski (2016). Flow in tube bundles is beginning to reach a level of understanding similar to that for flow in vertical tubes.

The flow conditions used in the CENS tests have been plotted on Kanizawa and Ribatski flow-regime maps in Fig 10-13. Note that the transition to churn flow corresponds well to the switch between bubbly flow and churn flow shown in Fig. 10-12. In addition, very few of the CENS flow conditions fall in the intermittent flow regime.

The flow conditions used in the CRL tests have also been plotted on Kanizawa and Ribatski maps in Fig. 10-13. Most of the 60° and 90° bundle CRL flow conditions fall within bubbly flow, with some high void fraction and high mass flux conditions falling in the intermittent flow region. In contrast, 30° bundle flow conditions have many data points in the intermittent flow regime.

The transition equations developed by Kanizawa and Ribatski (2016) use fluid properties, pitch-to-diameter ratio and tube diameter. As a result, transition lines shift for experimental programs with differing geometry and fluids. One finds that in air-water, a bundle with smaller P/D has a larger intermittent flow regime.

It should be noted that the steam-water and Freon flow-regime maps shown in Fig. 10-14 do not have an intermittent flow regime because the iterative transition line solution did not converge. This result is consistent with no intermittent flow being observed by the authors during visual observations of Freon flow tests. The reader will find papers and texts that refer to intermittent flow in Freon flow tests because those researchers have used flow maps by Grant (Fig. 3-12) or Ulbrich and Mewes (Figs. 3-14 and 3-16). These earlier flow maps did not distinguish between churn flow and intermittent flow.

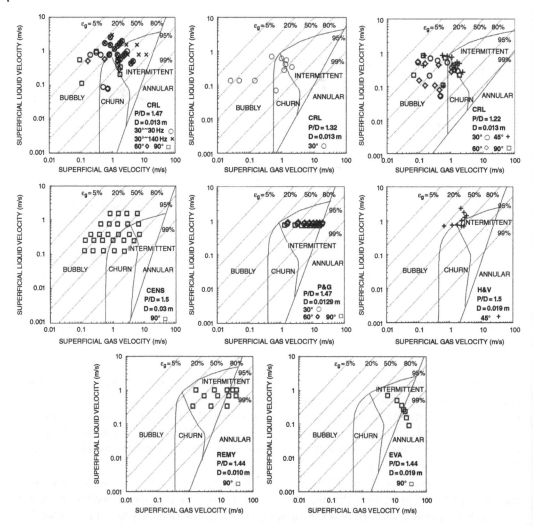

Fig. 10-13 Air-Water Flow Regime Maps Using Kanizawa and Ribatski (2016) Boundaries for Flow Conditions from CRL (Taylor et al, 1989 and Pettigrew et al, 2001), CENS (Taylor et al, 1996), P&G (Pettigrew and Gorman, 1973b), H&V (Heilker and Vincent, 1981), REMY (Remy, 1982) and EVA (Axisa et al, 1986).

The effects of various parameters on two-phase excitation forces in each flow regime are reviewed as dimensionless expressions for the PSDs are developed in Sections 10.5.2 through 10.5.8.

10.5.2 Effect of Void Fraction

The effect of void fraction in the bubbly-plug flow regimes is illustrated using 60° CRL data in Fig. 10-15 and can also be seen for lower void fractions in the CENS data in Fig. 10-12. The excitation forces increase with void fraction in the bubbly flow regime.

In the churn flow regime, the void-fraction effect is almost eliminated and may actually be reversed. This is illustrated in the high-void-fraction ratio data in Fig. 10-12.

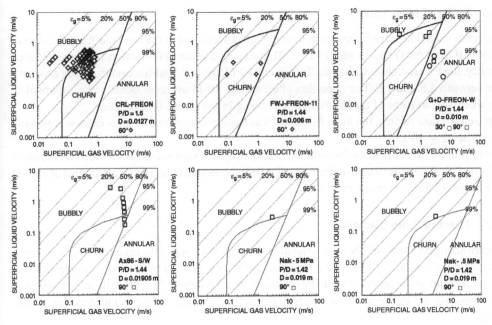

Fig. 10-14 Steam-Water and Freon Flow Regime Maps Using Kanizawa and Ribatski (2016) Boundaries for Flow Conditions from CRL (Pettigrew et al, 1995), FWJ (Feenstra et al, 1996), G&D (Gay et al, 1988 and Delenne et al, 1997), Ax86 (Axisa et al, 1986), and Nak (Nakamura et al, 1982).

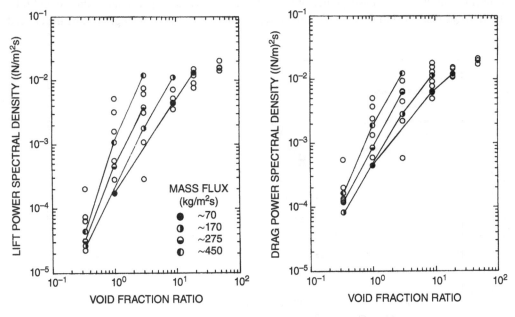

Fig. 10-15 CRL Air-Water Power Spectral Densities at fd_B/U_p = 0.1 for a 60° Rotated-Triangular Tube Bundle with P/D = 1.47 (Empty Circles Signify Mass Fluxes Other Than Those Noted in the Legend.).

10.5.3 Effect of Mass Flux

The effect of mass flux in the bubble-plug flow regime is illustrated in the low-void-fraction ratios of Fig. 10-12 (CENS data) and in Fig. 10-15 (CRL data). At a given void-fraction ratio and reduced frequency, the PSD increases with mass flux. In Fig. 10-12, the low-void-fraction mass flux exponent varies between 1.5 and 2.5. The mass-flux effect in Fig. 10-15 averages closer to an exponent of unity, likely due to some of the flow conditions being in the churn flow regime.

In the churn flow regime, the mass-flux effect was less significant and in some cases was actually reversed (see Fig. 10-12).

In intermittent flow, the mass-flux effect is consistently an exponent of two or more, as shown in Fig. 10-16. This figure provides data from a variety of sources that conducted experiments at a given void fraction while varying the mass flux.

10.5.4 Effect of Tube Diameter

Any force acting on a tube will be proportional to the projected area of the tube surface (tube length times tube diameter). Since the excitation forces have been presented as a force per unit length, the diameter remains as the other factor in the projected area. Therefore, the PSD of the excitation force will be proportional to the square of the tube diameter. This relationship is independent of flow regime.

10.5.5 Effect of Correlation Length

Each time the bundle geometry or the flow condition is altered, the correlation length of the excitation force field changes. Since the actual correlation length is not known, the correlation coefficient has been assumed to be unity. However, if the correlation length (or correlation coefficient) is sufficiently small, the joint acceptance of the excitation forces is proportional to the excited tube length, L_e (as discussed in Chapter 9). Therefore, an effect of correlation length can be introduced

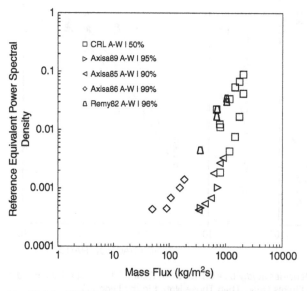

Fig. 10-16 Effect of Mass Flux on Reference Equivalent Power Spectral Density for Several Air-Water Tests.

by multiplying the PSD of the excitation forces by a dimensionless excited tube length. For convenience, the excited tube length is rendered dimensionless by dividing by a tube length equal to 1 m, $L_o = 1$ m.

Although the correlation length changes with each new flow condition, at a given flow condition it is assumed that the upstream correlation lengths were very similar in the CENS and CRL tests. This is a reasonable assumption because the CENS and CRL test sections have similar mixers a short distance upstream of the tube array. Similar mixers can be important, because the type of mixer influences the upstream flow structure and turbulence level, and because the correlation length depends upon the nature of the upstream flow.

A similar argument can be made for the correlation around the diameter of a tube. As a result, an effect of correlation around a tube can be introduced by multiplying the PSD of the excitation forces by a dimensionless diameter. For convenience, the excited diameter is rendered dimensionless by dividing by a reference diameter equal to 0.02 m, $D_o = 0.02$ m.

There is considerable disagreement regarding the need for these correlation length factors. Axisa et al (1990), Taylor (1994) and de Langre and Villard (1998) have introduced these factors, while Au-Yang (2001) argues that they are not necessary. In any case, it is recommended that the use of design guidelines produced in this chapter be restricted to span lengths and diameters that are in the same order of magnitude as those used in these test programs.

10.5.6   Effect of Bundle and Tube-Support Geometry

Unlike fluidelastic instability, random excitation forces acting on a tube in a tube bundle are not affected by the flexibility of the surrounding tubes. Random excitation forces measured in a rigid bundle (using pressure sensors), a flexible bundle (using strain gages) or a flexible tube in a rigid bundle are the same.

In 1990, Axisa et al (1990) stated that steam-water tube displacements in normal-triangular, parallel-triangular and normal-square bundles were quite similar. Pettigrew et al (2001) provide air-water data over a wide range of void fractions and mass fluxes showing that random excitation due to two-phase flow is similar for normal-triangular, parallel-triangular, normal-square and rotated-square tube bundle geometries. Pettigrew et al also found that pitch-to-diameter ratios between 1.22 and 1.47 have similar random excitation forces (see Fig. 10-17). Therefore, a single upper bound can be used to provide a design guideline for random excitation forces for all tube geometries.

As well as determining the appropriate number of tube supports, it is also necessary to consider the effectiveness of the support as it fits around or close to a heat exchanger tube. If the tube-support clearance is too large, the support will not be effective. In the U-bend region of steam generators, a common practice is the use of flat-bar restraints that do not provide any physical restraint in the in-plane direction. In these U-bends, the designer is relying on damping forces and misalignment of tubes in the clearances to ensure that the U-bend tubes do not develop large-amplitude in-plane motions. Research into U-bend tube-support effectiveness and potential tube wear issues in two-phase cross flow has been published, beginning in the early 1990s. Some examples include Taylor et al (1991), Taylor et al (1995), Janzen et al (2005), Feenstra et al (2009), Mohany et al (2012) and Feenstra et al (2017). The more recent U-bend test programs have taken place in Freon cross flow with similar results to the earlier tests in air-water. In all cases, concern was raised regarding the possibility of significant in-plane motion due to random excitation (and possibly fluidelastic instabilities, as noted in Chapter 8).

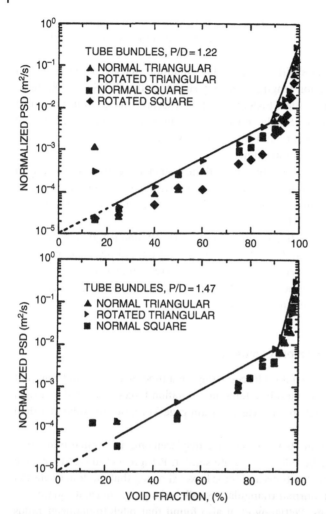

Fig. 10-17 Comparison of Normalized Random Excitation Power Spectral Densities for Different Bundle Geometries and Pitch-to-Diameter Ratios (Pettigrew et al, 2001).

10.5.7 Effect of Two-Phase Mixture

Thus far, this chapter has only used air-water data in the development of the PSD and frequency scaling factors. In the mid-1990s and early 2000s, CRL added a series of Freon tests as defined in Table 10-1. The results of the Freon-22 and Freon-134a rotated-triangular bundle test programs are presented in Figs. 10-18 and 10-19 and the Freon-22 normal-triangular bundle results are presented in Fig. 10-20. As in Figs. 10-12 and 10-15, a scaling length of d_B rather than D_B is used in these figures, with no effect on observed trends or conclusions.

The Freon test program results clearly show that vibration response due to turbulence-induced excitation forces is significantly lower in Freon than in air-water for the same void fraction and mass flow rate. Although the Freon tube bundle was slightly more flexible, the vibration response and, therefore, the excitation force levels were higher in the air-water mixture. Fluid properties such as surface tension, liquid-vapor density ratio and viscosity may play a role in explaining the differing force levels. Two-phase mixtures with lower surface tension and liquid-vapor density

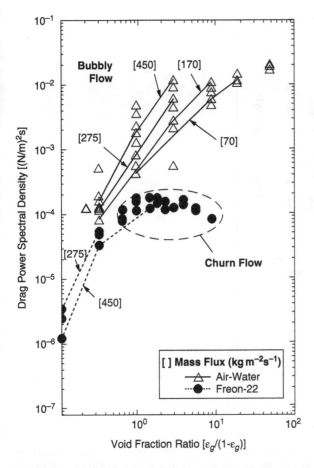

Fig. 10-18 Comparison of Air-Water and Freon-22 Drag Power Spectral Densities of Random Excitation Forces at Reduced Frequency $fd_B/U_p = 0.1$ for 60° Rotated-Triangular Tube Bundles with $PID \approx 1.5$ from Pettigrew et al (1995).

ratios tend to have smoother flows because the interaction forces between the phases are lower. Since Freon has lower surface tension and liquid-vapor density than air-water, it is not surprising that the excitation forces are also lower.

As listed in Table 10-1, CRL Freon tests were done with varying bundle geometry and fluids. These Freon test results were all very similar (see Figs. 10-18 to 10-20). There was no effect of tube bundle geometry or Freon type.

In Fig. 10-18, for lower void fractions, the Freon PSDs increase rapidly with increasing void fraction. Above 40% void fraction, the Freon PSDs do not change significantly with void fraction. This corresponds to the Freon transition from bubbly to churn flow shown in the CRL-Freon flow regime map (Fig. 10-14).

The Freon PSDs differ from the air-water PSDs with respect to the effect of increasing mass flux at a given void fraction. The CRL air-water results show a continuous increase in excitation force level with increasing mass flux. In contrast, the Freon results show almost no effect of increasing mass flux at void fractions above 40%. Again the flow regime maps in Fig. 10-13 show that very few of the CRL air-water tests fall into the churn regime, but rather remain in the bubbly flow regime or transition directly to the intermittent flow regime.

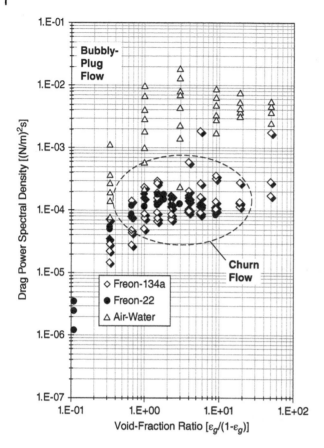

Fig. 10-19 Comparison of Air-Water, Freon-134a and Freon-22 Drag Power Spectral Densities of Random Excitation Forces at Reduced Frequency $fd_B/U_p = 0.1$ for 60° Rotated-Triangular Tube Bundles with $P/D \approx 1.5$ from Pettigrew et al (2002).

Comparing tests done in air-water and steam-water, the results are similar to the comparison between air-water and Freon. From the early tests carried out by Axisa et al (1990), it has been observed that random excitation forces due to steam-water are lower than those found in air-water. Axisa et al (1990) also note that the tube response to steam-water over a wide range of void fractions was almost constant, within a factor of two. The steam-water flow regime maps show that the steam-water tests were conducted in bubbly flow conditions close to the annular flow transition line.

A more recent paper by Mitra et al (2009) provides the results of tests done in bubbly flows in air-water and low-pressure steam-water cross flows. Mitra et al (2009) observed that air-water excitation forces increased quickly with increasing void fraction at low void fractions and that steam-water excitation forces also increased but at a lower rate. These results are consistent with the bubbly flow trends discussed in this chapter.

10.5.8 Effect of Nucleate Boiling

The effect of nucleate boiling is discussed in Chapter 3, Section 3.2.7. The experimental work noted in Chapter 3 consistently found that nucleate boiling does not increase random excitation forces for

Fig. 10-20 Comparison of Air-Water and Freon-22 Drag Power Spectral Densities of Random Excitation Forces at Reduced Frequency $fd_B/U_p = 0.1$ for 30° (Normal-Triangular) Tube Bundles With $P/D \approx 1.5$ from Pettigrew and Taylor (2009).

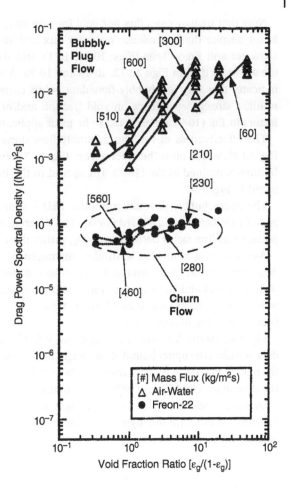

void fractions above 10%. Below 10% void fraction, two-phase effects are known to be very small and nucleate boiling effects were noticeable, but not large.

10.6 Dimensionless Power Spectral Density

Based on the discussion above and many iterations with possible scaling factors, the following dimensionless PSDs have been defined to collapse two-phase excitation forces:

$$\text{Churn/Bubbly Flow} \quad \widetilde{S}_F(f_R) = \frac{S_F(f)}{(\rho_\ell g D)^2 D^2} \frac{U_p}{D_B} \tag{10-9}$$

where ρ_ℓ is the liquid phase density and g is gravitational acceleration.

$$\text{Intermittent Flow} \quad \widetilde{S}_F(f_R) = \frac{S_F(f)}{\left(g\dot{m}_p^2 D\right) D^2} \frac{U_p}{D_B} \tag{10-10}$$

where $\dot{m}_p$ is the pitch mass flux.

Note that neither mass flux nor void fraction ratio were used in the churn/bubbly-flow scaling factor despite the dependence on mass flux and void fraction evident in the low-void-fraction air-water and Freon tests (Figs. 10-12, 10-18 and 10-19) and a strong mass-flux dependence in the air-water data in Figs. 10-12, 10-18 and 10-20. A review of other data (not CRL or CENS data) indicates that very few bubbly-flow data points correspond to void fractions less than 40%. As a result, a strong dependence on void fraction and/or mass flux was not observed. Thus, scaling factors in Eq. (10-9) are expected to be most applicable to void fractions above 40%.

The effectiveness of the churn/bubbly-flow dimensionless PSD (Eq. (10-9)) is illustrated in Fig. 10-21, which plots the spectra from Fig. 10-7 using this PSD scaling factor versus the reduced frequency defined in Eq. (10-8). The spread in the PSD data is significantly reduced when compared to Fig. 10-7.

The churn/bubbly flow dimensionless PSD is similar to the dimensionless scaling factor developed by de Langre and Villard (1998). The difference is that de Langre and Villard (1998) used a pressure scaling factor defined as $\rho_\ell g D_B$, rather than $\rho_\ell g D$.

As the scaling factors are so similar, one might ask why we decided to re-examine the available data. The reason is illustrated by plotting the data using the de Langre and Villard (1998) scaling factors and showing the results superimposed on the upper bound proposed by them, as in Figs. 10-22, 10-23 and 10-24. (A single figure could have been used but the symbols would have been impossible to decipher.)

Fig. 10-22 is the data from de Langre and Villard (1998). Fig. 10-23 shows that most of the CENS data is within the upper bound of de Langre and Villard (1998), but not all. In contrast, about half of the data in Fig. 10-24 is above the upper bound. It is clear that some experimental data is not within the de Langre and Villard (1998) upper bound.

One could speculate that the de Langre and Villard scaling factors do not work for all flow regimes. To examine this hypothesis, the experimental data has been collated by flow regime in Section 10.7.

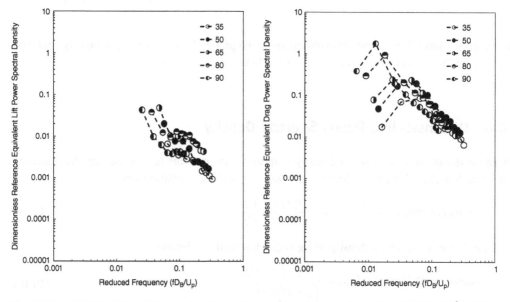

Fig. 10-21 CENS Air-Water Dimensionless Power Spectral Densities at a Mass Flux of 750 kg/(m²s) Collapsed Using Eq. (10-8) and Eq. (10-9).

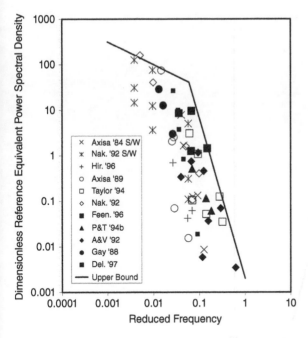

Fig. 10-22   Original Dimensionless Guideline with Data Points from de Langre and Villard (1998).

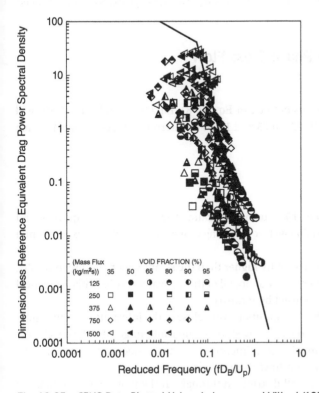

Fig. 10-23   CENS Data Plotted Using de Langre and Villard (1998) Scaling Factors and Upper Bound.

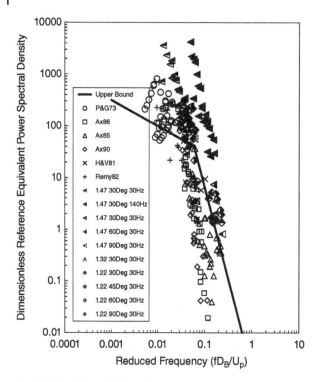

Fig. 10-24   CRL and Other Data Plotted Using de Langre and Villard (1998) Scaling Factors and Upper Bound.

10.7   Upper Bounds for Two-Phase Cross Flow Dimensionless Spectra

The proposed dimensionless scaling factors defined in Eq. (10-8), Eq. (10-9) and Eq. (10-10) will now be applied to the data sets in Figs. 10-22, 10-23 and 10-24 to determine upper bounds for each flow regime.

10.7.1   Bubbly Flow

All of the bubbly flow data is plotted in Fig. 10-25 using the scaling factors defined in Eq. (10-8) and Eq. (10-9). The data has collapsed reasonably well considering that a large number of experimental programs are included in the analysis.

Note that the air-water results tend to be slightly higher than the steam-water and Freon results. The upper bound has been chosen to cover most but not all of the air-water results. This is practical since air-water heat exchangers are rarely used by industry.

A few of the air-water data points with flow conditions very close to the intermittent flow regime displayed strong mass-flux dependence and were moved to the intermittent graph. The developers of flow regime maps have noted that transition to a different flow regime does not occur abruptly. Some flexibility in interpreting the transition lines is appropriate.

The steam-water and Freon results are well distributed throughout the plot suggesting that the scaling factors are suitable for collapsing the results for these fluids.

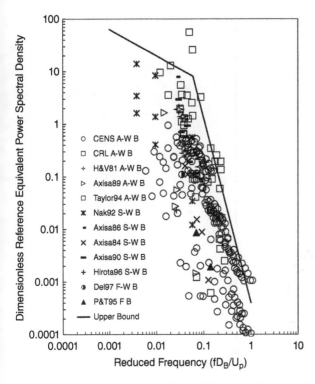

Fig. 10-25 Bubbly Flow Power Spectral Densities Collapsed Using Eq. (10-8) and Eq. (10-9).

The bubbly-flow upper bound (with the same shape used by de Langre and Villard (1998)) is defined as follows:

$$2.0 f_R^{-0.5}, \quad 0.001 < f_R < 0.06 \tag{10-11}$$

$$4.3 \times 10^{-4} f_R^{-3.5}, \quad 0.06 < f_R < 1 \tag{10-12}$$

10.7.2 Churn Flow

The churn data is plotted in Fig. 10-26 using the scaling factors defined in Eq. (10-8) and Eq. (10-9). The data collapse for churn flow is somewhat better than for bubbly flow. This result suggests that different fluid mixtures behave most similarly in this flow regime.

We have chosen to show the annular flow data in the churn flow plot. There is very little annular data and it is not clear whether this is the appropriate placement. For example, annular flow may have a mass-flux dependence similar to intermittent flow. However, we believe that forces caused by the largely air or vapor flow streaming by the tubes with a high superficial gas velocity will be relatively independent of mass flux.

The proposed upper bound shown in Fig. 10-26 is the same upper bound defined for bubbly flow. If a designer was very sure that a churn flow regime would exist in the tube bundle being designed, a lower upper bound for $f_R < 0.06$ could be chosen.

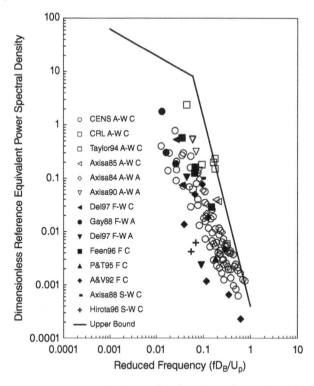

Fig. 10-26 Churn and Annular Flow Power Spectral Densities Collapsed Using Eq. (10-8) and Eq. (10-9).

10.7.3 Intermittent Flow

All of the intermittent flow data is plotted in Fig. 10-27 using the scaling factors defined in Eq. (10-8) and Eq. (10-10). The data collapse for intermittent flow (Fig. 10-27) is somewhat better than for bubbly flow (Fig. 10-25), but this result is not surprising since the intermittent data is only from air-water test programs.

This is the first time that an intermittent flow upper bound has been defined. It is only with the aid of the recent flow regime maps by Kanizawa and Ribatski (2016) that the churn and intermittent flow regimes have been clearly separated in an air-water flow regime map.

The intermittent flow upper bound (again using the same shape introduced by de Langre and Villard (1998) is defined as follows:

$$5.0 f_R^{-0.5}, \quad 0.001 < f_R < 0.06 \tag{10-13}$$

$$1 \times 10^{-3} f_R^{-3.5}, \quad 0.06 < f_R < 1 \tag{10-14}$$

Example 10-1 Random Excitation for Two-Phase Flow in SG U-Bend

Random Excitation Parameters:

$D_0 = 0.02$ m, $L_0 = 1.0$ m, $\phi_1^2 = 2.0$, $a_1 = 1.1$ (pinned-pinned condition from Table 9-1)

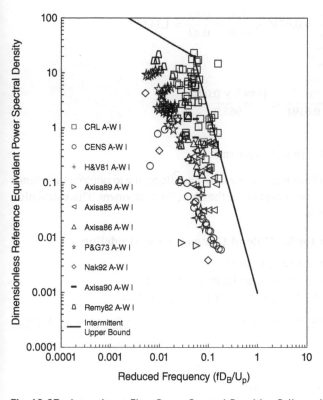

Fig. 10-27 Intermittent Flow Power Spectral Densities Collapsed Using Eq. (10-8) and Eq. (10-10).

From Examples 3-2, 4-2, 4-4 and 6-1 and Fig. 3-6: $D = 0.02$ m, $m = 0.791$ kg/m, $f = 118$ Hz, $\rho_g = 32$ kg/m^3, $\rho_\ell = 760$ kg/m^3 $U_p = 5.09$ m/s, $\varepsilon_g = 0.856$, $L_e = \ell_m = 0.59$ m, $\rho_{TP} = 137$ kg/m^3, $\zeta_T = 0.0191$

Two-Phase Random Excitation Calculation:

Using the steam-water map with $D = 0.019$ m from the flow regime maps in Fig. 10-14 and using Eq. (3-5) to calculate $U_g = 4.36$ m/s and $U_\ell = 0.73$ m/s, we can determine that this two-phase flow is in the bubbly flow regime.

Using Eqs. (10-7), 10-8 and 10-9,

$$D_B = \frac{0.1D}{\sqrt{1-\varepsilon_g}} = \frac{0.1 \times 0.02}{\sqrt{1-0.856}} = 0.0053 \text{ m}$$

$$p_o = \rho_\ell gD = 760 \times 9.81 \times 0.02 = 149.1 \text{ kg/m} \cdot \text{s}^2$$

$$f_o = U_p/D_B = 5.09/0.0053 = 965.9 \text{ Hz}$$

$$f_R = f/f_o = 118/965.9 = 0.1222$$

Using the Design Guideline Eq. (10-12),

$$\widetilde{S}_F(f_R)_e^o = 4.3 \times 10^{-4} f_R^{-3.5}, \quad 0.06 < f_R < 1$$

$$\widetilde{S}_F(f_R)_e^o = 4.3 \times 10^{-4}(0.1222)^{-3.5} = 0.6746$$

$$\tilde{S}_F(f_R)_e = \tilde{S}_F(f_R)_e^o \times \frac{L_o}{L_e} \times \frac{D}{D_o} = 0.6746 \times \frac{1.0}{0.59} \times \frac{0.02}{0.02} = 1.143$$

$$\overline{y^2}(x) = \frac{\tilde{S}_F(f_R)_e \phi_1^2 a_1}{64\pi^3 f_1^3 m^2 \zeta_1} \times \frac{(p_o D)^2}{f_o}$$

$$= \frac{1.143 \times 2.0 \times 1.1}{64\pi^3 \times 118^3 \times 0.791^2 \times 0.0191} \times \frac{(149.1 \times 0.02)^2}{965.9}$$

$$= 5.950 \times 10^{-10} \, \mathrm{m}^2$$

$$y(x)_{rms} = \left[\overline{y^2}(x)\right]^{0.5} = 2.44 \times 10^{-5} \, \mathrm{m} = 24.4 \, \mu\mathrm{m} \, \mathrm{rms}$$

The design guideline for maximum allowable rms vibration amplitude due to random turbulence excitation is based on calculated fretting-wear depth over the life of the component. This calculation is completed in Appendix A, Example A.2.

Example 10.2 Random Excitation for SG U-Bend Using an Early Guideline

Random Excitation Parameters (same as Example 10.1):

From previous calculations: $D = 0.02$ m, $m = 0.791$ kg/m, $f = 118$ Hz, $U_p = 5.09$ m/s, $\varepsilon_g = 0.856$, $\ell = L_e = 0.59$ m, $\rho_{TP} = 137$ kg/m^3, $\zeta_T = 0.0191$

Two-Phase Random Excitation Calculation (Early Empirical Approach from Pettigrew et al, 1991):

$$S_F(f) = NPSD_{g\ell}\left(\frac{\dot{m}^2 D^2}{L_e f^{1.25}}\right)$$

$$NPSD = 10^{(0.027\varepsilon_s - 3.45)}, \quad \text{for } 10\% < \varepsilon_g < 80\%$$

$$NPSD = 0.05, \quad \text{for } 80\% < \varepsilon_g < 99\%$$

$$S_F(f) = 0.05\left(\frac{(5.09 \times 137)^2 \times 0.02^2}{0.59 \times 118^{1.25}}\right) = 0.042 \, \mathrm{N}^2\mathrm{m}^{-2}\mathrm{s}$$

$$\overline{y^2}(0.5\ell) = \frac{S_F(f)\phi_1^2 a_1}{64\pi^3 f^3 m^2 \zeta_1}$$

$$= \frac{0.042 \times 2.0 \times 1.1}{64\pi^3 \times 118^3 \times 0.791^2 \times 0.0191} = 2.39 \times 10^{-9} \, \mathrm{m}^2$$

For two spans (ineffective supports) $f = 29.5$ Hz, $\zeta = 0.0294$ and $\overline{y^2}(0.5\ell) = 9.86 \times 10^{-8}$ m^2

For effective supports, $y(0.5\ell) = 48.9$ μm rms

For ineffective supports $y(0.5\ell) = 314$ μm rms

The rms vibration response at the U-bend tube midspan calculated with the earlier guideline (48.9 μm rms) is within a factor of two of that calculated with the current guideline (24.4 μm rms) (see Example 10-1).

10.8 Axial Flow Random Turbulence Excitation

Random turbulence excitation due to two-phase axial flow can also exist in process heat exchangers and nuclear components. In single-phase axial flow, significant far field pressure fluctuations are often transmitted to a structure as standing waves or convected turbulence. In two-phase flow, far-field pressure fluctuations are usually quickly attenuated by the inherently higher damping of two-phase mixtures. In two-phase flow, near-field pressure fluctuations are generated by turbulence enhanced by the presence of gas or vapor bubbles. This situation results in a more or less broadband random pressure field acting at the surface of a structure.

The analytical formulation for random excitation in axial flow is identical to that for cross flow. The mean-square response can be expressed using Eq. (10-1). However, experimental evidence suggests that the correlation length is much larger in axial flow than in cross flow (Kaneko et al, 2014).

Valuable pioneering experimental work was done by Gorman (1971) with a cylinder subjected to confined annular two-phase flow simulated by air-water mixtures. Pressure measurements were taken at several axial and peripheral locations on the wall of the test section. These measurements yielded useful information on the statistical properties of the random pressure field. For example, Fig. 10-28 shows the axial spatial correlation along the test section for two different simulated steam qualities. Using this information, Gorman (1971) had remarkable success in predicting the vibration response of the test cylinder. This success is demonstrated in Fig. 10-29, where predicted and measured vibration amplitudes are compared. The vibration response is maximum at a simulated steam quality of approximately 15%. Such behavior has also been observed in steam-water flow (Pettigrew and Gorman, 1973a; and Pettigrew, 1993). This behavior is explained in terms of flow regime changes, as discussed later in this section.

Measurements of vibration response in two-phase axial flow have largely been conducted as a part of nuclear fuel channel studies. As a consequence, a large proportion of the tests used steam-water flow and the test rig geometries were similar. Many flow and geometric parameters affect the magnitude of the excitation forces in axial two-phase flow. Because a significant range of geometric parameters and fluid properties have not been studied, our discussion herein concentrates on the changes for random excitation forces over a range of flow parameters in steam-water flow. Some of the most important flow parameters are flow regime, void fraction, temperature or pressure, and mass flux or velocity.

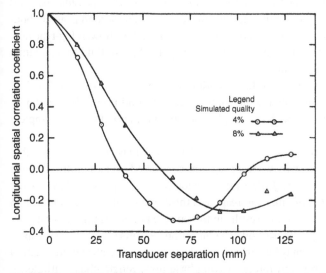

Fig. 10-28 Axial Spatial Correlation of Random Pressure Fluctuations in Two-Phase Annular Flow (Gorman, 1971).

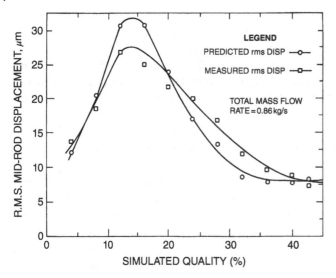

Fig. 10-29 Measured and Predicted Vibration Amplitude versus Simulated Steam Quality in Axial Two-Phase Flow (Gorman, 1971).

In fuel bundles and heat exchanger tube bundles, the flow area around each cylinder is an annulus described by an equivalent outer diameter, D_e. Flow regimes found in the annulus have been defined by several authors including Kelessidis and Dukler (1989). Flow patterns, such as bubbly, slug, churn, lumpy annular and annular can exist in an annulus.

The effect of flow regime on the void-fraction dependence of $S_F(f)$ is clearly illustrated in Fig. 10-30. This data corresponds to the lumpy-annular and annular flow regimes. In fact, the transition from lumpy annular to annular flow is clearly indicated by a sudden decrease in the magnitude of $S_F(f)$. A possible reason for the decrease in excitation force in annular flow is that a liquid wall formed by a layer of water on the cylinder surface does not transmit the random forces. Another possible reason is that the annular flow is less hydraulically noisy than lumpy-annular flow. The second peak at about 97% void fraction suggests that the annular flow may change to a mist or fog-type flow. A fog-type flow would occur when the void fraction is very high and there is no longer sufficient fluid to maintain a liquid film on the surface of the vibrating cylinder.

At lower void fractions, the flow is bubbly and the excitation forces increase with void fraction. However, complications due to changing flow regimes that depend on fluid properties, flow path geometries and flow conditions do not allow for a simple normalizing method for the complete void-fraction range.

In bubbly and lumpy-annular flow regimes, the effects of changes in mass flux are reasonably predictable. Figure 10-31 includes results from two test programs conducted over a significant range of mass fluxes. One set of data suggests that $S_F(f)$ depends on velocity to the exponent 2.7, while the other set indicates an exponent of 1.56. For vibration response, these exponents correspond to 1.35 and 0.78, respectively. Taking unity as an average results in the same exponent as that suggested by Pettigrew and Gorman (1973a). In single-phase flow, the vibration response tends to be proportional to velocity squared. This difference may be attributed to the predominant effect of interaction between phases in the generation of random forces in two-phase flow. Single-phase turbulence spectra do not apply in two-phase flow.

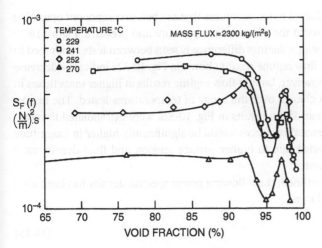

Fig. 10-30 Effect of Flow Regime on Void Fraction Dependence of $S_F(f)$ in Steam-Water Axial Flow (Data is from Pettigrew and Gorman, 1973a).

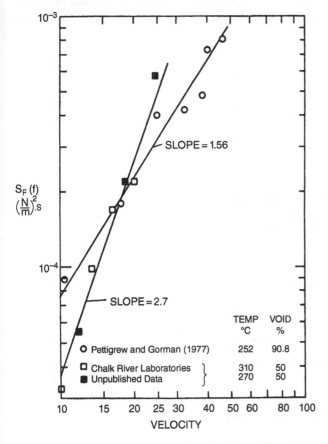

Fig. 10-31 Velocity Dependence of $S_F(f)$ in Steam-Water Axial Flow (Pettigrew and Taylor, 1994).

Since laboratory testing is often conducted at temperatures that are below component operating temperatures, it is important to understand the effects of temperature and pressure. Figure 10-32 illustrates the effect of temperature. Again, a distinct difference is seen between tests conducted in different flow regimes. Lumpy-annular flow regime results shown in Fig. 10-32a indicate a decrease in $S_F(f)$ with increasing temperature; however, bubbly-flow regime results at higher mass fluxes in Fig. 10-32b suggest that $S_F(f)$ does not change over the range of temperatures tested. The importance of knowing the flow regime is clear. If the results in Fig. 10-32a were extrapolated to room-temperature air-water tests, air-water excitation forces would be significantly higher in magnitude than steam-water results. This is expected due to higher surface tension and fluid-density-ratio values for air-water relative to steam-water.

Based on the parametric effects just outlined, the following power spectral density has been used to collapse axial two-phase excitation forces:

$$Normalized\ PSD = \frac{S_F(f)}{(\dot{m}D)^2} \tag{10-15}$$

In Fig. 10-33, axial high-temperature steam-water data from several sources are compared using Eq. (10-15). In most cases, the reference documents provided vibration response data that had to be converted to $S_F(f)$ using Eq. (10-1). An upper bound on the data provides an appropriate design guideline for use within the specified temperature range and flow regimes. Scatter was inevitable because of nonlinear flow regime transition effects. Much more research is required before the

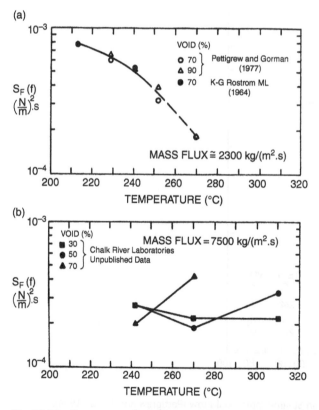

Fig. 10-32 Temperature Dependence of $S_F(f)$ in Steam-Water Axial Flow (Pettigrew and Taylor, 1994).

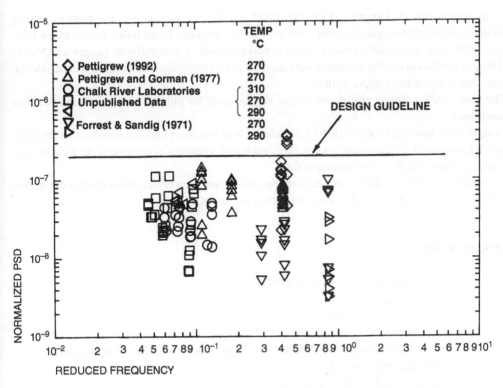

Fig. 10-33 Normalized Power Spectral Density Results from Several Researchers Plotted versus Reduced Frequency (Pettigrew and Taylor, 1994).

prediction of flow regimes will be reliable enough to assume lower excitation forces than those found at the top of the $S_F(f)$ peaks. When plotted against reduced frequency, the resulting guideline is a constant at *Normalized PSD* $= 2 \times 10^{-7}$. Other authors (e.g., Kaneko et al, 2014) point to the many single-phase correlations as an approach to calculate vibration amplitudes in two-phase axial flow. Figure 10-33 is based on steam-water tests done in several different laboratories and is our recommendation for two-phase axial flow in two-phase heat transfer equipment. An approach similar to that used in Example 10-2 could be used to determine the vibration amplitude in two-phase axial flow.

10.9 Conclusions

Random excitation force two-phase cross-flow data from many experimental programs conducted between 1970 and 2005 have been included in the development of suitable scaling factors and upper bounds. The primary purpose of this work has been to prepare guidelines that can be used by steam generator and process heat exchanger designers. Particular attention has been given to the effect of flow regime and the possibility of using a single upper bound.

The main conclusions are as follows:

- In agreement with Taylor (1994) and de Langre and Villard (1998), a characteristic void length is appropriate in defining the length scale for the frequency scaling factor.

- In agreement with de Langre and Villard (1998), gravity has been used in the pressure scaling factor, recognizing that gravity plays a key role in the buoyancy forces found in two-phase flow.
- Most of the data generated in many experimental programs is within the de Langre and Villard (1998) upper bound but the expanded data set used in this study shows that there is experimental data that is above their upper bound.
- The same scaling factors and upper bound can be used for bubbly, churn and annular flow conditions.
- In agreement with Taylor et al (1989), intermittent flow results in high-amplitude random vibration. This intermittent flow data is mass-flux dependent, resulting in separate scaling factors and a unique upper bound for intermittent flow.
- Recent advances in two-phase cross-flow regime maps have been critical in increasing our understanding of two-phase cross-flow random excitation forces.

Nomenclature

a_i	= modal correlation factor for Mode i
D	= tube diameter, m
D_B	= characteristic void length, m
D_o	= reference tube diameter, m
d_B	= characteristic void length defined in Taylor (1994) and Taylor et al (1996)
f_R	= reduced frequency, fD/U_p
g	= gravitational acceleration, m/s^2
f_i	= tube natural frequency for Mode i, Hz
f_o	= frequency scaling factor, Hz
J_i^2	= joint acceptance for Mode i
L_e	= excited tube length, m
L_o	= reference excited tube length, m
ℓ	= span length, m
M_i	= generalized mass per unit length for Mode i, kg/m
m	= total mass per unit length ($m_h + m_t$), kg/m
$\dot{m}_p$	= pitch mass flux, kg/(m^{2}s)
P	= pitch between tubes, m
p_o	= pressure scaling factor, Pa
$S_F(f)$	= force power spectral density per unit length, (N/m)^{2}s
$S_F(f)_e$	= equivalent force power spectral density, (N/m)^{2}s
$S_F(f)_e^o$	= reference equivalent force power spectral density, (N/m)^{2}s
$\tilde{S}_F(f_R)$	= dimensionless power spectral density
U_p	= homogenous pitch velocity, m/s
x	= distance along the tube, m
$\overline{y^2(x)_1}$	= first-mode mean-square of tube response, m^2
ε_g	= void fraction
ϕ_i^2	= normalized mode shape for Mode i
λ_c	= correlation length, m
ρ	= fluid density, kg/m^3
ρ_ℓ	= liquid phase density, kg/m^3
ζ_i	= generalized damping ratio for Mode i

References

Álvarez-Briceño, R., Kanizawa, F. T., Ribatski, G. and de Oliveira, L. P. R., 2018, "Validation of Turbulence Induced Vibration Design Guidelines in a Normal Triangular Tube Bundle During Two-Phase Cross Flow," *Journal of Fluids and Structures*, 76, pp. 301–318, 2017.10.013.

Álvarez-Briceño, R. and de Oliveira, L. P. R., 2020, "Combining Strain and Acceleration Measurements for Random Force Estimation via Karman Filtering on a Cantilevered Structure," *J. Sound and Vibration*, 469, 2019.115112.

Au-Yang, M. K., 2001, *Flow-Induced Vibration of Power and Process Plant Components*, Professional Engineering Publishing Ltd, New York, NY.

Axisa, F., and Villard, B., 1992, "Random Excitation of Heat Exchanger Tubes by Two-Phase Cross-Flows," ASME-WAM *1992 International Symposium on Flow-Induced Vibration and Noise*, (M. P. Païdoussis, principal editor), Vol. 1, Anaheim, Nov. 8-13, pp. 119–139.

Axisa, F., Villard, B., Gibert, R. J., Hetsroni, G. and Sundheimer, P., 1984, "Vibration of Tube Bundles Subjected to Air-Water and Steam-Water Cross-Flow: Preliminary Results on Fluidelastic Instability," *Proceedings of ASME Symposium on Flow-Induced Vibrations*, Vol. 2, New Orleans, Dec. 9-14, pp. 269–284.

Axisa, F., Boheas, M. A. and Villard, B., 1985, "Vibration of Tube Bundles Subjected to Steam-Water Cross-Flow: A Comparative Study of Square and Triangular Arrays," *8th International Conference on Structural Mechanics in Reactor Technology*, Brussels, Belgium, Aug. 19-23, Paper No. B1/2.

Axisa, F., Villard, B. and Sundheimer, P., 1986, "Flow-Induced Vibration of Steam Generator Tubes," Electric Power Research Institute, Palo Alto, CA, *Report No. EPRI NP-4559*.

Axisa, F., Villard, B. and Antunes, J., 1989, "Fluidelastic Instability of a Flexible Tube in a Rigid Normal Square Array Subjected to Uniform Two-Phase Flow," *Proceedings of the International Conference on Engineering Hydro-Aeroelasticity* EAHA, Prague, Czech Republic, Dec. 5-8.

Axisa, F., Antunes, J. and Villard, B., 1990, "Random Excitation of Heat Exchanger Tubes by Cross-Flows," *Journal of Fluids and Structures*, 4, No. 3, pp. 321–341.

Blevins, R. D., 1979, "Flow-Induced Vibration in Nuclear Reactors: A Review," *Progress in Nuclear Energy*, 4, pp. 25–49.

da Silva, B. L., Luciano, R. D., Utzig, J. and Meier, H. F., 2019, "Analysis of Flow Behavior and Fluid Forces in Large Cylinder Bundles By Numerical Simulations," *International Journal of Heat and Fluid Flow*, 75, pp. 209–226.

de Langre, E. and Villard, B., 1998, "An Upper Bound on Random Buffeting Forces Caused by Two-Phase Flows Across Tubes," *Journal of Fluids and Structures*, 12, pp. 1005–1023, fl980180.

Delenne, B., Gay, N., Campistron, R. and Banner, D., 1997, "Experimental Determination of Motion-Dependent Fluid Forces in Two-Phase Water-Freon Cross-Flow," *Proceedings of the Fourth International Symposium on Fluid-Structure Interactions, Aeroelasticity, Flow-Induced Vibration and Noise*, Vol. 2, Dallas, Texas, pp. 349–356.

Feenstra, P. A., 2000, "Modelling Two-Phase Flow-Excited Fluidelastic Instability in Heat Exchanger Tube Arrays," Ph.D. Thesis, McMaster University, Hamilton, Ontario, Canada.

Feenstra, P. A., Weaver, D. S., and Judd, R. L., 1996, "Stability Analysis of Parallel Triangular Tube Arrays Subjected to Two-Phase Cross Flow," *Proceedings of CSME Conference*, McMaster University, Hamilton, ON, May 7-9.

Feenstra, P. A., Weaver, D. S. and Nakamura, T, 2009, "Two-Phase Flow-Induced Vibration of Parallel Triangular Tube Arrays With Asymmetric Support Stiffness," *ASME Journal of Pressure Vessel Technology*, 131, 031301-1

Feenstra, P. A., Sawadoga, T, Smith, B., Janzen, V. and Cothron, H., 2017, "Investigations of In-Plane Fluidelastic Instability in a Multi-Span U-Bend Tube Bundle - Tests in Air Flow," *Proceedings of the ASME 2017 Pressure Vessels and Piping Conference*, Waikoloa, Hawaii, USA, July 16-20, PVP2017-66068.

Gay, N., Decembre, P. and Launay, J., 1988, "Comparison of Air-Water to Water-Freon Two-Phase Cross-Flow Effects on the Vibratory Behaviour of a Tube Bundle," ASME-WAM *1988 International Symposium on Flow-Induced Vibration and Noise*, (M. P. Païdoussis principal editor), Vol. 3, Chicago, Nov. 27-Dec. 2, pp. 139–158.

Gorman, D. J., 1971, "An Analytical and Experimental Investigation of the Vibration of Cylindrical Reactor Fuel Elements in Two-Phase Parallel Flow," *Nuclear Science and Engineering*, 44, pp. 277–290.

Goyder, H. G. D., 1987, "Two-Phase Buffeting of Heat Exchanger Tubes," *Proceedings of the International Conference on Flow-Induced Vibrations*, Bowness-on-Windermere, England, Paper E2.

Goyder, H. G. D., 1988, "Fluidelastic Instability and Buffeting of Heat Exchanger Tube Bundles due to Single and Two-Phase Flows," ASME-WAM *1988 International Symposium on Flow-Induced Vibration and Noise*, (M. P. Païdoussis principal editor), Vol. 2, Chicago, Nov. 27-Dec. 2, pp. 151–168.

Hara, F., 1982, "Two-Phase Cross Flow-Induced Forces Acting on a Circular Cylinder," ASME Symposium on *Flow-Induced Vibration of Circular Cylindrical Structures*, PVP-Vol. 63 (S. S. Chen, M. P. Païdoussis and M. K. Au-Yang editors), Orlando, pp. 9–17.

Hara, F., 1987a, "Vibration of a Single Row of Circular Cylinders Subjected to Two-Phase Bubble Cross-Flow," *Proceedings of the International Conference on Flow-Induced Vibrations*, Bowness-on-Windermere, England, May 12-14, Paper E1.

Hara, F., 1987b, "Vibrations of Circular Cylindrical Structures Subjected to Two-Phase Cross Flows," *JSME International Journal*, 30(263), pp. 711–722.

Hara, F. and Iijima, T, 1988, "Vibration of Two Circular Cylinders in Tandem Subjected to Two-Phase Bubble Cross Flows," ASME-WAM *1988 International Symposium on Flow-Induced Vibration and Noise*, (M. P. Païdoussis principal editor), Vol. 2, Chicago, Nov. 27-Dec. 2, pp. 63–78.

Heilker, W. J. and Vincent, R. Q., 1981, "Vibration in Nuclear Heat Exchangers due to Liquid and Two-Phase Flow," *ASME Journal of Engineering for Power*, 103, pp. 358–366.

Hirota, K., Nakamura, T., Mureithi, N., Kasahara, H., Kusakabe, H. and Takamatsu, H., 1996, "Dynamics of an In-Line Tube Array in Steam-Water Flow - Part 3: Fluidelastic Instability Tests and Comparison with Theory," *Flow-Induced Vibration - 1996 -Proceedings of the ASME PVP Conference*, Montreal, Canada, July 21-26, PVP-Vol.328, pp. 123–134.

Janzen, V. P., Hagberg, E. G., Pettigrew, M. J. and Taylor, C. E., 2005, "Fluidelastic Instability and Work-Rate Measurements of Steam Generator U-Tubes in Air-Water Cross-Flow," *ASME Journal of Pressure Vessel Technology*, 127, pp. 84–91.

Kaneko, S., Nakamura, T., Inada, F., Kato, M., Ishikara, K., Nishihara, T., Mureithi, N.W. and Langthjem, M.A., 2014, *Flow-Induced Vibrations: Classifications and Lessons from Practical Experiences*, 2nd Edition, Elsevier, Academic Press, London.

Kanizawa, F. T and Ribatski, G., 2016, Two-Phase Flow Patterns Across Triangular Tube Bundles for Air-Water Upward Flow," *International Journal of Multiphase Flow*, 80, pp. 43–56.

Kelessidis, V. C. and Dukler, A. E., 1989, "Modelling Flow Pattern Transitions for Upward Gas-Liquid Flow in Vertical Concentric and Eccentric Annuli," *International Journal of Multiphase Flow*, 15, pp. 173–191.

Lian, H. Y., Chan, A. M. C. and Kawaji, M., 1992, "Effects of Void Fraction on Vibration of Tube Bundles under Two-Phase Cross Flow," ASME-WAM *1992 International Symposium on Flow-Induced Vibration and Noise*, (M. P. Païdoussis principal editor), Vol. 1, Anaheim, Nov. 8-13, pp. 109–118.

Lian, H. Y., Noghrenkar, G., Chan, A. M. C. and Kawaji, M., 1997, "Effect of Void Fraction on Vibrational Behaviour of Tubes in Tube Bundle Under Two-Phase Cross Flow," *Journal of Vibration and Acoustics*, 119, pp. 457–463.

Mitra, D., Dhir, V. K. and Catton, I., 2009, "Fluid-Elastic Instability in Tube Arrays Subjected to Air-Water and Steam-Water Cross-Flow," *Journal of Fluids and Structures*, 25, pp. 1213–1235.

Mohany, A., Janzen, V. P., Feenstra, P. and King, S., 2012, "Experimental and Numerical Characterization of Flow-Induced Vibration of Multispan U-tubes," *ASME Journal of Pressure Vessel Technology*, 134, 011301-1.

Nakamura, T., Kanazawa, H. and Sakata, K., 1982, "An Experimental Study on Exciting Force by Two-Phase Cross-Flow," ASME Symposium on *Flow-Induced Vibration of Circular Cylindrical Structures*, PVP-Vol. 63 (S. S. Chen, M. P. Païdoussis and M. K. Au-Yang editors), Orlando, pp. 19–29.

Nakamura, T., Fujita, K., Kawanishi, K., Yamaguchi, N. and Tsuge, A., 1991, "Study on the Vibrational Characteristics of a Tube Array Caused by Two-Phase Flow—Part 1 - Random Vibration," ASME Symposium on *Flow-Induced Vibration and Wear 1991*, PVP-Vol. 206 (M. K. Au-Yang and F. Hara editors), San Diego, June 23-27 pp. 19–24.

Nishida, S., Kawakami, R., Hirota, K., Morita, H., Azuma, S., Kondo, Y. and Komuro, Y., 2019, "Random Excitation PSD Model Acting on Heat Exchanger Tube Bundle Under Two Phase Flow Condition," *Proceedings of the ASME 2019 Pressure Vessel and Piping Conference*, July 14-19, San Antonio, Texas, PVP2019-93458.

Païdoussis, M. P., 1982, "A Review of Flow-Induced Vibrations in Reactors and Reactor Components," *Nuclear Engineering Design*, 74, pp. 31–60.

Pettigrew, M. J., 1981, "Flow-Induced Vibration Phenomena in Nuclear Power Station Components," *Power Industry Research*, 1, pp. 97–133.

Pettigrew, M. J., 1993, "The Vibration Behaviour of Nuclear Fuel under Reactor Conditions," *Journal of Nuclear Science and Engineering*, 114, pp. 179–189.

Pettigrew, M. J. and Gorman, D. J., 1973a, "Experimental Studies on Flow-Induced Vibration to Support Steam Generator Design, Part 1: Vibration of a Heated Cylinder in Two-Phase Axial Flow," *Proceedings of the UKAEA/NPL International Symposium on Vibration Problems in Industry*, Keswick, UK, April 10-12, Paper 424, also *Atomic Energy of Canada Limited, Report No. AECL-4514*.

Pettigrew, M. J. and Gorman, D. J., 1973b, "Experimental Studies on Flow-Induced Vibration to Support Steam Generator Design, Part 3: Vibration of Small Tube Bundles in Liquid and Two-Phase Cross Flow," *Proceedings of the UKAEA/NPL International Symposium on Vibration Problems in Industry*, Keswick, UK, April 10-12, Paper 426, also *Atomic Energy of Canada Limited, Report No. AECL-5804*.

Pettigrew, M. J. and Gorman, D. J., 1978, "Vibration of Heat Exchange Components in Liquid and Two-Phase Cross-Flow," *Proceedings of the BNES International Conference on Vibration in Nuclear Plant*, Keswick, UK, May 9-12, Paper 2.3, also *Atomic Energy of Canada Limited, Report No. AECL-6184*.

Pettigrew, M. J. and Taylor, C. E., 1993, "Two-Phase Flow-Induced Vibration," *Technology for the 90's by ASME PVP Division*, New York, 1993, pp. 811–864.

Pettigrew, M. J. and Taylor, C. E., 1994, "Two-Phase Flow-Induced Vibration: An Overview," *ASME Journal of Pressure Vessel Technology*, 116, pp. 233–253.

Pettigrew, M. J. and Taylor, C. E., 2003, "Vibration Analysis of Shell-and-Tube Heat Exchangers: An Overview- Part 2: Vibration Response, Fretting-Wear, Guidelines," *Journal of Fluids and Structures*, 18, pp. 485–500.

Pettigrew, M. J. and Taylor, C. E., 2009 "Vibration of a Normal Triangular Tube Bundle Subjected to Two-Phase Freon Cross Flow," *ASME Journal of Pressure Vessel Technology*, 131, 051302 (1-7).

Pettigrew, M. J., Tromp, J. H. and Mastorakos, M., 1985, "Vibration of Tube Bundles Subjected to Two-Phase Cross-Flow," *ASME Journal of Pressure Vessel Technology*, 107, pp. 335–343.

Pettigrew, M. J., Carlucci, L. N., Taylor, C. E. and Fisher, N. J., 1991, "Flow-Induced Vibration and Related Technologies in Nuclear Components," *Nuclear Engineering and Design*, 131, pp. 81–100.

Pettigrew, M. J., Taylor, C. E., Jong, J. H., and Currie, I. G., 1995, "Vibration of a Triangular Tube Bundle in Two-Phase Freon Cross Flow," *ASME Journal of Pressure Vessel Technology*, 117(4), pp. 321–329.

Pettigrew, M. J., Taylor, C. E. and Kim, B. S., 2001, "The Effects of Bundle Geometry on Heat Exchanger Tube Vibration in Two-Phase Cross Flow," *Journal of Pressure Vessel Technology*, 123, pp. 414–420.

Pettigrew, M. J., Taylor, C. E., Janzen, V. P. and Whan, T., 2002, "Vibration Behavior of Rotated Triangular Tube Bundles in Two-Phase Cross Flows," *ASME Journal of Pressure Vessel Technology*, 124, pp. 144–153.

Remy, F. N., 1982, "Flow-Induced Vibration of Tube Bundles in Two- Phase Cross-Flow," Proceedings of the *UKAEA/BNES Third Keswick International Conference on Vibration in Nuclear Plants*, Keswick, UK, April 10-12, Paper 68.

Ribatski, G., Kanizawa, F. T., Álvarez-Briceño, R. and de Oliveira, L. P. R., 2019, "Flow Induced Vibration, Flow Patterns, Void Fraction and Pressure Drop for Two-Phase Flow Across Triangular Tube Bundles," *10th International Conference on Multiphase Flow, ICMF 2019*, Rio de Janeiro, Brazil, May 19-24.

Sim, W. G. and Mureithi, N. W., 2013, Drag Coefficient and Two-Phase Friction Multiplier on Tube Bundles Subjected to Two-Phase Cross Flow," *ASME Journal of Pressure Vessel Technology*, 135, 011302-1.

Taylor, C. E., 1994, "Random Excitation Forces in Tube Arrays Subjected to Water and Air-Water Cross Flow," Ph.D. Thesis, University of Toronto, Toronto, Canada.

Taylor, C. E., Pettigrew, M. J., Axisa, F. and Villard, B., 1988, "Experimental Determination of Single- and Two-Phase Cross-Flow-Induced Forces on Tube Rows," *ASME Journal of Pressure Vessel Technology*, 110, pp. 22–28.

Taylor, C. E., Currie, I. G., Pettigrew, M. J. and Kim, B. S., 1989, "Vibration of Tube Bundles in Two-Phase Cross-Flow: Part 3 - Turbulence-Induced Excitation," *ASME Journal of Pressure Vessel Technology*, 111, pp. 488–500.

Taylor, C. E., Pettigrew, M. J. and Tromp, J. H., 1991, "Vibration of Steam Generator U-Bend Tubes: Effectiveness of Flat-Bar Restraints," *Proceedings of 1991 Pressure Vessels and Piping Conference*, San Diego, California, USA, June, June 23-27, PVP-Vol. 206, pp. 1–8.

Taylor, C. E., Boucher, K. M. and Yetisir, M., 1995, "Vibration and Impact Forces Due to Two-Phase Cross Flow in U-Bend Region of Nuclear Steam Generators," *Flow-Induced Vibration, Proceedings Sixth International Conference on Flow-Induced Vibration*, London, UK, April, pp. 401–411.

Taylor, C. E., Pettigrew, M. J. and Currie, I. G., 1996, "Random Excitation Forces in Tube Bundles Subjected to Two-Phase Cross-Flow," *ASME Journal of Pressure Vessel Technology*, 188, pp. 265–277.

Van der Welle, R., 1985, "Void Fraction, Bubble Velocity and Bubble Size in Two-Phase Flow," *International Journal of Multiphase Flow*, 11, pp. 317–345.

Weaver, D. S., and Fitzpatrick, J. A., 1987, "A Review of Flow-Induced Vibrations in Heat Exchangers," *Proceedings of the International Conference on Flow-Induced Vibrations*, Bowness-on-Windermere, England, May 12-14, Paper A1.

Zhang, C., Pettigrew, M. J. and Mureithi, N. W., 2005, "Vibration Excitation Force Measurements in a Rotated Triangular Tube Bundle Subjected to Two-Phase Cross Flow," *Proceedings of PVP2005 ASME Pressure Vessels and Piping Division Conference*, Denver, Colorado, USA, July 17-21, PVP2005-71464.

11

Periodic Wake Shedding and Acoustic Resonance

David S. Weaver, Colette E. Taylor, and Michel J. Pettigrew

11.1   Introduction

Periodic wake shedding or more generally vortex shedding may be a problem when the shedding frequency coincides with the natural frequency of a cylindrical structure. This coincidence may lead to resonance and large vibration amplitudes. Periodic-wake-shedding resonance has been observed by Pettigrew and Gorman (1978) in tube bundles subjected to liquid flows (Fig. 11-1).

The vortex shedding phenomenon for an isolated single cylinder in cross flow has been studied extensively. See, for example, Griffin (1980). A typical example of the laminar-vortex-shedding phenomenon is shown in Fig. 11-2. The shedding frequency, f_S, may be formulated in terms of a Strouhal Number, S,

$$S = f_S D/U \tag{11-1}$$

where D is the tube diameter and U is the upstream flow velocity. The Strouhal Number is approximately 0.2 for Reynolds Numbers of 200 to 3×10^5, as shown in Fig. 11-3. The resulting periodic forces, F_L, may be formulated by analogy to steady fluid forces as

$$F_L = C_L D\rho U^2/2 \tag{11-2}$$

where ρ is the density of the fluid and C_L is the dynamic lift coefficient, which is approximately 0.5 for an isolated single cylinder as shown in Fig. 11-4.

The topic of vortex-shedding resonance of a single cylinder in cross flow will not be discussed further here as it is amply covered in the literature. More importantly, the discussion will focus on periodic wake shedding in tube bundles. The true nature of periodic wake shedding in tube bundles is not yet completely understood. For a very open tube bundle with a large *P/D*, say 2.5, the behavior may be similar to that of a single cylinder with a Strouhal number of about 0.2. For a closely spaced tube bundle, the behavior is governed by the available space between tubes to allow for the formation of coherent vortices. Figure 11-5 shows a flow visualization of vortex shedding in a rotated-square array with a pitch-to-diameter ratio of 1.7 (Weaver et al, 1993). Clearly, the coherent vortices scale with the space between the tubes which explains why smaller pitch ratio arrays exhibit smaller vortices associated with higher frequencies, larger Strouhal numbers, and smaller lift forces.

Periodic-wake-shedding resonance may be of concern in liquid cross flow where the flow is relatively uniform. It is not normally a problem at the entrance region of steam generators because the

Flow-Induced Vibration Handbook for Nuclear and Process Equipment, First Edition.
Michel J. Pettigrew, Colette E. Taylor, and Nigel J. Fisher.
© 2022 John Wiley & Sons, Inc. This Work is a co-publication between ASME Press and John Wiley & Sons, Inc.

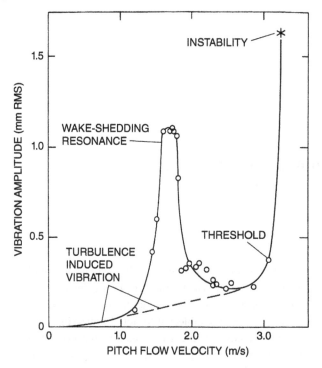

Fig. 11-1 Typical Vibration Response (Gorman, 1976).

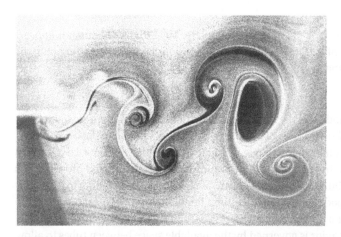

Fig. 11-2 Laminar Vortex Formation in the Wake of a Vibrating Cylinder at a Reynolds Number of 190. The Flow Pattern was Made Visible by Injecting a Sheet of Small Aerosol Particles into a Wind Tunnel Test Section (Griffin, 1980).

flow is non-uniform and quite turbulent (Pettigrew et al, 1973). Turbulence may inhibit periodic wake shedding in tube arrays (Cheung and Melbourne, 1983).

Acoustic resonance is a common problem in heat exchangers with shell-side gas flow. Acoustic resonance can occur when the periodic-wake-shedding frequency of the tubes coincides with the

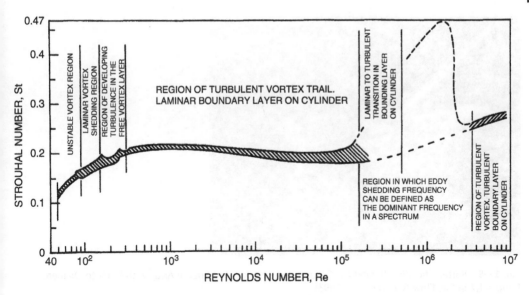

Fig. 11-3 Envelope of Strouhal Number versus Reynolds Number for Circular Cylinders (Lienhard, 1966).

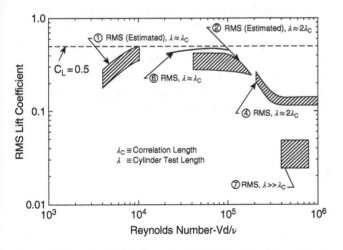

Fig. 11-4 Lift Coefficients for a Single Cylinder (Gerlach and Dodge, 1970).

acoustic frequency of the vessel containing the tube bundle. This problem is addressed separately later in this chapter.

Resonance of tubes due to periodic wake shedding is usually not a problem in gas heat exchangers. The gas density is usually too low to cause significant periodic forces at flow velocities close to resonance. Normal flow velocities in gas-flow heat exchangers are usually much higher than those required for resonance. However, resonance may be possible in high-pressure components with high-density gas and for higher modes of vibration with higher natural frequencies corresponding to higher flow velocities.

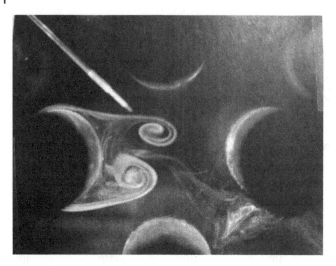

Fig. 11-5 Vortex Shedding Behind the Second Row in a Rotated-Square Array with Pitch-to-Diameter Ratio = 1.7 in Gas Flow (Weaver et al, 1993).

The early literature regarding vortex shedding in tube bundles can be confusing but by the late 1980s a much better understanding was emerging. Generally, there is less information on the magnitude of periodic-wake-shedding forces in tube bundles than for frequencies, since acoustic resonance in gas-flow heat exchangers has proven to be a more significant problem than tube vibration due to vortex shedding. It is important to note that periodic wake shedding is equally possible in finned-tube arrays (Mair et al, 1975; Chen, 1968; Bryce et al, 1978; and Kouba, 1986). In fact, in some cases, the phenomenon is better defined in finned-tube bundles (Bryce et al, 1978). Thus, periodic-wake-shedding resonance must be considered in finned-tube bundles.

11.2   Periodic Wake Shedding

11.2.1   Frequency: Strouhal Number

Periodic wake shedding in tube arrays is described in terms of a Strouhal Number, $S = f_s D/U_p$, where f_s is the vortex shedding frequency and U_p is the pitch velocity. The fluctuating lift coefficient, C_L, is used to estimate the periodic lift forces, F_L, due to wake shedding; thus, $F_L = C_L D\rho U_p^2/2$, where U_p is the pitch velocity at resonance, $U_p = f_s D/S$.

Figure 11-6 shows Strouhal numbers for tube bundles of various configurations and pitch-to-diameter ratios, P/D, in liquid flow (Pettigrew and Gorman, 1978). The Strouhal numbers based on the pitch flow velocity are generally between 0.33 and 0.67. For prominent resonance peaks in triangular bundles of P/D between 1.3 and 1.4, the Strouhal numbers are between 0.40 and 0.67.

Weaver et al (1987) reviewed the data for wake shedding in tube bundles in more detail. The data is summarised in Fig. 11-7 where the Strouhal numbers, St, are defined in terms of the approach velocity, U_∞. Although there is a lot of scatter in the data, expressions based on Owen's (1965) theory were proposed to formulate the average values, as shown by the curves in Fig. 11-7.

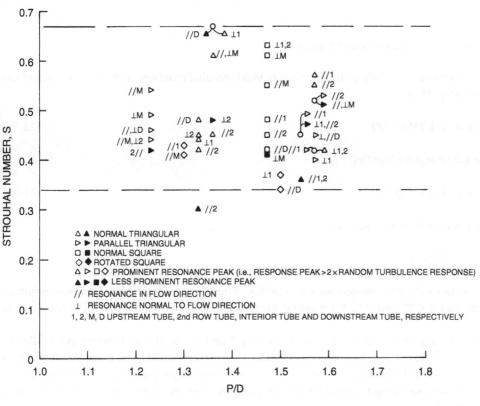

Fig. 11-6 Strouhal Numbers for Tube Bundles in Liquid Flow (Pettigrew and Gorman, 1978).

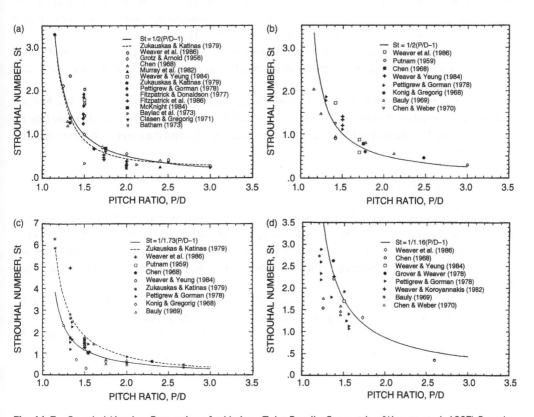

Fig. 11-7 Strouhal Number Expressions for Various Tube Bundle Geometries (Weaver et al, 1987) Based on Upstream Velocity, U_∞ : (a) In-Line-Square Arrays, (b) Rotated-Square Arrays, (c) Normal-Triangular Arrays and (d) Parallel-Triangular Arrays.

These expressions can easily be transformed to yield Strouhal numbers, S, defined in terms of the pitch velocity. Thus,

$$S = (1/1.731)(D/P) \tag{11-3a}$$

for normal-triangular bundles,

$$S = (1/1.161)(D/P) \tag{11-3b}$$

for rotated-triangular bundles, and

$$S = (1/2)(D/P) \tag{11-3c}$$

for both normal- and rotated-square bundles. These expressions give Strouhal numbers between about 0.32 and 0.70 for realistic heat exchanger tube bundles of P/D between 1.23 and 1.57.

This Strouhal number range is very similar to that found in Fig. 11-6 and corresponds to dimensionless pitch velocities U_p/fD between 1.5 and 3.0. Resonance should be assumed possible within this velocity range. In the case of a tube subjected to a time-varying velocity, only the fluctuating fluid forces corresponding to the region within the above range of dimensionless velocities need to be considered.

Other formulations have been developed for the prediction of Strouhal numbers in tube arrays. Several of these charts and empirical formula are summarized in the review by Weaver (1993). None of them have proven to be more accurate than the formulation proposed by Weaver et al (1987) based on Owen's hypothesis (1965) and given in Eq. (11-3).

The data on Strouhal numbers for finned-tube bundles is relatively sparse. Mair et al (1975) propose the use of an equivalent hydraulic diameter, D_{eff}, to estimate Strouhal numbers and fluid forces for finned tubes.

The fins may be approximated by a bare tube of equivalent hydraulic diameter, D_{eff}. The hydraulic diameter is based on the ratio, R_F, of the area occupied by the fins over the available area (the fin thickness divided by the fin pitch) between the root diameter, D_r, and the outer diameter of the fins, D_o, as follows:

$$D_{eff} = D_r + R_F(D_o - D_r) \tag{11-4}$$

Strouhal numbers for finned-tube arrays can then be estimated using data from bare-tube arrays, using the effective diameter for computing both the Strouhal number and the pitch velocity. While fins come in a variety of forms, numerous investigations on single finned tubes and tube arrays have demonstrated the efficacy of this approach, e.g., Ziada et al (2005) and Wang and Weaver (2012).

Kouba (1986) reported some results for normal-triangular finned-tube bundles (Fig. 11-8). He found $S = 0.53$ for $P/D = 1.51$ based on a root diameter, $D_r = 22.5$ mm.

Chen (1968) did experiments on staggered finned-tube bundles. For a nearly normal-triangular array of $P/D = 1.735$, he reported $S = 0.47$.

Fig. 11-8 Strouhal Numbers for Finned Tubes (Kouba, 1986): Dots are Experimental Points, Curve 1 Is from Chen (1977) and Curve 2 from Zukauskas and Katinas (1979).

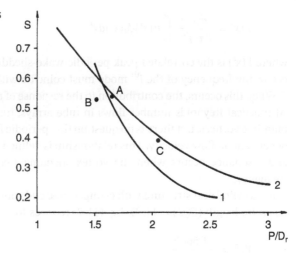

Most of the Strouhal number data in Figs. 11-6 through 11-8 is between 0.40 and 0.67. More weight has been given to the more relevant information, i.e., the average values of Weaver et al (1987) and the data deduced from actual tube resonances (Pettigrew and Gorman, 1978). The single point at $S = 0.20$ in Fig. 11-7(c) may not be significant, as there is no information on its relative magnitude.

In summary, the above results suggest a reference Strouhal number value of roughly $S = 0.50$ for heat exchanger tube bundles. Periodic-wake-shedding resonance is possible for Strouhal numbers between 0.40 and 0.67.

A thorough review of vortex shedding phenomena (Weaver, 1993) showed that the vorticity mechanisms in in-line and staggered tube arrays with practical pitch ratios are fundamentally different. As a result, reliable Strouhal number data for staggered arrays can be determined at resonant and off-resonant conditions; however, reliable Strouhal number data for in-line arrays should be obtained only under resonant conditions. In addition, acoustic resonance can lead to "lock-in" that increases the Strouhal number range for which periodic-wake-shedding resonance is possible (see Section 11.3.4).

11.2.2 Calculating Tube Resonance Amplitudes

If a vortex shedding resonance is predicted at the operating flow rates, the next step is to estimate the tube resonance amplitudes. From a structural dynamics perspective, a cylinder or similar structure can be considered as a beam with known end conditions. The tube response $y(x,t)$ at any point x and at any time t may be expressed as a normal-mode expansion in terms of the generalized coordinates $q_i(t)$ as follows:

$$y(x, t) = \sum_{i=1}^{\infty} \phi_i(x) q_i(t) \tag{11-5}$$

Using Lagrange's equation and assuming that the damping is small and that coupling between the modes does not occur, the equation of motion for the ith mode is

$$\ddot{q}_i(t) + 4\pi f_i \zeta \dot{q}_i(t) + 4\pi^2 f_i^2 q_i(t) = \int_0^L g(x, t) \phi_i(x) dx \tag{11-6}$$

where $g(x, t)$ is the forcing function.

Using Eqs. (11-5) and (11-6), it can be shown that the peak vibration response, $Y(x)$, of a uniform cylinder to periodic forces $g(x', t) = F(x') e^{j2\pi ft}$ at resonance in the ith mode is expressed by

$$Y(x) = \frac{\phi_i(x)}{8\pi^2 f^2 \zeta} \int_0^L F(x')\phi_i(x')dx' \tag{11-7}$$

where $F(x')$ is the correlated peak periodic-wake-shedding force per unit length. For resonance to occur, the frequency of the ith mode must coincide with the Strouhal frequency.

When this occurs, the contribution to the response of modes other than the ith mode is negligible. At practical Reynolds number flows in tube arrays, turbulence masks visualization of clear Karman-like vortices, but there is no question that periodic forces occur at a frequency which is linearly dependent on flow velocity. This relationship is defined using the Strouhal number as in Eq. (11-1) and resonance occurs when the vortex shedding frequency is equal to the structural natural frequency.

For a cylindrical structure with clamped-free end conditions subjected to uniform cross flow over its entire length ($F(x') = F(x)$), Eq. (11-7) reduces to

$$Y(\ell) = \frac{1.566F}{8\pi^2 mf^2 \zeta} \tag{11-8a}$$

and, for a cylindrical structure with pinned-pinned end conditions subjected to uniform cross flow over its entire length ($F(x') = F(x)$), Eq. (11-7) reduces to

$$Y(\ell/2) = \frac{F}{2\pi^3 mf^2 \zeta} \tag{11-8b}$$

where $Y(\ell/2)$ is the mid-span peak amplitude with pinned-pinned end conditions and $Y(\ell)$ is the peak amplitude at the free end with clamped-free end conditions. Cantilever tubes are not generally found in process heat exchangers, but Eq. (11-8a) was used to calculate the force coefficients in Table 11-5. A more complete discussion of the possible boundary conditions and mode shapes is found in Blevins (1990).

The wake shedding force may be formulated in terms of a dynamic coefficient, C_L, as shown earlier in Eq. (11-2),

$$F_L = C_L \left(\frac{1}{2}\rho DU^2 \right)$$

11.2.3 Fluctuating Force Coefficients in Single-Phase Flow

A review of the available data shows that fluctuating forces due to periodic wake shedding depend on several parameters such as bundle configuration, location within the bundle, Reynolds Number, turbulence level, fluid density and P/D. At the limit when P/D is large, fluctuating force coefficients should approach those for isolated single cylinders. On the other hand, when P/D is very small, the force coefficients are small since the fluid mass associated with the formation of vortices will be small as there is little space available within the bundle for large vortices.

Examples of single-phase vibration response curves for a range of P/D and bundle geometries are provided in Fig. 11-9a, b and c. Curves such as these with periodic response peaks are used to calculate fluctuating force coefficients.

The available data for fluctuating force coefficients in single-phase flow is compiled in Tables 11-1 and 11-2 and plotted in Fig. 11-10. The data in Table 11-2 was extracted from Chen (1979), Batham (1973), Shim et al (1988), Savkar and Litzinger (1982), and Pettigrew and Gorman (1978). The available data shows that the fluctuating lift coefficient, C_L, is very dependent on P/D up to $P/D = 2.5$. Unfortunately, there was no data for finned tubes. For heat exchanger tube bundles of $P/D < 1.6$, a

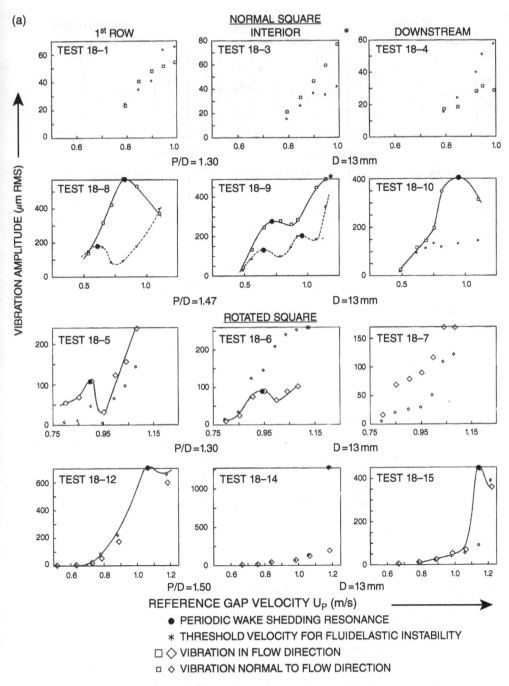

Fig. 11-9a Vibration Response to Single-Phase Forced Excitation (a) Normal-Square and Rotated-Square Arrays with D = 13 mm.

fluctuating force coefficient of 0.075 rms is recommended as a design guideline to calculate periodic-wake-shedding forces.

When resonance is considered possible, the maximum allowable tube vibration amplitude should not exceed 0.02D or 2% of the tube diameter. Below 0.02D, experience has shown that the vibration amplitudes are too small to cause long term damage due to fatigue or fretting wear.

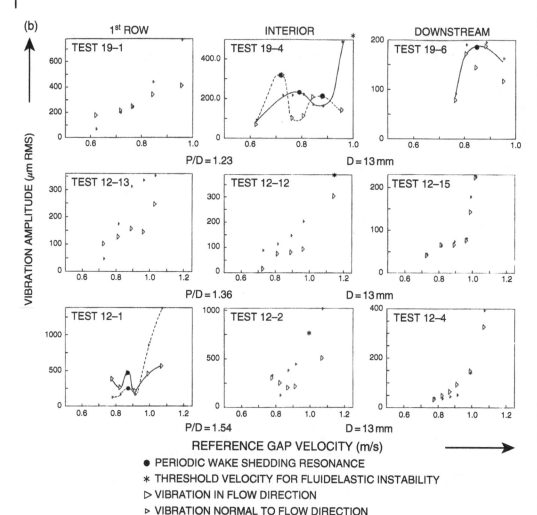

Fig. 11-9b Vibration Response to Single-Phase Forced Excitation (b) Rotated-Triangular Array with *D* = 13 mm.

11.2.4 Fluctuating Force Coefficients in Two-Phase Flow

It is sometimes difficult to distinguish between fluidelastic instability and periodic wake shedding in liquid and two-phase cross flow through arrays of cylinders, because the resonance velocity and stability threshold nearly coincide. The existence and characteristics of such flow and related vibration excitation mechanisms are discussed in this section. The spectral peaks in Fig. 11-11 are a clear indication that periodic forces may happen in two-phase cross flow. In fact, these periodic forces exist at 50% and 80% void fraction, not just at low void fractions. Thus, it is important to evaluate the characteristics of two-phase periodic forces. This evaluation may be done by deduction from the vibration response of test sections and other similar equipment.

There is not much available information on periodic wake shedding in tube bundles subjected to two-phase cross flow. This chapter has gathered data from test results at the Chalk River Laboratories (CRL) (Pettigrew et al, 1985) with the test loop and test sections shown in Figs. 8-4 and 8-5; at

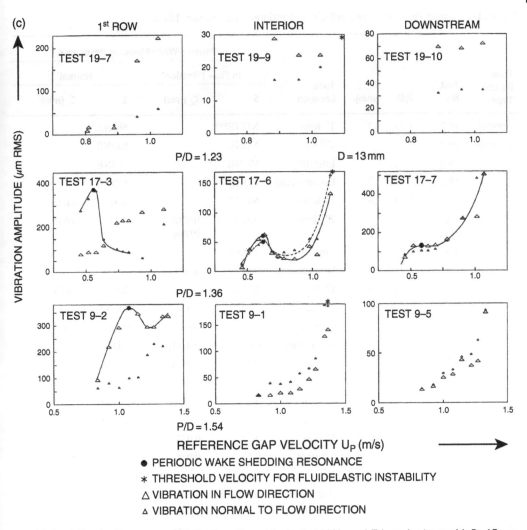

Fig. 11-9c Vibration Response to Single-Phase Forced Excitation (c) Normal-Triangular Array with $D = 13$ mm.

École Polytechnique (EP), Montreal (Pettigrew et al, 2005 and Senez et al, 2010); and at the Centre d'Études Nucléaires de Saclay (CENS), France (Taylor, 1994). The periodic forces were measured directly in the CENS and EP test sections, while the amplitude of motion was measured in the CRL test sections.

The CENS tests with $D = 30$ mm, $P/D = 1.5$ and normal-square array geometry had clear periodic forces only in the low-void-fraction regions, with the resulting lift coefficients listed in Table 11-3. The EP tests, with rotated-triangular array geometry, $D = 38$ mm and 17.5 mm, and $P/D = 1.5$ and 1.42, had clear periodic forces up to 80% void fraction, with some of the resulting lift coefficients listed in Table 11-4. Examples of the periodic frequency and periodic force versus pitch velocity provided in Fig. 11-12 clearly show that the periodic forces increase with velocity, while the behavior of the periodic frequency differs in the lift and drag directions. Very similar results for the Senez et al (2010) EP tests are observed in Figs. 11-13 and 11-14.

Table 11-1 Liquid Periodic Force Coefficients (Pettigrew and Gorman, 1978).

Tube Bundle Type	Test No.	P/D	D (mm)	Tube Location	Periodic-Wake-Shedding Resonance			
					In Flow Direction*		Normal**	
					S	C_L (rms)	S	C_L (rms)
Normal Triangle	19-7	1.23	13	1st Row	NONE***		NONE	
	19-8			2nd Row	NONE		NONE	
	19-9			Interior	NONE		NONE	
	19-10			Downstream	NONE		NONE	
	4-6	1.33	19	1st Row	NONE		0.44	0.046
	4-8			2nd Row	0.42/0.30	0.020/ (0.006)+	0.45	0.011
	4-4			Interior	NONE			
	4-12			Downstream	0.48	0.005		
	17-3	1.36	13	1st Row	NONE		0.67	0.064
	17-5			2nd Row	0.45	0.023	NONE	
	17-6			Interior	0.61	0.008	0.61	0.007
	17-7			Downstream	0.67	(0.023)	NONE	
	9-2	1.54	13	1st Row	0.36	(0.018)	NONE	
	9-3			2nd Row	0.37	(0.011)	NONE	
	9-1			Interior	NONE		NONE	
	9-5			Downstream	NONE		NONE	
	6-2	1.57	19	1st Row	0.57	0.027	0.42	0.019
	6-4			2nd Row	0.55	0.028	0.42	0.016
	6-8			Interior	NONE		NONE	
	6-9			Downstream	NONE		NONE	
Parallel Triangle	19-1	1.23	13	1st Row				
	19-2			2nd Row	0.42	(0.025)	0.44	0.064
	19-4			Interior	0.54/0.44	0.033/0.015	0.49	0.020
	19-6			Downstream	0.46	0.016	0.46	0.016
	5-2	1.33	19	1st Row			NONE	
				2nd Row				
	5-3			Interior	NONE		NONE	
	5-4			Downstream	NONE		NONE	
	12-13	1.36	13	1st Row	NONE		NONE	
	12-14			2nd Row	NONE		0.48	(0.012)
	12-12			Interior	NONE		NONE	
	12-15			Downstream	NONE		NONE	
	12-1	1.54	13	1st Row	0.45	0.033	0.45	(0.018)

Table 11-1 (Continued)

Tube Bundle Type	Test No.	P/D	D (mm)	Tube Location	Periodic-Wake-Shedding Resonance			
					In Flow Direction*		Normal**	
					S	C_L (rms)	S	C_L (rms)
	12-3			2nd Row	0.45	(0.03)		
	12-3			Interior				
	12-4			Downstream	NONE		NONE	
	7-2	1.57	19	1st Row	0.42	0.057	0.40	(0.021)
	7-3			2nd Row	0.52	0.023	NONE	
	7-4			Interior	0.52	(0.006)	0.52	(0.007)
	7-5			Downstream	0.45	0.01	0.45	0.014
Normal Square	18-1	1.3	13	1st Row	NONE		NONE	
	18-2			2nd Row	NONE		NONE	
	18-3			Interior	NONE		NONE	
	18-4			Downstream	NONE		NONE	
	18-8	1.47	13	1st Row	0.48	0.048	0.63	0.025
	18-11			2nd Row	0.45	0.047	0.63	0.018
	18-9			Interior	0.55	0.031	0.61/ 0.41	0.017/(0.012)
	18-10			Downstream	0.42	0.025	NONE	
Rotated Square	18-5	1.30	13	1st Row	0.43	0.007	NONE	
				2nd Row				
	18-6			Interior	0.41	0.006	NONE	
	18-7			Downstream	NONE		NONE	
	18-12	1.5	13	1st Row			0.37	0.35
	18-13			2nd Row	NONE		NONE	
	18-14			Interior	NONE		NONE	
	18-15			Downstream	0.34	0.019	NONE	

* Resonant vibration in flow direction,

** Resonant vibration normal to flow direction,

*** NONE indicates that no resonance was observed. A blank space indicates that the tests were not done or were inconclusive.

+Parentheses indicate less prominent resonance peak (i.e., Resonance peak < 2 x Random turbulence response).

In the 1980s, the possibility of periodic wake shedding at low void fraction was raised. Tests at CRL were conducted to simulate the entrance region of steam generators where low void fraction flow may exist. Example vibration response curve results are shown in Figs. 11-15a, b, and c and the resulting fluctuating force coefficients are listed in Table 11-5. Figure 11-15 shows that there is little evidence of periodic wake shedding above 20% void fraction. However, the question remained for higher void fraction (70 to 99%) two-phase flow in the U-bend region. Consequently, the EP experimental program was implemented to study flow-induced vibration in two-phase cross flow.

Table 11-2 Dynamic Lift Coefficients for Tube Bundles.

Source	P/D	Re	C_L rms	Comments
Chen	1.4		0.05	
(1979)	2.32		0.436	
Batham	1.25	$\sim 7 \times 10^4$	0.18	From Pressure
(1973)	2.0	$\sim 10^5$	0.18	From Pressure
Shim et al	2.67	$\sim 10^5$	0.3-0.7	
(1988)				
Savkar &	1.2	3×10^5	0.075	Upstream
Litzinger (1982)	1.2	3×10^5	0.05	Interior
	1.5	1.5×10^5	0.178	Upstream
	1.5	1.5×10^5	0.128	Interior
	1.71	1.2×10^5	0.136	Upstream
	1.71	1.2×10^5	0.136	Interior
Pettigrew &	1.33	$\sim 2 \times 10^4$	0.046	Upstream
Gorman (1978)	1.36	$\sim 2 \times 10^4$	0.064	Upstream
	1.36	$\sim 2 \times 10^4$	0.023	Upstream
	1.57	$\sim 2 \times 10^4$	0.028	Interior

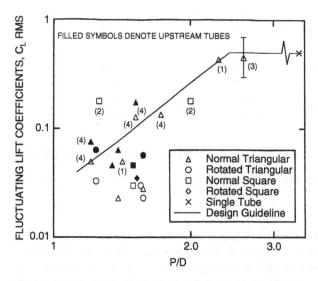

Fig. 11-10 Fluctuating Force Lift Coefficients for Tube Bundles in Single-Phase Flow: (1) Chen (1979) in Air, (2) Batham (1973) in Air, (3) Shim et al (1988) in Air, (4) Savkar and Litzinger (1982) in Air and all other Points are Pettigrew and Gorman (1978) in Water.

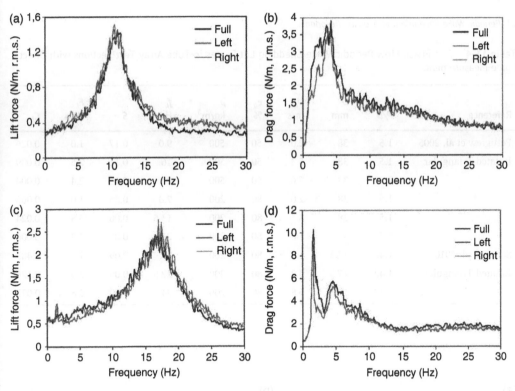

Fig. 11-11 Force Spectra at 80% Void Fraction: (a) and (b) Pitch Flow Velocity 5 m/s, and (c) and (d) Pitch Flow Velocity 10 m/s (Zhang et al, 2007).

Table 11-3 Single and Two-Phase Periodic Wake Shedding: Dynamic Coefficients and Strouhal Numbers for a Normal-Square Tube Bundle with P/D = 1.5 (Taylor, 1994).

e_g (%)	Tube Row	Drag		Lift	
		S	C_D	S	C_L
0	First	0.48	0.072	0.44	0.044
	Second	0.44	0.058	0.44	0.040
	Interior	0.62	0.057	0.47	0.029
	Downstream	0.51	0.037	0.44	0.042
5	First	0.46	0.061	0.46	-
	Second	0.42	0.043	0.40	0.013
	Interior	0.42	0.034	0.46	0.041
	Downstream	0.42	0.029	0.42	0.016
10	First	0.43	0.033	-	-
	Second	0.43	0.024	0.43	0.012
	Interior	0.41	0.038	0.41	0.020
	Downstream	0.43	0.031	0.43	0.013
15	First	0.40	0.043	0.40	0.015
	Second	0.36	0.022	0.36	0.032
	Interior	0.41	0.055	0.41	0.023
	Downstream	0.39	0.052	0.39	0.025

Table 11-4   Two-Phase Flow Periodic-Wake-Shedding Lift Forces in Tube Array Test Sections with Direct Force Measurement.

Reference	P/D	D mm	U_p m/s	ε_g %	ρ kg/m³	f_L Hz	S	F_L N/m	C_L
Pettigrew et al, 2005	1.5	38	2.0	50	500	9.0	0.17	1.0	0.026
Rotated Triangular	1.5	38	5.3	50	500	16	0.11	2.5	0.009
	1.5	38	7.6	50	500	20	0.1	2.4	0.004
	1.5	38	2.0	80	200	7.0	0.26	1.0	0.066
	1.5	38	8	80	200	13.5	0.06	5.8	0.024
	1.5	38	14	80	200	21	0.06	4.8	0.006
Senez et al, 2010	1.42	17.5	1.5	80	200	8	0.09	1.9	0.483
Rotated Triangular	1.42	17.5	4.2	80	200	22.5	0.09	3.9	0.126
	1.42	17.5	6.8	80	200	34	0.09	5.8	0.072

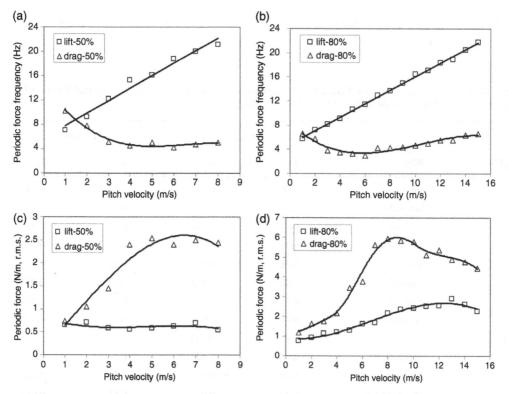

Fig. 11-12   Periodic Force Frequency and Periodic Force versus Pitch Velocity at 50% and 80% Void Fraction (Pettigrew et al, 2005).

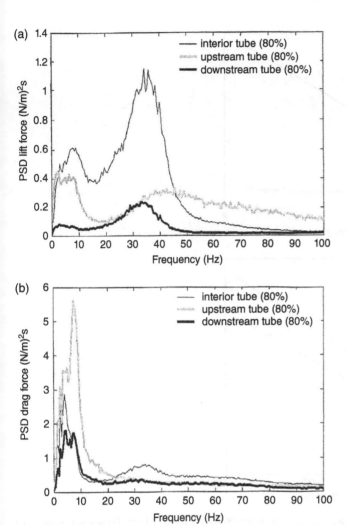

Fig. 11-13 Force Power Spectral Density (PSD) at 80% Void Fraction and 6.8 m/s Pitch Velocity: (a) Lift and (b) Drag (Senez et al, 2010).

In the EP program, both Pettigrew et al (2005) and Senez et al (2010) state that a wake between in-line cylinders was observed visually and periodic forces were measured, even though periodic wake shedding was not expected at such high void fractions. The Strouhal numbers attained in the EP tests (see Table 11-4) are between 0.06 and 0.26, as opposed to between 0.4 and 0.7 observed in the CRL and CENS low-void-fraction tests. Low-void-fraction wake shedding is similar to single-phase wake shedding, whereas the high-void-fraction mechanism producing fluctuating forces is different.

The two-phase force coefficient data from Tables 11-3 through 11-5 is shown in Fig. 11-16 under the single-phase design guideline from Fig. 11-10. These two-phase forces are similar in magnitude to single-phase data.

In Fig. 11-16 there is one data point from the rotated-triangular bundle geometry at 80% void fraction that is well above the other data points. A closer look at the source document (Senez et al, 2010) shows that the periodic force spectrum for this point (U_p = 1.5 m/s) does not have a

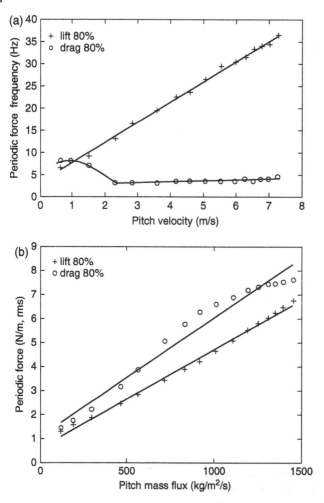

Fig. 11-14 Periodic Force Frequency and Periodic Force at 80% Void Fraction a) Force Frequency versus Pitch Velocity and (b) Periodic Force versus Mass Flux (Senez et al, 2010).

well-defined resonance peak. In fact, the periodic-wake-shedding peak is not distinguishable from a low-frequency peak. (A low-frequency peak occurs in the spectrum for each pitch velocity.) It is likely that the large force coefficient is due to low-frequency forces that are not related to periodic wake shedding. As the Senez et al (2010) periodic-wake-shedding flow conditions largely fell within the intermittent flow regime, these low-frequency forces may have been the result of flow regime, rather than periodic wake shedding. If we ignore the Senez data point that is well above the other data points, the single-phase design guideline can be used as an upper bound for two-phase periodic force coefficients.

11.2.5 The Effect of Bundle Orientation and *P/D* on Fluctuating Force Coefficients

Bundle orientation has no clear effect on the magnitude of the force coefficients in the single-phase data in Fig. 11-10. The two-phase data in Fig. 11-16, suggest that the rotated-triangular orientation may produce larger force coefficients than the other bundle orientations. The rotated-triangular

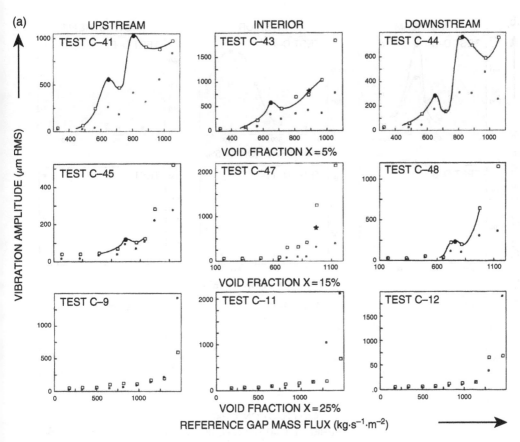

Fig. 11-15a Vibration Response to Air-Water Forced Excitation (Solid Circle: Periodic Wake Shedding Peak; Large Symbols: Lift Direction; Small Symbols: Drag Direction): (a) Normal-Square Tube Bundle of *P/D* = 1.47 (Pettigrew et al, 1985).

orientation was also observed to be the most susceptible to fluidelastic instability (see Chapters 7 and 8).

The effect of pitch-to-diameter ratio was clear in the single-phase data shown in Fig. 11-10. The force coefficients were found to increase with larger *P/D*. The two-phase results do not show a clear dependence on *P/D*. Additional data at varying pitch-to-diameter ratios is required to confirm that fluctuating force coefficients depend on *P/D* in two-phase flow.

The Senez et al (2010) test section had tubes that were about half the diameter of the Pettigrew et al (2005) test section, but the *P/D* ratios were kept similar. One might expect the fluctuating force coefficients to be lower in the Senez et al (2010) data, but this was not the case.

11.2.6   The Effect of Void Fraction and Flow Regime on Fluctuating Force Coefficients

Most of the available two-phase test results from laboratories around the world state that periodic wake shedding was not evident above 5% to 15% void fraction. This was also found in the Freon test results of Feenstra et al (2003). However, Pettigrew et al (2005) and Senez et al (2010) have shown that significant fluctuating forces can occur at high void fraction.

One could speculate that flow regime may have an effect on the formation of periodic wakes. Classic periodic wake shedding with expected Strouhal numbers has been found only in

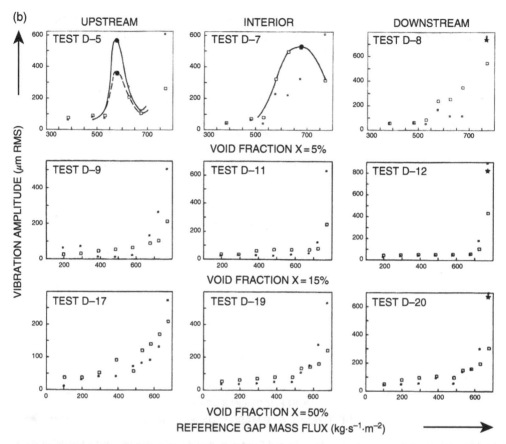

Fig. 11-15b Vibration Response to Air-Water Forced Excitation (Solid Circle: Periodic Wake Shedding Peak; Large Symbols: Lift Direction; Small Symbols: Drag Direction): (b) Normal-Square Tube Bundle oF *P/D* = 1.22.

low-void-fraction bubbly flow. The flow-regime boundaries in Fig. 11-17 illustrate that flow-regime boundaries change with changes in tube diameter and pitch-to-diameter ratio. As the tube diameter decreases and P/D is kept about the same, the bubbly flow region reduces in size. This behavior suggests that classic periodic wake shedding would be less likely in tube bundles with smaller diameter tubes. As P/D decreases, the bubbly flow region actually increases slightly, which could be interpreted as the likelihood of seeing wake shedding forces at slightly higher void fractions increasing as P/D decreases.

The Kanizawa and Ribatski (2016) flow maps in Fig. 11-17 were largely borrowed from Chapter 10 and, therefore, show many flow conditions that did not result in periodic wake shedding. However, it is simple to remember that, except for the EP tests, all of the periodic-wake-shedding data is from tests between 5% and 15% void fraction and, therefore, all in bubbly flow. The EP flow maps were prepared for this chapter and only show the flow conditions that are plotted in Figs. 11-12 and 11-14. The Pettigrew et al (2005) flow conditions all fall in the bubbly flow regime, despite the high void fractions. It is likely that intermittent flow and churn flow would inhibit classic periodic wake shedding, but the Senez et al (2010) tests that fall in the intermittent flow regime showed a fluctuating force that they believe is due to a type of wake shedding. This is unexpected and needs more study.

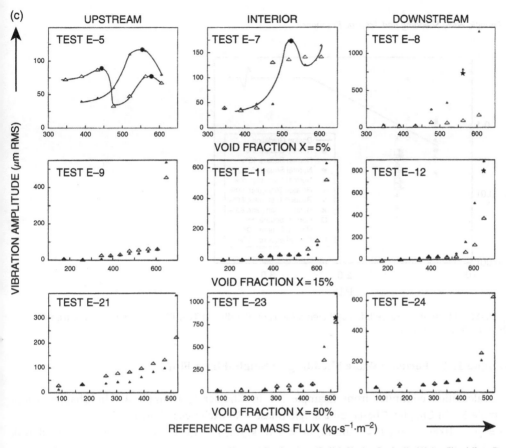

Fig. 11-15c Vibration Response to Air-Water Forced Excitation (Solid Circle: Periodic Wake Shedding Peak; Large Symbols: Lift Direction; Small Symbols: Drag Direction): (c) Normal-Triangular Tube Bundle of $P/D = 1.22$.

Table 11-5 Two-Phase Flow Periodic Wake Shedding Lift Forces in Tube Array Test Sections with Amplitude Measurement.

Reference	P/D	D mm	Mass Flux kg/m²s	ε_g %	ρ kg/m³	f_L Hz	m kg/m	ζ %	S	Y_{rms} mm	F_L N/m	C_L
Pettigrew et al, 1985	1.47	13	650	5	950	26.1	0.512	2.3	0.50	0.25	0.101	0.035
Normal Square	1.47	13	800	5	950	26.1	0.512	2.3	0.40	0.4	0.162	0.037
	1.47	13	650	5	950	26.1	0.512	2.3	0.50	0.3	0.121	0.042
	1.47	13	650	5	950	26.1	0.512	2.3	0.50	0.16	0.0646	0.022
	1.47	13	725	15	850	26.1	0.512	3.2	0.40	0.09	0.0506	0.013
	1.47	13	700	15	850	26.1	0.512	3.2	0.40	0.1	0.0562	0.015
CRL	1.22	13	580	5	950	26.3	0.508	2.6	0.56	0.57	0.262	0.114
Normal Square	1.22	13	580	5	950	26.3	0.508	2.6	0.56	0.22	0.101	0.044
	1.22	13	580	5	950	26.3	0.508	2.6	0.56	0.16	0.0736	0.032
CRL	1.22	13	560	5	950	26.6	0.496	2.2	0.59	0.115	0.0447	0.021
Normal Triangular	1.22	13	520	5	950	26.6	0.496	2.2	0.63	0.17	0.0661	0.036
	1.22	13	475	5	950	26.6	0.496	2.2	0.69	0.25	0.0972	0.063

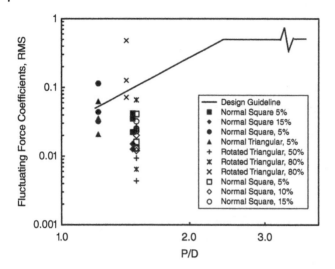

Fig. 11-16 Fluctuating Force Lift Coefficients for Tube Bundles in Two-Phase Flow with Data from Tables 11-3, 11-4 and 11-5.

Example 11-1   Periodic Wake Shedding in Single-Phase Flow

This calculation continues from Example 3-1 in Chapter 3, Example 4-1 in Chapter 4 and Example 5-1 in Chapter 5 based on a single-phase heat exchanger shown in Fig. 3.5.,

$\rho = \rho_i = 1000 \text{ kg/m}^3$, $D = 0.02$ m, $m = 1.16$ kg/m, $f = 135.6$ Hz, $\zeta = 1.96\%$ (in windows), $U_p = 1.5$ m/s.

Remembering that periodic wake shedding is possible in tube bundles for Strouhal numbers between 0.40 and 0.67, we can calculate the Strouhal number for this bundle using Eq. (11-1).

$$S = \frac{f_S D}{U_p} = \frac{135.6 \times 0.02}{1.5} = 1.8$$

This Strouhal number is outside of the Strouhal number range for periodic wake shedding.

If we now consider the longer span lengths within the window region of the heat exchanger, the tube frequency is much lower.

$$S = \frac{f_S D}{U_p} = \frac{33.9 \times 0.02}{1.5} = 0.45$$

This Strouhal number is very close to the lower limit of the possible Strouhal number range.

A calculation of the periodic forces (Eq. (11-2)), assuming a lift coefficient, $C_L = 0.075$, can be used to estimate a maximum amplitude for pinned-pinned end conditions using Eq. (11-8b).

$$F_L = \frac{C_L \rho \, D U_P^2}{2} = \frac{0.075 \times 1000 \times 0.02 \times 1.5^2}{2} = 1.69 \text{ N/m}$$

$$Y_{(\ell/2)} = \frac{F_L}{2\pi^3 m f^2 \zeta} = \frac{1.69}{2\pi^3 \times 1.16 \times 33.9^2 \times 0.0196}$$
$$= 0.0010 \text{ m} = 1.0 \text{ mm}$$

$\dfrac{Y_{(\ell/2)}}{D} = \dfrac{1.0}{20} = 0.0520 = 5.2\% > 2.0\%$ Possibly could result in fatigue or fretting-wear damage.

This heat exchanger may be susceptible to periodic wake shedding in the window region.

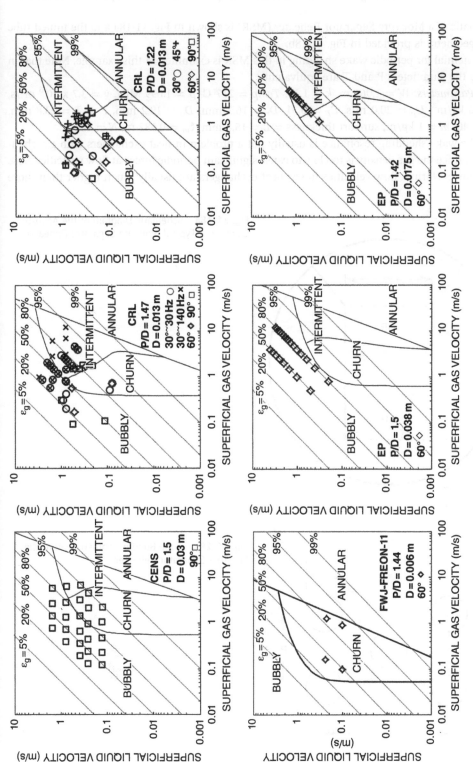

Fig. 11-17 Flow Regime Maps for the Two-Phase Fluctuating Force Data Sources.

Example 11-2 Calculating Periodic Wake Shedding in a Moisture Separator Reheater

A schematic of a Moisture Separator Reheater (MSR) is shown in Fig. 11-18a and the finned-tube bundle geometry is provided in Fig. 11-18b.

The potential for periodic wake shedding in this MSR is examined in this example. Note that in this case the parameters P and T are equivalent.

MSR Parameters: $W = 1.064$ m, $k = 1.33$, $Temp = 217°C$, $p = 1.099$ MPa, $\nu = 3.7 \times 10^{-6}$ m^2/s, $\rho = 5.06$ kg/m^3, $U_p = 20.0$ m/s, $R_F = 0.31$, $D_r = 16.3$ mm, $D_o = 19.2$ mm, $P = T = 23.8$ mm, $L/D = 2.40$, $m = 1$ kg/m, support thickness $= L = 15.8$ mm, $\ell_m = 0.682$ m, $N = 18$.

Periodic-wake-shedding resonance is usually not a problem in heat exchangers with gas shell-side flow, since the gas density is too low to cause significant periodic forces at flow velocities close to resonance. Normal flow velocities in gas heat exchangers are usually much higher than those

(a)

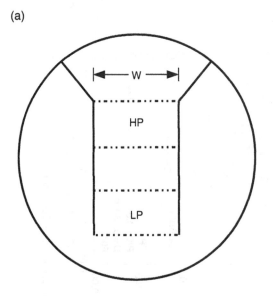

Fig. 11-18a Schematic Diagram of the Acoustic Cavity in a Typical Moisture Separator Reheater.

(b)

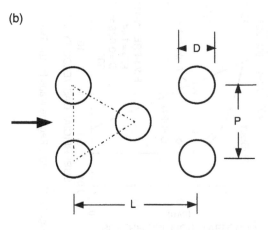

Fig. 11-18b Tube Bundle Geometry within the Moisture Separator Reheater.

required for resonance. However, it may be possible in high-pressure components such as the MSR, with higher-density gas on the shell side. It could also happen for higher modes of vibration with higher frequencies corresponding to higher flow velocities. Thus, periodic wake shedding cannot be ignored in MSRs.

Effective Diameter

Using Eq. (11-4), an effective diameter for finned tubes can be calculated, as follows:

$$D_{eff} = D_r + R_F(D_o - D_r) = 17.2 \text{ mm}$$

This diameter is used in the calculation of vibration excitation parameters such as Strouhal No., dimensionless flow velocity and random excitation. The equivalent pitch-to-diameter ratio, P/D, becomes P/D_{eff} for finned tubes. Thus, $P/D = P/D_{eff} = 1.38$ in this case.

Strouhal Numbers in this MSR

From Fig. 11-6, we find that around P/D of 1.38, the Strouhal numbers are between 0.4 and 0.67. Examining Fig. 11-7c for normal-triangular bundles (with Strouhal number based on the upstream velocity), the Strouhal number based on pitch velocity is defined by Eq. (11-3a) as

$$S = (1/1.73)(D/P) = (1/1.73)(1/1.38) = 0.418$$

The data on Strouhal numbers in finned-tube bundles is sparse, but using Fig. 11-8 we find that the experimental Point B (based on a tube with $P/D_r = 1.51$ and $D_r = 22.5$ mm) has $S = 0.53$. Based on this MSR with $P/D_r = 1.46$, and using Curve 2, this Strouhal number becomes $S = 0.63$.

Therefore, this MSR can be assumed to fall in the same Strouhal number range as a typical heat exchanger with $P/D = 1.5$, with a reference value of $S = 0.5$ and periodic wake shedding possible for $0.40 < S < 0.67$.

Calculating the Periodic-Wake-Shedding Frequencies

The Strouhal number range for this tube bundle configuration is extended to allow for possible "lock-in" by Blevins and Bressler (1987) (see Section 11.3.4) resulting in

$$0.4 (1 - 0.2) < S < 0.67 (1 + 0.3)$$

Thus, resonance is possible over the range $0.32 < S < 0.87$.

Based on $D_{eff} = 17.2$ mm. and the average flow velocity $U_p = 20.0$ m/s, the corresponding periodic-wake-shedding frequency range with $f_s = S U_p/D_{eff}$ is

$$372 \text{ Hz} < f_s < 1012 \text{ Hz}$$

Tube Damping

The dominant damping mechanism in heat exchangers with gas on the shell-side is friction between tubes and tube supports. Referring to Chapter 5, we find the following design recommendation for estimating the friction damping ratio, ζ_F, in percent:

$$\zeta_F = 5\left(\frac{N-1}{N}\right)\left(\frac{L}{\ell_m}\right)^{0.5} = 5\left(\frac{18-1}{18}\right)\left(\frac{0.0158}{0.682}\right)^{0.5} = 0.72\%$$

which takes into account the effect of support thickness, L, span length, ℓ_m, and the number of spans, N.

Hartlen and Barnstaple (1971) reported the only relevant damping data for finned tubes. When their minimum damping values are normalized to MSR geometry, they correspond to $\zeta_F = 0.57\%$. Since this value is somewhat lower than that for bare tubes, it will be used in this analysis.

Determining the MSR Lift Coefficients

Using the lift coefficient data provided in Fig. 11-10, we can infer that for $P/D_{eff} = 1.38$, the average value or reference value is about $C_L = 0.05$ for troubleshooting.

For use in designing MSRs with this geometry, the design guideline would be $C_L = 0.075$ as per Section 11.2.3.

Calculation of the Periodic Forces

A calculation of the periodic force per unit length (Eq. (11-2)), assuming a lift coefficient, $C_L = 0.075$, can be used to estimate a maximum amplitude for pinned-pinned end conditions using Eq. (11-8b).

$$F_L = \frac{C_L \rho\, D_{eff} U_p^2}{2} = \frac{0.075 \times 5.06\ \text{kg/m}^3 \times 0.0172\ \text{m} \times (20.0\ \text{m/s})^2}{2}$$
$$= 1.30\ \text{kg/s}^2$$

Vibration Amplitude

$$Y_{(\ell/2)} = \frac{F_L}{2\pi^3 m f^2 \zeta} = \frac{1.30\ \text{kg/s}^2}{2\pi^3 \times 1.0\ \text{kg/m} \times 372^2\ 1/\text{s}^2 \times 0.0057} = 0.0266\ \text{mm}$$

Note that the lowest frequency in the periodic-wake-shedding frequency range was chosen to give the largest amplitude result.

$$\frac{Y_{(\ell/2)}}{D_{eff}} = \frac{0.0266}{17.2} = 0.0015 = 0.15\% < 2.0\%$$

This MSR is unlikely to be susceptible to fatigue or fretting-wear damage due to periodic wake shedding.

11.3 Acoustic Resonance

Acoustic resonance may take place in heat exchanger tube bundles when vortex or periodic-wake-shedding frequencies coincide with a natural frequency for acoustic standing waves within a heat exchanger cavity. The flow-excited acoustic standing waves are generally transverse to both the axis of the tubes and to the direction of flow. Such resonances normally cause intense acoustic noise with sound pressure amplitudes that can exceed 160 dB. Tube, baffle and duct damage from fatigue may also occur in time if nothing is done to eliminate the noise problem. Acoustic resonance is possible in gas heat exchangers with both finned and bare tubes.

Acoustic resonance requires two conditions: 1) a necessary condition is the coincidence of shedding and acoustic frequencies, and 2) sufficient acoustic energy to overcome the acoustic damping in order that sustained acoustic standing wave resonance occurs.

11.3.1 Acoustic Natural Frequencies

Acoustic resonance in heat exchangers is relatively common so it is not surprising that a substantial effort has been directed toward predicting its occurrence. However, given our inability to scale or model acoustic energy production and acoustic damping in tube arrays, no completely reliable method for predicting acoustic resonance has been found. In discussing the relative merits of various criteria, Païdoussis (1980) lamented the lack of quantitative comparisons that would assist in determining which criterion was most successful in predicting acoustic resonance. To some extent, this situation has been redressed by Fitzpatrick (1986), Blevins and Bressler (1987), Ziada et al (1989a and 1989b), Blevins (1992), and Eisinger et al (1994). Blevins (1992) compared the criteria of Grotz and Arnold (1956), Chen and Young (1974), Fitzpatrick (1986), and Ziada et al (1989a and 1989b) for 22 different tube arrays. None of the criteria were 100% successful in predicting the occurrence of resonance. Blevins (1992), and Blevins and Bressler (1993) proposed predicting sound pressure level instead. Eisinger et al (1994) used operating experience with 66 steam generators with in-line tube arrays to evaluate the damping criteria of Grotz and Arnold (1956), Chen (1968), Chen and Young (1974), Fitzpatrick (1986), Ziada et al (1989b), and Blevins and Bressler (1987). This work suggests that none of the criteria are completely satisfactory and underscores the importance of developing damping criteria from field data rather than primarily from laboratory studies. The various approaches are summarized briefly in the following sub-sections.

11.3.2 Equivalent Speed of Sound

The presence of tubes in a duct is known to reduce the effective speed of sound by an amount that is proportional to the ratio of the volume of the tubes to the volume of the duct, σ. Parker (1978) showed that the effective speed of sound in a tube array, C_e, is related to the speed of sound in the fluid, C, by

$$C_e = C/\sqrt{1+\sigma} \qquad (11\text{-}9)$$

Experiments indicate that the actual speed of sound through a tube array is somewhere between C and C_e. Blevins and Bressler (1987) suggest that this is the result of the sound field spilling out into a region of the duct not occupied by tubes. Ziada et al (1989a and 1989b) indicate that the effective speed of sound correlates well with the ratio of the depth of the tube array in the flow direction, D_d, to the width of the flow duct in the direction of the sound waves, W. This result is shown in Fig. 11-19, which provides a reasonable estimate of the speed of sound in closely spaced tube arrays as demonstrated, for example, by Feenstra et al (2006).

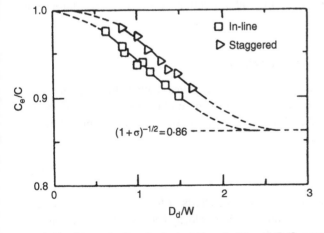

Fig. 11-19 Dimensionless Equivalent Speed of Sound C_e/C versus D_d/W of the Array (Ziada et al, 1989a).

11.3.3 Acoustic Natural Frequencies $(f_a)_n$,

Acoustic standing wave frequencies, $(f_a)_n$, in rectangular ducts are defined by:

$$(f_a)_n = nC_e/2W \tag{11-10}$$

where W is the dimension of the heat-exchanger tube-bundle cavity in the direction normal to the flow and the tube axes. The integer, n, is the mode order and C_e is the effective speed of sound.

In general, the speed of sound C is obtained from

$$C = \sqrt{kp/\rho} \tag{11-11}$$

where k is the specific heat ratio ($k = 1.33$ for steam), p is the shell-side pressure and ρ is the shell-side fluid density.

The speed of sound is affected by the presence of the tubes. This effect is related to the solidity ratio, σ, which is the ratio of the volume occupied by the tubes (with tube diameter, D) over the volume of the tube bundle, as noted above. For a triangular tube bundle

$$\sigma = \pi D^2 / \left(2\sqrt{3}P^2\right) \tag{11-12a}$$

and, for a square tube bundle

$$\sigma = \pi D^2 / P^2 \tag{11-12b}$$

The acoustic standing wave natural frequencies should be calculated for the first few acoustic modes (i.e., the first five modes should suffice). If one or more acoustic mode frequencies fall within the range of periodic-wake-shedding frequency, acoustic resonance can occur.

As previously noted, the vast majority of acoustic resonance problems in heat exchangers are associated with acoustic modes that are transverse to both the tube axes and the direction of flow. Defining this dimension as the duct width, W, the lowest acoustic natural frequency, $(f_a)_1$, will have a wavelength $2W$. Taking the speed of sound as lying between C and C_e, the nth acoustic natural frequency will lie in the range

$$\frac{nC_e}{2W} < (f_a)_n < \frac{nC}{2W} \quad n = 1, 2, 3, \dots \tag{11-13}$$

If the duct shape is such that Eq. (11-13) will not give a reasonable estimate of acoustic natural frequencies, more general formulas or even numerical procedures may be used (see, for example, Blevins (1986)).

11.3.4 Frequency Coincidence – Critical Velocities

In staggered arrays, acoustic resonance may occur for flow velocities at which the frequency of vortex shedding, f_s, is approximately equal to one of the acoustic natural frequencies $(f_a)_n$. These velocities of frequency coincidence are called critical velocities $(U_{cp})_n$ and may be determined from the Strouhal number(s), S, for a given array:

$$\left(U_{cp}\right)_n = \frac{(f_a)_n D}{S} \quad n = 1, 2, 3, \dots \tag{11-14}$$

The Strouhal number can be obtained from the various charts or formulas in Section 11.2.1.

In general, resonance will be excited over a range of flow velocity because of frequency entrainment or "lock-in" and may occur for several frequencies within the velocity range of a heat exchanger unit. This is shown in Fig. 11-20 from Blevins (1992), where the first critical velocity occurs at a reduced gap velocity of about 3. After a period of lock-in in the first mode, lock-in in the second occurs at a reduced velocity of about 4.75. Note that the highest sound pressure levels occur at and just beyond frequency coincidence in both modes. Because of this lock-in effect, Barrington (1973), Rogers and Penterson (1977), and Ziada et al (1989b) recommend a range of critical velocity $(U_{cp})_n$ ±20%. Blevins and Bressler (1987) suggest that in some cases the lock-in range may be even larger, $(U_{cp})_n$ -20% and $(U_{cp})_n$ +30%.

Given that acoustic resonance in in-line arrays is not triggered by flow periodicity observed at off-resonant conditions, prediction of critical velocities for such in-line arrays must be carried out using Strouhal numbers obtained under acoustic resonant conditions. Regardless of the conditions required for obtaining acoustic Strouhal numbers, all of the methods use the approach just described.

Fitzpatrick (1986) recommends such an approach for in-line tube arrays, i.e., the Strouhal number computed from velocity and frequency under conditions of acoustic resonance. Indeed, this design guide proposal involves determining a range of flow velocities and, through the acoustic Strouhal number, a range of acoustic frequencies which may be excited.

Fitzpatrick's proposal takes a similar approach by defining a critical velocity range based on empirical data from the literature. Ziada et al (1989b) corroborated Fitzpatrick's method and Blevins (1992) found it to be the best (albeit not perfect) predictor of those available for in-line arrays.

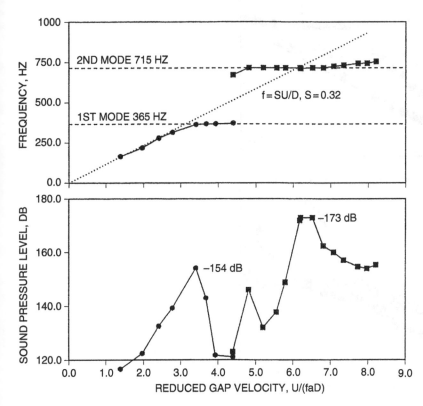

Fig. 11-20   Acoustic Resonance in First and Second Mode Excited by Vortex Shedding in a Staggered Array (Blevins, 1992).

11.3.5 Damping Criteria

The coincidence of a flow periodicity frequency with a duct acoustic natural frequency is a necessary but not sufficient condition for acoustic resonance. The flow energy must also be sufficiently high to overcome the acoustic damping in the system. This situation is shown in Fig. 11-21, Ziada et al (1989), where the first acoustic resonance mode excited is the second mode. The natural vortex shedding frequency passes the first duct mode as the flow velocity is increased. In many practical cases, the lowest mode of resonance may be well above the fundamental acoustic mode (see for example, Eisinger et al (1994)). When the flow energy is sufficient to overcome damping, an acoustic mode is excited and the acoustic particle velocity feedback can completely organize the flow structure within the tube array. Such effects of sound have been shown, for example, by Welsh et al (1990).

Blevins and Bressler (1987) suggest that resonance is less likely for closely spaced tube bundles probably because acoustic damping is higher. Also closely spaced tube bundles, have higher

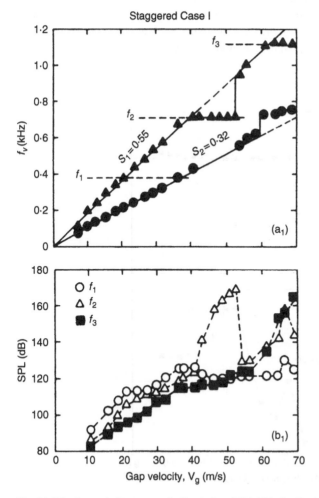

Fig. 11-21 Acoustic Resonance in Second and Third Modes Excited by Vortex Shedding in a Staggered Array (Ziada et al, 1989a).

Strouhal numbers and smaller gaps between the tubes for coherent vortex structures to develop. Thus, the acoustic excitation energy is smaller and more easily suppressed by damping. From experimental data for first-mode acoustic resonance, Blevins and Bressler (1987) show that resonance is unlikely for $T/D < 1.6$ and $L/D < 3.0$ for staggered (triangular) tube bundles, and for $L/D < 1.4$ for in-line (square) tube bundles (see Fig. 11-22). Here, T is the transverse pitch and L is the longitudinal pitch, as shown in Figs. 11-18b and 11-22. However, they do not discuss the applicability of their criterion to higher-order acoustic modes.

Given frequency coincidence, predicting acoustic resonance may be reduced to establishing a suitable damping criterion. Grotz and Arnold (1956) produced a criterion for acoustic resonance

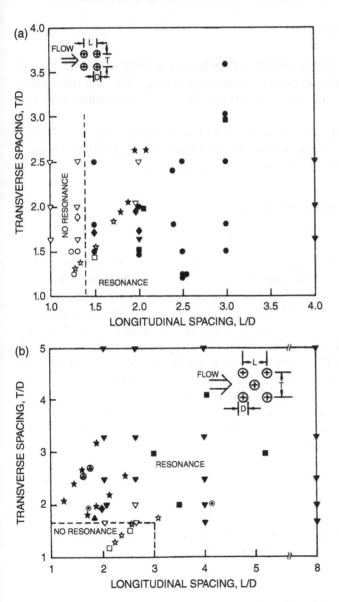

Fig. 11-22 Acoustic Resonance Criterion for a) In-Line (Square) Bundles and b) Staggered (Triangular) Bundles (Blevins and Bressler, 1987).

in in-line arrays. **Fitzpatrick (1986)** argued that this did not allow for geometric scaling and developed a modified damping criterion for in-line arrays, Δ^*, given by

$$\Delta^* = \frac{(Re)^{0.5}}{MS}\left[\frac{1}{2(X_L - 1)X_T}\right] \tag{11-15}$$

where Re is the Reynolds number, M is the Mach number, and S is the Strouhal number, all based on gap velocity at resonance. The array geometry is accounted for by the longitudinal and transverse pitch ratios, $X_L = L/D$ and $X_T = T/D$, respectively. From Fitzpatrick's graph of experimental data for resonance, resonance is predicted to occur if Δ^*, is in the range of approximately:

$$\frac{8200}{X_L} - 3000 < \Delta^* < \frac{8200}{X_L} - 700 \tag{11-16}$$

As noted, this criterion appears to work reasonably well for laboratory models. However, Eisinger et al (1994) found that for in-service steam generators, the resonant range of Δ^*, was substantially greater than indicated by Eq. (11-16) and that this criterion predicted resonances where none were observed.

Chen (1968) reported a damping criterion for in-line arrays based on wind tunnel tests. This criterion was later modified for heat exchangers in service (**Chen and Young, 1974**). Resonance is predicted to occur if the damping criterion, ψ exceeds an empirical number:

$$\psi = \frac{Re}{S}\left(1 - \frac{1}{X_L}\right)^2 \frac{1}{X_T} > 600 \text{ for wind tunnels} \tag{11-17}$$

$$> 2000 \text{ for heat exchangers}$$

Fitzpatrick (1982) has criticized this criterion because it fails to account for geometric scaling. (See also the discussions of Païdoussis (1980) and Ziada et al (1989b).) Blevins (1992) found that this criterion predicted acoustic resonance in a number of cases when it was not, in fact, observed. Note that, for staggered arrays, X_L in Eq. (11-17) must be replaced by $X_L/2$ because of the definition of longitudinal pitch used in this chapter (see Figs. 11-18b and 11-22b). Indeed, it should be noted that while Chen's criterion, Eq. (11-17), has been widely used for staggered arrays, it was not developed for that purpose. Eisinger et al (1994) found the Chen and Young (1974) criterion to be the most practically useful of the various criteria they evaluated. Using Strouhal numbers from Fitz-Hugh (1973), Eisenger et al (1994) reported no acoustic resonance for $\psi < 1300$, a mixed region of resonant and non-resonant cases for $1300 < \psi < 2700$, and all cases resonant for $\psi > 2700$.

Ziada et al (1989b) also developed damping criteria for in-line and staggered arrays. For in-line arrays, the resonance parameter, G_i is given by

$$G_i = \frac{(R_c)^{0.5}}{R_a}X_T \tag{11-18}$$

where
  R_c = the critical Reynolds number based on the gap velocity at resonance
  R_a = the acoustic Reynolds number based on the speed of sound in the array
For staggered arrays, the resonance parameter, G_s, is given by

$$G_s = \frac{(R_c)^{0.5}}{R_a}\frac{[X_L(X_T - 1)]^{0.5}}{(X_L - 1)} \tag{11-19}$$

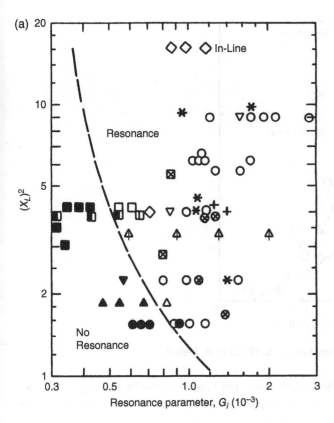

Fig. 11-23a Damping Criteria for In-Line Arrays, G_i (Ziada et al, 1989b).

Resonance is predicted to occur when the value of the resonance parameter exceeds an array geometry scale parameter on charts of resonance data as shown in Fig. 11-23. The resonance parameter, G_s, is correlated against the flow path parameter, L/h, in Fig. 11-23b. The parameter L/h represents the ratio between the length, L, of the flow jet emerging between the tubes and the minimum flow gap or thickness, h, of the jet ($h = (T-D)/2$). It is a measure of the stability of the flow jet and, hence, the propensity to resonance. Price and Zahn (1991) reported that this method worked well for predicting acoustic resonance in their laboratory study of a normal-triangular array with a pitch-to-diameter ratio of 1.375. For in-line arrays, Eisinger et al (1994) found that Eq. (11-18) did not satisfactorily separate resonant from non-resonant cases for in-service steam generators.

11.3.6 Sound Pressure Level

As noted in the previous sub-sections, none of the available predictive methods are completely reliable. It is noteworthy that none of the so-called damping criteria contain any direct measure of acoustic damping in the tube arrays explicitly. The implicit assumption is that the acoustic damping is included in the empirical bounds for the criteria. Sound wave attenuation may be considered to be at least partially accounted for through the inclusion of tube spacing ratios and Reynolds number, which are related to viscous flow losses. However, several authors have observed that a relatively small amount of fouling of the tubes of heat exchangers can reduce or eliminate acoustic resonance

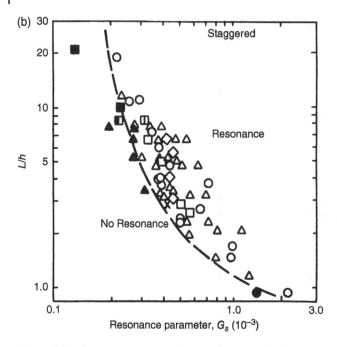

Fig. 11-23b Damping Criteria for Staggered Arrays, G_S (Ziada et al, 1989b).

(Baird (1959), Blevins and Bressler (1987), and Rogers and Penterson (1977)). Fouling of tubes will likely increase acoustic damping and pressure drop, and cleaning the tubes may cause the acoustic resonance to return.

In an attempt to overcome the problem of finding a reliable damping criterion, Blevins (1992) has proposed an alternative approach. If frequency coincidence occurs, he computes the maximum sound pressure level, noting that acoustic resonance has never been observed for reduced velocities less than 2. If

$$\frac{U_g}{f_a D} > 2 \quad SPL_{max} = 12.5 \frac{U_g}{C} \Delta p \tag{11-20}$$

where U_g is the average velocity in the minimum tube gap, SPL_{max} is the maximum sound pressure, C is the speed of sound in the array, and Δp is the total pressure drop through the array. Thus, this criterion specifies that the maximum sound pressure level is proportional to the product of the Mach number of the flow through the tube gaps and the total pressure drop across the array. This criterion is a measure of the flow energy generating sound, but does not consider the size of the tube array, nor the acoustic damping.

Based on laboratory results, Blevins provided maps of *SPL* for in-line and staggered arrays as functions of array geometry as shown in Fig. 11-24. Note that the longitudinal pitch for staggered (triangular) arrays defined in Fig. 11-24 is actually equal to one-half of the longitudinal pitch definition used in this chapter.

Blevins' approach has the advantage that it provides an indication of the severity of an acoustic resonance problem. However, the lack of a model for scaling array size and inclusion of acoustic damping reduces the usefulness of Blevin's maps. Eisinger et al (1994) plotted their 66 industrial cases on a Blevins-type map showing Reynolds number and sound pressure levels for modes up to the fifth acoustic mode. They conceded that the chart was informative, but concluded that much

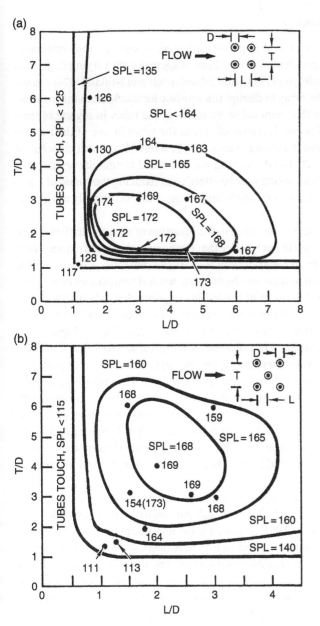

Fig. 11-24 Maximum Sound Pressure Levels as a Function of Tube Pattern and Spacing for a) In-Line (Square) Arrays and b) Staggered (Triangular) Arrays (Blevins, 1992).

more practical data was required before the benefits of such a map could be realized. Of particular concern are the design predictions of no acoustic resonance for small pitch-to-diameter ratio arrays based on Blevins and Bressler maps but which actually generated high resonance sound pressure levels when built. The study of Feenstra et al (2006) demonstrated this concern and showed that increasing the width of the tube array increased the sound pressure levels of the resonance. It is concluded that laboratory studies are very useful for establishing the necessary conditions for acoustic resonance in industrial-sized heat exchangers, but that sound pressure levels, i.e., the severity of the resonance, are not accurately predicted by such small-scale studies.

11.3.7 Elimination of Acoustic Resonance

If severe acoustic resonance is predicted or found in a heat exchanger in service, some method of elimination is required. A number of modifications have been suggested with varying degrees of success. Walker and Reising (1968), Barrington (1973), and Zdravkovich and Nuttall (1974) discuss the removal of selected tubes from a tube array to disrupt the acoustic feedback mechanism. However, Blevins and Bressler (1987) found that removal of 3% to 6% of the tubes in several of their arrays had a negligible effect on sound levels. Removal of 16% of the tubes in one of their arrays did reduce the sound level appreciably but this does not appear to be a reliable or practical solution to the problem. Blevins and Bressler (1987) also discuss the use of a tuned Helmholtz resonator in order to increase acoustic damping. This approach may be effective in reducing the sound of a specific acoustic mode if the resonator volume is sufficiently large. Again, this is probably not a practical solution in many cases.

The best approach to eliminating acoustic resonance is the insertion of acoustic baffles. These are inserted parallel to the direction of flow in such a way that the acoustic natural frequencies of the channels created are well removed from the excitation frequencies. Basically, the value of the duct width, W, in Eq. (11-13) is decreased to increase the acoustic natural frequencies beyond the excitation frequencies. The number and location of the baffles required depends on the lowest acoustic natural frequencies needed to prevent resonance. Such baffles not only modify the acoustic natural frequencies, but also may significantly increase the acoustic damping. Eisinger (1992) discussed the installation of baffles and outlined the drawbacks of limited service life due to baffle vibration and thermal effects. To overcome these problems, he suggested a novel approach of attaching fore and aft fins to the tubes in a column parallel to the flow. The effect is essentially a perforated baffle. Byrne (1983) suggested increasing the acoustic damping using a porous baffle of ceramic or glass fiber. However, Blevins and Bressler (1987) noted that solid baffles were more effective than perforated baffles. Feenstra et al (2005) showed that a rigid baffle located at the antinode effectively eliminated the first acoustic mode but did not entirely eliminate the higher modes. Similarly, two baffles entirely eliminated the first two acoustic modes but did not entirely eliminate the third and higher modes.

For in-line arrays, the insertion of solid plate baffles is relatively simple. For staggered arrays, the problem is more complicated because of the lack of a clear path parallel to the flow for baffle insertion. To overcome this difficulty, Eisinger (1992) patented a flexible corrugated acoustic baffle which can be used to retrofit staggered arrays to eliminate excessive noise. Most authors agree that acoustic baffles are the most effective means of eliminating acoustic resonance in heat exchangers. They have the distinct advantage that they can usually be inserted into existing arrays as a retrofit for excessively noisy heat exchangers in service. This approach has even been used as an effective retrofit for a large staggered-array economizer where the baffles were not parallel to the flow and the pressure drop created was not too large.

Example 11-3 Calculating Acoustic Resonance in a Moisture Separator Reheater

Parameters: Staggered (triangular) bundle, $W = 1.064$ m, $k = 1.33$, $Temp = 217°C$, $p = 1.099$ MPa, $v = 3.7 \times 10^{-6}$ m^2/s, $\rho = 5.06$ kg/m^3, $U_{cp} = 20.0$ m/s, $D_{eff} = 17.2$ mm, $T/D = 1.4$, $L/D = 2.4$, dimensions are defined in Fig. 11-18b.

<u>Calculating the Acoustic Frequencies</u>

It was shown in Example 11-2 that the periodic-wake-shedding frequency range is

$$372 \text{ Hz} < f_S < 1012 \text{ Hz}$$

Acoustic standing wave frequencies are defined using (Eq. (11-10)), $(f_a)_n = nC_e/2W$, where W is the dimension of the MSR tube bundle cavity in the direction normal to the flow and the tube axes (Fig. 11-18b).

The speed of sound, C, is obtained from Eq. (11-11)

$$C = \sqrt{kp/\rho} = \sqrt{\frac{1.33 \times 1.099\,\text{MPa}}{5.06\,\text{kg/m}^3}} = 537.5\,\text{m/s}$$

The speed of sound is affected by the solidity ratio. The solidity ratio can be calculated for a staggered bundle using Eq. (11-12a)

$$\sigma = \frac{\pi D^2}{2\sqrt{3}P^2} = \frac{\pi \times 17.2^2}{2\sqrt{3}(1.4 \times 17.2)^2} = 0.46$$

The effective speed of sound C_e, is defined in Eq. (11-9)

$$C_e = C/\sqrt{1+\sigma} = \frac{537.5}{\sqrt{1+0.46}} = 445.0\,\text{m/s}$$

The acoustic standing wave frequencies can now be calculated using Eq. (11-10). Thus, for the first-mode acoustic frequency

$$(f_a)_1 = nC_e/2W = \frac{445\,\text{m/s}}{2 \times 1.064\,\text{m}} = 209\,\text{Hz}$$

Similarly, higher-mode acoustic frequencies are calculated as

$$(f_a)_2 = 418\,\text{Hz}, \quad (f_a)_3 = 627\,\text{Hz},$$
$$(f_a)_4 = 836\,\text{Hz}, \quad (f_a)_5 = 1045\,\text{Hz}$$

Recalling, the periodic-wake-shedding frequency range of 372 Hz $< f_S <$ 1012 Hz, coincidence of acoustic and shedding frequency leading to resonance is

1) unlikely for the 1st and 5th acoustic modes,
2) possible for the 2nd and 4th acoustic modes because of possible "locking in" of the wake shedding frequency, and
3) most likely for the 3rd mode, since it corresponds closely to the mid-range reference Strouhal number, $S = 0.5$, which has a corresponding Strouhal Frequency of $f_S = 581$ Hz, which is reasonable close to $(f_a)_3 = 627$ Hz.

Therefore, a resonance condition is possible.

Susceptibility to Resonance

1) Using Fig. 11-22b from Blevins and Bressler (1987), we find that resonance is unlikely for T/D < 1.6 and L/D < 3.0. For the MSR, both $T/D = 1.4$ and $L/D = 2.4$ fall in the unlikely range. Thus, acoustic resonance is predicted to be unlikely.
2) A second approach to determining the likelihood of acoustic resonance is found with the Ziada et al (1989b) definition of a resonance parameter, G_s, for staggered arrays (Eq. (11-19)) and the resonance criterion plotted in Fig. 11-23b. To calculate the resonance parameter, the Reynolds number based on critical flow, R_c, must be calculated.

$$R_c = \frac{U_{cp}D}{\nu} = \frac{20 \text{ m/s} \times 0.0172 \text{ m}}{3.7 \times 10^{-6} \text{ m}^2/\text{s}} = 93.0 \times 10^3$$

$$R_a = \frac{CD}{\nu} = \frac{445 \text{ m/s} \times 0.0172 \text{ m}}{3.7 \times 10^{-6} \text{ m}^2/\text{s}} = 2.07 \times 10^6$$

$$G_s = \frac{(R_c)^{0.5}}{R_a} \frac{[L/D(T/D-1)]^{0.5}}{(L/D-1)}$$

$$= \frac{(93.0 \times 10^3)^{0.5}}{2.07 \times 10^6} \frac{[2.4(1.4-1)]^{0.5}}{(2.4-1)} = 1.03 \times 10^{-4}$$

From Fig. 11-23b, the flow path parameter, L/h, with $h = (T\text{-}D)/2$

$$\frac{L}{h} = \frac{L}{((T-D)/2)} = \frac{2L/D}{(T/D-1)} = \frac{2 \times 2.4}{(1.4-1)} = 12.0$$

The MSR parameters $L/h = 12.0$ and $G_s = 1.03 \times 10^{-4}$ are clearly in the no-resonance region of Fig. 11-23b. Thus, acoustic resonance is predicted to be unlikely.

Both Fitzpatrick (1986) and Chen and Young (1974) developed acoustic resonance criterion similar to that of Ziada et al (1989b). Unfortunately, these criteria are limited to in-line tube bundles and, therefore, not applicable to this case.

11.4 Conclusions and Recommendations

The state of knowledge of vortex shedding and acoustic resonance in heat exchanger tube arrays has been reviewed. Emphasis has been placed on the underlying basic excitation mechanisms because only through an understanding of these can the often conflicting and apparently scattered data in the literature be properly interpreted and reliable design guidelines be developed. The following are principal conclusions and recommendations for future research.

- The vorticity excitation mechanisms in in-line and staggered tube arrays with practical pitch-to-diameter ratios are fundamentally different.
- Alternate vortex shedding develops in the first few tube rows of staggered tube arrays. Multiple Strouhal numbers may exist for such arrays depending on the array geometry, pitch ratio, Reynolds number, and location of measurement. Under conditions of acoustic resonance, the standing waves are coherent throughout the array and spill out into ducting upstream and downstream of the tube arrays. Reliable Strouhal data for staggered arrays can be determined at resonant and off-resonant conditions.
- The acoustic excitation mechanism in in-line arrays over the range of practical pitch-to-diameter ratios is a jet instability that continues to develop through about the first five tube rows. This instability may be coupled with tube motion or acoustic resonance and, being a type of self-excitation, is not sustained at off-resonant conditions. Thus, reliable Strouhal data must be obtained under resonance conditions. The constant Strouhal number periodicity observed in in-line arrays under non-resonant conditions is due to symmetric jet-instability and is not related to the resonance problems usually observed. Alternate vortex shedding may occur in in-line arrays with pitch-to-diameter ratios greater than about 3.

- Vorticity phenomena may cause tube vibration in liquid flows because the higher fluid density substantially increases the fluid dynamic forces. Resonance conditions may be predicted using available Strouhal data and vibration amplitudes estimated using the appropriate force coefficients.
- For cases of tube resonance in liquid flows, more data is required for force coefficients as functions of array geometry and pitch-to-diameter ratio so that more accurate predictions of maximum response amplitudes may be made.
- In two-phase flow, vortex shedding has been shown to exist and force coefficients similar to those found in single-phase flow have been measured. These periodic forces have been found at void fractions as high as 50% to 80% in some laboratory test sections.
- Vorticity phenomena may cause acoustic resonance in gas flows. Acoustic natural frequencies can be reasonably well predicted with existing theory and excitation frequencies can be estimated from available Strouhal data. The latter must be acoustic Strouhal numbers for in-line arrays. Given frequency coincidence, the available criteria provide reasonable predictions for acoustic resonance. Unfortunately, the lack of reliable models for scaling acoustic source strength and acoustic damping means that resonance observed in laboratory models may not be found in full scale equipment, i.e., the severity of predicted resonance is inadequately modeled. More detailed technical data from service experience is necessary for criteria development and evaluation. Until more reliable criteria become available, useful guidance can be obtained from existing criteria, especially when used to check against the various predictions.
- More research is required using systematic, parametric studies to improve the Strouhal maps for staggered arrays (especially parallel-triangular arrays) so that excitation frequencies may be predicted with greater precision. In cases where there are multiple Strouhal numbers, it is important to determine their relative importance in exciting resonance.
- For cases of acoustic resonance, more reliable criteria for predicting the existence and severity of resonance are required. Essential to this development is a foundation on the underlying physics of the phenomena and the use of data from heat exchangers in service. It would be extremely useful to establish a data bank from the latter that would include sufficient technical information for full and proper assessment.

Nomenclature

C	= speed of sound, m/s
C_e	= effective speed of sound, m/s
C	= dynamic coefficient (C_L in the lift direction)
D	= tube outside diameter, m
D_d	= depth into the tube array in the flow direction, m
D_{eff}	= effective diameter, m
D_o	= outer diameter at fin tip for finned tubes, m
D_r	= root diameter for finned tubes, m
D_s	= shell inside diameter, m
F_L	= periodic force per unit length in the lift direction, N/m
F	= periodic force per unit length, N/m
$F(x)$	= force per unit length at position x, N/m

f	= vibration frequency, Hz
f_S	= Strouhal frequency, Hz
$(f_a)_n$	= acoustic standing wave frequency for Mode n, Hz
G_i, G_s	= resonance parameters for in-line and staggered arrays
g	= gravitational constant (i.e., 9.81 m/s^2)
$g(x,t)$	= periodic forces per unit length, N/m
h	= minimum width of the jet flowing between the tubes, m
k	= specific heat ratio
L	= longitudinal pitch, m, or length of the jet in L/h
L	= support thickness, m
ℓ_m	= span length, m
M	= Mach number
m	= tube mass per unit length, kg/m
N	= number of spans
n	= integer mode order
P	= tube pitch, m
p	= shell-side pressure, Pa
Δp	= total pressure drop through the array, N/m^2
$q_i(t)$	= generalized coordinates, m
Re	= Reynolds number, $Re = \dfrac{\rho U W}{\mu}$ for channel flow and $Re = \dfrac{\rho U_p D}{\mu}$ for cross flow
R_c	= critical Reynolds number based on the pitch velocity at resonance
R_a	= acoustic Reynolds number based on speed of sound in the array
R_F	= ratio of the area occupied by the fins over the available area (fin width divided by the fin pitch)
S	= Strouhal number
St	= Strouhal number defined using the free stream velocity
SPL_{max}	= maximum sound pressure, N/m^2
T	= transverse pitch, m
U	= free stream velocity (also U_∞), m/s
U_{cp}	= critical pitch velocity, m/s
U_g	= average velocity in the minimum tube gap, m/s
U_p	= pitch velocity, m/s
W	= path width, m, also bundle width in the direction of the sound waves, m
X_L	= longitudinal pitch divided by the tube diameter
X_T	= transverse pitch divided by the tube diameter
$Y(x)$	= peak vibration response at position x, m
$y(x,t)$	= amplitude of motion at position x and time t, m
ζ	= damping ratio
μ	= dynamic viscosity, kg/(m-s)
ν	= kinematic viscosity, m^2/s
ρ	= shell-side fluid density, kg/m^3
Δ^*	= modified damping criterion for in-line arrays (Fitzpatrick, 1986)
σ	= solidity ratio
$\phi_i(x)$	= modal shape (i^{th} mode) at position x
ψ	= damping criterion from Chen and Young (1974)

References

Baird, R. C., 1959, "Discussion of Putnam (1959)," *ASME J. Eng. For Power*, 81, pp. 420.

Barrington, E. A., 1973, "Acoustic Vibrations in Tubular Heat Exchangers" *Chemical Engineering Progress*, 69, pp. 62–68.

Batham, J. P., 1973, "Pressure Distribution on In-Line Arrays in Cross Flow," *Proceedings of the International Symposium on Vibration Problems in Industry*," Keswick, UK, Paper No. 412.

Blevins, R. D., 1986, "Acoustic Modes of Heat Exchanger Tube Bundles," *J. Sound and Vibration*, 109, pp. 19–31.

Blevins, R. D., 1990, *Flow-Induced Vibration*, 2nd Edition, Van Nostrand Reinhold Co., New York,

Blevins, R. D., 1992, "Experiments on Acoustic Resonance in Heat Exchangers Tube Bundles: II. Prediction and Suppression of Resonance," *ASME J. Pressure Vessel Technology*, 109, pp. 282–288.

Blevins, R. D and Bressler, M. M., 1987, "Acoustic Resonance in Heat Exchanger Tube Bundles. (1. Physical Nature of the Phenomenon; 2. Prediction and Suppression of Resonance.)," *ASME Journal of Pressure Vessel Technology*, 109(3), pp. 275–283.

Blevins, R. D. and Bressler, M. M., 1993, "Experiments on Acoustic Resonance in Heat Exchanger Tube Bundles," *J. of Sound and Vibration*, 164(30), pp. 503–533.

Bryce, W. B., Wharmby, J. S. and Fitzpatrick, J., 1978, "Duct Acoustic Resonances Induced by Flow over Coiled and Rectangular Heat Exchanger Test Banks in Plain and Finned Tubes," *Proceedings of BNES Symposium on Vibration in Nuclear Plant*, Keswick, UK, Paper No. 3.5.

Bryne, K. B., 1983, "The Use of Porous Baffles to Control Acoustic Vibrations in Cross Flow Tubular Heat Exchangers" *ASME J. Heat Transfer*, 105, pp. 751–758.

Chen, S. S., 1987, *Flow-Induced Vibration of Circular Cylindrical Structures*, Hemisphere Publishing Corporation, Washington.

Chen, Y. N., 1968, "Flow-Induced Vibration and Noise in Tube Bank Heat Exchangers due to von Karman Streets," *ASME J. Engineering for Industry*, 90, pp. 134–146.

Chen, Y. N., 1977, "The Sensitive Tube Spacing Region of Tube Bank Heat Exchangers for Fluid-Elastic Coupling in Cross-Flow," *ASME Fluid Structure Interaction Phenomena in Pressure Vessel Piping Systems*, pp. 1–18.

Chen, Y. N., 1979, "Flow-Induced Vibrations of Plates, Single Cylinders and Tube Bundles," *Practical Experiences with Flow-Induced Vibrations*, Paper A17, pp. 201–211.

Chen, Y. N. and Young, W.C., 1974, "Damping Capability of the Tube Band Against Vortex Excited Sonic Vibrations," *ASME J. Engineering for Industry*, 96, pp. 1072–1075.

Cheung, J. C. K. and Melbourne, W. H., 1983, "Turbulence Effects on Some Aerodynamic Parameters of a Circular Cylinder at Supercritical Reynolds Numbers," *Journal of Wind Engineering and Industrial Aerodynamics*, 14, p. 399.

Eisenger, F. L., 1992, *Personal communication*.

Eisenger, F. L., Sullivan, R. E., and Francis, J. T., 1994, "A Review of Acoustic Vibration Criteria Compared to Inservice Experience with Steam Generator In-Line Tube Banks," *ASME Journal of Pressure Vessel Technology*, 116, pp. 17–23.

Feenstra, P. A., Weaver, D. S. and Nakamura, T., 2003, "Vortex Shedding and Fluidelastic Instability in a Normal Square Tube Array Excited by Two-Phase Cross Flow," *Journal of Fluids and Structures*, 17, pp. 793–811.

Feenstra, P. A., Weaver, D. S. and Eisenger, F. L., 2005, "Acoustic Resonance in a Staggered Tube Array: Tube Response and Effect of Baffles," *Journal of Fluids and Structures*, 21, pp. 89–101.

Feenstra, P. A., Weaver, D. S. and Eisenger, F. L., 2006, "A Study of Acoustic Resonance in a Staggered Tube Array," *ASME Journal of Pressure Vessel Technology*, 128, pp. 533–540.

Fitz-Hugh, J. S., 1973, "Flow-Induced Vibrations in Heat Exchangers," *Proceedings UKAEA.NPL Int. Symposium on Vibration Problems in Industry*, Keswick, England, Paper 427.

Fitzpatrick, J. A., 1982, Letter to the Editor, "Acoustic Resonance in In-Line Tube Tanks," *J. Sound and Vibration*, 85, pp. 435–437.

Fitzpatrick, J. A., 1986, "A Design Guide Proposal for Avoidance of Acoustic Resonances in In-Line Heat Exchangers," *ASME J. Vibration, Acoustics, Stress & Reliability in Design*, 108, pp. 296–300.

Gerlach, C. R. and Dodge, F. T., 1970, "An Engineering Approach to Tube Flow-Induced Vibrations," *Proceedings of the Conference on Flow- Induced Vibrations in Reactor System Components*, May 14-15, ANL-7685, Argonne National Laboratory, Argonne, Illinois, pp. 205–224.

Gorman, D. J., 1976, "Experimental Development of Design Criteria to Limit Liquid Cross-Flow-Induced Vibration in Nuclear Reactor Heat Exchange Equipment," *Journal of Nuclear Science and Engineering*, 61, pp. 324–336.

Griffin, O. M., 1980, "Observations of Vortex Streets and Patterns -in the Atmosphere -in the Oceans -in the Laboratory," *Keynote Paper for the Symposium on Vortex Flows, ASME Winter Annual Meeting*, Chicago, Illinois.

Grotz, B. J. and Arnold, F. R., 1956, "Flow-Induced Vibration in Heat Exchangers," Dept. of Mech. Eng., Stanford University, California, Report No. 31.

Hartlen, R. T and Barnstaple, S. G., 1971, *Unpublished Report*.

Kanizawa, F.T and Ribatski, G., 2016, "Two-Phase Flow Patterns Across Triangular Tube Bundles for Air-Water Upward Flow," *International Journal of Multiphase Flow*, 80, pp. 43–56.

Kouba, J., 1986, "Vortex Shedding and Acoustic Emission in Finned Tube Banks Exposed to Cross Flow," *Proceedings of the ASME PVP Symposium on Flow-Induced Vibration*, Vol. 104, Chicago, Illinois, July, pp. 213–217.

Lienhard, J. H., 1966, "Synopsis of Lift, Drag and Vortex Frequency Data for Rigid Circular Cylinders," Bulletin 300, Washington State University, Pullman, Washington.

Mair, W. A., Jones, P. D. F. and Palmer, R. K. W., 1975, "Vortex Shedding from Finned Tubes," *Journal of Sound and Vibration*, 39, pp. 293–296.

Owen, P. R., 1965, "Buffeting Excitation of Boiler Tube Vibration," *Journal of Mechanical Engineering Science*, 7, pp. 431–439.

Païdoussis, M. P., 1980, "Flow-Induced Vibration in Nuclear Reactors and Heat Exchangers: Practical Experiences and State of Knowledge," *IUTAM-IAHR Symposium on Practical Experiences with Flow- Induced Vibrations*. Ed. E Naudascher and D. Rockwell, pp. 1–81. Berlin, Springer Verlag.

Parker, R., 1978, "Acoustic Responses in Passages Containing Banks of Heat Exchanger Tubes," *J. Sound and Vibration*, 57, pp. 245–260.

Pettigrew, M. J., and Gorman, D. J., 1973, "Experimental Studies on Flow-Induced Vibration to Support Steam Generator Design, Part III: Vibration of Small Tube Bundles in Liquid and Two-Phase Cross Flow," *International Symposium on Vibration Problems in Industry*, Paper No. 424, Keswick, UK, also, *Atomic Energy of Canada Limited Report, AECL-5804*.

Pettigrew, M. J. and Gorman, D. J., 1978, "Vibration of Heat Exchange Components in Liquid and Two-Phase Cross-Flow," Proceedings of the *BNES International Conference on Vibration in Nuclear Plant*, Keswick, UK, Paper 2.3, (also *AECL-6184*).

Pettigrew, M. J., Tromp, J. H. and Mastorakos, M., 1985, "Vibration of Tube Bundles Subjected to Two-Phase Cross Flow," *ASME Journal of Pressure Vessel Technology*, 107, pp. 335–343.

Pettigrew, M. J., Taylor, C. E. and Kim, B. S., 1989a, "Vibration of Tube Bundles in Two-Phase Cross Flow - Part 1: Hydrodynamic Mass and Damping," *ASME Journal of Pressure Vessel Technology*, 111, pp. 466–477.

Pettigrew, M. J. and Taylor, C. E., 1993, "Two-Phase Flow-Induced Vibration," *Technology for the 90s by ASME PVP Division*, New York, 1993, pp. 811–864.

Pettigrew, M. J., Zhang, C., Mureithi, N.W. and Pamfil, D., 2005, "Detailed Flow and Force Measurements in a Rotated Triangular Tube Bundle Subjected to Two-Phase Cross Flow," *Journal of Fluids and Structures*, 20, pp. 567–575.

Price, S. J. and Zahn, M. L., 1991, "Fluidelastic Behaviour of a Normal Triangular Array Subject to Cross Flow," *J. Fluids and Structures*, 5, pp. 259–278.

Rogers, J. D. and Penterson, C. A., 1977, "Predicting Sonic Vibration in Cross Flow Heat Exchangers: Experience and Model Testing," *ASME Paper 77-FE-7*, New York, ASME.

Savkar, S. D. and Litzinger, T A., 1982, "Buffeting Forces Induced by Cross Flow Through Staggered Arrays of Cylinders," *General Electric Report No. 82CRD238*.

Senez, H., Mureithi, N. W. and Pettigrew, M. J., 2010, "Vibration Excitation Forces in a Rotated Triangular Tube Bundle Subjected to Two-Phase Cross Flow," Proceedings of *FEDSM-ICNMM2010*, Aug 1-5, Montreal, Paper No. 30528, pp. 565–573.

Shim, K. C., Hill, R. C. and Lewis, R. I., 1988, "Fluctuating Lift Forces and Pressure Distributions due to Vortex Shedding in Tube Banks," *International Journal Heat and Fluid Flow*, 9(2), pp. 131–146.

Taylor, C. E., 1994, "Random Excitation Forces in Tube Arrays Subjected to Water and Air-Water Cross Flow," *Ph.D. Thesis*, University of Toronto, Toronto, Canada.

Violette, R., Pettigrew, M. J. and Mureithi, N.W., 2006, "Fluidelastic Instability of an Array of Tubes Preferentially Flexible in the Flow Direction Subjected to Two-Phase Cross Flow," *ASME Journal of Pressure Vessel Technology*, 128, pp. 148–159.

Walker, E. M. and Reising, G. F. S., 1968, "Flow-Induced Vibrations in Cross Flow Heat Exchangers," *Chemical Process Engineering*, 49, pp. 95–103.

Wang, J. and Weaver, D. S., 2012, "Fluidelastic Instability in Normal and Parallel Triangular Tube Arrays of Finned Tubes," *ASME Journal of Pressure Vessel Technology*, 134, 021302-1-7.

Weaver, D. S., 1993, "Chapter 6: Vortex Shedding and Acoustic Resonance in Heat Exchanger Tube Arrays," *Technology for the '90s*, (Principal Editor M. K. Au-Yang), ASME, New York, pp. 777–810.

Weaver, D. S., Fitzpatrick, J. A. and El Kashlan, M., 1987, "Strouhal Numbers for Heat Exchanger Tube Arrays in Cross Flow," *ASME Journal of Pressure Vessel Technology*, 109, pp. 219–223.

Weaver, D. S., Lian, H. Y. and Huang, X. Y., 1993, "Vortex Shedding in Rotated Square Tube Arrays," *J. Fluids and Structures*, 7, pp. 107–121.

Welsh, M. C., Hourigan, K., Welch, L. W., Downie, R. J., Thompson, M. C. and Stokes, A. N., 1990, "Acoustic and Experimental Methods: The Influence of Sound on Flow and Heat Transfer," *Experimental Thermal & Fluid Sciences*, 3, pp. 138–152.

Zdravkovich, M. M. and Nuttall, J. A., 1974, "On the Elimination of Aerodynamic Noise in a Staggered Tube Bank," *J. Sound and Vibration*, 34, pp, 173–177.

Zhang, C., Pettigrew, M. J. and Mureithi, N. W., 2007, "Vibration Excitation Force Measurements in a Rotated Triangular Tube Bundle Subjected to Two-Phase Cross Flow," *Journal of Pressure Vessel Technology*, pp. 21–27, *also PVP2005-71464*.

Ziada, S., Oengoren, A. and Buhlmann, E. T., 1989a, "On Acoustical Resonances in Tube Arrays, Part I: Experiments," *Journal of Fluids and Structures*, 3, pp. 293–314.

Ziada, S., Oengoren, A. and Buhlmann, E. T., 1989b, "On Acoustical Resonances in Tube Arrays, Part II: Damping Criteria," *Journal of Fluids and Structures*, 3, pp. 315–324.

Ziada, S., Jebodhsingh, D., Weaver, D. S. and Eisenger, F. L., 2005, "The Effect of Fins on Vortex Shedding from a Cylinder in Cross Flow," *Journal of Fluids and Structures*, 21, pp. 689–705.

Zukauskas, A. and Katinas, V., 1979, "Flow-Induced Vibration in Heat Exchanger Tube Banks," *Proceedings of the Symposium on Practical Experiences with Flow-Induced Vibrations, IUTAM*, Karlsruhe, pp.188–198.

12

Assessment of Fretting-Wear Damage in Nuclear and Process Equipment

Michel J. Pettigrew, Metin Yetisir, Nigel J. Fisher, Bruce A. W. Smith, and Victor P. Janzen

12.1 Introduction

Generally, the problems caused by excessive vibration are fatigue cracks and fretting-wear damage. This chapter treats the problem of fretting-wear damage between a vibrating structure and its supports. Typical components of concern are piping systems and pipe supports, multi-span heat exchanger tubes and tube supports, and nuclear fuel bundles and fuel channels. Fatigue problems, on the other hand, can be analysed using classical methods that are readily available in the literature.

Fretting-wear damage is related to the dynamic interaction between a structure and its supports. This is conveniently formulated in terms of a parameter called "work-rate" to predict fretting-wear damage. Work-rate is simply the integral of contact force over sliding distance per unit time. Work-rate can be estimated by performing a time domain simulation of the dynamics of a structure vibrating within its supports (see Fisher et al, 1992). However, this is not a trivial exercise. Another approach is to look at the problem from an energy point of view. It is essentially the mechanical energy or power dissipated through contact forces and sliding that causes fretting-wear damage at the supports.

Vibration energy in a structure is related to its vibration amplitude, frequency, mode shape, mass and damping. This relationship led to the development of a semi-empirical formulation to estimate work-rate based on experimental data and time-domain simulations (Yetisir et al, 1998). However, it soon became clear that an exact formulation based on energy considerations could be derived simply from vibration theory. The development of this simple formulation to relate tube vibration response and fretting-wear damage at the supports is presented in this chapter together with the time-domain simulation approach. Applications to components such as heat exchangers, piping, and nuclear fuels are discussed. It should be noted that the energy approach has already been mentioned in a practical book by Au-Yang (2001), where he refers to the original publication by Yetisir et al (1998).

Generally, energy methods are classical. However, the application to fretting-wear predictions is somewhat novel and is much simpler than existing methodologies based on time-domain simulations (Yetisir and Fisher, 1996). Some interesting early thinking on fretting-wear damage prediction may be found in Connors (1980) and Blevins (1978).

Flow-Induced Vibration Handbook for Nuclear and Process Equipment, First Edition.
Michel J. Pettigrew, Colette E. Taylor, and Nigel J. Fisher.
© 2022 John Wiley & Sons, Inc. This Work is a co-publication between ASME Press and John Wiley & Sons, Inc.

12.2 Dynamic Characteristics of Nuclear Structures and Process Equipment

The dynamic behavior of process and nuclear structures is generally very complex. These structures often have ill-defined boundary conditions, and are non-linear because of necessary clearances at the supports. They are sometimes non-stationary and subjected to random excitation forces. They can only be defined in terms of both deterministic parameters such as diameter, length and fluid properties and statistical parameters such as straightness, alignment and contact forces at the supports.

12.2.1 Heat Exchangers

For example, from a dynamic point of view, a heat exchanger such as a nuclear steam generator is essentially a multitude (1000s) of multi-span U-bend tubes supported by clearance-supports of more-or-less complex geometries. This is effectively a highly non-linear system because of the clearances between the tube and tube supports. At a given support location, the tube may be typically: not touching the support, contacting lightly, or in contact with a significant preload depending on tube straightness, support alignment and hydraulic drag forces. Thus, the system is considerably ill-defined.

Furthermore, the tube contact loading at the support may change with time, since hydrodynamic forces may vary due to thermal power changes, and tube straightness and alignment may be affected by thermal expansion. Crudding over time can also affect the tube loading conditions at the support. Thus, the system is non-stationary.

The geometry of the support can be a further complication. For example, in a tri-lobar broached-hole support there are twelve principal contact points, as shown in Fig. 12-1. The tube may be contacting any one of six edges, sitting on any one of three lands between two edges, or be jammed between two edges of neighboring lands. Similarly, in a lattice-bar type of support (see Fig. 12-1) there are eight principal contact points. A large number of contact points increases the number of possible boundary conditions dramatically. Considering that there are some twenty supports along a steam generator tube, there may be some 10^{30} different combinations of possible contact situations in a given steam generator tube, as shown schematically in Fig. 12-2.

It is obviously impossible to carry out time-domain simulations for all of these tube contact combinations. Rather, the approach taken is to model a statistically representative sample of the population of contact combinations. Such modelling may be done for example to predict the effect on

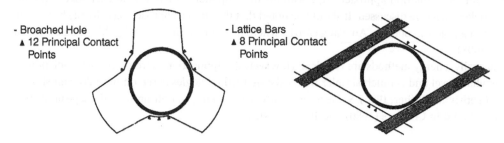

- Broached Hole
  ▲ 12 Principal Contact Points

- Lattice Bars
  ▲ 8 Principal Contact Points

Fig. 12-1 Complex Tube-Support Geometry: Possible Contact Points (Pettigrew et al, 1998).

Fig. 12-2   Steam Generator Tube and Support Contact Combinations (Pettigrew et al, 1998).

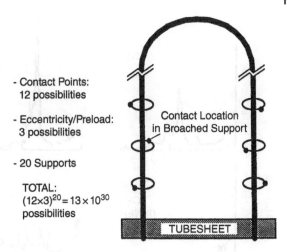

- Contact Points:
  12 possibilities

- Eccentricity/Preload:
  3 possibilities

- 20 Supports

TOTAL:
$(12 \times 3)^{20} = 13 \times 10^{30}$
possibilities

Contact Location
in Broached Support

TUBESHEET

fretting-wear damage of increasing the clearance between the tubes and tube supports. Enlarged diametrical clearances may result from chemical cleaning or from manufacturing difficulties.

12.2.2   Nuclear Structures

Nuclear fuel is typically in the form of bundles or clusters containing some 30 to 200 fuel elements. The fuel elements are supported by end plates, support grids, bearing pads, etc. The fuel bundles are contained in some hundreds of fuel channels in a typical reactor core.

Depending on geometry, flexural rigidity, clearances and straightness, fuel elements may be in contact with the associated fuel channels or support grids at several locations. There may be a pre-load or small clearance at the support locations. Thus, fuel elements and bundles are highly non-linear and ill-defined systems.

The dynamic stiffness of fuel assemblies depends on end plate and support system stiffness and on the flexural rigidity of the fuel elements. There can be great variation in stiffness due to ill-defined boundary conditions and clearances. The vibration response of otherwise identical fuel elements can be very different, as shown in Fig. 12-3 (Pettigrew, 1993).

The flexural rigidity of fuel elements is affected by thermal expansion of the uranium fuel pellets, and by creep of the enveloping fuel sheath under reactor power. Fuel pellet cracking, densification due to neutron radiation, and swelling due to fission gas production are time-dependent phenomena that also affect the mechanical characteristics of fuel elements (Pettigrew, 1993). These phenomena lead to considerable differences in vibration response over time as shown for example in Fig. 12-4 for prototype CANDU[1] fuel.

Thus, the vibration response of nuclear fuel is significantly non-stationary. The dynamic behavior of nuclear fuel can only be described in statistical terms within a bounding range, as shown in Fig. 12-5.

The dynamic behavior of other components such as piping systems could be similarly described. These components can either be analyzed using a statistical approach or by understanding the dynamic behavior within the bounding limits of the dynamic parameters, as discussed in the following sections.

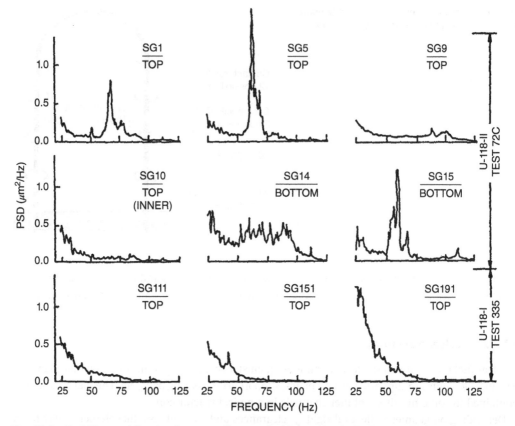

Fig. 12-3 Fuel Element Vibration Response Compared with Location in Fuel String (Pettigrew, 1993).

12.3 Fretting-Wear Damage Prediction

Fretting-wear damage occurs between a vibrating structure and its supports. The dynamic interaction between structure and supports is conveniently formulated in terms of a parameter called "work-rate", $\dot{W}$. Work-rate is simply the integral of contact force, $F(t)$, over sliding distance, S, per unit time (Frick et al, 1984), or:

$$\dot{W} = \frac{1}{T}\int_0^S F(t)\,ds \tag{12-1}$$

The fretting-wear damage volume rate, $\dot{V}$, can be calculated from:

$$\dot{V} = K_{FW}\dot{W} \tag{12-2}$$

where K_{FW} is a wear coefficient obtained experimentally, as discussed in Chapter 13.

12.3.1 Time-Domain Approach

To predict fretting-wear damage in a multi-span structure it is necessary to evaluate work-rate at every support location. Work-rate can be measured experimentally or calculated with a time-domain

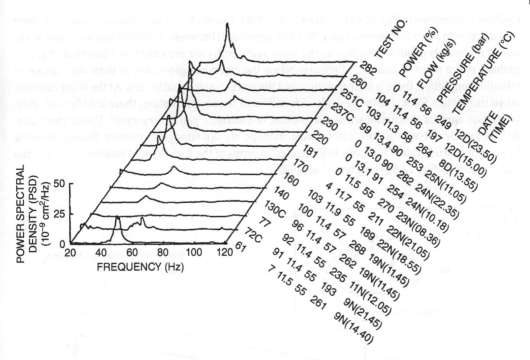

Fig. 12-4 Effect of History on Fuel Element Vibration Response (Pettigrew, 1993).

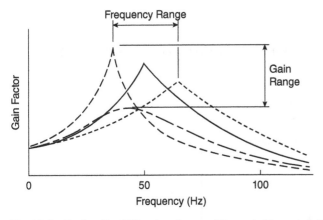

Fig. 12-5 Nuclear Fuel Vibration: Range of Dynamic Characteristics (Pettigrew, 1993).

computer model such as the VIBIC code (Yetisir and Fisher, 1996). VIBIC simulates the vibration response of multi-span structures, and, in particular, the dynamic interaction between structure and supports, which yields the work-rate. VIBIC can model deterministic fluid forces and random turbulence excitation, and realistic support geometry with appropriate contact forces.

The approach has been validated in several benchmark tests where work-rate was measured experimentally and compared against model predictions. For example, the dynamic interaction between a CANDU nuclear fuel element and a section of fuel channel was measured for the configuration shown in Fig. 12-6 (Pettigrew et al, 1998). The work-rate measurements are compared to prediction in Fig. 12-7 (Fisher et al, 1996). The agreement between measured and predicted work-rate is reasonably good. In this figure, the work-rate results are presented as a function of gap or preload at the point of contact. Obviously when the clearance gap is larger than the maximum vibration amplitude, there is no interaction and the work-rate is equal to zero. At the other extreme, when the friction force due to the preload is sufficient to prevent motion, there is little work-rate. Work-rate appears maximum when the preload or clearance gap is very small. Under these conditions, contact forces are sufficient to cause damage but not enough to prevent sliding, resulting in maximum work-rate. The above is illustrated in terms of the fuel element motion relative to the fuel channel in Fig. 12-8 (Pettigrew et al, 1998).

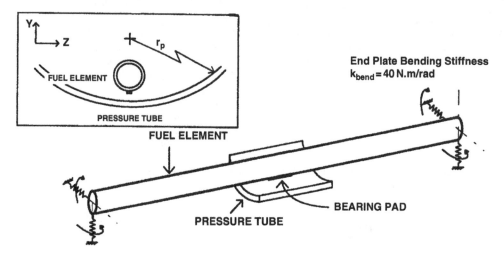

Fig. 12-6 Model of Fuel Element Bearing Pad and Fuel Channel Contact (Pettigrew et al, 1998).

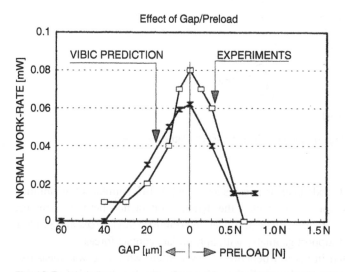

Fig. 12-7 Work-Rate versus Gap/Preload for a Fuel Element Vibration Response of 25 μm rms (Fisher et al, 1996).

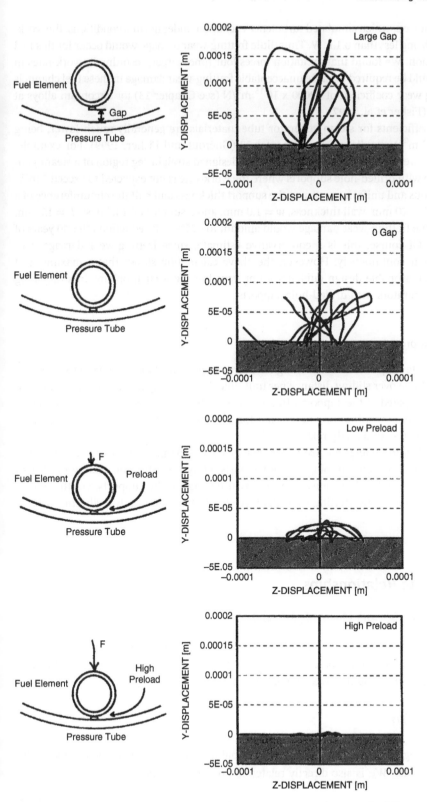

Fig. 12-8 Effect of Clearance or Preload on Dynamic Interaction between Fuel Element and Fuel Channel (Pettigrew et al, 1998).

For this particular case, which simulated turbulence excitation under normal conditions, the work-rate was very low, being less than 0.1 mW. Thus, little fretting-wear damage would occur for this fuel element configuration. We found that excitation forces four times larger, resulting in work-rates in excess of 1 mW, would be required to cause unacceptable fretting-wear damage in these fuel channels based on a fretting-wear coefficient of ~2000 x 10^{-15} m^2/N (see Chapter 13) for zirconium alloys at reactor conditions (Fisher et al, 1997).

Fretting-wear coefficients for steam generator tube materials are generally much lower, being typically 20 x 10^{-15} m^2/N, as discussed in Chapter 13 (Guérout and Fisher, 1999). For example, we can predict tube wear depth quite simply for a well-designed straight-leg region of a steam-generator U-bend tube with drilled-hole supports where the work-rate is not expected to exceed 5 mW. Assuming continuous and uniform wear over the support thickness and half the circumference of a tube of diameter, $D = 20$ mm, wall thickness, $w = 1.0$ mm, and a support of thickness, $L = 15$ mm, we can show that the fretting-wear damage would amount to ~25% wall reduction after 40 years of reactor operation. Of course, this is a conservative estimate, since fretting-wear damage does not necessarily occur continuously. However, the above calculation shows that a maximum of 5 mW would be a reasonable design guideline to prevent fretting-wear problems in straight-leg regions of steam generators with drilled-hole supports.

12.3.2 Energy Approach

Alternately, work-rate may be estimated using an energy approach. Work-rate, being simply the integral of contact force over sliding distance per unit time, is essentially a measure of mechanical energy or power dissipated at the supports. In air or gases, the energy absorbed by the structure from the fluid, the structure vibration energy, and the energy dissipated at the supports through fretting-wear damage are all directly related.

In liquids, the vibration energy is also dissipated by viscous damping between the structure and fluid and by squeeze-film damping at the supports. Only the vibration energy dissipated through contact forces and sliding causes fretting-wear damage at the supports. On the whole, there is an energy balance and fretting-wear damage should be related to the vibration energy in the tube.

12.4 Work-Rate Relationships

12.4.1 Shear Work Rate and Mechanical Power

Although fretting-wear predictions are usually made in terms of the normal work-rate, $\dot{W}_N$, the true work-rate or mechanical power dissipated at a support is the shear work-rate, $\dot{W}_S$, which is the integral of shear force, F_S, over sliding distance per unit time:

$$\dot{W}_S = \frac{1}{T}\int_0^S F_S(t)ds \tag{12-3}$$

Since, during sliding, the shear force equals the normal force times the dynamic friction coefficient, μ, shear work-rate, $\dot{W}_S$, is also directly related to normal work-rate, $\dot{W}_N$:

$$\dot{W}_S = \dot{W}_N\mu \tag{12-4}$$

We have found during fretting-wear experiments that the dynamic friction coefficient is often around 0.5 for the material combinations of concern here (Ko, 1985; Pendlebury et al, 1985). Eq. (12-4) conveniently becomes:

$$\dot{W}_N = 2\dot{W}_S \tag{12-5}$$

12.4.2 Vibration Energy Relationship

In a vibrating structure under steady-state conditions, damping dissipates vibration excitation energy. Thus, vibration excitation energy is equal to vibration damping energy. There are usually several energy dissipation mechanisms contributing to the overall damping. It is convenient to formulate damping in terms of the damping ratio. In heat exchanger tubes, for example, the overall damping ratio, ζ_T, includes fluid viscous damping, ζ_V, and damping between tube and supports, ζ_S. Damping at the supports includes, both friction damping, ζ_F, and squeeze-film damping, ζ_{SF}. Since they cannot be easily isolated, they are often considered together as support damping. Hence,

$$\zeta_T = \zeta_V + \zeta_S \tag{12-6}$$

Similarly, the total mechanical power dissipated by damping, P_T, is equal to:

$$P_T = P_V + P_S \tag{12-7}$$

However, only the mechanical power dissipated at the supports causes fretting-wear damage. Thus:

$$\dot{W}_S = P_S \tag{12-8}$$

12.4.3 Single Degree-of-Freedom System

From Thomson (1993), the energy dissipated per cycle, E_d, due to damping for a single-degree of freedom system under harmonic excitation at resonance is:

$$E_d = 2\pi\zeta_T kY^2 \tag{12-9}$$

where k is the stiffness and Y is the vibration amplitude (0 to peak) at resonance. The natural frequency, f, is:

$$f^2 = k/(4\pi^2 M)$$
$$k = 4\pi^2 M f^2 \tag{12-10}$$

where M is the equivalent dynamic mass of the system. The available vibration energy per unit time or mechanical power, P_T, is:

$$P_T = E_d f \tag{12-11}$$

Inserting Eqs.(12-9) and (12-10) into (12-11)

$$P_T = 8\pi^3 M f^3 Y^2 \zeta_T \tag{12-12}$$

This expression can be used to predict the mechanical power or work-rate dissipated by damping for a single-degree of freedom system. Since $\dot{W}_N = 2\dot{W}_S$, and, in terms of the root-mean-square (rms) response, y_{rms}, $Y^2 = 2(y_{rms}^2)$, then the normal work-rate may be formulated by:

$$\dot{W}_N = 32\pi^3 f^3 M y_{rms}^2 \zeta_S \tag{12-13}$$

where support damping, ζ_S, is substituted for overall (total) damping, ζ_T, as only mechanical power dissipated at the supports causes fretting-wear damage.

12.4.4 Multi-Span Beams Under Harmonic Excitation

Process equipment structures such as heat exchanger tubes, nuclear fuels and piping systems are often uniform multi-span beams of equal span length, ℓ. Since the number of spans and number of supports is relatively large, it can be assumed that the vibration energy of one span is dissipated at one support. For the limiting case where the clearances at the supports are large and the vibration response at the supports is nearly the same as at mid-span, then the beam is essentially vibrating as a rigid body and the equivalent mass in Eq. (12-13) is effectively the mass of the beam over one span or:

$$M = m\ell \tag{12-14}$$

where m is the mass of the beam per unit length including the hydrodynamic mass.

Thus, in terms of the normal work-rate at the support only, Eq. (12-13) becomes:

$$\dot{W}_N = 32\pi^3 f^3 m\ell y_{rms}^2 \zeta_S \tag{12-15}$$

Eq. (12-15) is effectively an upper bound.

For the other limiting case, where the clearances are small and the vibration response is much larger at mid-span than at the supports (i.e., the beam is essentially pinned at the support), the vibration mode shape of the beam is essentially sinusoidal. In this case, the vibration response, $y(x)$, at any point, x, along a given span of the beam may be formulated by:

$$y_{(x)} = y_{(\ell/2)} \sin\left(\frac{\pi x}{\ell}\right) \tag{12-16}$$

where $y(\ell/2)$ is the mid-span vibration response.

Discretizing Eq. (12-15) along x:

$$d\dot{W}_N = 32\pi^3 f^3 m y_{(x)_{rms}}^2 \zeta_S dx \tag{12-17}$$

Inserting Eq.(12-16) in (12-17)

$$\dot{W}_N = 32\pi^3 f^3 m \zeta_S y_{(\ell/2)_{rms}}^2 \int_0^\ell \sin\left(\frac{\pi x}{\ell}\right) dx$$

and integrating:

$$\dot{W}_N = 16\pi^3 f^3 m\ell y_{(\ell/2)rms}^2 \zeta_S \tag{12-18}$$

In this case, the effective dynamic mass, $M = m\ell/2$. In reality, the beam dynamic behavior is somewhere between Eqs. (12-15) and (12-18) and probably closer to the latter in well-performing components. Eq. (12-18) is a practical lower bound. For example, the tubes in well-designed heat exchangers do not normally vibrate back and forth within the available clearance. If this were the case, extensive fretting-wear damage would occur, as discussed later in this chapter.

12.4.5 Response to Random Excitation

It can be shown (Blake, 1986) that the total mechanical power absorbed in the form of vibration energy by a single degree-of-freedom system subjected to random turbulence excitation may be formulated by:

$$P_T = \frac{S_F(f)}{4M} \tag{12-19}$$

where $S_F(f)$ is the power spectral density (PSD) of the random forces.

From Thompson (1993), the vibration response of the system is

$$y_{rms}^2 = \frac{f\pi S_F(f)}{4k^2\zeta_T} \tag{12-20}$$

Since $k = 4\pi^2 M f^2$,

$$y_{rms}^2 = \frac{S_F(f)}{64\pi^3 f^3 M^2 \zeta_T} \tag{12-21}$$

$$S_F(f) = 64\pi^3 f^3 M^2 y_{rms}^2 \zeta_T \tag{12-22}$$

Inserting Eq. (12-22) in (12-19),

$$P_T = 16\pi^3 f^3 M y_{rms}^2 \zeta_T \tag{12-23}$$

Substituting support damping, ζ_S, for overall (total) damping, ζ_T, as before, in terms of normal work-rate:

$$\dot{W}_N = 32\pi^3 f^3 M y_{rms}^2 \zeta_S \tag{12-24}$$

which is identical to Eq. (12-13).

Similarly, for the first mode of a single-span beam with pinned supports:

$$P_T = \frac{S_F(f_1)J_1^2 \ell}{4m} \tag{12-25}$$

where $S_F(f_1)$ is the PSD per unit length of the forcing function in the vicinity of the first-mode natural frequency, f_1, and J_1^2 is the joint acceptance for the first mode.

From Taylor and Pettigrew (1999), the mid-span tube vibration response, $y_{(\ell/2)rms}$, may be formulated by:

$$y_{(\ell/2)rms}^2 = \frac{S_F(f_1)\phi_1^2(\ell/2)J_1^2}{(64\pi^3 f^3 m^2 \zeta_T)} \tag{12-26}$$

or

$$S_F(f_1)J_1^2 = \frac{(64\pi^3 f^3 m^2 \zeta_T)y_{(\ell/2)rms}^2}{\phi_1^2(\ell/2)} \tag{12-27}$$

where ϕ_1^2 is the first-mode normalized mode-shape factor. Considering that $\phi_1^2(\ell/2) = 2$ for a pinned-pinned beam and inserting Eq. (12-27) into (12-25) yields:

$$P_T = 8\pi^3 f^3 m\ell y_{(\ell/2)rms}^2 \zeta_T \tag{12-28}$$

Substituting support damping, ζ_S, for overall (total) damping, ζ_T, Eq. (12-28) can be rearranged to estimate the normal work-rate at the supports of a multi-span beam of equal span length vibrating in the first mode, assuming the energy of one span is dissipated at one support:

$$\dot{W}_N = 2P_S = 16\pi^3 f^3 m\ell y_{(\ell/2)rms}^2 \zeta_S \tag{12-29}$$

Equation (12-29) is identical to (12-18) for harmonic motion. This is not surprising since it is the same quantity of energy whether harmonic or random. Equation (12-29) is a practical lower bound.

12.4.6 Work-Rate Estimate: Summary

As already mentioned, process equipment structures are often multi-span beams of N spans of equal length, ℓ. The total mechanical power, assuming similar vibration response in each span may be formulated with the aforementioned equations. Considering that there is essentially one support per span and assuming that the energy is dissipated equally between supports, a realistic normal work-rate at each support may be estimated with Eq. (12-15) or (12-29). The mechanical energy formulated by these equations is the maximum energy available to cause fretting-wear damage at one support on the average. In reality, the work-rate may not be distributed evenly among the supports depending on support conditions, i.e., preloads or clearances, as discussed in Pettigrew et al (1999). A maldistribution factor of two or three has been observed in experiments and in simulations (using VIBIC) for similar situations. A factor of three could be used to obtain an upper bound to take into account maldistribution. However, the total work-rate for all the supports could never be larger than the total estimated for each span with the above equations.

12.5 Experimental Verification

An experimental program was conducted to understand damping of multi-span heat exchanger tubes (Taylor et al, 1998). A single heat exchanger tube, clamped at one end and supported at intermediate locations by three realistic supports, was used in these experiments. The supports were fully instrumented to measure work-rate directly, as outlined in Fig. 12-9. The tube was excited by random vibration and the energy input was measured. The tests were done in air and in water at different temperatures to study the effect of viscosity on damping.

An energy approach was used to investigate damping. In these experiments, the energy dissipated at each support was measured in the form of a shear work-rate. The vibration excitation energy was similarly measured in the form of an input work-rate defined as the average mechanical power the shaker provided to the tube.

In air, we found that the excitation energy was approximately equal to the sum of the mechanical energies dissipated at the supports as shown in Fig. 12-10a.

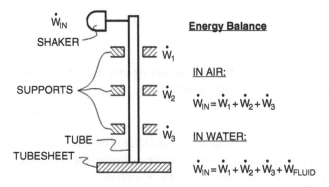

Energy Balance

IN AIR:

$$\dot{W}_{IN} = \dot{W}_1 + \dot{W}_2 + \dot{W}_3$$

IN WATER:

$$\dot{W}_{IN} = \dot{W}_1 + \dot{W}_2 + \dot{W}_3 + \dot{W}_{FLUID}$$

Fig. 12-9 Work-Rate Balance in Multi-Span Heat Exchanger Tube Test.

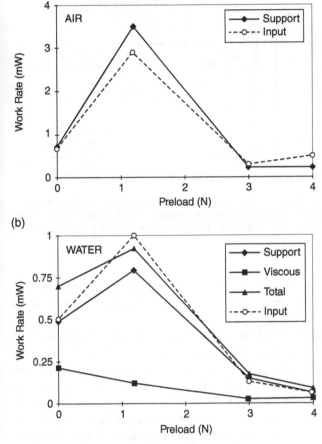

(a)

(b)

Fig. 12-10 Input and Dissipated Work-Rates for Multi-Span Heat Exchanger Tube Tests a) in Air and b) in Water (Taylor et al, 1998).

We found similar results in water. However, the energy dissipated through viscous damping between tube and fluid had to be taken into account separately since it is not measured as work-rate at the supports. The total dissipated work-rate is remarkably close to the input work-rate, as shown in Fig. 12-10b. Internal or material damping is usually small in such structures and is, thus, neglected. Although energy conservation is not surprising, its experimental verification in multi-span heat exchanger tubes was an interesting challenge.

In summary, vibration excitation energy is directly related to work-rate dissipated at the supports since there must be an overall energy balance. Thus, the prediction of work-rate from vibration excitation energy and vibration response is entirely logical.

12.6 Comparison to Time-Domain Approach

Yetisir et al (1998) conducted thousands of work-rate calculations on realistic heat exchanger tube configurations to confirm the form of Eq. (12-29). The simulations were done using the VIBIC code. The parameters that were investigated were: number of spans, excitation levels, support

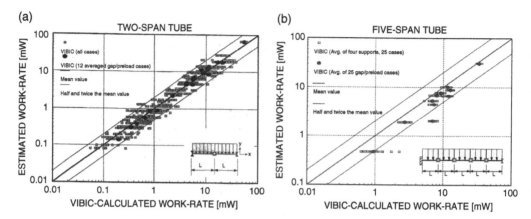

Fig. 12-11 Estimated and VIBIC-Calculated Work-Rates for: a) Two-Span Simulations (Total Number of Simulations Is 1144; Only 27 (3%) of the Data Points are Outside the Factor-of-Two Bounds.): b) Five-Span Simulations (Total Number of Simulations is 300; 35 (12%) of the Data Points are Outside the Factor-of-Two Bounds) (Yetisir et al, 1998).

clearances, preload at the supports, tube flexural rigidity, mass per unit length, span length, modal damping, etc. The results were compared to those obtained with Eq. (12-29), as shown in Figs. 12-11a and 12-11b.

The agreement between VIBIC simulations and the simple calculations using the proposed criterion was remarkably good, being mostly within a factor of ±2. The range reflects the statistical nature of this problem and, in particular, the effect of the statistical parameters governing the boundary conditions at the supports. Not surprisingly, the exponents of the parameters in Eq. (12-29) were found to be a best fit to the data from the simulations. The proposed approach was tested for a frequency range of 8 to 136 Hz, maximum-vibration-amplitude range of 7 to 1860 μm, span-length range of 0.5 to 2.0 m, mass per unit length range of 0.3 to 1.2 kg/m and a damping-ratio range of 0.01 to 0.05. The calculated work-rate varied from 0.08 to 65 mW, which encompasses most heat exchanger situations.

12.7 Practical Applications: Examples

The practical application of this approach to U-bend heat exchanger tubes and to other structures such as nuclear fuels and piping systems has been discussed in papers by Yetisir and Pettigrew (1999), Fluit and Pettigrew (2001) and by Pettigrew et al (1999). Simple examples are outlined below.

Example 12-1 Heat Exchanger Tubes

Take, for example, the typical multi-span heat exchanger tube shown in Fig. 12-12. The resultant mid-span vibration response of such a tube to turbulence-induced excitation would usually be less than 25 μm rms under normal operating conditions. The mechanical characteristics of the tube are taken to be typically $D = 20$ mm, $m = 1.0$ kg/m, $\ell = 0.75$ m, $L = 16$ mm, $f = 60$ Hz and $\zeta = 2\%$. Assuming viscous damping between the tube and fluid to be 0.5%, damping at the supports may be estimated to be $\zeta_S = 1.5\%$. Using Eq. (12-29):

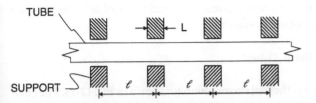

Fig. 12-12 Hypothetical Multi-Span Heat Exchanger Tube.

$$\dot{W}_N = 16\pi^3 f^3 m \ell y^2_{(\ell/2)rms} \zeta_S$$
$$\dot{W}_N = 16\pi^3 (60 \text{ Hz})^3 1.0 \text{ kg/m} \times 0.75 \text{ m} \times (25 \text{ } \mu m)^2 \times 0.015$$

we obtain a normal work-rate:

$$\dot{W}_N = 0.75 \text{ mW}$$

The total fretting-wear damage volume, V, after a typical component life, T_S, of 40 years may be calculated to be 1.9 x 10^{-8} m^3 using Eq. (12-2) and a wear coefficient $K_{FW} = 20$ x 10^{-15} m^2/N:

$$\dot{V} = K_{FW} \times \dot{W}_N$$
$$V = \dot{V}T_S = K_{FW} \times \dot{W}_N \times T_S$$
$$V = 20 \times 10^{-15} \text{ m}^2/\text{N} \times 0.753 \times 10^{-3} \text{ W} \times 40 \text{ y} \times 365 \text{ d/y} \tag{12-30}$$
$$\times \ 24 \text{ h/d} \times 3600 \text{ s/h}$$
$$V = 19 \times 10^{-9} \text{ m}^3 \text{ or } 19 \text{ mm}^3$$

The resulting tube wall wear depth, d_w, can be calculated from the wear volume. This calculation requires the knowledge of the relationship between d_w and V. For example, for a tube within a circular hole, it may be assumed that the wear is occurring uniformly over the support thickness, L, and half the circumference, πD, of the support. Thus,

$$d_w = 2V/(\pi DL) \tag{12-31}$$

For a tube diameter, $D = 20$ mm, and a support thickness, $L = 16$ mm, a life-time tube-wall wear depth:

$$d_w = 2 \times 19 \times 10^{-9} \text{ m}^3/(\pi \times 0.02 \text{ m} \times 0.016 \text{ m})$$
$$d_w = 37.8 \text{ } \mu m$$

is calculated.

Taking a typical tube wall thickness of 1.0 mm and a maldistribution factor of three, an upper bound wear depth of 113 μm or 11.3% is obtained. This amount of tube wall loss is reasonable for a well-designed heat exchanger.

On the other hand, if the tube were assumed to vibrate back and forth within a typical tube-to-support diametral clearance of 0.4 mm (equivalent vibration response of 400 $\mu m/(2\sqrt{2}) = 141$ μm rms), a work-rate of 48 mW would be estimated using Eq. (12-15). Such a work-rate would lead to 25% tube wall reduction within approximately four years without accounting for maldistribution. This would be unacceptable in practice. In general, the operating experience with real heat exchangers has shown that this situation is unlikely. However, it could be possible in the presence of large vibration amplitudes due to fluidelastic instability.

Example 12-2   Steam Generator U-Bend Tubes

Consider now the U-bend region of the hypothetical steam generator shown in Fig. 12-13. The characteristics of the largest radius tube, which is supported by vertical flat-bar supports, are: $D = 16$ mm, $w = 1.0$ mm, $m = 0.6$ kg/m, average $\ell = 0.6$ m, $f = 100$ Hz, $L = 25$ mm, and $\zeta_S = 1.0\%$. The mid-span vibration response of the tube to turbulence excitation is assumed to be a realistic $10\ \mu m$ rms. Assuming that the supports are effective (i.e., the tube-to-support clearances are small and the vibration response at the supports is small relative to that at the mid-span), a work-rate of 0.18 mW is calculated for this case using Eq. (12-29)

$$\dot{W}_N = 16\pi^3 (100\ Hz)^3 0.6\ kg/m \times 0.6\ m \times (10\ \mu m)^2 \times 0.010 = 0.18\ mW$$

Assuming fretting wear is taking place continuously for 40 years and taking the wear coefficient to be 20×10^{-15} m²/s, the wear volume may be calculated with Eq. (12-30);

$$V = 20 \times 10^{-15}\ m^2/N \times 0.1786 \times 10^{-3}\ W \times 40\ y \times 365\ d/y \times 24\ h/d$$
$$\times\ 3600\ s/h$$
$$V = 4.506 \times 10^{-9}\ m^3 = 4.5\ mm^3$$

For flat-bar-type supports, it is assumed that wear is occurring on one side and only on the tubes. Thus, the bars and the wear scars remain flat (see Fig. 12-14a). This assumption leads to:

$$V = \frac{LD^2}{8}(2\theta - \sin 2\theta) \tag{12-32}$$

where θ is the intersection angle,

$$\theta = arc\cos\left[(D - 2d_w)/D\right] \tag{12-33}$$

Thus, the tube wall fretting-wear depth, d_w, may be estimated using Eqs. (12-32) and (12-33). However, these equations need to be solved iteratively. A curve of d_w versus V for $L = 25$ mm and $D = 16$ mm is shown in Fig. 12-14b. The wear volume of 4.5 mm³ for this geometry corresponds

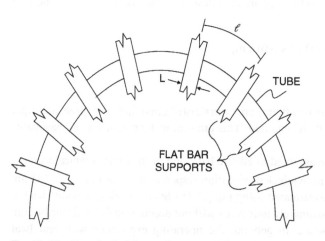

TUBE

FLAT BAR
SUPPORTS

Fig. 12-13   Hypothetical Steam Generator U-Bend Tube with Flat-Bar Supports.

(a)

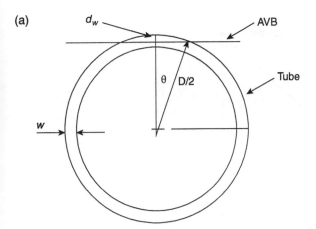

Fig. 12-14a Contact between a Steam Generator Tube and Flat-Bar-Type (AVB) Support.

to a wear depth of approximately 100 μm or 10% tube wall reduction for a wall thickness of 1.0 mm, which is acceptable from an operation point of view.

On the other hand, if for some reason the tube-to-support diametral clearances became larger (say 1.0 mm) at several adjacent support locations, then the supports could become ineffective. Under such conditions, the tubes would be susceptible to amplitude-limited fluidelastic instability and could possibly rattle within the available clearance. Using Eq. (12-15) and taking the vibration response corresponding to the diametral clearance as 354 μm rms, a work-rate of 447 mW may be estimated:

$$\dot{W}_N = 32\pi^3 (100 \text{ Hz})^3 \times 0.6 \text{ kg/m} \times 0.6 \text{ m} \times (354 \,\mu\text{m})^2 \times 0.010 = 447 \text{ mW}$$

From Fig. 12-14b, the wear volume corresponding to through-wall wear ($d_w = 1.0$ mm in this case) is 131 mm³. From Eq. (12-30), the time taken to reach this wear volume is:

$$T_S = V/(K_{FW} \dot{W}_N)$$
$$T_S = 131 \times 10^{-9} \text{ m}^3 / (20 \times 10^{-15} \text{ m}^2/\text{N} \times 0.4465 \text{ W})$$
$$T_S = 14.65 \times 10^6 \text{ s or } 0.46 \text{ year}$$

which is less than half a year. This is obviously an unacceptable situation. Although a very pessimistic example was chosen here, it does illustrate the importance of maintaining small clearances to ensure effective supports.

Example 12-3 Nuclear Fuels

Take, for example, the CANDU nuclear fuel element shown in Fig. 12-6. The resultant vibration response of such a fuel element is less than 25 μm rms under normal operating conditions. The mechanical characteristics of such fuel elements are typically: $m = 1.5$ kg/m, $\ell = 0.5$ m, $f = 30$ Hz and $\zeta_T = 3\%$. It would be reasonable to assume that half the vibration energy is dissipated at the support (bearing pad/pressure tubes contact) while the other half is absorbed internally by the fuel pellets and by viscous damping between the fuel element and the fluid; thus, $\zeta_S = 1.5\%$. Using Eq. (12-29) to calculate the normal work-rate;

$$\dot{W}_N = 16\pi^3 (30 \text{ Hz})^3 \times 1.5 \text{ kg/m} \times 0.5 \text{ m} \times (25 \,\mu\text{m})^2 \times 0.015$$
$$\dot{W}_N = 0.094 \text{ mW}$$

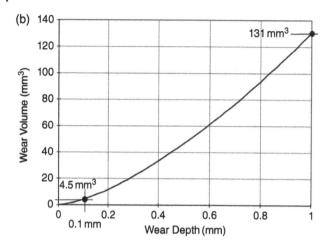

Fig. 12-14b Wear Volume versus Wear Depth.

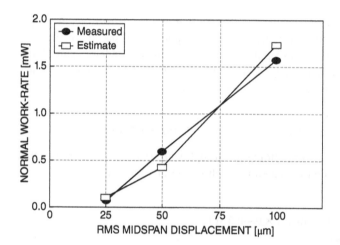

Fig. 12-15 Estimated and Measured Work-Rate for Fuel Element Subjected to Turbulent Excitation.

This is acceptable and would not cause excessive fretting-wear damage. However, it is easy to see with Eq. (12-29) that a vibration response of 100 μm rms would result in $\dot{W}_N$ =1.5 mW, which could cause excessive fretting-wear damage. Estimated and measured work-rates for a range of fuel element vibration responses are compared in Fig. 12-15. They are remarkably close, giving further credibility to the proposed energy approach.

Pressurized water reactor (PWR) fuel rods are another interesting example. They are typically 5 m long, supported every 0.5 m by grid-supports, have a mass per unit length, $m = 0.7$ kg/m, and a natural frequency of typically 27.5 Hz. Using a realistic damping ratio of 2% and vibration response level of 15 μm rms, the normal work-rate can be calculated using Eq. (12-29):

$$\dot{W}_N = 16\pi^3(27.5 \text{ Hz})^3 \times 0.7 \text{ kg/m} \times 0.5 \text{ m} \times (15 \,\mu\text{m})^2 \times 0.02$$
$$\dot{W}_N = 0.016 \text{ mW}$$

Could this level of vibration energy cause a problem? Very unlikely if the fuel rods are tight within the grids. Then no sliding occurs and the work-rate is nil. If for some reason, such as creep, clearances develop between the fuel rod and grids, then the maximum work-rate available at the grids would be 0.016 mW. Would this be sufficient to cause a problem? Assuming the total contact area at one grid to be 10 mm^2 and the fretting-wear coefficient to be 1000 x 10^{-15} m^2/N, the wear depth after a fuel residency, T, of 10,000 hours can be estimated with Eq. (12-30):

$$V = \dot{V}T = K_{FW} \times \dot{W}_N \times T$$
$$V = 1000 \times 10^{-15} \text{ m}^2/\text{N} \times 16.2 \times 10^{-6} \text{ W} \times 10{,}000 \text{ h} \times 3600 \text{ s/h}$$
$$V = 0.585 \times 10^{-9} \text{ m}^3 = 0.59 \text{ mm}^3$$
$$d_w = V/A = 0.585 \text{ mm}^3/10 \text{ mm}^2 = 59 \text{ } \mu\text{m}$$

Such a wear depth is probably tolerable. However, it is easy to see that a vibration response of 75 μm rms would not be acceptable since it would cause an estimated wear depth of 1.5 mm (i.e., 25 times greater).

Example 12-4   Piping Systems

A typical example of recommended vibration limits for average piping systems is shown in Fig. 12-16 (Wachel, 1982). It is interesting to compare this recommended practice to the energy approach with a simple example. Take a multi-span steel pipe of 100 mm diameter, 5 mm wall, containing flowing water and supported at 3.0 m intervals (see Fig. 12-17). Using a modulus of elasticity $E = 0.2 \times 10^{12}$ N/m^2,

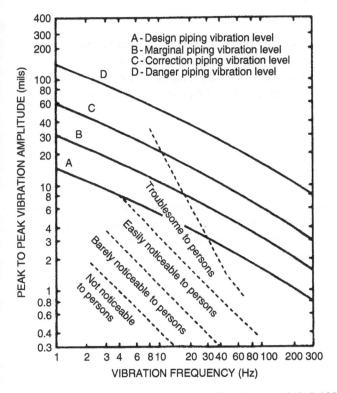

Fig. 12-16   Typical Vibration Limits for Piping System (Wachel, 1982).

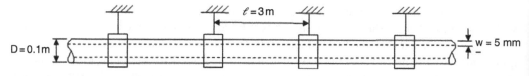

Fig. 12-17 Multi-Span Steel Pipe.

water density, $\rho_w = 1000 \, \text{kg/m}^3$, and a pipe material density, $\rho_t = 8000 \, \text{kg/m}^3$, a total mass per unit length, $m = 18.3 \, \text{kg/m}$, the first-mode natural frequency, $f = 23.7 \, \text{Hz}$, can easily be calculated as discussed in Chapter 4. The recommended vibration limit from Fig. 12-16 for such a pipe is 3.3 mils (p-p) or approximately 30 μm rms. Assuming a realistic damping ratio for pipes of $\zeta_S = 2\%$ and using Eq. (12-29) leads to

$$\dot{W}_N = 16\pi^3 (23.7 \, \text{Hz})^3 \times 18.3 \, \text{kg/m} \times 3.0 \, \text{m} \times (30 \, \mu\text{m})^2 \times 0.02$$
$$\dot{W}_N = 6.53 \, \text{mW}$$

Now, if we assume that the above mentioned multi-span pipe is supported by typical 25 mm wide circular steel bands, that fretting wear is taking place over half the circumference of the band, and that the fretting-wear coefficient for this material combination is $K_{FW} = 50 \times 10^{-15} \, \text{m}^2/\text{N}$, we can calculate with Eqs. (12-30) and (12-31) a pipe wall reduction after 40 years of:

$$V = 50 \times 10^{-15} \times \text{m}^2/\text{N} \times 6.53 \, \text{mW} \times 40 \, \text{y} \times 365 \, \text{d/y} \times 24 \, \text{h/d} \times 3600 \, \text{s/h}$$
$$V = 412 \times 10^{-9} \, \text{m}^3 \text{ or } 412 \, \text{mm}^3$$
$$d_w = 2 \times 411.6 \times 10^{-9} \, \text{m}^3/(\pi \times 0.1 \, \text{m} \times 0.025 \, \text{m})$$
$$d_w = 0.105 \times 10^{-3} \, \text{m or } 0.11 \, \text{mm}$$

which corresponds to a 2.1% pipe wall reduction. This is acceptable in most practical cases. Of course, it is assumed that fretting wear is the damage mechanism in this case.

Similarly, the danger vibration level of 33 mils (p-p) or ~300 μm rms, from Fig. 12-16 would result in a 210% pipe wall reduction after 40 years. This amount of damage is certainly not acceptable.

12.8 Concluding Remarks

In summary, a simple relationship based on vibration energy has been formulated to predict work-rate and fretting-wear damage in process equipment. The proposed approach has been verified. It provides an alternative to comprehensive time-domain simulations for simple cases such as multi-span heat exchanger tubes and nuclear fuels.

Nomenclature

d_w	= wear depth, m
D	= tube outside diameter, m
E	= elastic modulus, N/m^2
E_d	= energy dissipated per cycle due to damping, J

f	= natural frequency, Hz
$F, F(t)$	= contact force, N
$F_S, F_S(t)$	= shear component of contact force, N
J^2	= joint acceptance
k	= stiffness, N/m
K_{FW}	= fretting-wear coefficient, m^2/N
ℓ	= span length, m
L	= thickness of support, m
m	= mass per unit length, kg/m
M	= dynamic mass, kg
N	= number of spans in a multi-span tube
P_S	= mechanical power dissipated at supports, W
P_T	= total mechanical power. W
P_V	= mechanical power dissipated by viscous damping,
S	= sliding distance, m
$S_F(f)$	= power spectral density, $(N/m)^2s$
T, t	= time, s
T_S	= component life, s
V	= wear volume, m^3
$\dot{V}$	= wear volume rate, m^3/s
w	= wall thickness, m
$\dot{W}$	= work-rate, W
$\dot{W}_N$	= normal work-rate, W
$\dot{W}_S$	= shear work-rate, W
$y(\ell/2)$	= mid-span vibration response, m
x	= position along beam, m
y_{rms}	= vibration amplitude, m
Y	= vibration amplitude (0 to peak) at resonance, m
ρ_t	= tube material density, kg/m^3
ρ_w	= fluid density, kg/m^3
θ	= intersection angle, rads
ϕ^2	= normalized mode-shape factor
ζ_F	= friction damping ratio
ζ_{SF}	= squeeze-film damping ratio
ζ_S	= support damping ratio
ζ_T	= total (overall) damping ratio
ζ_V	= viscous damping ratio
μ	= dynamic friction coefficient

Note

1 CANDU is a registered trademark of Atomic Energy of Canada Limited.

References

Au-Yang, M. K., 2001, *Flow-Induced Vibration of Power and Process Plant Components: A Practical Workbook*, ASME Press, New York.

Blake, W. K., 1986, *Mechanics of Flow-Induced Sound and Vibration: Volume I-General Concepts and Elementary Sources*, Academic Press, Orlando.

Blevins, R. D., 1978, "Fretting-Wear of Heat Exchanger Tubes," *ASME Paper No. 78-JPGC-NE-9, Joint Power Generation Conference*, September 10-14, Dallas, Texas.

Connors, H. J., 1980, "Flow-Induced Vibration and Wear of Steam Generator Tubes," *Westinghouse R&D Report 80-IE7-TUBES-P1*, SG-09-041.

Fisher, N. J., Ing, J. G., Pettigrew, M. J. and Rogers, R. J., 1992, "Tube-to-Support Dynamic Interaction for a Multispan Steam Generator Tube," *Proceedings of ASME International Symposium on Flow-Induced Vibration and Noise*, Anaheim, California, November 8-13, Vol. 2, pp. 301–316.

Fisher, N. J., Tromp, J. H. and Smith, B. A. W., 1996, "Measurement of Dynamic Interaction Between a Vibrating Fuel Element and Its Support," *Proceedings, ASME-PVP Symposium on Flow-Induced Vibration-1996*, Montreal, Canada, July 21-26, ASME Publication PVP-Vol. 328, pp. 271–283.

Fisher, N. J., Weckwerth, M. K., Grandison, D. A. E. and Cotnam, B. M., 1997, "Fretting-Wear of Zirconium Alloys," *Transactions of the 14th International Conference on Structural Mechanics in Reactor Technology (SMiRT14)*, Lyon, France, August 17-22, pp. 183–191.

Fluit, S. M. and Pettigrew, M. J., 2001, "Simplified Method for Predicting Vibration and Fretting-Wear in Nuclear Steam Generators," *Proceedings, ASME-PVP Symposium on Flow-Induced Vibration-2001*, Atlanta, Georgia, July 22-26, ASME Publication PVP Vol. 420-1, pp. 7–16.

Frick, T. M., Sobek, T. E. and Reavis, J. R., 1984, "Overview on the Development and Implementation of Methodologies to Compute Vibration and Wear of Steam Generator Tubes," *Proceedings, Int. Symposium on Flow-Induced Vibrations: Volume 3: Vibration in Heat Exchangers*, ASME Special Publication, pp. 149–161.

Guérout, F. M. and Fisher, N. J., 1999, "Steam Generator Fretting-Wear Damage: A Summary of Recent Findings," *Proceedings, ASME-PVP Symposium on Flow-Induced Vibration - 1999*, Boston, Massachusetts, August 1-5, ASME Publication PVP-Vol. 389, pp. 227–234.

Ko, P. L., 1985, "The Significance of Shear and Normal Force Components on Tube Wear due to Fretting and Periodic Impacting," *AECL Report No 8845*, October 1985.

Pendlebury, R. E., Aldham, D. and Warburton, J., 1985, "The Unlubricated Fretting Wear of Mild Steel in Air," *Wear*, 106, pp. 177–201.

Pettigrew, M. J., 1993, "The Vibration Behaviour of Nuclear Fuel under Reactor Conditions," *ANS Journal of Nuclear Science and Engineering*, 114, pp. 179–189.

Pettigrew, M. J., Taylor, C. E., Fisher, N. J., Yetisir, M., and Smith, B. A. W., 1998, "Flow-Induced Vibration: Recent Findings and Open Questions," *J. of Nuclear Engineering and Design*, 185(2-3), pp. 249–276.

Pettigrew, M. J., Yetisir, M., Fisher, N. J., Smith, B. A. W. and Taylor, C. E., 1999, "Prediction of Vibration and Fretting-Wear Damage: An Energy Approach," *Proceedings, ASME-PVP Symposium on Flow-Induced Vibration - 1999*, Boston, Massachusetts, August 1-5, pp. 283–290.

Taylor, C. E. and Pettigrew, M. J., 1999, "Random Excitation Forces in Heat Exchanger Tube Bundles," *Proceedings, ASME-PVP Symposium on Flow-Induced Vibration - 1999*, Boston, Massachusetts, August 1-5, pp. 35–42.

Taylor, C. E., Pettigrew, M. J., Dickinson, T. J., Currie, I. G. and Vidalou, P., 1998, "Vibration Damping in Multispan Heat Exchanger Tubes," *ASME Journal of Pressure Vessel Technology*, 120(3), pp. 283–289.

Thomson, W. T., 1993, *Theory of Vibration with Applications*, Fourth Edition, Prentice Hall, Englewood Cliffs, New Jersey.

Wachel, J. C., 1982, "Field Investigation of Piping System for Vibration-Induced Stress and Failures," *ASME Bound Vol. #H00219*, Pressure Vessel and Piping Conference, June 1982.

Yetisir, M. and Fisher, N. J., 1996, "Fretting-Wear Prediction in Heat Exchanger Tubes: The Effect of Chemical Cleaning and Modelling Ill-Defined Support Conditions," *Proceedings, ASME-PVP Symposium on Flow-Induced Vibration - 1996*, Montreal, Canada, July 21-26, ASME Publication PVP-Vol. 328, pp. 359–368.

Yetisir, M. and Pettigrew, M. J., 1999, "A Simple Approach to Estimate Heat Exchanger Tube Fretting-Wear: Application to U-bend Tubes," *Proceedings ASME-PVP Flow-Induced Vibration Symposium*, Boston, Massachusetts, August 1-5, pp. 273–282.

Yetisir, M., McKerrow, E. and Pettigrew, M. J., 1998, "Fretting-Wear Damage of Heat Exchanger Tubes: A Proposed Criterion Based on Tube Vibration Response," *ASME Journal of Pressure Vessel Technology*, 120(3), pp. 297–305.

13

Fretting-Wear Damage Coefficients
Nigel J. Fisher and Fabrice M. Guérout

13.1   Introduction

Excessive flow-induced vibration may cause damage by fatigue or by fretting wear. Fatigue is a relatively well understood damage mechanism. It is the subject of a number of books and design guidelines. Maximum dynamic stresses can be estimated from the vibration response and the number of dynamic stress cycles can be easily calculated from the response frequency. These two parameters are then used to estimate acceptability or fatigue life of components. Since fatigue damage information is already available and well documented, this chapter will focus on the much less well understood fretting-wear damage mechanisms.

13.2   Fretting-Wear Damage Mechanisms

Fretting wear occurs as a result of low-amplitude relative motion between contacting components. This low-amplitude motion distinguishes fretting wear from the more general sliding wear (see Fig. 13-1). The amplitude of motion for sliding wear is large in comparison to the area of contact, so wear particles are easily removed from between the contacting parts. Examples of sliding wear are journal bearings and nuclear reactor fuel handling systems. "Classical" fretting wear occurs in components where contacting parts are held together with normal loads or tight tolerance fits. It is similar to sliding wear in that the contact loads are large enough to hold the wear surfaces together. However, because the relative motion is small, wear particles are trapped between the surfaces and can participate in the wear process. An example of "classical" fretting wear is electrical contacts, where trapped wear debris can increase the electrical resistance across contacts.

13.2.1   Impact Fretting Wear

In components such as shell-and-tube heat exchangers (HX) and nuclear steam generators (SG), where vibration occurs as a result of flow-induced excitation forces and where clearances exist between contacting parts, wear can be a combination of "classical" fretting wear and impact-sliding (oblique impact) wear. We call this type of wear "impact fretting wear" to differentiate it from "classical" fretting wear, as shown in Fig. 13-1. Like "classical" fretting wear, impact fretting wear is characterized by small amplitude motion. However, the static normal loads (preloads) are much smaller, such that separation between the contacting surfaces can occur. This periodic separation

Flow-Induced Vibration Handbook for Nuclear and Process Equipment, First Edition.
Michel J. Pettigrew, Colette E. Taylor, and Nigel J. Fisher.

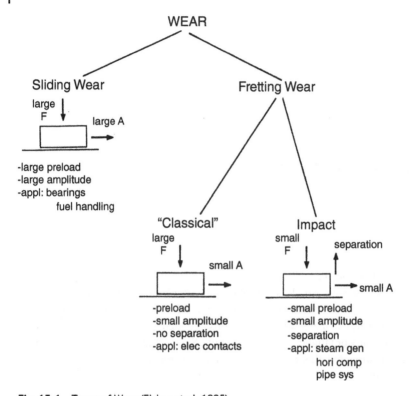

Fig. 13-1   Types of Wear (Fisher et al, 1995).

affects the wear behavior in two important ways: 1) wear particles can be removed (flushed) from the wear interface, and 2) impact forces exist between the contacting surfaces. In addition to SG and other HX, impact fretting wear can occur in reactor internals and piping systems.

Vingsbo and Soderberg's (1987) fretting map is shown in Fig. 13-2, where four wear regimes are defined based on the amplitude of relative motion between the contacting parts. Low-damage fretting occurs in the stick regime, where relative motions are in the order of several μm and can be accommodated by elastic deflection of the contacting parts. In this regime, the bodies experience little damage due to wear, and no fatigue crack growth. Fretting fatigue occurs at motion amplitudes of approximately 5 to 10 μm. This is the mixed stick and slip regime. Here, wear and corrosion effects are small, but accelerated crack growth can lead to reduced component fatigue life. Fretting wear occurs in the gross-slip regime, where amplitudes of motion range from approximately 20 to 300 μm. In the gross-slip regime, severe fretting-wear damage due to corrosion-assisted wear can occur, but limited fatigue crack growth occurs, because wear is progressing at a faster rate. At larger amplitudes of motion, wear damage is characteristic of unidirectional sliding.

Amplitudes of motion for impact fretting wear are in the order of 10 to 100 μm. Thus, impact fretting wear is in the gross-slip regime.

13.2.2   Trends

Early impact fretting-wear studies conducted by Ko (1979a, 1980, 1984, 1985a) focused on the relative effects of environmental and geometrical parameters on tube material wear rates. He found

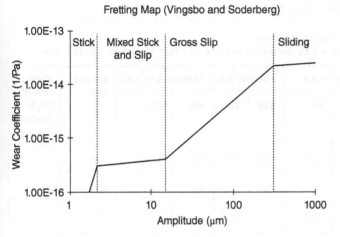

Fig. 13-2 Fretting Map (adapted from Vingsbo and Soderberg, 1987).

that material combination, test environment, tube-to-support motion, dynamic force level, support geometry, and test duration were all important parameters. In general, the effects of these parameters are interrelated, so it is sometimes difficult to attribute changes in wear rate to variations in one parameter only. However, Ko's studies did reveal general trends in fretting-wear rate due to these parameters.

Tube and support material combination is the most important parameter affecting fretting wear. Materials are generally chosen because of their resistance to chemical and environmental conditions, rather than their resistance to fretting-wear damage. Fretting wear is caused more by sliding motion than impacting. The compatibility of contacting materials is more important than hardness, as high surface affinity in identical or similar materials tends to significantly increase the development of strong micro-welds between contacting surface asperities, even for short-time loaded-contact conditions. These micro-welds lead to more damaging adhesive wear mechanisms and, therefore, result in significantly greater fretting-wear rates. Fretting-wear rates between the same or similar materials can be an order of magnitude greater than rates between dissimilar materials, as shown in Table 13-1 (Ko, 1984) for Incoloy[®][1] 800 or Inconel[®] 600 tubing and Inconel 600 supports.

The effect of test environment includes the effects of both temperature and water chemistry. Temperature can have a great effect on wear rate. In general, fretting-wear rates increase with increasing temperature, as shown in Fig. 13-3 (Ko, 1980). However, the underlying mechanism responsible for the increased wear rate is unclear.

The increase could be due to either decreased material strength or increased oxidation rates at higher temperatures. Both of these effects probably contribute. Fretting-wear studies of CANDU[®][2] fuel channel materials (Zircaloy-4 bearing pads versus Zirconium 2.5% Niobium pressure tubes) showed increasing fretting-wear rates from room temperature to 250°C and then decreasing wear rates from 250 to 315°C, as shown in Fig. 13-4 (Fisher et al, 1990).

In HX and SG, tube-to-support relative motion is a major parameter affecting fretting-wear rates. Motion type can be broadly differentiated into three classifications: impacting, sliding, and combined impact-sliding. To assess the effect of tube motion, Ko (1979a) conducted tests with motions ranging from primarily impacting to circumferential sliding. Ko's results showed that rubbing

Table 13-1 Ko's Short-Term Relative Fretting-Wear Data (Ko, 1984).

Fretting Wear of Tubing Materials in 265°C Pressurized Water
Excitation Frequency: 28 Hz, Force ratio, f_y/f_x = 3 Equivalent Depth of Wear of Tube Specimens, $\mu m/10^6$ cycles

Tubing Specimens \ Support Specimens	A36 Carbon Steel	Austenitic S.S.		Ferritic S.S.		Martensitic S.S.		Inconel 600	Two Other High Nickel Alloys
		347	304	405	430	403	410		
Incoloy 800	0.33	0.23	0.19	0.22	0.07	0.15	0.28	2.80	
Inconel 600	0.16	0.11	0.17	0.11	0.14	0.10	0.11	2.90	>1.50
Monel 400	0.11		0.20		0.15		0.23	0.60	

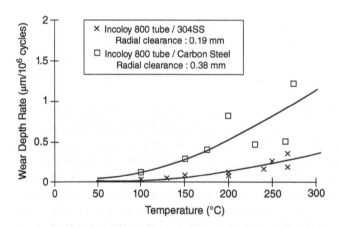

Fig. 13-3 Effect of Temperature on Fretting-Wear Rates (Ko, 1980).

motion caused more fretting wear than impacting motion. Therefore, fretting wear is primarily due to shear. Ko's later studies (Ko and Basista, 1984; and Ko, 1985b) focused on the importance of shear forces due to rubbing or oblique impact on fretting-wear rates.

In most HX and SG, the motion between a tube and its supports is unidirectional, perpendicular to the tube axis. However, in CANDU fuel channels it can be bi-directional, having components both perpendicular and parallel to the fuel element axis. Studies of fretting-wear damage in nuclear fuels have shown that fretting-wear coefficients for bi-directional motion may be an order of magnitude larger than for unidirectional motion (Guérout and Grandison, 1997).

The dynamic force level between the structure and its supports has a major effect on fretting-wear rates. Ko (1979a) showed that the resultant force, defined as the vector sum of the root-mean-square (rms) values of the support reaction force components, was an adequate parameter for comparing fretting-wear damage, as long as other parameters, such as motion type and contact geometry, were

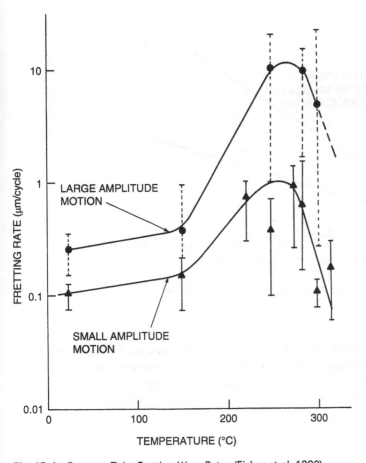

Fig. 13-4 Pressure Tube Fretting-Wear Rates (Fisher et al, 1990).

maintained constant. For short-duration tests, where the tube-to-support clearance does not change significantly, fretting-wear rates increase fairly linearly with contact force, as shown in Fig. 13-5 for Incoloy 800 tubing versus carbon steel supports at room temperature (Ko, 1979a). The motion type for these tests was combined impact-sliding.

In HX and SG, support geometry affects fretting-wear rates in three ways. First, relative tube-to-support motion is dependent upon support geometry. For example, rubbing motion can occur in drilled-hole supports, while combined impact-sliding or impacting motion is more likely in broached-hole supports. Second, the fretting-wear penetration rate is dependent on the tube-to-support contact area. Equal volume fretting-wear rates result in greater penetration rates for broached-hole supports, because wear is concentrated on the land areas. The third effect is an extension of the second. For moderate differences in contact area (e.g., drilled-hole versus broached-hole), volume fretting-wear rates are equal for comparable motion types and dynamic force levels. However, volume fretting-wear rates are affected by extreme contact areas. Tests with severe support geometries to simulate loose parts trapped in steam generator tube bundles have shown that severe geometries (such as knife-edges or threaded rods) result in lower fretting-wear rates but greater wear-depth rates (Guérout et al, 1993). This effect is probably due to differences in contact pressure (Lim and Ashby, 1987).

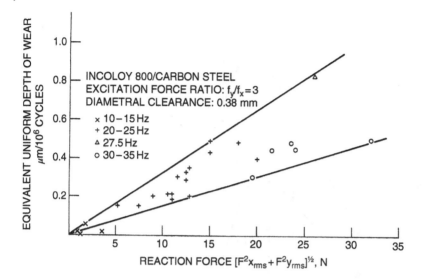

Fig. 13-5   Effect of Contact Force Level on Fretting-Wear Rates (Ko, 1979a).

Fretting-wear rates vary with test duration. Early in a fretting-wear test, the fretting-wear rate can change significantly as the contacting surfaces "wear-in" and steady-state wear is achieved. The fretting-wear rate then remains constant for a period before changing again, as the amount of fretting wear reaches the point where the clearance gap has increased sufficiently to affect other parameters, such as motion type and contact force.

13.2.3   Work-Rate Model

Archard's wear equation for sliding wear (Archard, 1953; Suh, 1986) states that wear volume, V, is linearly related to the product of applied force, F, and sliding distance, S, divided by material hardness, H, thus:

$$V = kFS/3H \tag{13-1}$$

The dimensionless constant of proportionality, k, is called the dimensionless wear coefficient and, ideally, is a constant property for a given material combination and environment.

The theoretical basis for Archard's wear equation is adhesion, whereby wear is postulated to result from adhesion of surface asperities during sliding. Wear occurs under this scenario when a given pair of asperities remains welded together and one of the pair is torn from its original surface and transferred to the other surface. The factor 3 in Archard's equation is a shape factor resulting from the assumption that welded junctions between asperities are circular and that wear fragments are hemispherical.

Similar forms of Archard's equation can be derived for other types of wear. Ko (1986) summarizes various equations and notes that all have the same format, with wear relating linearly to the product of normal force and sliding distance divided by material hardness.

In practice, Archard's equation is adequate for sliding where, to a rough approximation, wear volume relates linearly to sliding distance and normal force. However, the equation implies that harder materials wear less, which is not always true. For example, soft, commercially pure copper can be much more resistant to wear than AISI 1045 steel, which is much harder (Suh, 1986). Hardness, along with the shape factor, can be incorporated into the dimensionless wear coefficient, resulting in a dimensional wear coefficient, K_{FW}. The units of the wear coefficient are m^2/N or Pa^{-1}. In this form, the wear equation is

$$V = K_{FW}FS \tag{13-2}$$

As mentioned previously, Ko (1979a) showed that the resultant force was adequate for comparing fretting-wear rates of various material combinations, as long as other conditions, such as geometry and motion, were maintained constant. In later studies, Ko showed that rubbing motion caused more fretting-wear than impacting motion, and he focused on the importance of shear forces due to rubbing or oblique impact. By combining force and motion, Ko (1985b) developed a new correlating parameter that he called the "force function". To compute the force function, Ko used the instantaneous relative motion to resolve the instantaneous contact force into normal and shear components, and assembled force histograms for each. The force function represented the integrated area under the histogram curve for the shear force component. Fretting-wear data from tests with differing motions correlated well with this force function, exhibiting a power relationship between fretting-wear rate and force function.

Frick et al (1984) defined a similar parameter for correlating wear data called work-rate, $\dot{W}$. Work-rate is defined as the normal component of contact force, F_n, integrated over the real sliding distance, s, per unit time, T

$$\dot{W} = \frac{1}{T}\int_0^S F_n(t)ds \tag{13-3}$$

Work-rate, like Ko's force function, combines both contact force and relative motion, and differentiates shear and normal force components by integrating over the sliding distance. For impacting motion, work-rate is small because, although the contact force level may be high, the sliding distance is small. The work-rate for sliding motion is moderate to high, because both force level and sliding distance are moderate to high. Expressing Archard's wear equation, Eq. (13-2), in rate form and replacing the product of force and sliding distance with work-rate results in

$$\dot{V} = K_{FW}\dot{W} \tag{13-4}$$

where $\dot{V}$ is the volumetric wear rate.

Archard's equation in rate form (Eq. (13-4)) implies that wear rate is linearly related to work-rate through a constant wear coefficient. This linearity is not necessarily valid for all types of wear. In some instances, and especially over broad ranges of contact conditions encompassing different wear regimes, the wear coefficient may be dependent on work-rate, resulting in a nonlinear relationship between wear rate and work-rate. Consistent with Ko's (1985b) experience of correlating wear versus his "force function", a power relationship between wear rate and work-rate would not be unreasonable.

Thus far, work-rate (Eq. (13-3)) has been defined in terms of the normal component of contact force, F_n, because that definition is consistent with Archard's original formulation. However, work-rate could also be defined in terms of the shear component, F_S. This latter definition makes sense as the so-defined work-rate would directly represent the mechanical energy dissipated at the point of contact.

13.3 Experimental Considerations

13.3.1 Experimental Studies

Fretting wear of process and nuclear component materials has been studied experimentally at Canadian Nuclear Laboratories (CNL) for more than four decades. These studies date back to fretting-wear failures that occurred in a SG at the Douglas Point prototype CANDU power station (Pettigrew et al, 1991) (see Chapter 1). Tubes in the U-bend region of this SG vibrated with sufficient amplitude to cause tube-to-support and tube-to-tube fretting wear. One tube failure occurred in 1971 and a second in 1977, about 10 years after station start-up. These failures were attributed to inadequate support combined with high-velocity two-phase cross flow. More examples of fretting-wear damage are given in Chapter 1.

Steam generator fretting wear remains an important issue. According to Dow's (1996) survey of SG tube performance during the 1992/93 period, approximately 60 power stations reported fretting-wear-related leaking and 500 to 600 tubes were plugged each year.

Tests at CNL are conducted in machines that simulate SG environmental conditions and tube-to-support dynamic interactions. Two types of test machines are used. Simple room-temperature machines are used to conduct exploratory tests that show trends within the range of interest of relevant parameters. Results from these tests provide insights into fretting-wear mechanisms. More complicated high-temperature machines are used to test material combinations at typical operating conditions. Fretting-wear coefficients from these tests are used to predict damage in actual components.

13.3.2 Room-Temperature Test Data

A room-temperature fretting-wear test program (Fisher et al, 1995) was conducted to verify that work-rate is a suitable correlating parameter for impact fretting wear. Eleven impact fretting-wear tests, each approximately 140 hours in duration, were conducted in distilled water in a room-temperature test machine.

The room-temperature machine (see Fig. 13-6) consists of a single-span cantilevered SG tube, a clearance support plate and supporting structure. Tube motion is induced by a vibration generator attached to the cantilevered end of the tube. The tube can be forced to oscillate with elliptical motions of various amplitude ratios, ranging from straight-line impacting motion to circular rubbing motion.

The support geometry for these tests was drilled-hole. The tube outside diameter was 14.6 mm and the support diametral clearance was 0.38 mm. The material combination was Incoloy 800 tubing and Inconel 600 supports, chosen because the high fretting-wear rate would result in readily measurable wear during a short-duration test.

Contact forces and relative motions were monitored continually throughout each test. Contact forces were measured using a force transducer assembly (see Fig. 13-7) consisting of four miniature piezoelectric force transducers located at 90-deg intervals around the circumference of the support (Ko, 1985a; and Fisher and Ingham, 1989). Relative tube motions within the drilled-hole support were measured using a pair of high-sensitivity eddy-current displacement probes mounted on the support platform.

Contact forces and motions in the two principal directions were simultaneously recorded periodically throughout each test. Work-rates were computed from the four recorded signals which were digitized at a sampling frequency of 16 kHz. The digital records were combined and resolved to compute normal and shear force components, sliding distances, and work-rates.

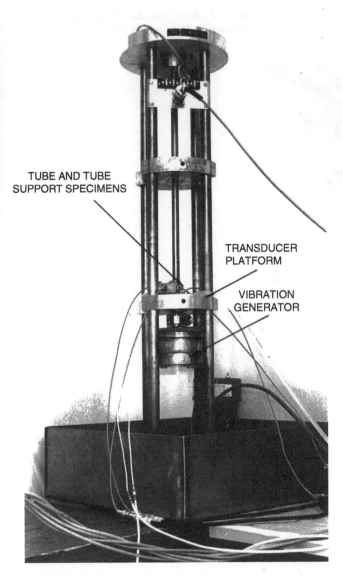

TUBE AND TUBE
SUPPORT SPECIMENS

TRANSDUCER
PLATFORM

VIBRATION
GENERATOR

Fig. 13-6   Room-Temperature Fretting-Wear Machine.

Excitation frequencies for the eleven tests were approximately 25 Hz. Resultant force levels ranged from 10 to 17 N rms. Tube motions ranged from pure impacting to circular rubbing: two impacting, six combined, and three rubbing. Tube fretting-wear rates are plotted versus work-rate in Fig. 13-8. Work-rates for the two impacting motion tests were small and the fretting-wear damage was also small. Work-rates and wear damage for the other tests were greater. In general, greater work-rates resulted in greater wear damage, although experimental scatter is evident at larger work-rates. Over this large range of work-rate, the fretting-wear rate appears to follow a power relationship with respect to work-rate. For smaller ranges (i.e., less than 100 mW), a linear correlation between fretting-wear rate and work-rate would be appropriate.

Fig. 13-7 Force Transducer Assembly for Room-Temperature Machine (Ko, 1985a).

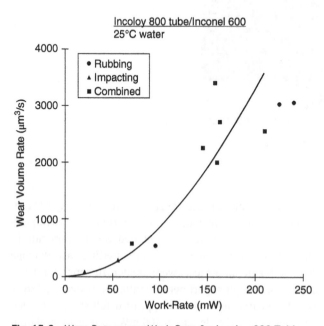

Fig. 13-8 Wear Rate versus Work-Rate for Incoloy 800 Tubing and Inconel 600 Supports at Room Temperature (Fisher et al, 1995).

13.3.3 High-Temperature Experimental Facility

The CNL high-temperature impact fretting-wear test facility consists of test machines connected to a water storage tank and accumulator. Temperature and pressure inside the test machines can be independently controlled. Either pressurized water or saturated steam conditions can be achieved. The operating temperature and pressure limits are 320°C and 11.7 MPa.

Each machine, see Fig. 13-9, consists of an autoclave, excitation tube, instrument platform, vibration generator, and supporting structure. The tube wear specimen is attached to the excitation tube

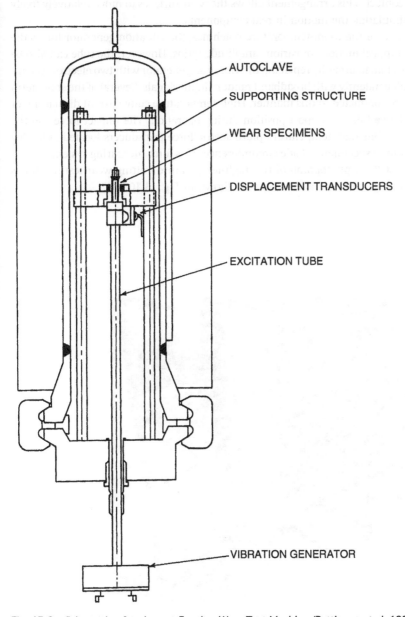

AUTOCLAVE

SUPPORTING STRUCTURE

WEAR SPECIMENS

DISPLACEMENT TRANSDUCERS

EXCITATION TUBE

VIBRATION GENERATOR

Fig. 13-9 Schematic of an Impact Fretting-Wear Test Machine (Pettigrew et al, 1991).

and the support wear specimen is mounted on the instrument platform. Two additional control specimens are placed within the autoclave to monitor material oxidation. These specimens are mounted on the instrument platform next to the support wear specimen. The relative position of the tube wear specimen with respect to the support wear specimen is controlled by adjustment screws that clamp the excitation tube to the autoclave base.

The vibration generator is attached to the end of the excitation tube that protrudes outside the autoclave through a flexible sleeve. The test rigs simulate the dynamic interaction characteristic of components such as nuclear fuel channels, SG and HX. Unlike conventional wear testing machines, where the contact motion is strictly controlled, the excitation in these rigs is remote from the point of specimen contact. This arrangement allows the wear surfaces to move relatively freely against one another, simulating the motion in real components.

As described previously for the room-temperature machines, the vibration generator forces the tube to oscillate with elliptical motions of various amplitude ratios. However, the tube can also be forced to vibrate in a random manner by replacing the vibration generator with two electromechanical vibrators driven independently with broadband random input signals. Several of the short-term tests described later were conducted in this manner. High-temperature eddy-current displacement probes monitor the relative tube-to-support position during operation. The impact forces on the support are measured by four high-temperature piezoelectric force transducers located at 90-deg intervals around the support specimen outside circumference. This force measuring system is capable of operation up to 350°C. A photograph of two high-temperature fretting-wear test machines connected to a water storage tank and accumulator is shown in Fig. 13-10.

Fig. 13-10 Two High-Temperature Machines Used at AECL-CNL (Guérout and Fisher, 1999).

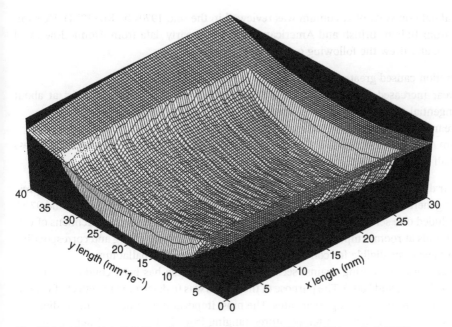

Fig. 13-11 Isometric 3-D Wear Profile (Specimen C297 PT-124) (Fisher et al, 1997).

13.3.4 Wear Volume Measurements

Fretting wear is usually a long-term wear mechanism resulting in failures in years or decades rather than days. Of necessity, fretting-wear testing is much shorter, 500 hours being a typical test duration. Thus, the wear volume to be measured may be very small and must be measured very accurately. Two techniques are used to measure wear. Wear may be estimated from the weight loss of the fretting-wear specimens. Extraneous effects such as corrosion are compensated for by the use of reference specimens. Very accurate weighing scales in an environmentally controlled room are required.

Another technique to measure wear volume is three-dimensional (3-D) profilometry. This technique is often more accurate since wear volume is measured directly. An isometric 3-D profile for a zirconium nuclear fuel channel specimen from a 1000-hour duration test is shown in Fig. 13-11. The maximum depth of the wear mark is 100 μm.

13.4 Fretting Wear of Zirconium Alloys

13.4.1 Introduction

Zirconium alloys are widely used in the core of nuclear reactors. Components such as fuel channels and nuclear fuel assemblies are made of zirconium alloys for neutron economy. Thus, it is important to understand the fretting-wear behavior of zirconium alloys, particularly when it may lead to fuel failures or affect the integrity of pressure boundaries such as fuel channels. To relate vibration response to fretting-wear damage, the wear coefficients of materials at realistic levels of dynamic interaction and under realistic environmental conditions must be known.

International data on wear of zirconium was reviewed in the mid-1970s by Ko (1974). He summarized data from Italian, British and American sources and early data from Atomic Energy of Canada Limited, and drew the following qualitative conclusions:

1) Rubbing motion caused greater fretting wear than impacting motion.
2) Fretting wear increased with amplitude. A threshold amplitude appeared to exist at about 200 μm tangential motion, above which fretting wear increased dramatically.
3) Fretting wear increased approximately proportionally with normal load.
4) Results from American sources indicated that the fretting-wear rate of Zircaloy-2 at 315°C is less than one-half of that at 260°C, which itself is only one-tenth of that at 232°C.

Fretting-wear data related to CANDU fuel channels (i.e., fuel bearing pad interaction with pressure tube) was generated by Ko (1979b) in short-term tests at temperatures between 23 and 97°C. These tests included one series in which Zircaloy-2 tubing was worn against flat specimens of various other materials at room temperature and 97°C. Wear rates for both tubing and corresponding flat-bar supports were negligible for combinations using platinum and palladium.

Testing of bearing pads fretting against sections of pressure tube has been conducted at CNL (Fisher et al, 1990, 1997, and 2002). The purpose of these tests was to determine the effect of various parameters on pressure-tube fretting-wear rates. The most important parameter to be studied was temperature, with tests conducted at temperatures ranging from 25 to 315°C. Other parameters studied included motion type, vibration amplitude, pH control additive, dissolved oxygen content and pressure tube pre-oxidation.

13.4.2 Experimental Set-Up

The bearing pad wear specimen, consisting of a short length of fuel sheath with a bearing pad, is attached to the central tube, as shown in Fig. 13-12. The pressure tube wear specimen is mounted on the instrument platform. Additional bearing pad and pressure tube specimens are mounted on the instrument platform to monitor material oxidation. All of the tests in this program were conducted in pressurized water.

Prior to starting the fretting-wear test program, scoping tests were conducted to identify motion types that were repeatable and that caused measurable wear within a reasonable test duration. The intent of the program was to simulate the interaction occurring between fuel elements and the pressure tube due to flow turbulence. Some elements are postulated to be close to or lightly in contact with the pressure tube, such that contact could occur as a result of low levels of turbulence-induced vibration (Yetisir and Fisher, 1995). The vibration level under these conditions would be in the order of 25 μm rms, and would be in the normal and tangential directions. Accordingly, the scoping tests were conducted at fairly low levels of dynamic interaction and did not include axial motion or large preloads. Four stable but different motion types were identified:

1) Large-amplitude impacting at 15 Hz (LAI),
2) Large-amplitude rubbing at 15 Hz (LAR),
3) Small-amplitude impacting at 25 Hz (SAI), and
4) Small-amplitude rubbing at 25 Hz (SAR).

The amplitudes of the two motion types at 15 Hz were typically three times greater than the amplitudes at 25 Hz. For both frequencies, the tangential components of vibration were equal, while the normal components were nearly zero for rubbing motion and almost equal to the tangential components for impacting motion. The contact force magnitudes for impacting motion

Fig. 13-12 Bearing Pad and Pressure Tube Specimens Installed in the Wear Machine (Fisher et al, 1997).

Fig. 13-13 Schematic of the SAI Motion Type (Fisher et al, 1990).

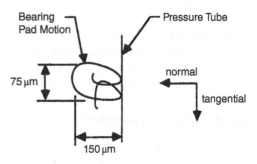

at both frequencies were equal and about two or three times greater than the magnitudes for rubbing motion.

The motion type for the majority of the test program was SAI, as this is most typical of turbulence-induced interaction between fuel elements and the pressure tube. This motion is shown schematically in Fig. 13-13. Typical displacements, forces and work-rates for this motion type were 65 to 100 μm rms, 5 to 15 N peak, and 0.5 to 4 mW, respectively.

The test program consisted of more than 125 tests, most of 250 hours duration but some as long as 1000 hours. The effects of environmental factors such as temperature, pH control additive and dissolved oxygen content were investigated. As well, the effects of other factors such as pressure-tube pre-oxidation, motion type, and vibration amplitude were studied.

At least three tests were conducted at each test condition to assess test repeatability and increase data reliability. The smallest variation in test results was for tests conducted at 25°C, where wear coefficient standard deviations were less than 50%. However, at higher temperatures above 150°C, the standard deviations were as great as 100%. To increase data reliability, more than six tests were conducted at some high temperature conditions where large variability was apparent.

Wear coefficients for the long duration tests were comparable to those from shorter duration tests conducted at the same conditions. Therefore, there was no apparent effect of test duration so long as the duration exceeded 250 hours.

13.4.3 Effect of Vibration Amplitude and Motion Type

Fretting-wear rates for all four types of motion (SAI, LAI, SAR and LAR) were compared at two temperatures: 25 and 286°C. Greater fretting-wear rates were observed for large amplitude motion than for small amplitude motion. However, work-rates for large amplitude motion were also greater, resulting in similar wear coefficients. Fretting-wear rates for rubbing and impacting motions at the same amplitude were approximately equal. As work-rates for rubbing motion were greater, wear coefficients were generally lower.

13.4.4 Effect of Pressure-Tube Pre-Oxidation and Surface Preparation

In the normal fabrication process, a thin black oxide layer is formed on the pressure-tube surface for wear protection. To investigate the effectiveness of this layer, some tests were conducted with pressure-tube specimens from an unoxidized tube. The presence of the oxide reduced the fretting-wear rate at lower temperatures slightly, where the material oxidation rate was low. However, at higher temperatures, the presence of the layer had no effect on the fretting-wear rate.

The surface finish of early pressure tubes was the result of grit blasting, while later tubes were honed. The grit-blasted surfaces are much rougher than the honed surfaces. However, comparison of tests conducted with both types of surface at nominally identical conditions did not reveal any significant differences in fretting-wear rates.

Overall, the effects of pressure-tube pre-oxidation and surface preparation were insignificant in comparison to the effects of other parameters, such as vibration amplitude and temperature.

13.4.5 Effect of Temperature

The effect of temperature on fretting-wear damage is very significant. Average results for 78 SAI tests are plotted in Fig. 13-14. Error bars on each data point indicate the range of results of individual tests at each temperature. More than ten individual tests are included in the average and range at some temperatures.

The range in results at lower temperatures (25 and 150°C) is a factor of two to three, which is typical repeatability for wear testing. At higher temperatures, the range in results is much greater, more than an order of magnitude at temperatures above 225°C. This increased scatter is thought to be characteristic of the fretting-wear behavior at these higher temperatures. The average wear coefficient decreases by an order of magnitude between 250 and 315°C. Therefore, it is likely that the increased scatter is due to slight variations in the test conditions resulting in changes in the fretting-wear regime.

The maximum fretting-wear coefficient was observed in the 225 to 286°C temperature range. Less damage occurred at lower temperatures (25 and 150°C) and at higher temperatures (300 and 315°C). In particular, the fretting-wear coefficient for SAI motion at 300 to 315°C is 300×10^{-15} m^2/N, ten times lower than the value of 3000×10^{-15} m^2/N for the 225 to 286°C range, and similar to the value of 150×10^{-15} m^2/N for room temperature. This observed decrease in

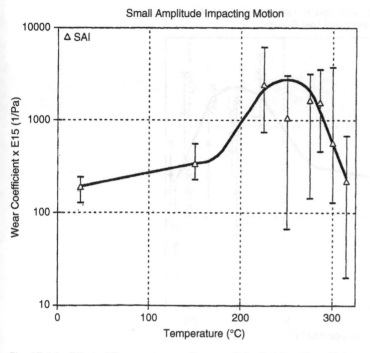

Fig. 13-14 Effect of Temperature on Pressure Tube Fretting Wear (Fisher et al, 1997).

fretting-wear coefficient with increasing temperature above 250°C is similar to that observed in early tests from American sources (Ko, 1974).

13.4.6 Effect of pH Control Additive and Dissolved Oxygen Content

Two pH control additives were used in the program to maintain the water pH between 9.5 and 10.5. Hydrazine was used in early tests because it also acted as an oxygen scavenger, ensuring that dissolved oxygen levels were maintained below 10 ppb. Later, when testing at higher levels of dissolved oxygen was required, lithium hydroxide was used. Lithium hydroxide is used in-reactor.

Early tests conducted at 250 and 315°C appeared to indicate that less fretting wear was occurring with lithium hydroxide. However, additional tests at those temperatures and at 225, 275 and 300°C showed that any difference is less than the scatter between individual tests and, thus, is insignificant.

Fretting wear is a combination of both wear and corrosion processes, with material oxidation affecting the wear process. Initially, it was postulated that the decrease in pressure tube wear coefficient at higher temperatures resulted from the development of a protective oxide layer on the worn surfaces at these temperatures. Tests were conducted with high levels (>500 ppb) of dissolved oxygen to test this hypothesis. Test results for high and low levels of dissolved oxygen are compared in Fig. 13-15.

Fretting-wear coefficients at 250°C were equal. Average fretting-wear coefficients at 300 and 315°C were greater for high levels of dissolved oxygen, while they were lower at 225 and 275°C. The scatter in test results was as great as the differences at all temperatures. Within the accuracy of the test results, there was no consistent difference in the fretting-wear coefficient for low and high levels of dissolved oxygen content. Therefore, it was concluded that increased oxidation at

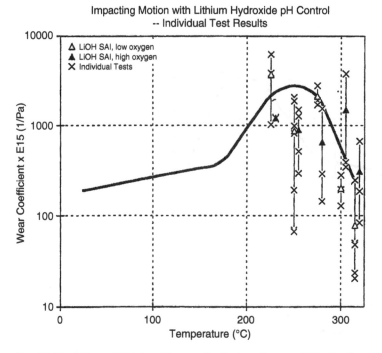

Fig. 13-15 Effect of Dissolved Oxygen Content on Pressure Tube Fretting Wear (Fisher et al, 2002).

higher test temperatures could not account for the decrease in the fretting-wear coefficient. Note that the fretting-wear coefficients presented in Fig. 13-14 include the results for both low and high levels of dissolved oxygen.

13.4.7 Discussions

Other researchers have observed similar effects of temperature on wear coefficients. Saito and Mino (1995) conducted unlubricated sliding tests of Alloy X40, a cobalt-based alloy, at various temperatures and sliding speeds in a conventional pin-on-disk wear machine. Their results are shown in Fig. 13-16.

They observed peak wear rates between 200 and 300°C, with much lower rates at temperatures above 500°C. Saito and Mino explained their results in terms of wear regimes. For low sliding speeds, they identified three regimes:

1) Mild metallic wear below 200°C,
2) Severe metallic wear between 200 and 400°C, and
3) Oxide wear above 400°C.

As discussed in more detail in Section 13.5.3, Guérout et al (1996) studied impact fretting wear of steam generator materials in pressurized water at various temperatures. Their results for work-rates less than 30 mW show that the fretting-wear coefficient for Incoloy 800 tubing material decreased by an order of magnitude between 215 and 315°C. The fretting-wear coefficient at 315°C was approximately equal to that at room temperature as shown in Fig. 13-17. They performed scanning electron microscopy (SEM) examinations of the worn tube samples to compare the wear process at

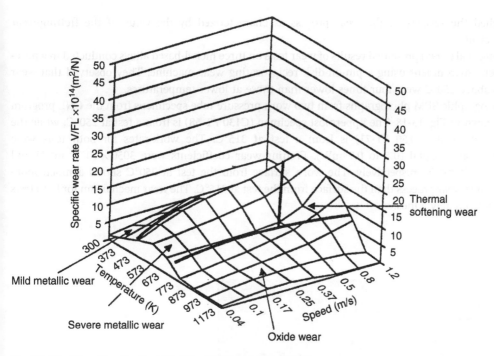

Fig. 13-16 Wear Map for the X40 Alloy (Saito and Mino, 1995).

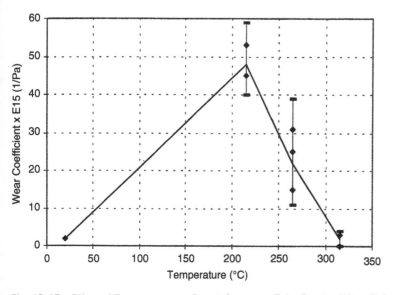

Fig. 13-17 Effect of Temperature on Steam Generator Tube Fretting Wear (Fisher et al, 2002).

each temperature. The wear process at 215°C was very aggressive, exhibiting wear damage by the removal of large wear sheets. In contrast, the wear process at 265 and 315°C was less severe, characterized by the removal of small plastically deformed edges of superimposed lap structures. Guérout et al (1996) concluded that the severity of the observed surface damage at each temperature matched the severity of the wear process as characterized by the value of the fretting-wear coefficient.

Jiang et al (1995) presented results of wear tests on three nickel-based alloys conducted in a pure-oxygen environment using a pin-on-disk reciprocating wear machine. They observed that wear rates above 250°C were four times lower than those at lower temperatures.

Topographic SEM photographs from two worn pressure tube specimens from the CNL program are shown in Fig. 13-18. The uppermost specimen (C130 PT-58) is from a test at 250°C, while the lower specimen (C150 PT-71) is from a test at 315°C. The work-rates for both tests were approximately equal (0.6 to 0.7 mW). Fretting-wear coefficients were 3050×10^{-15} m^2/N and 660×10^{-15} m^2/N, respectively. The worn surface from the test at 250°C shows a much more aggressive wear process than the surface from the test at 315°C. The wear mechanism for both tests is predominately adhesion.

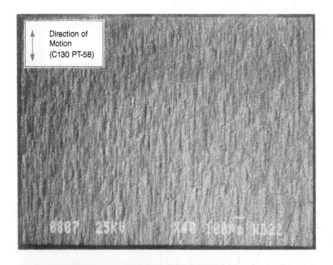

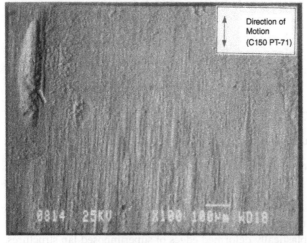

Fig. 13-18 Topographic SEM Photographs of Fret Marks (Fisher et al, 2002).

13.5   Fretting Wear of Heat Exchanger Materials

Fretting-wear studies of HX and SG material combinations have been conducted by several researchers. Early work by Blevins (1978 and 1985) and Cha et al (1987) was done at low temperatures. Later work by Hofmann et al (1986 and 1992) was done at temperatures up to 200°C. None of these studies presented results in terms of work-rates and fretting-wear coefficients.

13.5.1   Work-Rate Model and Wear Coefficient

A series of short-term fretting-wear tests was conducted at CNL to determine the wear coefficient for the material combination of Incoloy 800 tubing versus Type 410 stainless-steel supports (Fisher et al, 1995). The outside diameter of the tubing was nominally 15.9 mm. The environment for all tests was pressurized water at 265°C with typical CANDU secondary-side chemistry: pH 8-9, 50 ppb hydrazine, 5 ppm morpholine, and <5 ppb dissolved oxygen.

Tests were conducted for both drilled- and broached-hole supports. Figure 13-19 shows a typical pair of wear specimens. Both random and sinusoidal force excitations were used. Test durations ranged from 200 to 800 hours. The support surface finish, thickness, eccentricity, and diametral clearance were also varied.

Test results are summarized in Table 13-2 and plotted in Fig. 13-20. The test point identified as Table 1 is from Ko's table of relative material fretting-wear rates (Table 13-1). The work-rate for this data point has been estimated by measuring work-rate at an excitation similar to that reported by Ko. Work-rates computed from pre-test and post-test force measurements are listed in Table 13-2. The data is plotted versus the post-test work-rate in Fig. 13-20, as that work-rate is believed to be more representative of the average test conditions.

Two post-test work-rates are reported for Data Point C6-1. The pre-test and first post-test work-rate were measured with room temperature water completely covering the test specimens. The second post-test work-rate was measured with the water level lowered to just below the test specimens. This second work-rate is believed to be more representative of the work-rate at 265°C, because the tube motion was very similar. The tube motion with room-temperature water covering the test specimens was much less than at 265°C, due to greater squeeze-film damping within the drilled-hole support at room temperature. This effect was not observed in the broached-hole support tests.

A best-fit straight line through the data of Fig. 13-20 results in a derived fretting-wear coefficient of 40×10^{-15} m^2/N for Incoloy 800 tubing versus Type 410 stainless-steel supports at typical CANDU secondary-side conditions. The tests conducted with random excitation (C6-4 and C6-5) were at lower levels of interaction (work-rate) and resulted in lower fretting-wear rates, indicating that wear damage depends on the relative motion and contact force between the tube and support during impact, rather than the overall tube motion and type of excitation. Within a repeatability of ± 50 percent, work-rate is an effective correlating parameter for this short-term impact fretting-wear data.

Fig. 13-19   Incoloy 800 Tubing and Type 410 Stainless Steel Support Specimens (Fisher et al, 1995).

Table 13-2 Short-Term Fretting-Wear Data for Incoloy 800 Tubing and Type 410 Stainless Steel Supprts (Fisher et al, 1995).

Test	Test Type Excitation	Test Type Eccentricity	Support Type	Support Finish	Support Thickness (mm)	Support Clearance (mm)	Duration (hours)	Work-Rate (mW) Pre-Test	Work-Rate (mW) Post-Test	Wear Volum Rate (μm³/
Table 1	s	c	d	d	6.35	0.38	144	40 (estimated)		2030
C6-1	s	c	d	d	22.2	0.38	492	18	19/54	957
C6-2	s	c	b	d	22.2	0.38	456	29	31	1139
C6-4	r	c	b	b	22.2	0.38	505	- -	3	156
C6-5	r	e	b	b	22.2	0.38	503	0.4	- -	131
C6-7	s	e	b	b	22.2	0.38	500	8	7	348
C6-8	s	e	b	b	22.2	0.38	672	9	12	135
C6-9	s	c	b	b	6.35	0.38	827	12	24	867
C6-10	s	c	b	b	22.2	0.76	264	7	24	1104

Notes:
1) Excitation: s = sinusoidal; r = random
2) Eccentricity: c = concentric ; e = eccentric
3) Support Type: d = drilled-hole ; b = broached-hole
4) Finish: d = drilled ; b = broached

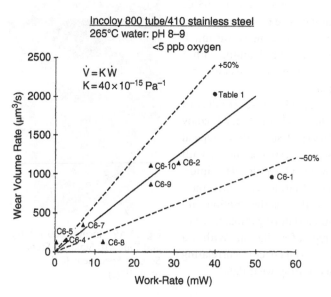

Fig. 13-20 Incoloy 800 Tubing and Type 410 Stainless Steel Fretting-Wear Curve (Fisher et al, 1995).

13.5.2   Effect of Test Duration

Two long-term fretting-wear tests were conducted to determine the time-dependency of fretting-wear data and to provide long-term fretting-wear test data (Fisher et al, 1995). The material combination for both tests was Inconel 600 tubing versus carbon steel supports. The outside diameter of the tubing was nominally 13.0 mm. The support geometries for the tests were 6.35-mm thick drilled-hole, and 25.4-mm thick broached-hole. The environment was pressurized water at 265°C with typical CANDU secondary-side conditions.

The procedure for the two long-term tests was similar to that used in the previous short-term tests. Forces were measured and work-rates were computed at room temperature before and after each test. The tests were interrupted every 500 to 1000 hours to measure fretting-wear damage and work-rate.

The initial diametral clearance for the drilled-hole test was 0.31 mm. The relative tube-to-support position was concentric. The work-rate measured at the beginning of the test was 55 mW. The motion at the start of the test was combined impact-sliding (see Fig. 13-21a). After 1100 hours, the clearance gap had increased to the point where the motion began to change from impact-sliding to predominantly impacting. After 1500 hours, the motion stabilized at predominantly impacting. The end-of-test motion (4922 hours) is shown in Fig. 13-21b.

The drilled-hole test results are shown in Fig. 13-22 in terms of apparent tube wear depth versus time. To calculate apparent wear depth, wear was assumed to occur uniformly over the full tube circumference. The fretting-wear rate increased for the first 500 hours of the test and then stabilized for the period from 500 to 1500 hours. The measured work-rate over this period varied from 55 to 75 mW. After 1500 hours, the fretting-wear rate decreased as the motion changed from combined impact-sliding to predominantly impacting. The work-rate measured at the end of the test was 60 mW.

The initial diametral clearance for the broached-hole test was 0.29 mm. The work-rate measured at the beginning of the test was 20 mW, corresponding to impacting motion as shown in Fig. 13-23a. The relative tube-to-support position was eccentric, with the tube resting against one land. The motion throughout the broached-hole test was impacting. The broached-hole geometry appeared to inhibit sliding motion. The end-of-test motion (3021 hours) is shown in Fig. 13-23b. The broached-hole results are shown in Fig. 13-24 in terms of apparent tube wear depth versus time.

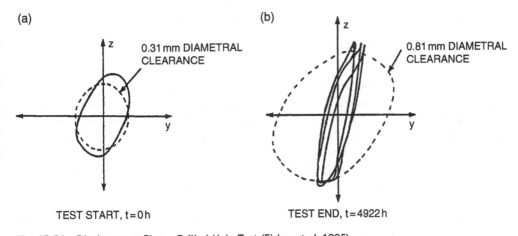

Fig. 13-21   Displacement Plots - Drilled-Hole Test (Fisher et al, 1995).

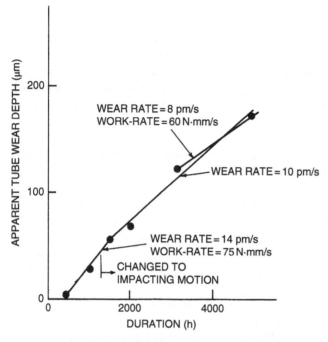

Fig. 13-22 Drilled-Hole Test Results (Fisher et al, 1995).

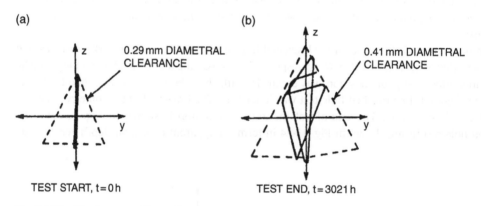

Fig. 13-23 Displacement Plots - Broached-Hole Test (Fisher et al, 1995).

To calculate apparent wear depth for the broached-hole support, wear was assumed to occur uniformly over the full tube circumference and was normalized for a 6.35 mm thick support. As the actual support was four times thicker (25.4 mm), with a total land area of approximately 18% of the tube circumferential area, the actual average land wear rates were approximately 35% greater than the apparent wear rates reported here.

The fretting-wear rate for the broached-hole test decreased over the first 120 hours of the test and then stabilized for the remaining period. The measured work-rate over this period varied from 10 to 40 mW.

Fretting-wear rates from the two long-term tests are plotted versus work-rate in Fig. 13-25. The test point identified as Table 1 is from Ko's table of relative material fretting-wear rates (Table 13-1).

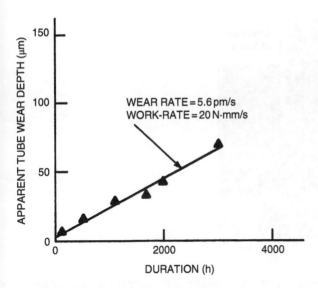

Fig. 13-24 Broached-Hole Test Results (Fisher et al, 1995).

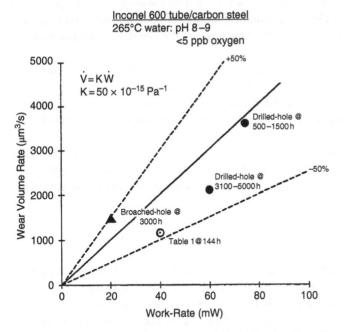

Fig. 13-25 Inconel 600 Tubing and Carbon Steel Fretting-Wear Curve (Fisher et al, 1995).

The work-rate for this data point has been estimated by measuring work-rate at an excitation similar to that reported by Ko. A best-fit straight line through the data of Fig. 13-25 results in a derived fretting-wear coefficient of 50 x 10^{-15} m^2/N for Inconel 600 tubing versus carbon steel supports at typical CANDU secondary side conditions. Within a repeatability of ± 50 percent, work-rate is an effective correlating parameter for this long-term impact fretting-wear data.

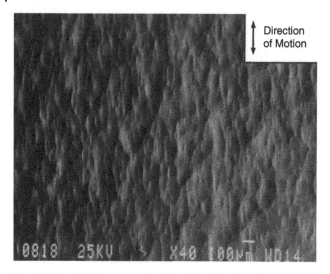

Fig. 13-26 SEM Photograph at 40x Magnification of the Worn Surface of the Inconel 600 Tubing Wear Specimen (Fisher et al, 1995).

A SEM photograph of the worn surface of the tube specimen from the drilled-hole support test at 40 times magnification is shown in Fig. 13-26. Elliptical dimples of approximately 100 to 200 μm length and 3:1 aspect ratio are clearly visible. The longer axes of the dimples are aligned with the direction of motion.

In summary, tests with different support geometries (i.e., drilled-hole and broached-hole), varying motion types, and different test durations showed that work-rate is an effective correlating parameter. Fretting-wear rates were shown to be linearly related to work-rate, within a repeatability of ± 50 percent. Therefore, material wear coefficients derived in this manner appear to be independent of support geometry, motion type, and time.

Attia and Magel (1999) obtained somewhat similar results for the steam generator materials they tested except at very high work-rates (in the order of 100 mW), where they measured higher fretting-wear coefficients, probably due to different wear mechanisms occurring at such high work-rates.

13.5.3 Effect of Temperature

Tests were performed at CNL for four temperatures (25, 215, 265, and 315°C) to investigate the effect of temperature on SG material combinations (Guérout et al, 1996, and Guérout and Fisher, 1999). Identical chemistry parameters such as pH, oxygen content, and hydrazine content were maintained for all tests performed. A pH of 9 was controlled by the addition of morpholine. The oxygen level was maintained below 5 ppb. Incoloy 800 tubing was used for the SG tube specimens. The typical outside diameter was 15.9 mm. The support material was Type 410 stainless steel. Two types of supports were tested: flat bar and broached hole. The support thickness was 25.4 mm for the flat-bar supports and 27.2 mm for the broached-hole supports. Tube-to-support diametral clearances were 420 and 360 μm, respectively. Impacting motion was dominant for all tests. Typical tube motion is shown in Fig. 13-27 for both types of support (Guérout and Fisher, 1999). The flat-bar support configuration is typical of the U-bend region of modern CANDU steam generators. The broached-hole support configuration is typical of the straight leg of earlier steam generators.

(a) (b)

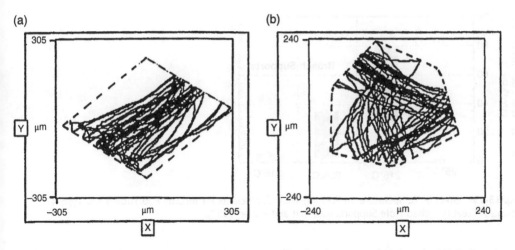

Fig. 13-27 Typical Tube-to-Support Relative Motion: a) Flat-Bar Support and b) Broached-Hole Support (Guérout and Fisher, 1999).

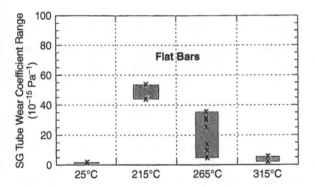

Fig. 13-28 Effect of Temperature on Fretting-Wear Coefficients for Incoloy 800 Tubing and Type 410 Stainless Steel Flat-Bar Supports (Guérout and Fisher, 1999).

Fretting-wear coefficients are plotted in Fig. 13-28 for the flat-bar support configuration (for work-rates ranging from 12 to 25 mW), and in Fig. 13-29 for the broached-hole support configuration (for work-rates ranging from 12 to 40 mW). For both types of support geometry, the wear coefficient increases between 25 and 215°C, and decreases between 215 and 315°C.

The results are consistent over a large range of work-rate values. For the tube-to-flat-bar support configuration, and for work-rates in the order of 20 mW, the comparison of wear coefficients between elevated temperatures and room temperature is as follows: the wear coefficient is approximately 30 times higher at 215°C than at 25°C, 20 times higher at 265°C than at 25°C, and two times higher at 315°C than at 25°C. Therefore, temperature has a significant effect on fretting wear and, in order to be relevant, tests must be performed at operating temperatures.

Examination of worn specimens (Guérout et al, 1996) showed that the fretting-wear mechanism is dependent on temperature. At 265°C, the worn surface exhibits a structure of superimposed laps that wear on the edge by removal of small plastically deformed tongues. The wear damage is similar at the higher temperature of 315°C except that the laps are smaller. At the lower temperature of

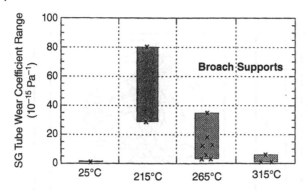

Fig. 13-29 Effect of Temperature on Fretting-Wear Coefficients for Incoloy 800 Tubing and Type 410 Stainless Steel Broached-Hole Supports (Guérout and Fisher, 1999).

215°C where the wear coefficient is greater, the fretting-wear process is much more severe, characterized by the removal of large scale-like layers (wear sheets).

13.5.4 Effect of Water Chemistry

Tests were performed to study the effect of normal steam generator water chemistry on tube fretting wear (Guérout and Fisher, 1999). The objectives were to compare normal water chemistries in use at various CANDU stations from a fretting-wear perspective (effect of using ammonia versus morpholine control, effect of hydrazine, effect of phosphates, effect of boric acid addition), and also to provide baseline wear coefficients prior to the study of the more corrosive environments resulting from faulted water or ion concentration conditions. The test program was divided into five series of three identical tests.

The first test series (DNGS) was performed using ammonia and hydrazine (pH = 9.8, dissolved oxygen (O_2) < 5 ppb). The second test series (G-2) used morpholine only (pH = 9.3, O_2 < 5 ppb). The third test series (PLNGS) was performed with congruent phosphate treatment (morpholine, hydrazine, and phosphate with pH = 9.5 and O_2 < 5 ppb). Flat-bar supports made of Type 410 stainless-steel stock bar were used for all tests of the three series. Incoloy 800 tubing was used for the tube specimens. The test duration was 500 hours for each test and work-rate was maintained in the 5 to 10 mW range. This level of work-rate was chosen to be as representative as possible of real steam generator tube excitation, while still generating measurable wear volumes at the end of a 500 hour test.

Even though low work-rates were used, the results showed relatively high wear-rate-to-corrosion-rate ratios. These ratios indicate the level of possible interaction of wear and corrosion. Work by Batchelor and Stachowiak (1988) indicates that this interaction is likely to affect the wear results if the ratio is less than 10:1. For the current tests, work rates of 5 to 10 mW resulted in wear-rate-to-corrosion-rate ratios in the order of 200:1 for the tube specimens and 50:1 for the supports. Even for realistic SG tube-to-support interaction levels of 1 to 5 mW, the ratios would still be higher than the limit ratio of 10:1, below which wear-corrosion interaction occurs.

The fretting-wear coefficients for the chemistries corresponding to the three test series are shown in Fig. 13-30. Average wear coefficients of 28×10^{-15} m^2/N, 20×10^{-15} m^2/N, and 11×10^{-15} m^2/N were derived for the first, second and third test series, respectively.

The fretting-wear coefficients obtained for each series confirm that, within the scatter of results observed for each series (between ±30% and ±70%), the type of normal SG water treatment used has no significant effect on fretting-wear damage so long as the chemistry parameters are kept within

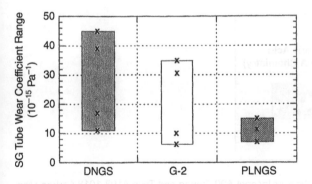

Fig. 13-30 Effect of Steam Generator Chemistry Control on Fretting Wear of Incoloy 800 Tubing and Type 410 Stainless Steel Flat-Bar Supports (Guérout and Fisher, 1999).

Fig. 13-31 Tube-to-Drilled-Hole-Support Relative Motion (Guérout and Fisher, 1999).

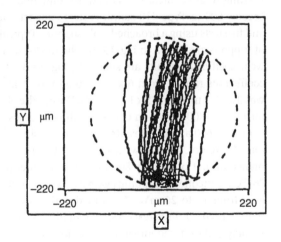

normal specifications. The average fretting-wear coefficient obtained for the third test series is slightly lower than that for the first and second test series, but additional testing would be required to confirm this trend.

A fourth and a fifth test series of three 500-hour tests were performed to study the effect of adding boric acid to normal SG water chemistry. Morpholine and hydrazine (pH = 9.7, O_2 < 5 ppb) was used for the fourth test series, whereas boric acid was added for the fifth test series. The tube-to-support configuration used for these two series was Inconel-600-to-A108-1018-carbon-steel drilled-hole. The drilled-hole specimens were used to simulate scallop bar geometry. Typical tube-to-support relative motion for this test series is shown in Fig. 13-31. The fretting-wear coefficient results are shown in Fig. 13-32 for both types of water chemistry. For the fourth test series, fretting-wear coefficients ranged from 5×10^{-15} to 19×10^{-15} m^2/N, averaging 13×10^{-15} m^2/N. For the fifth test series (boric acid added), fretting-wear coefficients ranged from 14×10^{-15} to 26×10^{-15} m^2/N, averaging 19×10^{-15} m^2/N. Considering the scatter in the results for each series, no significant effect of boric acid addition on tube wear coefficient was observed: the scatter of the results was 12×10^{-15} m^2/N with boric acid and 12×10^{-15} m^2/N without. The difference in average coefficients with the two different types of chemistry was only 6×10^{-15} m^2/N.

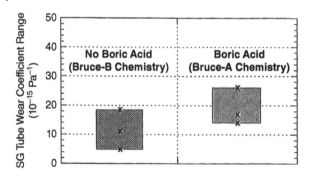

Fig. 13-32 Effect of Boric Acid on Fretting Wear of Inconel 600 Tubing and Type A108-1018 Carbon Steel Scallop Bars (Guérout and Fisher, 1999).

13.5.5 Effect of Tube-Support Geometry and Tube Materials

Fretting-wear coefficient results were compared for ten 500-hour tests performed at 265°C for the Incoloy-800-to-Type-410-stainless-steel support configuration: five tests using a flat-bar support and five tests using a broached-hole support. Typical tube-to-support relative motions for both types of supports are shown in Fig. 13-27. The work-rate range investigated was 8 to 27 mW. The fretting-wear coefficient results are shown in Fig. 13-33. For the flat-bar supports, the tube fretting-wear coefficients ranged from 6×10^{-15} to 35×10^{-15} m^2/N. For the broached-hole supports, the range was 3×10^{-15} m^2/N to 35×10^{-15} m^2/N. Within the observed scatter, average fretting-wear coefficients for Incoloy 800 tubing in contact with both types of support are similar.

Inconel 690 and Incoloy 800 are used as tubing material in most modern nuclear steam generators. Therefore, fretting-wear testing of Inconel 690 tubing was conducted to compare its performance to that of Incoloy 800 tubing under identical test conditions. Six tests were performed at 265°C for Inconel 690 tubing against a Type 410 stainless steel flat-bar support with work-rates ranging from 10 to 26 mW. The test duration was 500 hours. The typical tube-to-support relative motion used for these tests is shown in Fig. 13-27a. The results were compared to several tests using Incoloy 800 under identical test conditions.

Fretting-wear coefficients are compared in Fig. 13-34 for work-rates varying from 8 to 21 mW. For Inconel 690, fretting-wear coefficients ranged from 9×10^{-15} to 49×10^{-15} m^2/N, with the average at

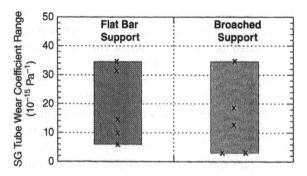

Fig. 13-33 Effect of Support Geometry on Fretting Wear of Incoloy 800 Tubing and Type 410 Stainless Steel Supports (Guérout and Fisher, 1999).

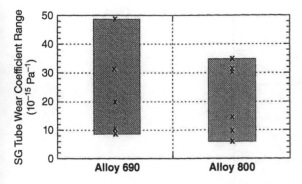

Fig. 13-34 Effect of Tube Material on Fretting Wear for Type 410 Stainless Steel Flat-Bar Supports (Guérout and Fisher, 1999).

24×10^{-15} m^2/N. For Incoloy 800, fretting-wear coefficients ranged from 6×10^{-15} to 35×10^{-15} m^2/N with the average at 25×10^{-15} m^2/N. Therefore, for predominantly impacting motion, the fretting-wear behaviour of SG tubes against Type 410 stainless steel flat-bar supports is similar regardless of whether Inconel 690 or Incoloy 800 is used as tubing material.

13.5.6 Discussion

The test results show that SG tube wear coefficients do not vary significantly (within the scatter of results), whether a normal SG chemistry is controlled with ammonia or morpholine, whether phosphates or boric acid are added, whether a flat-bar support or a broached-hole support is used, or whether Incoloy 800 or Inconel 690 is used for the SG tubing. At 265°C, for work-rates ranging from 6 to 18 mW, fretting-wear coefficients range from 5×10^{-15} to 45×10^{-15} m^2/N for the various types of test conditions described above. The test results are shown in terms of wear rate versus work-rate in Fig. 13-35. The average SG tube fretting-wear coefficient value in the 6 to 18 mW work-rate range is 20×10^{-15} m^2/N. A conservative value for design purposes is 40×10^{-15} m^2/N.

For identical test conditions, the scatter in the data was slightly greater when tests were performed at lower excitation levels (6 to 10 mW). This raises the question of which SG tube fretting-wear coefficient values to use for realistic SG tube excitation levels (i.e., 1 to 2 mW). The results presented earlier are considered short-term test results in view of the long-term wear process taking place in a steam generator. Most tests were performed for slightly higher excitation levels (5 to 20 mW) in order to generate measurable wear within a 500-hour test duration. The use of a large number of short-term tests is required whenever the effect of various parameters on the wear process needs to be investigated within a reasonable amount of time. These effects are now better understood and the wear coefficient average values for work rates of 6 to 18 mW are well defined.

However, there is a need for long-term tests (5000 to 10 000 hours) at tube-to-support interaction levels of 1 to 2 mW. There is already some indication in the literature that SG tube fretting-wear coefficients will not vary significantly. Results published by Hofmann et al (1996) are shown in Figs. 13-35 and 13-36 for impacting motion of Incoloy 800 tubing against Type 321 stainless steel anti-vibration bars. Ten tests were performed at 200°C for tube-to-support interaction levels ranging from 1 to 10 mW. The fretting-wear coefficient for tests with work-rates from 1 to 3 mW ranged from 9×10^{-15} to 54×10^{-15} m^2/N, with an average of 28×10^{-15} m^2/N. This range of fretting-wear coefficients is in good agreement with that found for higher levels of interaction. Therefore, until

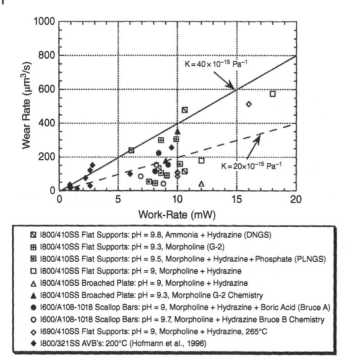

Fig. 13-35 Fretting-Wear Results at High Temperature for Various Types of Normal Steam Generator Chemistry, Support Geometry, Tube Material And Support Material (Guérout and Fisher, 1999).

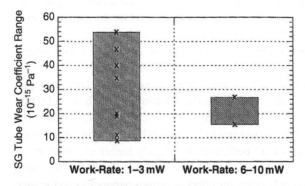

Fig. 13-36 Fretting-Wear Results of Incoloy 800 Tubing versus Type 321 Stainless Steel at 200°C with Data from Hofmann et al, 1996 (Guérout and Fisher, 1999).

the results of long-term tests are available, it is recommended that, for the prediction of SG tube damage resulting from impacting motion, a conservative value of 40×10^{-15} m^2/N for the SG tube fretting-wear coefficient be used. The recommended value for an average fretting-wear coefficient is 20×10^{-15} m^2/N.

The fretting-wear coefficients presented in this chapter can be applied to typical impact-sliding and impacting motions. Wear damage resulting from pure sliding motion is generally less severe than that resulting from the combination of impacting and sliding. Therefore, the fretting-wear coefficients reported herein would likely be conservative for pure sliding motion.

13.6   Summary and Recommendations

Generally, the "work-rate" model is appropriate to analyse fretting-wear data and to predict fretting-wear damage.

There is a significant effect of temperature on fretting-wear rates for both SG tube materials and zirconium components. For equivalent SG tube-to-support interaction levels, the fretting-wear coefficient was maximum at 215°C, where it was found to be 30 times higher than at 25°C. At 265°C, the fretting-wear coefficient was 20 times higher than at 25°C. On the other hand, the fretting-wear coefficient at 315°C was approximately equal to that at room temperature. Similar trends were found for zirconium alloys, where the maximum wear coefficient was observed in the 225 to 286°C temperature range, and the average wear coefficient decreases by an order of magnitude between 250 and 315°C.

Within the range of normal operating environment, the effect of chemistry (i.e., pH, additives, dissolved oxygen, etc.) is not significant compared to the variability of test results. As discussed earlier, for normal levels of interaction, it is unlikely that corrosion would interfere with the fretting-wear damage process.

For impact fretting wear, SG tube wear coefficients are similar for both Inconel 690 and Incoloy 800 tubing materials, both flat-bar and broached-hole supports, and both Type 410 stainless steel and Type 321 stainless steel support materials. The fretting-wear coefficients for tube and support of the same material can be more than an order of magnitude larger.

For work-rates ranging from 1 to 18 mW, a conservative tube fretting-wear coefficient of 40 x 10^{-15} m^2/N, and an average tube fretting-wear coefficient of 20 x 10^{-15} m^2/N may be used for various types of HX and SG normal water chemistry, support geometry, tube material, and support material (where tube and support materials are dissimilar).

The fretting-wear coefficient for zirconium alloys is much larger being in the order of 1000 x 10^{-15} m^2/N for realistic operating conditions.

Nomenclature

F	= applied force, N
F_n	= normal component of applied force, N
F_S	= shear component of applied force, N
H	= material hardness, Pa
k	= dimensionless wear coefficient
K_{FW}	= fretting-wear coefficient, m^2/N
S	= sliding distance, m
T	= time, s
V	= wear volume, m^3
$\dot{V}$	= wear volume rate, m^3/s
$\dot{W}$	= work-rate, W

Notes

1 Incoloy and Inconel are registered trademarks of International Nickel Company.
2 CANDU is a registered trademark of Atomic Energy of Canada Limited.

References

Attia, M. H. and Magel, E., 1999, "Experimental Investigation of Long-Term Fretting Wear of Multispan Steam Generator Tubes with U-bend Sections," *Wear*, 225, pp. 563–574.

Archard, J. F., 1953, "Contact and Rubbing of Flat Surfaces," *Journal of Applied Physics*, 24, pp. 981–988.

Batchelor, A. W. and Stachowiak, G. W., 1988, "Predicting Synergism Between Corrosion and Abrasive Wear," *Wear*, 123, pp. 281–291.

Blevins, R. D., 1978, "Fretting-Wear of Heat Exchanger Tubes, Part 1: Experiments," ASME Paper 78-JPG-NE-8, *Joint Power Generation Conference*, Dallas, Texas, USA.

Blevins, R. D., 1985, "Vibration-Induced Wear of Heat Exchanger Tubes," *ASME Journal of Engineering Materials and Technology*, 107, pp. 61–67.

Cha, J. H., Wambsganss, M. W., and Jendrzejczyk, J. A., 1987, "Experimental Study on Impact/Fretting Wear in Heat Exchanger Tubes," *ASME Journal of Pressure Vessel Technology*, 109, pp. 265–274.

Dow, B. L. Jr., 1996, "Steam Generator Progress Report: Revision 12," *Electrical Power Research Institute Report* EPRI TR-106365.

Fisher, N. J. and Ingham, B., 1989, "Measurement of Tube-to-Support Dynamic Forces in Fretting-Wear Rigs," *ASME Journal of Pressure Vessel Technology*, 111, pp. 385–393.

Fisher, N. J., Taylor, C. E., and Pettigrew, M. J., 1990, "Fuel-Element Vibration and Bearing Pad to Pressure Tube Fretting," *Atomic Energy of Canada Report AECL-10164*.

Fisher, N. J., Chow, A. B., and Weckwerth, M. K., 1995, "Experimental Fretting-Wear Studies of Steam Generator Materials," *ASME Journal of Pressure Vessel Technology*, 117, pp. 312–320.

Fisher, N. J., Weckwerth, M. K., Grandison, D. A. E., and Cotnam, B.M., 1997, "Fretting-Wear of Zirconium Alloys," Transactions, *14th International Conference on Structural Mechanics in Reactor Technology (SMiRT 14)*, Lyon, France, August 17–22.

Fisher, N. J., Weckwerth, M. K., Grandison, D. A. E., and Cotnam, B.M., 2002, "Fretting-Wear of Zirconium Alloys," *Nuclear Engineering and Design*, 213, pp. 79–90.

Frick, T. M., Sobek, T. E. and Reavis, J. R., 1984, "Overview on the Development and Implementation of Methodologies to Compute Vibration and Wear of Steam Generator Tubes," *Proceedings, Int. Symposium on Flow-Induced Vibrations: Volume 3: Vibration in Heat Exchangers*, ASME Special Publication, pp. 149–161.

Guérout, F. M., and Grandison, D. A., 1997, "Effect of Fuel Element Motion on Pressure Tube Fretting-Wear of CANDU Reactors," *14th International Conference on Structural Mechanics in Reactor Technology (SMiRT 14)*, Lyon, France, August 17–22.

Guérout, F. M., and Fisher, N. J., 1999, "Steam Generator Fretting-Wear Damage: A Summary of Recent Findings," *ASME Journal of Pressure Vessel Technology*, 121, pp. 304–310.

Guérout, F. M., Zbinden, M., Fisher, N. J., and Weckwerth, M. K., 1993, *Unpublished Report*.

Guérout, F. M., Fisher, N. J., Grandison, D. A., and Weckwerth, M. K., 1996, "Effect of Temperature on Steam Generator Fretting-Wear," Proceedings, *ASME Symposium on Flow-Induced Vibration-1996*, PVP-Vol. 328, Book No. H01056-1996, pp. 233–246.

Hofmann, P. J., Schettler, T., and Steininger, D. A., 1986, "Pressurized Water Reactor Steam Generator Tube Fretting and Fatigue Wear Characteristics," *Pressure Vessels and Piping Conference*, 86-PVP-2, Chicago, IL., USA.

Hofmann, P. J., Schettler. T., and Steininger, D. A., 1992, "PWR Steam Generator Tube Fretting and Fatigue Wear Phenomena and Correlation," *FSI/FVI in Cylinder Arrays in Cross-Flow*, ASME Publication HTD-Vol. 230, pp. 211–236.

Hofmann, P. J., Friedrich, B. C. and Schröder, H. J., 1996, "Prediction of PWR Steam Generator Tube Material Loss Caused by Perpendicular Impacting," Proceedings, *ASME Symposium on Flow-Induced Vibration-1996,* PVP-Vol. 328, *Flow-Induced Vibration,* ASME Book No. H01056-1996, pp. 219–231.

Jiang, J., Stott, F. H. and Stack, M. M., 1995, "A Mathematical Model for Sliding Wear of Metals at Elevated Temperatures," *Wear,* 181-183, pp. 20–31.

Ko, P.L., 1974, *Unpublished Information.*

Ko, P. L., 1979a, "Experimental Studies of Tube Fretting in Steam Generators and Heat Exchangers," *ASME Journal of Pressure Vessel Technology,* 101, pp. 125–133.

Ko, P. L., 1979b, "Wear of Zirconium Alloys due to Fretting and Periodic Impacting," *Wear of Materials - 1979,* edited by K.C. Ludema et al., ASME, New York, pp. 388–395.

Ko, P. L., 1980, "Fretting-Wear Studies of Heat Exchanger Tubes," *Flow-Induced Heat Exchanger Tube Vibration -1980, ASME* Publication HTD-Vol. 9, pp. 11–18.

Ko, P. L., 1984, "A Brief Review of Tube Fretting-Wear in Heat Exchangers," *Atomic Energy of Canada Report AECL-8272.*

Ko, P. L., 1985a, "Heat Exchanger Tube Fretting-Wear: Review and Application to Design," *ASME Journal of Tribology,* 107, pp. 149–156.

Ko, P. L., 1985b, "The Significance of Shear and Normal Force Components on Tube Wear due to Fretting and Periodic Impacting." *Wear,* Vol. 106, pp. 261–281.

Ko, P. L., 1986, "Metallic Wear-A Review with Special References to Vibration-Induced Wear in Power Plant Components," ASME Paper No. 86-PVP-1, *Pressure Vessels and Piping Conference,* Chicago, IL, USA.

Ko, P. L. and Basista, H., 1984, "Correlation of Support Impact Force and Fretting-Wear for a Heat Exchanger Tube," *ASME Journal of Pressure Vessel Technology,* 106, pp. 69–77.

Lim, S. C. and Ashby, M. F., 1987, "Wear Mechanism Maps," *Acta Metallurgica,* 35, pp. 1–24.

Pettigrew, M. J., Carlucci, L. N., Taylor, C. E., and Fisher, N. J., 1991, "Flow-Induced Vibration and Related Technologies in Nuclear Components," *Nuclear Engineering and Design,* 131, pp. 81–100.

Saito, Y. and Mino, K., 1995, "Elevated Temperature Wear Maps of X40 and Mar-M247 Alloys," *Journal of Tribology,* 117, pp. 524–528.

Suh, N. P., 1986, *Tribophysics,* Prentice-Hall Inc., Englewood Cliffs, NJ, USA.

Vingsbo, O. and Soderberg, S., 1987, "On Fretting Maps," *Wear of Materials,* 2, pp. 885–894.

Yetisir, M. and Fisher, N. J., 1995, "Prediction of Pressure Tube Fretting-Wear Damage due to Fuel Vibration," Proceedings, *16th Annual Canadian Nuclear Association/ Canadian Nuclear Society Conference,* Saskatoon, Saskatchewan, Canada, June 4–7.

Appendix A

Component Analysis

A.1 Introduction

Excessive flow-induced vibration causing failures by fatigue or fretting wear must be avoided in process and nuclear components. As stated in the preface, that is the purpose of this handbook. Herein, the term "process components" is used generally to describe all shell-and-tube heat exchangers, including nuclear steam generators; power plant condensers, boilers and coolers; nuclear fuels and reactor internals; and piping systems.

The emphasis in this handbook is on two-phase flow-induced vibration in cross flow.

Although half of all heat exchanger equipment operates in two-phase flow, existing flow-induced vibration texts have provided limited guidance regarding vibration induced by two-phase flow, and in estimating fretting-wear damage caused by random turbulence excitation.

Design guidelines for flow-induced vibration analysis are presented in this handbook. These guidelines are based on extensive analysis of the literature and, in particular, on experimental and field data obtained during 40 years of research and development at Canadian Nuclear Laboratories in Chalk River (formerly, Atomic Energy of Canada Ltd., Chalk River Laboratories).

This handbook is for the practicing engineer who is designing or troubleshooting nuclear and process system components. As such, use of the design guidelines is illustrated within each chapter by example calculations, as listed below:

Chapter 3 - Calculation of Flow Velocities

Example 3-1: Flow in a Process Heat Exchanger
Example 3-2: Flow in a Nuclear Steam Generator (SG)

Chapter 4 - Calculation of Total Tube Mass and Natural Frequency of Vibration

Example 4-1: Single-Phase Total Tube Mass in a Process Heat Exchanger
Example 4-2: Two-Phase Total Tube Mass in a Nuclear Steam Generator
Example 4-3: Calculation of Frequency in a Process Heat Exchanger
Example 4-4: Calculation of Frequency in a Nuclear Steam Generator U-Bend

Chapter 5 - Calculation of Damping in Single-Phase Flow

Example 5-1: Calculation of Damping in a Gas Heat Exchanger
Example 5-2: Calculation of Damping in a Process Heat Exchanger

Chapter 6 - Calculation of Damping in Two-Phase Flow

Example 6-1: Calculation of Damping in a Nuclear Steam Generator U-Bend

Flow-Induced Vibration Handbook for Nuclear and Process Equipment, First Edition.
Michel J. Pettigrew, Colette E. Taylor, and Nigel J. Fisher.
© 2022 John Wiley & Sons, Inc. This Work is a co-publication between ASME Press and John Wiley & Sons, Inc.

Note that the above list identifies the chapter where each example is presented but does not provide the corresponding chapter title. A general description of the chapter subject is used instead.

The purpose of presenting the examples within the handbook in this order is to allow the reader to use and better understand the equations associated with a particular guideline immediately after reading the sections describing the development and usage of that guideline. Unfortunately, this approach makes it more difficult for the inexperienced reader to understand the overall approach and order of steps required to undertake a complete flow-induced vibration analysis of a particular component.

To show the required approach, most of the example calculations are repeated in this appendix in the order that should be followed when conducting a flow-induced analysis of a component. Thus, this appendix is divided into sections corresponding to the analysis of two components, as follows:

Section A.2 - Analysis of a Process Heat Exchanger

Section A.3 - Analysis of a Nuclear Steam Generator U-Bend

Example 4-2: Two-Phase Total Tube Mass in a Nuclear Steam Generator
Example 4-4: Calculation of Frequency in a Nuclear Steam Generator U-Bend
Example 6-1: Calculation of Damping in a Nuclear Steam Generator U-Bend
Example 8-1: Fluidelastic Instability in a Steam Generator U-Bend
Example 10-1: Random Excitation for Two-Phase Flow in Steam Generator U-Bend
Example 10-2: Random Excitation for SG U-Bend Using an Early Guideline
Example A-2: Fretting-Wear Damage Prediction in a Steam Generator U-Bend

Note that two additional examples (Examples A-1 and A-2) are included to complete the analysis of Section A.2 (process heat exchanger) and Section A.3 (nuclear steam generator U-bend), respectively.

Within each section, necessary figures, tables and equations are repeated such that the reader does not have to refer back to the handbook chapters for this information. Numbers for these repeated figures, tables and equations are kept the same as in the handbook so the reader can refer back to the chapter version, if desired.

Additional description beyond that given in the handbook examples is provided within each section to clarify the analysis steps for inexperienced readers. Also, additional lines of intermediate calculations are included for calculations involving complex equations (e.g., two-phase damping calculations).

Note that italic and non-italic "ℓ" should be considered equivalent in this Appendix. Script "ℓ" is used as a symbol for span length and subscript for liquid phase.

A.2 Analysis of a Process Heat Exchanger

The principal dimensions of a process heat exchanger are shown in Fig. 3-5.

The heat exchanger has two sets of baffles, one set on each side of the shell, offset by 0.5 m, as shown in Fig. 3.5. Tubes in the center of the bundle are supported by both sets of baffles. Tubes in the periphery of the bundle are only supported by one set of baffles. The baffle thickness, L, is 15 mm.

Thus, tubes in the center of the bundle have ten spans, $N = 10$, (i.e., nine baffles spaced 0.5 m apart). The two outer spans are 1.0 m long. The other eight spans are 0.5 m long. For the purposes of this example, $\ell = 0.5$ m. Note that a more complex analysis of this tube could be done with a vibration analysis code where the outer spans are 1.0 m long and have clamped boundary conditions at the outer ends, as described in Chapter 4, Section 4.3.3. In that case, the first-mode frequency would be slightly different than that calculated in Example 4-3.

Tubes on the periphery of the bundle near the shell-side inlet have six spans, $N = 6$, (i.e., five baffles spaced at 1.0 m). Because these tubes are directly exposed to the inlet flow over the left-hand span, we term this region the "window". The span length in the window region, $\ell_{window} = 1.0$ m.

The tube outside diameter, $D = 0.020$ m and inside diameter, $D_i = 0.018$ m. The tube array is triangular with a pitch, $P = 0.030$ m. Therefore, the pitch-to-diameter ratio, P/D, is $P/D = 0.030/0.020 = 1.5$.

The elastic (Young's) modulus of the tube material, $E = 200 \times 10^9$ Pa. The density of the tube material, $\rho_t = 8000$ kg/m^3.

The internal fluid density, $\rho_i = 1000$ kg/m^3, and the shell-side fluid density, $\rho = 1000$ kg/m^3. The kinematic viscosity of the shell-side fluid, $\nu = 1 \times 10^{-6}$ m^2/s.

The shell-side volume flow rate, $\dot{V} = 0.25$ m^3/s.

Example 3-1 Flow in a Process Heat Exchanger

The minimum free-stream cross-sectional area, A, on the shell side is determined by the diameter of the shell (1.0 m) and the baffle spacing (0.5 m) (see Fig. 3-5), which form an approximately rectangular area (approximate because the shell is cylindrical). Thus, the minimum flow area is slightly less than,

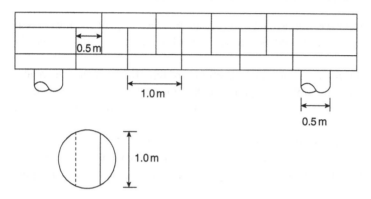

Fig. 3-5   Example 3-1 - Process Heat Exchanger Schematic.

$A \sim 0.5 \text{ m} \times 1.0 \text{ m} = 0.5 \text{ m}^2$

The free stream velocity, U, is calculated from Eq. (3-4),

$U = \dot{V}/A$ (3-4)

$U = 0.250/0.5$

$U = 0.5 \text{ m/s}$

The pitch velocity, U_p, is calculated from Eq. (3-6),

$U_p = U \times P/(P - D)$ (3-6)

$U_p = 0.5 \times 0.03/(0.03 - 0.020)$

$U_p = 1.5 \text{ m/s}$

Example 4.1 Single-Phase Total Tube Mass in a Process Heat Exchanger

As discussed in Chapter 4, Section 4.2, the effective mass of a tube with internal fluid and vibrating in a dense external fluid (i.e., not air) is greater than the tube mass alone. The effective or total tube mass per unit length, m, is calculated from Eq. (4-1)

$m = m_t + m_h + m_i$ (4-1)

where m_t is the tube mass per unit length, m_h is the hydrodynamic mass per unit length due to the external fluid, and m_i is the mass per unit length of the internal fluid.

The tube mass per unit length is calculated from Eq (4-2),

$m_t = \dfrac{\pi}{4}\rho_t\left(D^2 - D_i^2\right)$ (4-2)

$$m_t = (\pi/4)\rho_t\left(D^2 - D_i^2\right)$$

$$m_t = (0.785)(8000)(0.020^2 - 0.018^2)$$

$$m_t = (0.785)(0.608)$$

$$m_t = 0.478 \text{ kg/m}$$

The internal fluid mass per unit length is also calculated from Eq. (4-2),

$$m_i = \frac{\pi}{4}\rho_i D_i^2$$

$$m_i = (\pi/4) \times 1000 \times (0.018^2)$$

$$m_l = 0.254 \text{ kg/m}$$

The ratio of effective diameter, D_e, to tube diameter, D, for a triangular array is calculated using Eq. (4-9),

$$D_e/D = (0.96 + 0.5P/D)P/D \tag{4-9}$$

$$D_e/D = (0.96 + 0.5(1.5)) \times (1.5)$$

$$D_e/D = 2.565$$

Therefore,

$$D_e = (2.565) \text{ x } (0.020)$$

$$D_e = 0.0513 = 51.3 \text{ mm}$$

The hydrodynamic mass per unit length is calculated from Eq. (4-8),

$$m_h = \left(\frac{\pi}{4}\rho D^2\right)\left[\frac{(D_e/D)^2 + 1}{(D_e/D)^2 - 1}\right] \tag{4.8}$$

The confinement coefficient for hydrodynamic mass per unit length at $D_e/D = 2.565$ is,

$$(2.565^2 + 1)/(2.565^2 - 1) = 1.36$$

Therefore, the hydrodynamic mass per unit length is,

$$m_h = (\pi/4) \times 1000 \times (0.020)^2 \times (1.36)$$

$$m_h = (0.3142) \times (1.36)$$

$$m_h = 0.427 \text{ kg/m}$$

As per Eq. (4-1), the total tube mass per unit length is the algebraic sum of the tube, hydrodynamic and internal fluid masses per unit length,

$$m = m_t + m_h + m_i \tag{4-1}$$

$$m = 0.478 + 0.427 + 0.254$$

$$m = 1.16 \text{ kg/m}$$

Example 4-3   Calculation of Frequency in a Process Heat Exchanger

Free vibration analysis of tubes is discussed in Chapter 4. Natural frequencies, f_i, for single-span tubes with pinned end conditions are calculated from Eq. (4-26),

$$f_i = \frac{\beta_i^2}{2\pi} \left(\frac{EI}{m\ell^4}\right)^{\frac{1}{2}} \qquad (4\text{-}26)$$

where $\beta = n\pi$ is the eigenvalue and n takes on positive integer values greater than zero (see Chapter 4, Section 4.3.1). Thus, $\beta_1 = \pi$ for the first mode with pinned-pinned boundary conditions, f_1. Note that higher-order modes (i.e., f_2, f_3, f_4, etc.) corresponding to higher eigenvalues will be simply π multiples of the first mode, $f_2 = \pi f_1$, $f_3 = 2\pi f_1$, $f_4 = 3\pi f_1$, etc.

The area moment of inertia, I, for a tube with a circular cross section is,

$$I = \frac{\pi}{64}\left(D^4 - D_i^4\right)$$

$$I = (\pi/64) \times \left(0.020^4 - 0.018^4\right)$$

$$I = 2.7 \times 10^{-9}\ \mathrm{m}^4$$

For this example, we will calculate first-mode frequencies for two tubes: a tube in the center of the bundle, and a tube in the window region.

Therefore, the first modal frequency for a tube in the center of the bundle, f_1, is,

$$f_1 = \left(\pi^2/2\pi\right) \times \left\{\left[(200 \times 10^9) \times (2.7 \times 10^{-9})\right]/\left[(1.16) \times (0.5)^4\right]\right\}^{1/2}$$

$$f_1 = (\pi/2) \times \left\{[540]/[0.0725]\right\}^{1/2}$$

$$f_1 = 135.6\ \mathrm{Hz}$$

where $\ell = 0.5\mathrm{m}$, and for a tube in the window region, $f_{1window}$ is,

$$f_{1window} = \left(\pi^2/2\pi\right) \times \left\{\left[(200 \times 10^9) \times (2.7 \times 10^{-9})\right]/\left[(1.16) \times (1.0)^4\right]\right\}^{1/2}$$

$$f_{1window} = (\pi/2) \times \left\{[540]/[1.16]\right\}^{1/2}$$

$$f_{1window} = (\pi/2) \times (465.5)^{1/2}$$

$$f_{1window} = 33.9\ \mathrm{Hz}$$

where $\ell = \ell_{window} = 1.0\ \mathrm{m}$. Note that $f_{1window}$ is a factor of four less than f_1 because the span length is double and f is proportional to ℓ^{-2} (see Eq. (4-26)).

Example 5-2 Calculation of Damping in a Process Heat Exchanger

Damping in single-phase flow is discussed in Chapter 5. Calculation of total damping, ζ_T, in single-phase flow involves computing contributions from viscous damping in the shell-side fluid, ζ_V, and both squeeze-film damping, ζ_{SF}, and friction damping, ζ_F, between the tube and its supports, as per Eq. (5-38) and (5-39),

$$\zeta_T = \zeta_V + \zeta_{SF} + \zeta_F \qquad (5\text{-}38)$$

$$\zeta_T = \frac{100\pi}{\sqrt{8}} \left\{\frac{\left[1 + \left(\frac{D}{D_e}\right)^3\right]}{\left[1 - \left(\frac{D}{D_e}\right)^2\right]^2}\right\} \left(\frac{\rho D^2}{m}\right)\left(\frac{2v}{\pi f D^2}\right)^{1/2} + \left(\frac{N-1}{N}\right)\left[\frac{(1460)}{f}\left(\frac{\rho D^2}{m}\right)\left(\frac{L}{\ell_m}\right)^{1/2} + 0.5\left(\frac{L}{\ell_m}\right)^{1/2}\right]$$

$$(5\text{-}39)$$

where the first, second and third terms are viscous, squeeze-film and friction damping contributions, respectively. Note that all three terms in Eq. (5-39) calculate damping in percent (%), as that is how they were derived from the experimental data (see Chapter 5).

In this example, we will calculate damping for the two tubes described above.

Viscous damping (in percent) in the shell-side fluid, ζ_V, from the first term of Eq. (5-39), is,

$$\zeta_V = \frac{100\pi}{\sqrt{8}}\left(\frac{\rho D^2}{m}\right)\left(\frac{2v}{\pi f D^2}\right)^{0.5}\left\{\frac{\left[1+\left(\frac{D}{D_e}\right)^3\right]}{\left[1-\left(\frac{D}{D_e}\right)^2\right]^2}\right\} \tag{5-5}$$

The confinement coefficient for viscous damping is dependent solely on $D/D_e = 1/2.565$ (see Example 4-1). Thus, the confinement ratio for both tubes is,

$$= [1+(1/2.565)^3]/\left\{[1-(1/2.565)^2]^2\right\}$$

$$= [1+0.0593]/\{1-0.1520]^2\}$$

$$= [1.0593]/\{[0.8480]^2\}$$

$$= [1.0593]/\{0.7191\}$$

$$= 1.47$$

For a tube in the center of the bundle where $f_1 = 135.6$ Hz, ζ_V is,

$\zeta_V = [100\,\pi/(8^{0.5})]\,(1000\times0.02^2/1.16)\,\{[(2)(1.0\times10^{-6})]/[\pi\,(135.6)(0.02^2)]\}^{0.5}(1.47)$

$\zeta_V = [111.1]\,(0.3448)\,\{[(2.0\times10^{-6})/(0.1704)]\}^{0.5}(1.47)$

$\zeta_V = [111.1]\,(0.3448)\,(0.003426)\,(1.47)$

$\zeta_V = 0.193\%$

For a tube in the window region where $f_{1window} = 33.9$ Hz, ζ_V is,

$\zeta_V = [100\,\pi/(8^{0.5})]\,(1000\times0.02^2/1.16)\,\{[(2)(1.0\times10^{-6})]/[\pi\,(33.9)(0.02^2)]\}^{0.5}(1.47)$

$\zeta_V = [111.1]\,(0.3448)\,\{[(2.0\times10^{-6})/(0.04260)]\}^{0.5}(1.47)$

$\zeta_V = [111.1]\,(0.3448)\,(0.006852)\,(1.47)$

$\zeta_V = 0.386\%$

Note that ζ_V in the window region is a factor of two greater than in the center of the bundle because $f_{1window}$ is a factor of four less than f_1 and ζ_V is proportional to $f^{-0.5}$ (see Eq. (5-5)).

Squeeze-film damping (in percent) between the tube and its supports, ζ_{SF}, from the second term of Eq. (5-39), is,

$$\zeta_{SF} = [(N-1)/N]\zeta_{SFn} \tag{5-10}$$

Where ζ_{SFn} is the normalized squeeze-film damping,

$$\zeta_{SFn} = \left(\frac{1460}{f}\right)\left(\frac{\rho D^2}{m}\right)\left(\frac{L}{\ell_m}\right)^{0.5} \tag{5-34}$$

The characteristic span length, ℓ_m, is discussed in Chapter 5, Section 5.6.3. In most cases ℓ_m, is calculated as the average length of the three longest spans of the tube under analysis. However, special cases apply, as discussed in Section 5.6.3.

For a tube in the center of the bundle, $(N-1)/N = (9/10)$, $f_1 = 135.6$ Hz, and $\ell_m = \ell = 0.5$ m. Thus, ζ_{SF} is,

$$\zeta_{SF} = (9/10)[1460/135.6]\left[(1000\times0.02^2)/1.16\right][0.015/0.5]^{0.5}$$

$$\zeta_{SF} = (0.9)(10.77)(0.3448)(0.1732)$$

$$\zeta_{SF} = 0.579\%$$

For a tube in the window region, $(N\text{-}1)/N = (5/6)$, $f_1 = 33.9$ Hz, and $\ell_m = \ell_{window} = 1.0$ m. Thus, ζ_{SF} is,

$$\zeta_{SF} = (5/6)[1460/33.9]\left[(1000 \times 0.02^2)/1.16\right][0.015/1.0]^{0.5}$$

$$\zeta_{SF} = (0.833)(43.07)(0.3448)(0.1225)$$

$$\zeta_{SF} = 1.52\%$$

Note that ζ_{SF} for a tube in the window region is a factor of 2.6 greater than for a tube in the center of the bundle because $f_{1window}$ is a factor of four less than f_1 and ζ_{SF} is proportional to f^{-1}, while ℓ_{window} is double ℓ and ζ_{SF} is proportional to $\ell_m^{-0.5}$, and $(N\text{-}1)/N$ for the window tube is 93% of that for the bundle center tube (see Eq. (5-34)).

Friction damping (in percent) between the tube and its support, ζF, from the third term of Eq. (5-39), is,

$$\zeta_F = [(N-1)/N]\zeta_{Fn} \tag{5-10}$$

where ζ_{Fn} is the normalized friction damping (analogous to the normalized squeeze-film damping),

$$\zeta_{Fn} = 0.5(L/\ell_m)^{0.5} \tag{5-35}$$

For a tube in the center of the bundle, $(N\text{-}1)/N = (9/10)$ and $\ell_m = \ell = 0.5$ m Thus, ζ_F is,

$$\zeta_F = (9/10)\ (0.5)\ [0.015/0.5]^{0.5}$$

$$\zeta_F = (0.45)\ (0.1732)$$

$$\zeta_F = 0.078\%$$

For a tube in the window region, $(N\text{-}1)/N = (5/6)$ and $\ell_m = \ell_{window} = 1.0$ m. Thus, ζ_F is,

$$\zeta_F = (5/6)\ (0.5)\ [0.015/1.0]^{0.5}$$

$$\zeta_F = (0.417)\ (0.1225)$$

$$\zeta_F = 0.051\%$$

Note that ζ_F for a tube in the window region is 66% of that for a tube in the center of the bundle because ℓ_{window} is double ℓ and ζ_F is proportional to $\ell_m^{-0.5}$, while $(N\text{-}1)/N$ for the window tube is 93% of that for the bundle center tube (see Eq. (5-35)).

Finally, the total damping (in percent), ζ_T, is the sum of ζ_V, ζ_{SF} and ζ_F, as per Eq. (5-38). For a tube in the center of the bundle, ζ_T is,

$$\zeta_T = 0.193 + 0.579 + 0.078$$

$$\zeta_T = 0.85\%$$

and, for a tube in the window region, ζ_T is,

$$\zeta_T = 0.386 + 1.52 + 0.051$$

$$\zeta_T = 1.96\%$$

Example 7-2   FEI Calculation in a Single-Phase Process Heat Exchanger with Window Region

Calculation of fluidelastic instability in single-phase flow is discussed in Chapter 7. The critical pitch velocity for FEI, U_{pc}, is calculated using Eq. (7-5),

$$\frac{U_{pc}}{fD} = K\left(\frac{2\pi\zeta m}{\rho D^2}\right)^{0.5} \tag{7-5}$$

where the instability constant, $K = 3.0$ (see Chapter 7, Section 7.8). Note that damping in Eq. (7-5) is **not** formulated in percent, but as a ratio.

For a tube in the center of the bundle, where $f_1 = 135.6$ Hz, and $\zeta_T = 0.85\%$, U_{pc} is,

$U_{pc} = (135.6)\,(0.02)\,(3.0)\,\{[2\pi\,(0.0085)\,(1.16)]/[(1000)\,(0.02^2)]\}^{0.5}$

$U_{pc} = (135.6)\,(0.02)\,(3.0)\,\{[0.06195]/[0.4]\}^{0.5}$

$U_{pc} = (135.6)\,(0.02)\,(3.0)\,\{0.1549\}^{0.5}$

$U_{pc} = (135.6)\,(0.02)\,(1.18)$

$U_{pc} = 3.2$ m/s

Comparing U_{pc} to the pitch velocity, U_p, in the process heat exchanger (see Example 3-1),

$U_p/U_{pc} = 1.5/3.2 = 0.47 < 1.0$

Therefore, this process heat exchanger should not experience FEI in the center of the bundle.

For a tube in the window region where $f_{1window} = 33.9$ Hz, and $\zeta_T = 1.96\%$, U_{pc} is,

$U_{pc} = (33.9)\,(0.02)\,(3.0)\,\{[2\pi\,(0.0196)\,(1.16)]/[(1000)\,(0.02^2)]\}^{0.5}$

$U_{pc} = (33.9)\,(0.02)\,(3.0)\,\{[0.1429]/[0.4]\}^{0.5}$

$U_{pc} = (33.9)\,(0.02)\,(3.0)\,\{0.3571\}^{0.5}$

$U_{pc} = (33.9)\,(0.02)\,(1.79)$

$U_{pc} = 1.22$ m/s

Comparing U_{pc} to the pitch velocity, U_p, in the process heat exchanger (see Example 3-1),

$U_p/U_{pc} = 1.5/1.22 = 1.23 > 1.0$

As U_p is greater that U_{pc}, this process heat exchanger may experience FEI in the window region.

Example 9-1   Random Excitation in Process Heat Exchanger Interior Flow

Random turbulence excitation in single-phase flow is discussed in Chapter 9. The formulation for the product of power spectral density of the input force per unit length, $S_F(f)$ and correlation length, λ_c, is given in Eq. (9-22) and Eq. (9-19),

$$S_F(f)\lambda_c = \widetilde{S}_F(f_R)_e \times \frac{\left(\frac{1}{2}\rho U_p^2 D\right)^2 D}{U_p} \times L_e \tag{9-22}$$

$$\widetilde{S}_F(f_R)_{e\,@\,L_o\,=\,1m} = \widetilde{S}_F(f_R)_e \times \frac{L_e}{1} \tag{9-19}$$

where the reduced frequency, f_R, is defined in Chapter 9, Section 9.4 as,

$$f_R = fD/U_p$$

For tubes in the center of the tube bundle, $f = f_1 = 135.6$ Hz (see Example 4-3) and $U_p = 1.5$ m/s (see Example 3.1). Thus,

$$f_R = (135.6) \times (0.02)/(1.5) = 1.808$$

The design guideline of dimensionless equivalent power spectral density of the excitation force per unit length, $\widetilde{S}_F(f_R)_e\,@\,L_o\,=\,1m$, for interior row tubes is shown in Eq. (9-20):

$$4 \times 10^{-4}\left(fD/U_p\right)^{-0.5}, \quad 0.01 < fD/U_p < 0.5 \tag{9-20a}$$

$$5 \times 10^{-5}\left(fD/U_p\right)^{-3.5}, \quad 0.5 < fD/U_p \tag{9-20b}$$

As $f_R = 1.808 > 0.5$, Eq. (9-20b) applies for this tube. Thus,

$$\tilde{S}_F(f_R)_{e \ @ \ L_o \ = \ 1m} = 5 \times 10^{-5}(1.808)^{-3.5} = 6.292 \times 10^{-6}$$

Combining Eq. (9-22) and Eq. (9-19), the L_e terms cancel out. Thus,

$S_F(f)\lambda_c = (6.292 \times 10^{-6}) \ [(0.5) \ (1000) \ (1.5)^2 \ (0.02)]^2(0.02)/(1.5)$

$S_F(f)\lambda_c = (6.292 \times 10^{-6}) \ [22.5]^2(0.02)/(1.5)$

$S_F(f)\lambda_c = (6.292 \times 10^{-6}) \ (506.3) \ (0.02)/(1.5)$

$S_F(f)\lambda_c = (6.292 \times 10^{-6}) \ (6.75)$

$S_F(f)\lambda_c = 4.247 \times 10^{-5} \ (\text{N/m})^2\text{s m}$

The maximum mean square of vibration response, $\overline{y^2}$, for a single-span tube is calculated from the power spectral density of the excitation force using Eq. (9-15).

$$\overline{y^2(x)}_1 = \frac{[S_F(f)\lambda_c]\phi_1^2(x)a_1}{64\pi^3 f_1^3 m^2 \zeta_1 L_e} \tag{9-15}$$

where, $\phi_1^2(x)$ is the normalized mode shape for the first mode, a_1 is the numerical coefficient for the first mode, f_1 is the tube first-mode natural frequency, m is the total tube mass, ζ_1 is the damping ratio for the first mode, and L_e is the excited tube length (the length of the portion of the tube that is subjected to flow).

Table 9-1 contains values of $\phi_{1 \ max}^2$ and a_1 for various end conditions. From Table 9-1, for pinned-pinned end conditions, $\phi_{1 \ max}^2 = 2.0$ and $a_1 = 1.1$.

For tubes in the center of the tube bundle, $f = f_1 = 135.6$ Hz (see Example 4-3), $m = 1.16$ kg/m (see Example 4-1), $\zeta_1 = \zeta_T = 0.85\%$ (see Example 5-2) and $L_e = \ell = 0.5$ m. Thus, $\overline{y^2}$ at midspan, $\ell/2$, is,

$$= (4.247 \times 10^{-5}) \ (2) \ (1.1)/[64\pi^3(135.6^3) \ (1.16^2) \ (0.0085) \ (0.5)]$$

$$= (4.247 \times 10^{-5}) \ (2) \ (1.1)/[64\pi^3(2.493 \times 10^6) \ (1.346) \ (0.0085) \ (0.5)]$$

$$= (4.247 \times 10^{-5}) \ (2) \ (1.1)/[64\pi^3(1.426 \times 10^4)]$$

$$= (4.247 \times 10^{-5}) \ (2) \ (1.1)/[2.830 \times 10^7]$$

$$= 3.302 \times 10^{-12} \ \text{m}^2$$

Therefore, the root-mean-square (rms) vibration response at the midspan is.

$y(\ell/2) = (3.302 \times 10^{-12})^{0.5}$

$y(\ell/2) = 1.82 \times 10^{-6}$ m rms

$y(\ell/2) = 1.82 \ \mu\text{m rms}$

The design guideline for maximum allowable rms vibration amplitude due to random turbulence excitation is based on calculated fretting-wear depth over the life of the component. This calculation is conducted in Example A-1.

Example 11-1 Periodic Wake Shedding in Single-Phase Flow

Periodic wake shedding (vortex shedding) is discussed in Chapter 11. Periodic wake shedding is possible in single-phase flow for Strouhal numbers, S, between 0.40 and 0.67 (see Chapter 11, Section 11.2.1). The Strouhal number is calculated using Eq. (11-1),

$$S = f_s D/U \tag{11-1}$$

where f_S is the wake shedding frequency and $U = U_p = 1.5$ m/s for the process heat exchanger (see Example 3.1).

The first-mode natural frequency, $f_1 = 135.6$ Hz for a tube in the center of the process heat exchanger bundle (see Example 4-3). For wake shedding to occur at this frequency (remembering that the phenomenon occurs when wake shedding and natural frequencies coincide),

$$S = (135.6)\,(0.02)/(1.5) = 1.8$$

As S is not within the 0.40 to 0.67 range, periodic wake shedding is not a concern for tubes in the center of the process heat exchanger bundle.

The first-mode natural frequency, $f_{1window} = 33.9$ Hz for a tube in the window region of the process heat exchanger bundle (see Example 4-3). For wake shedding to occur at this frequency,

$$S = (33.9)\,(0.02)/(1.5) = 0.45$$

As S is within the 0.40 to 0.67 range, periodic wake shedding is possible for tubes in the window region of the process heat exchanger bundle.

The fluctuating fluid forces, F_L, for a tube in the window region at resonance due to periodic wake shedding are calculated using Eq. (11-2),

$$F_L = C_L\left(\tfrac{1}{2}\rho D U^2\right) \tag{11-2}$$

where, C_L is the fluctuating force lift coefficient (see Chapter 11, Section 11.2.3) and $U = U_p = 1.5$ m/s.

Fluctuating force lift coefficients for tube bundles in single-phase flow are shown versus P/D in Fig. 11-10. The design guideline for heat exchanger tube bundles of $P/D < 1.6$ is $C_L = 0.075$ rms (see Chapter 11, Section 11.2.3). As $P/D = 1.5$ for this case (see Example 3.1), F_L is,

$$F_L = (0.075)\,(0.5)\,(1000)\,(0.02)\,(1.5^2) = 1.69 \text{ N/m}$$

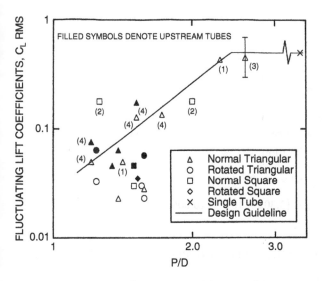

Fig. 11-10 Fluctuating Force Lift Coefficients for Tube Bundles in Single-Phase Flow.

The peak vibration amplitude at the mid-span, $Y_{(\ell/2)}$, for a cylindrical structure with pinned-pinned end conditions subjected to uniform cross flow over its entire length (see Eq. (11-8b)) is,

$$Y_{(\ell/2)} = F_L / \left(2\pi^3 m f^2 \zeta\right) \tag{11-8b}$$

Note that damping in Eq. (11-8b) is **not** formulated in percent, but as a ratio.

For a tube in the window region, $m = 1.16$ kg/m (see Example 4-1), $f = f_{1window} = 33.9$ Hz (see Example 4-3), and $\zeta = \zeta_T = 1.96\%$ (see Example 5-2). Thus, $Y_{(\ell/2)}$ for a tube in the window region due to periodic wake shedding is,

$$Y_{(\ell/2)} = (1.69) / \left[2\pi^3 (1.16)\left(33.9^2\right)(0.0196)\right]$$

$$Y_{(\ell/2)} = 0.0010 \text{ m} = 1.0 \text{ mm}$$

The design guideline for maximum allowable vibration amplitude when resonance is possible is $0.02D$, or 2% of tube diameter (see Chapter 11, Section 11.2.3).

$$Y_{(\ell/2)}/D = 0.0010/0.02 = 5.2\% > 2\%$$

Therefore, tubes in the window region of the process heat exchanger may be susceptible to fatigue or fretting-wear damage due to periodic wake shedding.

Example A-1 Fretting-Wear Damage Prediction in a Process Heat Exchanger

Fretting-wear damage prediction is discussed in Chapter 12. The amount of fretting-wear damage due to random turbulence excitation can be predicted using Eq. (12-29),

$$\dot{W}_N = 16\pi^3 f^3 m \ell y_{(\ell/2)rms}^2 \zeta_S \tag{12-29}$$

where the support damping, ζ_S, is calculated using Eq. (12-6),

$$\zeta_S = \zeta_T - \zeta_V \tag{12-6}$$

From Example 5-2,

$$\zeta_T = 0.85\%$$

$$\zeta_V = 0.193\%, \text{ and}$$

$$\zeta_S = \zeta_{SF} + \zeta_F = 0.579 + 0.078 = 0.657\%$$

for a tube in the center of the bundle.

From Example 9-1, the mid-span vibration response for a tube in the center of the bundle, $y_{(\ell/2)}$, is 1.82 μm rms. For this level of vibration, $\dot{W}_N$ is,

$$\dot{W}_N = 16\pi^3 \left(135.6^3\right)(1.16)(0.5)\left(1.82 \times 10^{-6}\right)^2 (0.00657)$$

$$\dot{W}_N = 16\pi^3 \left(2.493 \times 10^6\right)(1.16)(0.5)\left(3.302 \times 10^{-12}\right)(0.00657)$$

$$\dot{W}_N = 15.6 \times 10^{-6} \text{ W} = 0.0156 \text{ mW}$$

where $f = f_1 = 135.6$ Hz (see Example 4-3), $m = 1.16$ kg/m (see Example 4-1) and $\ell = 0.5$ m for a tube in the center of the bundle.

The fretting-wear damage resulting from this level of work-rate is calculated using Eq. (12-2),

$$\dot{V} = K_{FW}\dot{W}_N \tag{12-2}$$

Where K_{FW} is wear coefficient obtained experimentally as discussed in Chapter 13.

For a wear coefficient of $K_{FW} = 20 \times 10^{-15}\text{m}^2/\text{N}$, the volume wear rate, $\dot{V}$, is,

$$\dot{V} = (20 \times 10^{-15})(15.6 \times 10^{-6})$$
$$\dot{V} = 3.121 \times 10^{-19} \text{ m}^3/\text{s}$$

At this volume wear rate, the total worn volume, V, after a typical component lifetime, $T_s = 40$ years is,

$$V = (3.121 \times 10^{-19})(40)(365)(24)(3600)$$
$$V = 3.94 \times 10^{-10} \text{ m}^3$$

The resulting tube wall depth, d_w, can be calculated from the wear volume. This calculation requires knowledge of the relationship between d_w and V. For example, for a tube within a circular hole, it may be assumed that the wear is occurring uniformly over the support thickness and half of the support circumference. Thus,

$$d_w = 2V/(\pi DL) \tag{12-31}$$

For $D = 20$ mm and $L = 15$ mm, d_w is,

$$d_w = (2)(3.94 \times 10^{-10})/[\pi (0.02)(0.015)]$$
$$d_w = 8.35 \times 10^{-7}\text{m} = 0.084 \text{ }\mu\text{m}$$

For a typical tube wall thickness of 1.0 mm, and assuming a maldistribution factor of three for the wear around the support to be conservative, an upper bound wear depth of 2.5 μm, or 0.25%, is determined. This amount of tube wall loss is negligible.

A.3   Analysis of a Nuclear Steam Generator U-Bend

A sketch of the U-bend region of a typical nuclear steam generator is shown in Fig. 3-6.

The tube bundle outside diameter, $D_{BO} = 3.0$ m. In the U-bend region, the tubes are supported by a number of anti-vibration bars (AVBs). The span length, $\ell = 0.59$ m. The total number of spans in the U-bend, $N = 10$.

Fig. 3-6   Example 3-2 - U-Bend Schematic.

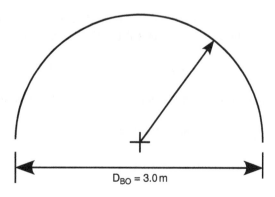

$$D_{BO} = 3.0\text{ m}$$

The tube outside diameter, $D = 20$ mm, and the pitch-to-diameter ratio, $P/D = 1.5$. Thus, the tube array pitch, $P = (1.5)(20) = 30$ mm. The tube inside diameter, $D_i = 18$ mm.

The elastic (Young's) modulus of the tube material, $E = 200 \times 10^9$ Pa. The density of the tube material, $\rho_t = 8000$ kg/m^3.

The internal fluid density, $\rho_i = 1000$ kg/m^3.

The shell-side mixture is two-phase steam-water at temperature, $T = 275$°C. The gas (steam) density, $\rho_g = 32$ kg/m^3 and the liquid (water) density, $\rho_e = 760$ kg/m^3. The gas phase kinematic viscosity, $v = 8.0 \times 10^{-7}$ m^2/s and the liquid phase kinematic viscosity, $v = 1.26 \times 10^{-7}$ m^2/s.

The feedwater volume flow rate, $\dot{V}_{FW} = 2.4$ m^3/s. The circulation ratio, CR = total mass flow/exit steam flow = total mass flow/feedwater flow = 5.

Example 3-2 Flow in a Nuclear Steam Generator (SG)

All of the flow in a steam generator tube bundle must exit through the hemispherical area formed by the largest U-tubes. Thus, the flow cross-sectional area outside the bundle, A, is

$$A = (\pi/2)D_{BO}^2 = (\pi/2)(3.0)^2 = 14.41 \text{ m}^2$$

The tube-bundle cross-sectional area, A_p, is calculated from Eq. (3-8),

$$A_p = A\frac{(P-D)}{P} \tag{3-8}$$

$$A_p = (14.41)(0.030 - 0.020)/(0.030) = 4.71 \text{ m}^2$$

The steam quality at the exit of the U-bend, X, in a recirculating steam generator is,

$$X = 1/CR = 1/5 = 0.2 \text{ or } 20\%$$

Fluid parameters for a two-phase mixture are defined in Chapter 3, Section 3.2.2. The two-phase mixture density, ρ_{TP}, is calculated by combining Eq. (3-2) and Eq. (3-5),

$$\rho_{TP} = \frac{1}{\dfrac{X}{\rho_g} + \dfrac{(1-X)}{\rho_\ell}} = \frac{1}{\dfrac{0.20}{32} + \dfrac{(1-0.20)}{760}} = 136.9 \text{ kg/m}^3$$

Similarly, the void fraction, ε_g, is calculated by combining Eq. (3-1) and Eq. (3-5),

$$\varepsilon_g = \frac{X/\rho_g}{X/\rho_g + (1-X)/\rho_\ell} = \frac{0.2/32}{0.2/32 + (1-0.2)/760} = 0.856 = 85.6\%$$

The pitch mass flux at the exit of the U-bend, $\dot{m}_p$, is calculated from Eq. (3-4), using feedwater volumetric flow multiplied by CR to calculate the total volumetric flow,

$$\dot{m}_p = \rho(\dot{V}_{FW} \cdot CR)/A_p = 136.9 \times 2.4 \times 5/4.71 = 349 \text{ kg/m}^2\text{s} \tag{3-4}$$

The pitch velocity in the bundle, U_p, is calculated from Eq. (3-7),

$$U_p = \frac{\dot{m}_p}{\rho_{TP}} = 349/137 = 2.54 \text{ m/s} \tag{3-7}$$

A maldistribution factor is applied in the U-bend region to account for differences in pressure drop and void fraction along a U-tube. Originally, a value of 2 was used based on experience. Results of thermalhydraulics analyses conducted later yielded similar values. Thus, the maximum pitch velocity, $(U_p)_{max}$, is double U_p,

$$(U_p)_{max} = U_p \times \text{maldistribution factor} = 2.54 \times 2 = 5.09 \text{ m/s}$$

Example 4-2 Two-Phase Total Tube Mass in a Nuclear Steam Generator

The total mass per unit length, m, is calculated from Eq (4-1),

$$m = m_t + m_h + m_i \tag{4-1}$$

where m_t is the tube mass per unit length, m_h is the hydrodynamic mass per unit length due to the external fluid, and m_i is the mass per unit length of the internal fluid.

The tube mass per unit length is calculated from Eq (4-2),

$$m_t = \frac{\pi}{4}\rho_t\left(D^2 - D_i^2\right) \tag{4-2}$$

$$m_t = (\pi/4)\rho_t\left(D^2 - D_i^2\right)$$

$$m_t = (0.785)\,(8000)\left(0.020^2 - 0.018^2\right)$$

$$m_t = (0.785)\,(0.608)$$

$$m_t = 0.478 \text{ kg/m}$$

The internal fluid mass per unit length is also calculated from Eq. (4-2),

$$m_i = \frac{\pi}{4}\rho_i D_i^2. \tag{4-2}$$

$$m_i = (\pi/4)\,(1000)\left(0.018^2\right)$$

$$m_i = 0.254 \text{ kg/m}$$

The ratio of effective diameter, D_e, to tube diameter, D, for a triangular array is calculated using Eq. (4-9),

$$D_e/D = (0.96 + 0.5P/D)P/D \tag{4-9}$$

$$D_e/D = (0.96 + 0.5(1.5)) \times (1.5)$$

$$D_e/D = 2.565$$

Therefore,

$$D_e = (2.565) \times (0.020)$$

$$D_e = 0.0513 = 51.3 \text{ mm}$$

The hydrodynamic mass per unit length is calculated from Eq. (4-8),

$$m_h = \left(\frac{\rho\pi D^2}{4}\right)\left[\frac{(D_e/D)^2 + 1}{(D_e/D)^2 - 1}\right] \tag{4-8}$$

where, $\rho = \rho_{TP}$.

The confinement coefficient for $D_e/D = 2.565$ is,

$$(2.565^2 + 1)/(2.565^2 - 1) = 1.36$$

Therefore, the hydrodynamic mass per unit length is,

$m_h = (\pi/4)(137)(0.020)^2(1.36)$

$m_h = (0.0430)(1.36)$

$m_h = 0.0585 \text{ kg/m}$

As per Eq. (4-1), the total tube mass per unit length is the algebraic sum of the tube, hydrodynamic and internal fluid masses per unit length,

$$m = m_t + m_h + m_i \tag{4-1}$$

$m = 0.478 + 0.0585 + 0.254$

$m = 0.791 \text{ kg/m}$

Example 4-4   Calculation of Frequency in a Nuclear Steam Generator U-Bend

Assuming the U-tube curved spans can be approximated as straight spans, natural frequencies, f_i, are calculated from Eq. (4-26),

$$f_i = \frac{\beta_i^2}{2\pi}\left(\frac{EI}{m\ell^4}\right)^{\frac{1}{2}} \tag{4-26}$$

where the first-mode eigenvalue, $\beta_1 = \pi$ for pinned-pinned boundary conditions.

The area moment of inertia, I, for a tube with a circular cross section is,

$I = \frac{\pi}{64}\left(D^4 - D_i^4\right)$

$I = (\pi/64) \times \left(0.020^4 - 0.018^4\right)$

$I = 2.7 \times 10^{-9} \text{m}^4$

The first modal frequency for the tube, f_1, is,

$f_1 = (\pi^2/2\pi) \times \left\{[(200 \times 10^9) \times (2.7 \times 10^{-9})]/[(0.791) \times (0.59)^4]\right\}^{1/2}$

$f_1 = (\pi/2) \times \{[540]/[0.0958]\}^{1/2}$

$f_1 = (\pi/2) \times (5634)^{1/2}$

$f_1 = 118 \text{ Hz}$

If every second AVB in the U-bend were ineffective, then $\ell_{\text{ineffective}} = 2 \times \ell = 1.18$ m, and the first modal frequency, $f_{1 ineffective}$ is,

$f_{1 ineffective} = (\pi^2/2\pi) \times \left\{[(200 \times 10^9) \times (2.7 \times 10^{-9})]/[(0.791) \times (1.18)^4]\right\}^{1/2}$

$f_{1 ineffective} = (\pi/2) \times \{[540]/[1.534]\}^{1/2}$

$f_{1 ineffective} = (\pi/2) \times (352.1)^{1/2}$

$f_{1ineffective} = 29.5 \text{ Hz}$

Note that $f_{1ineffective}$ is a factor of four less than f_1 because the span length is double and f is proportional to ℓ^{-2} (see Eq. (4-26)).

Example 6-1   Calculation of Damping in a Nuclear Steam Generator

Damping in two-phase flow is discussed in Chapter 6. Calculation of total damping, ζ_T, in two-phase flow involves computing contributions from support damping, ζ_S, between the tube and its supports, viscous damping in the shell-side fluid, ζ_V, and two-phase damping, ζ_{TP}, as per Eq. (6-14),

$$\zeta_T = \zeta_S + \zeta_V + \zeta_{TP} \tag{6-14}$$

Note that all three terms in Eq. (5-39) calculate damping in percent (%), as that is how they were derived from the experimental data (see Chapter 5 and Chapter 6).

In this example, we will calculate damping for two void fractions, $\varepsilon_g = 40\%$ and 85%. Two-phase damping is calculated using Eq. (6-23),

$$\zeta_{TP} = A\left(\frac{\rho_\ell D^2}{m}\right)\left(f(\varepsilon_g)\right)\left\{\frac{\left[1 + (D/D_e)^3\right]}{\left[1 - (D/D_e)^2\right]^2}\right\} \tag{6-23}$$

where, the coefficient, A, is equal to 4.0 when ζ_{TP} is expressed in percent and the void fraction function, $f(\varepsilon_g)$, is derived from Eq. (6-21). Note that two-phase damping is proportional to ρ_ℓ, as discussed in Chapter 6, Section 6.6.

The design guideline for the void-fraction function, $f(\varepsilon_g)$, is given in Eq. (6-21) as,

$$\begin{aligned} f(\varepsilon_g) &= 1 \text{ for } \varepsilon_g = 40\% \text{ to } 70\% \\ f(\varepsilon_g) &= \varepsilon_g/40 \text{ for } \varepsilon_g < 40\% \\ f(\varepsilon_g) &= 1 - (\varepsilon_g - 70)/30 \text{ for } \varepsilon_g > 70\% \end{aligned} \tag{6-21}$$

where ε_g is expressed in percent.

The ratio of effective diameter, D_e, to tube diameter, D, for a triangular array is calculated using Eq. (4-9),

$$D_e/D = (0.96 + 0.5P/D)P/D \tag{4-9}$$
$$D_e/D = (0.96 + 0.5(1.5)) \times (1.5)$$
$$D_e/D = 2.565$$

Therefore,

$$D_e = (2.565) \times (0.020)$$
$$D_e = 0.0513 = 51.3 \text{ mm}$$

The confinement function, $f(D_e/D)$, for two-phase damping (see Eq. (6-22)) is the same as the confinement coefficient for viscous damping (dependent solely on $D/D_e = 1/2.565$).

Thus, the confinement function for the steam generator U-tube is,

$$= \left[1 + (1/2.565)^3\right]/\left\{\left[1 - (1/2.565)^2\right]^2\right\}$$

$$= [1 + 0.0593]/\{[1 - 0.1520]^2\}$$
$$= [1.0593]/\{[0.8480]^2\}$$
$$= [1.0593]/\{0.7191\}$$
$$= 1.47$$

The total mass per unit length for the U-tube, m, includes the hydrodynamic mass, m_h, of the two-phase steam-water mixture on the shell side.

The hydrodynamic mass per unit length is calculated from Eq. (4-8) (see Example 4-2),

$$m_h = \left(\frac{\rho \pi D^2}{4}\right)\left[\frac{(D_e/D)^2 + 1}{(D_e/D)^2 - 1}\right] \tag{4-8}$$

where, $\rho = \rho_{TP}$, the homogeneous density of the two-phase water-steam mixture, and the confinement coefficient (see Example 4-2) is,

$$(2.565^2 + 1)/(2.565^2 - 1) = 1.36$$

The two-phase mixture density, $\rho_{TP} = \rho$, is calculated from Eq. (3-2),

$$\rho = \rho_\ell(1 - \varepsilon_g) + \rho_g \varepsilon_g \tag{3-2}$$

where, at $T = 275°C$, $\rho_\ell = 760$ kg/m^3 and $\rho_g = 31.2$ kg/m^3

In Example 3-2, ρ_{TP} was calculated for a void fraction, $\varepsilon_g = 85.6\%$. For the two void fractions of interest in this example, $\varepsilon_g = 40\%$ and 85%,

for $\varepsilon_g = 40\%$ $\quad \rho_{TP} = 760\,(1 - 0.4) + 31.2\,(0.4) = 468.5$ kg/m^3, and
for $\varepsilon_g = 85\%$ $\quad \rho_{TP} = 760\,(1 - 0.85) + 31.2\,(0.85) = 140.5$ kg/m^3

In Example 4-2, m_h was calculated for a void fraction, $\varepsilon_g = 85.6\%$. For the two void fractions of interest in this example, $\varepsilon_g = 40\%$ and 85%, m_h is,

for $\varepsilon_g = 40\%$ $\quad m_h = (\pi/4) \times 468.5 \times (0.020)^2 \times (1.36) = 0.200$ kg/m, and
for $\varepsilon_g = 85\%$ $\quad m_h = (\pi/4) \times 140.5 \times (0.020)^2 \times (1.36) = 0.060$ kg/m

As per Eq. (4-1), the total tube mass per unit length, m, is the algebraic sum of the tube, hydrodynamic and internal fluid masses per unit length (see Example 4-2),

$$m = m_t + m_h + m_i \tag{4-1}$$

In Example 4-2, m was calculated for a void fraction, $\varepsilon_g = 85.6\%$. For the U-tube, $m_t = 0.478$ kg/m and $m_i = 0.254$ kg/m (see Example 4-2). Thus, for the two void fractions of interest in this example, $\varepsilon_g = 40\%$ and 85%, m is,

for $\varepsilon_g = 40\%$ $\quad m = 0.478 + 0.200 + 0.254 = 0.932$ kg/m, and
for $\varepsilon_g = 85\%$ $\quad m = 0.478 + 0.060 + 0.254 = 0.792$ kg/m

For the two void fractions of interest in this example, $\varepsilon_g = 40\%$ and 85%, the void-fraction function, $f(\varepsilon_g)$, derived from Eq. (6-21) is,

for $\varepsilon_g = 40\%$ $\quad f(\varepsilon_g) = 1$, and
for $\varepsilon_g = 85\%$ $\quad f(\varepsilon_g) = 1 - (\varepsilon_g - 70)/30 = 0.5$

Therefore, for the two void fractions of interest in this example, $\varepsilon_g = 40\%$ and 85%, ζ_{TP} as per Eq. (6-23) is,

$$\text{for } \varepsilon_g = 40\% \quad \zeta_{TP} = 4\left\{\left[(760)\,(0.020^2)\right]/(0.932)\right\}\,(1)\,(1.47)$$

$$\zeta_{TP} = 4\,\{0.304/0.932\}\,(1)\,(1.47)$$

$$\zeta_{TP} = 4\,\{0.326\}\,(1)\,(1.47)$$

$$\zeta_{TP} = 1.92\%$$

$$\text{for } \varepsilon_g = 85\% \quad \zeta_{TP} = 4\left\{\left[(760)\,(0.020^2)\right]/(0.792)\right\}\,(0.5)\,(1.47)$$

$$\zeta_{TP} = 4\,\{0.304/0.792\}\,(0.5)\,(1.47)$$

$$\zeta_{TP} = 4\,\{0.384\}\,(0.5)\,(1.47)$$

$$\zeta_{TP} = 1.13\%$$

The kinematic viscosity for a two-phase mixture, ν_{TP}, is calculated from Eqs. (3-5) and (3-10), and with the knowledge that $\mu = \rho\nu$. For $\varepsilon_g = 40\%$, ν_{TP} is

$$\nu_{TP} = \cfrac{\nu_\ell}{1 + \varepsilon_g\left(\cfrac{\nu_\ell}{\nu_g} - 1\right)} = \cfrac{1.26 \times 10^{-7}}{1 + 0.40\left(\cfrac{1.26 \times 10^{-7}}{8.0 \times 10^{-7}} - 1\right)} = 1.9 \times 10^{-7}\,\text{m}^2/\text{s}$$

Similarly, for $\varepsilon_g = 85\%$, $\nu_{TP} = 4.44 \times 10^{-7}\,\text{m}^2/\text{s}$.

Viscous and support damping in two-phase flow use the same formulations as derived for single-phase flow (see Chapter 5) with two-phase values for the shell-side fluid parameters.

Viscous damping (in percent) in the shell-side fluid, ζ_V, from Eq. (5-5) is,

$$\zeta_V = \frac{100\pi}{\sqrt{8}}\left(\frac{\rho D^2}{m}\right)\left(\frac{2\nu}{\pi f D^2}\right)^{0.5}\left\{\frac{\left[1 + \left(\frac{D}{D_e}\right)^3\right]}{\left[1 - \left(\frac{D}{D_e}\right)^2\right]^2}\right\} \qquad (5\text{-}5)$$

The confinement coefficient for viscous damping = 1.47, as calculated above.

For $\varepsilon_g = 40\%$, $\rho = \rho_{TP} = 468.5\,\text{kg/m}^3$, $m = 0.932\,\text{kg/m}$, and $\nu = \nu_{TP} = 1.9 \times 10^{-7}\,\text{m}^2/\text{s}$. Thus, ζ_V is,

$$\zeta_V = \left[100\pi/(8^{0.5})\right](468.5 \times 0.02^2/0.932)\left\{\left[(2)(1.9 \times 10^{-7})\right]/\left[\pi(f)(0.02^2)\right]\right\}^{0.5}(1.47)$$

$$\zeta_V = [111.1](0.2011)\left\{[3.8 \times 10^{-7}]/\left[\pi(f)(0.02^2)\right]\right\}^{0.5}(1.47)$$

When the AVBs in the U-bend are effective, $f = f_1 = 118\,\text{Hz}$, and ζ_V is,

$$\zeta_V = [111.1](0.2011)\left\{[3.8 \times 10^{-7}]/\left[\pi(118)(0.02^2)\right]\right\}^{0.5}(1.47)$$

$$\zeta_V = [111.1](0.2011)\left\{[3.8 \times 10^{-7}]/[0.1483]\right\}^{0.5}(1.47)$$

$$\zeta_V = [111.1](0.2011)\left\{2.563 \times 10^{-6}\right\}^{0.5}(1.47)$$

$$\zeta_V = [111.1](0.2011)\{0.001601\}(1.47)$$

$$\zeta_V = 0.053\%$$

When every second AVB in the U-bend is ineffective, then, $f = f_{1ineffective} = 29.5\,\text{Hz}$, and ζ_V is,

$$\zeta_V = [111.1](0.2011)\left\{[3.8 \times 10^{-7}]/\left[\pi(29.5)(0.02^2)\right]\right\}^{0.5}(1.47)$$

$$\zeta_V = [111.1](0.2011)\{[3.8 \times 10^{-7}]/[0.03707]\}^{0.5}(1.47)$$

$$\zeta_V = [111.1](0.2011)\{1.025 \times 10^{-5}\}^{0.5}(1.47)$$

$$\zeta_V = [111.1](0.2011)\{0.003202\}(1.47)$$

$$\zeta_V = 0.105\%$$

For $\varepsilon_g = 85\%$, $\rho = \rho_{TP} = 140.5$ kg/m^3, $m = 0.792$ kg/m, and $v = v_{TP} = 4.44 \times 10^{-7}$ m^2/s. Thus, ζ_V is,

$$\zeta_V = [100\pi/(8^{0.5})](140.5 \times 0.02^2/0.792)\{[(2)(4.44 \times 10^{-7})]/[\pi(f)(0.02^2)]\}^{0.5}(1.47)$$

$$\zeta_V = [111.1](0.07096)\{[8.88 \times 10^{-7}]/[\pi(f)(0.02^2)]\}^{0.5}(1.47)$$

When the AVBs in the U-bend are effective, $f = f_1 = 118$ Hz, and ζ_V is,

$$\zeta_V = [111.1](0.07096)\{[8.88 \times 10^{-7}]/[\pi(118)(0.02^2)]\}^{0.5}(1.47)$$

$$\zeta_V = [111.1](0.07096)\{[8.88 \times 10^{-7}]/[0.1483]\}^{0.5}(1.47)$$

$$\zeta_V = [111.1](0.07096)\{5.989 \times 10^{-6}\}^{0.5}(1.47)$$

$$\zeta_V = [111.1](0.07096)\{0.002477\}(1.47)$$

$$\zeta_V = 0.028\%$$

When every second AVB in the U-bend is ineffective, then $f = f_{1ineffective} = 29.5$ Hz, and ζ_V is,

$$\zeta_V = [111.1](0.07096)\{[8.88 \times 10^{-7}]/\pi(29.5)[(0.02^2)]\}^{0.5}(1.47)$$

$$\zeta_V = [111.1](0.07096)\{[8.88 \times 10^{-7}]/[0.03707]\}^{0.5}(1.47)$$

$$\zeta_V = [111.1](0.07096)\{2.395 \times 10^{-5}\}^{0.5}(1.47)$$

$$\zeta_V = [111.1](0.07096)\{0.004894\}(1.47)$$

$$\zeta_V = 0.057\%$$

Support damping (in percent) includes squeeze-film and friction damping terms. Squeeze-film damping (in percent) between the U-tube and its supports, ζ_{SF}, from Eq. (5-10) and Eq. (5-34), is

$$\zeta_{SF} = [(N-1)/N]\zeta_{SFn} \tag{5-10}$$

where ζ_{SFn} is the normalized squeeze-film damping,

$$\zeta_{SFn} = \left(\frac{1460}{f}\right)\left(\frac{\rho D^2}{m}\right)\left(\frac{L}{\ell_m}\right)^{0.5} \tag{5.34}$$

For the U-tube, $L = 15$ mm.

For squeeze film damping in two-phase flow, we assume the supports are wet in well-designed recirculating steam generators (see Eq. (2-20)). Thus, $\rho = \rho_\ell = 760$ kg/m^3.

For $\varepsilon_g = 40\%$, $m = 0.932$ kg/m. Therefore, ζ_{SF} is

$$\zeta_{SF} = (N-1)/N[1460/f][(760 \times 0.02^2)/0.932][0.015/\ell_m]^{0.5}$$

When the AVBs in the U-bend are effective, $(N-1)/N = (9/10)$, $f = f_1 = 118$ Hz, and $\ell_m = \ell = 0.59$ m. Therefore, ζ_{SF} is,

$$\zeta_{SF} = (9/10)[1460/118][(760 \times 0.02^2)/0.932][0.015/0.59]^{0.5}$$

$\zeta_{SF} = (0.9)(12.37)(0.3262)(0.1594)$

$\zeta_{SF} = 0.579\%$

When every second AVB in the U-bend is ineffective, $(N\text{-}1)/N = (4/5)$, $f = f_{1ineffective} = 25.9$ Hz, and $\ell_m = \ell_{ineffective} = 2 \times \ell = 1.18$ m Therefore, ζ_{SF} is,

$\zeta_{SF} = (4/5) [1460/29.5] \left[(760 \times 0.02^2)/0.932\right] [0.015/1.18]^{0.5}$

$\zeta_{SF} = (0.8) (49.49) (0.3262) (0.1127)$

$\zeta_{SF} = 1.46\%.$

For $\varepsilon_g = 85\%$, $m = 0.792$ kg/m. Therefore, ζ_{SF} is

$\zeta_{SF} = (N-1)/N[1460/f]\left[(760 \times 0.02^2)/0.792\right][0.015/\ell_m]^{0.5}$

When the AVBs in the U-bend are effective, $(N\text{-}1)/N = (9/10)$, $f = f_1 = 118$ Hz, and $\ell_m = \ell = 0.59$ m. Therefore ζ_{SF} is,

$\zeta_{SF} = (9/10)[1460/118]\left[(760 \times 0.02^2)/0.792\right][0.015/0.59]^{0.5}$

$\zeta_{SF} = (0.9)(12.37)(0.3838)(0.1594)$

$\zeta_{SF} = 0.68\%$

When every second AVB in the U-bend is ineffective, $(N\text{-}1)/N = (4/5)$, $f = f_{1ineffective} = 25.9$ Hz, and $\ell_m = \ell_{ineffective} = 2 \times \ell = 1.18$ m. Therefore, ζ_{SF} is,

$\zeta_{SF} = (4/5)[1460/29.5]\left[(760 \times 0.02^2)/0.792\right][0.015/1.18]^{0.5}$

$\zeta_{SF} = (0.8)(49.49)(0.3838)(0.1127)$

$\zeta_{SF} = 1.71\%$

Friction damping (in percent) between the U-tube and its supports, ζ_F from Eq. (5-10) and Eq. (5-35), is

$$\zeta_F = [(N-1)/N]\zeta_{Fn} \tag{5-10}$$

where ζ_{Fn} is the normalized friction damping (analogous to the normalized squeeze-film damping),

$$\zeta_{Fn} = 0.5(L/\ell_m)^{0.5} \tag{5-35}$$

Friction damping is independent of the shell-side fluid properties. Thus, ζ_F is equal for both void fractions of interest.

When the AVBs in the U-bend are effective, $(N\text{-}1)/N = (9/10)$, and $\ell_m = \ell = 0.59$ m. Therefore, ζ_F is,

$\zeta_F = (9/10) (0.5) [0.015/0.59]^{0.5}$

$\zeta_F = (0.45) (0.1594)$

$\zeta_F = 0.072\%$

When every second AVB in the U-bend is ineffective, $(N\text{-}1)/N = (4/5)$, $f = f_{1ineffective} = 25.9$ Hz, and $\ell_m = \ell_{ineffective} = 2 \times \ell = 1.18$ m. Therefore, ζ_F is,

$\zeta_F = (4/5) (0.5) [0.015/1.18]^{0.5}$

$\zeta_F = (0.4) (0.1127)$

$$\zeta_F = 0.045\%$$

Finally, the total damping (in percent), ζ_T, for the U-tube is calculated from Eq. (6-14) as the algebraic sum of the support damping, ζ_S, between the U-tube and its supports, viscous damping in the shell-side fluid, ζ_V, and two-phase damping in the shell-side fluid, ζ_{TP}. The support damping is the algebraic sum of the squeeze-film damping, ζ_{SF}, and the friction damping, ζ_F,

$$\zeta_S = \zeta_{SF} + \zeta_F$$

Therefore, ζ_T at $\varepsilon_g = 40\%$ is

$$\zeta_T = 0.579 + 0.072 + 0.053 + 1.92$$
$$\zeta_T = 2.62\% \text{ with effective U-bend supports,}$$

and,

$$\zeta_T = 1.46 + 0.045 + 0.105 + 1.92$$
$$\zeta_T = 3.53\% \text{ with every second AVB ineffective,}$$

and, ζ_T at $\varepsilon_g = 85\%$ is,

$$\zeta_T = 0.68 + 0.072 + 0.028 + 1.13$$
$$\zeta_T = 1.91\% \text{ with effective U-bend supports,}$$

and,

$$\zeta_T = 1.71 + 0.045 + 0.057 + 1.13$$
$$\zeta_T = 2.94\% \text{ with every second AVB ineffective.}$$

Example 8-1   Fluidelastic Instability in a Steam Generator U-Bend

Fluidelastic instability in two-phase flow is discussed in Chapter 8. The critical pitch velocity for FEI, U_{pc}, is calculated using Eq. (8-5),

$$U_{pc}/fD = \left(K\sqrt{2\pi\zeta}\right)\left(m/\rho D^2\right)^{1/2} \tag{8-5}$$

where the instability constant, $K = 3.0$ (see Chapter 8, Section 8.10.1). Note that damping in Eq. (8-5) is **not** formulated in percent, but as a ratio.

Eq. (8-5) is the same formulation derived for single-phase flow (see Chapter 7) with two-phase values for the shell-side fluid parameters.

Parameter values for the U-bend tube in two-phase steam-water flow have been calculated in earlier examples.

From Example 3-2, $\rho = \rho_{TP} = 137 \text{ kg/m}^3$ at $\varepsilon_g = 85.6\%$.

From Example 4-2, $m = 0.791 \text{ kg/m}$

From Example 6-1, $\zeta = \zeta_T = 1.91\%$ at $\varepsilon_g = 85\%$ with effective U-bend supports, and $\zeta = \zeta_T = 2.94\%$ with every second AVB ineffective. Note that for this example, we will use the damping values at $\varepsilon_g = 85\%$ to approximate those at $\varepsilon_g = 85.6\%$.

From Example 4-4, $f = f_1 = 118 \text{ Hz}$ with effective U-bend supports, and $f = f_{1 ineffective} = 29.5 \text{ Hz}$ with every second AVB ineffective.

For the U-tube with effective U-bend supports, U_{pc} is,

$$U_{pc} = (118)(0.02)(3.0)[2\pi(0.0191)]^{0.5}\{(0.791)/[(137)(0.02^2)]\}^{0.5}$$
$$U_{pc} = (118)(0.02)(3.0)[0.1200]^{0.5}\{(0.791)/[0.0548]\}^{0.5}$$

$$U_{pc} = (118)\,(0.02)\,(3.0)\,(0.3464)\,\{14.43\}^{0.5}$$

$$U_{pc} = (118)\,(0.02)\,(3.0)\,(0.3464)\,(3.799)$$

$$U_{pc} = (118)\,(0.02)\,(3.948)$$

$$U_{pc} = 9.32 \text{ m/s}$$

Comparing U_{pc} to the pitch velocity, U_p, in the steam generator U-tube bundle (see Example 3-2),

$$U_p/U_{pc} = 5.09/9.32 = 0.55 < 1.0$$

Since the ratio is less than 1.0, this U-tube will not experience fluidelastic instability when the AVBs are effective.

For the U-tube with every second AVB ineffective, U_{pc} is,

$$U_{pc} = (29.5)(0.02)(3.0)[2\pi(0.0294)]^{0.5}\{(0.791)/[(137)(0.02^2)]\}^{0.5}$$

$$U_{pc} = (29.5)\,(0.02)\,(3.0)\,[0.1847]^{0.5}\{(0.791)/[0.0548]\}^{0.5}$$

$$U_{pc} = (29.5)\,(0.02)\,(3.0)\,(0.4298)\,\{14.43\}^{0.5}$$

$$U_{pc} = (29.5)\,(0.02)\,(3.0)\,(0.4298)\,(3.799)$$

$$U_{pc} = (29.5)\,(0.02)\,(4.898)$$

$$U_{pc} = 2.89 \text{ m/s.}$$

Comparing U_{pc} to the pitch velocity, U_p, in the steam generator U-tube bundle (see Example 3-2),

$$U_p/U_{pc} = 5.09/2.89 = 1.76 > 1.0$$

Fluidelastic instability is likely to occur for this U-tube where every second AVB is ineffective since the pitch velocity is significantly greater than the critical pitch velocity.

Example 10-1   Random Excitation for Two-Phase Flow in Steam Generator U-Bend

Random turbulence excitation in two-phase flow is discussed in Chapter 10. The formulation for the dimensionless power spectral density of the input force per unit length, $\widetilde{S}_F(f_R)$ is given in Eq. (10-9) for churn/bubbly flow and Eq. (10-10) for intermittent flow,

$$\widetilde{S}_F(f_R) = \frac{S_F(f)}{(\rho_\ell g D)^2 D^2}\frac{U_p}{D_B} \tag{10-9}$$

$$\widetilde{S}_F(f_R) = \frac{S_F(f)}{\left(g\dot{m}_p^2 D\right)D^2}\frac{U_p}{D_B} \tag{10-10}$$

where the reduced frequency, f_R, is defined in Eq. (10-8) as,

$$f_R = fD_B/U_p \tag{10-8}$$

and where the characteristic void length, D_B, is defined in Eq. (10-7) as,

$$D_B = 0.1D/\sqrt{1-\varepsilon_g} \tag{10-7}$$

The flow regime for the steam generator U-bend tube is determined from the flow regime maps in Fig. 10-14 based on the superficial gas and liquid velocities, U_g and U_ℓ, respectively. The superficial phase velocities for the U-tube shell-side steam-water mixture are calculated using Eq. (3-5).

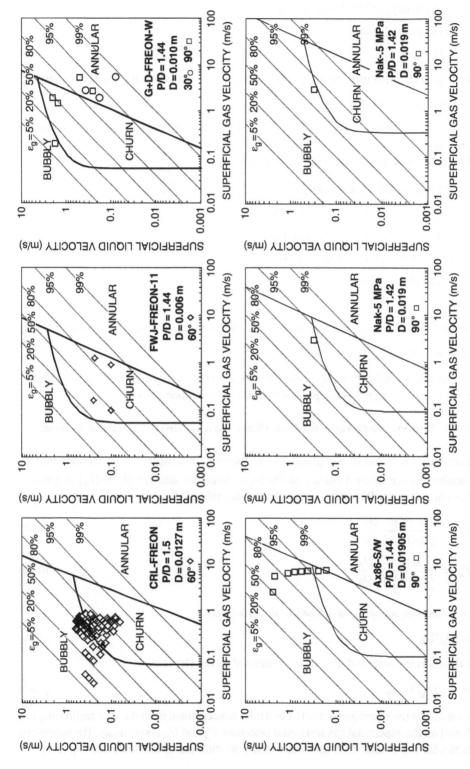

Fig. 10-14 Steam-Water and Freon Flow Regime Maps.

$$U_g = \frac{\varepsilon_g \dot{m}}{\rho} = \frac{X\dot{m}}{\rho_g}, U_\ell = \frac{(1-\varepsilon_g)\dot{m}}{\rho} = \frac{(1-X)\dot{m}}{\rho_\ell} \tag{3-5}$$

Properties for the shell-side steam-water mixture were calculated in Example 3-2. For $X = 20\%$, $\dot{m} = 349$ kg/m²s, $\rho_g = 32$ kg/m³, and $\rho_\ell = 760$ kg/m³,

$$U_g = (0.20)(349)/(32) = 2.18 \text{ m/s, and}$$
$$U_\ell = (1-0.20)(349)/(760) = 0.37 \text{ m/s}$$

Accounting for maldistribution, $U_g = 2 (2.18) = 4.36$ m/s, and $U_\ell = 2 (0.37) = 0.73$ m/s.

The steam-water maps for $D = 19$ mm (bottom row in Fig. 10-14) are most relevant for the U-bend tube. From these, the two-phase flow in the steam generator U-bend is within the bubbly regime. Therefore, Eq. (10-9) is the appropriate random turbulence excitation formulation.

For $\varepsilon_g = 85.6\%$, D_B calculated from Eq. (10-7) is,

$$D_B = [(0.1) (0.02)]/(1 - 0.856)^{0.5} = 0.00527 \text{ m}$$

For $f = f_1 = 118$ Hz (see Example 4-4) and $U_p = (U_p)_{max} = 5.09$ m/s (see Example 3-2), f_R calculated from Eq. (10-8) is,

$$f_R = (118) (0.00527)/(5.09) = 0.1222$$

The design guideline of dimensionless reference equivalent power spectral density of the excitation force per unit length, $\hat{S}_F(f_R)_e^o$, based on a reference excited tube length, L_o, and a reference diameter, D_o, for bubbly flow is given in Eq. (10-11) and Eq. (10-12),

$$\widetilde{S}_F(f_R)_e^o = 2.0 f_R^{-0.5}, \quad 0.001 < f_R < 0.06 \tag{10-11}$$

$$\widetilde{S}_F(f_R)_e^o = 4.3 \times 10^{-4} f_R^{-3.5}, \quad 0.06 < f_R < 1 \tag{10-12}$$

As $f_R = 0.1222 > 0.06$, Eq. (10-12) applies for the steam generator U-bend. Thus,

$$\widetilde{S}_F(f_R)_e^o = 4.3 \times 10^{-4}(0.1222)^{-3.5} = 0.6746$$

The design guideline was derived for $L_o = 1.0$ m, and $D_o = 20$ mm (see Chapter 10, Section 10.5.5). The dimensionless equivalent power spectral density is calculated following Eq. (10-5),

$$S_F(f)_e^o = S_F(f)_e \times \frac{L_e}{L_o} \times \frac{D_o}{D} \tag{10-5}$$

Note that Eq. (10-5) is for dimensional power spectral densities; however, the scaling between reference equivalent and equivalent power spectral densities is the same for the dimensionless spectra.

For the U-tube, $L_e = \ell = 0.59$ m, and $D = 20$ mm. Thus, the dimensionless equivalent power spectral density = (0.6746) x $(1.0 / 0.59)$ x $(20 / 20) = 1.143$.

The dimensionless power spectral density is rendered dimensional using Eq. (10-2),

$$\widetilde{S}_F(f_R) = \frac{S_F(f)}{(p_o D)^2} f_o \tag{10-2}$$

where p_o is a pressure scaling factor and f_o is a frequency scaling factor.

The frequency scaling factor is defined as, $f_o = f / f_R$ (see Chapter 10, Section 10.3). Thus, for the U-tube with effective supports, where, $f = f_1 = 118$ Hz, $f_o = (118) / (0.1222) = 965.9$ Hz.

The pressure scaling factor for churn/bubbly flow is defined as $p_o = \rho_{\ell}gD$ (see Chapter 10, Section 10.6), where g is the acceleration due to gravity ($g = 9.81$ m/s^2). Thus, for the U-tube, where

$$\rho_{\ell} = 760 \text{ kg/m}^3, \ p_o = (760)(9.81)(0.02) = 149.1 \text{ kg/m·s}^2$$

The maximum mean square vibration response, $\overline{y^2}$, for a tube in two-phase flow is calculated from Eq. (10-4).

$$\overline{y^2(x)}_1 = \frac{\widetilde{S}_F(f_R)_e \phi_1^2 a_1}{64\pi^3 f_1^3 m^2 \zeta_1} \frac{(p_o D)^2}{f_o}$$

where, $\phi_1(x)$ is the normalized mode shape for the first mode, a_1 is the numerical coefficient for the first mode, f_1 is the tube first-mode natural frequency, m is the total tube mass and ζ_1 is the damping ratio for the first mode.

Table 9-1 contains values of $\phi_1^2{}_{max}$ and a_1 for various end conditions. From Table 9-1, for pinned-pinned end conditions, $\phi_1^2{}_{max} = 2.0$ and $a_1 = 1.1$.

For the steam generator U-tube with effective supports, $f = f_1 = 118$ Hz (see Example 4-4), $m = 0.791$ kg/m (see Example 4-2), $\zeta_1 = \zeta_T = 1.91\%$ (Note that $\varepsilon_g = 85.6\%$, but for simplicity we use the total damping calculated in Example 6-1 for $\varepsilon_g = 85\%$). Thus, $\overline{y^2}$ at midspan, $\ell/2$, is

$$= \{(1.143)(2)(1.1)/[64\pi^3(118^3)(0.791^2)(0.0191)]\} \times [(149.1)(0.02)]^2/(965.9)$$

$$= \{(1.143)(2)(1.1)/[64\pi^3(1.964 \times 10^4)]\} \times [2.982]^2/(965.9)$$

$$= \{(1.143)(2)(1.1)/[64\pi^3(1.964 \times 10^4)]\} \times (8.892)/(965.9)$$

$$= \{(1.143)(2)(1.1)/[3.896 \times 10^7]\} \times (0.009216)$$

$$= \{6.457 \times 10^{-8}]\} \times (0.009216)$$

$$= 5.950 \times 10^{-10} \text{ m}^2$$

Therefore, the rms vibration response at the midspan is,

$$y_{(\ell/2)} = (5.950 \times 10^{-10})^{0.5}$$

$$y_{(\ell/2)} = 24.4 \times 10^{-6} \text{ m rms} = 24.4 \ \mu\text{m rms}$$

The design guideline for maximum allowable rms vibration amplitude due to random turbulence excitation is based on calculated fretting-wear depth over the life of the component. This calculation is completed in Example A.2.

Example 10-2   Random Excitation for SG U-Bend Using an Early Guideline

An earlier design guideline for random turbulence excitation in two-phase flow is presented in Pettigrew et al, 1991.

In this formulation, the power spectral density of the input force per unit length, $S_F(f)$, is given in terms of a normalized power spectral density (NPSD) as follows,

$$S_F(f) = NPSD\left(\frac{\dot{m}^2 D^2}{L_e f^{1.25}}\right)$$

The earlier design guideline of NPSD is,

$$NPSD = 10^{(0.027\varepsilon_g - 3.45)}, \quad \text{for } 10\% < \varepsilon_g < 80\%$$
$$NPSD = 0.05, \quad\quad\quad\ \ \text{for } 80\% < \varepsilon_g < 99\%$$

For the steam generator U-tube, $\varepsilon_g = 85.6\%$. Therefore, the NPSD = 0.05.

For the steam generator U-tube, $\dot{m} = 349$ kg/m²s (see Example 3-2). Accounting for maldistribution, $\dot{m} = 2 \times 349 = 698$ kg/m²s for this calculation. Also, $L_e = \ell = 0.59$ m and $f = f_1 = 118$ Hz for the steam generator U-tube with effective supports (see Example 4-4). Therefore, $S_F(f)$ is,

$$S_F(f) = (0.05) \left[(698^2)\, (0.02^2) \right] / \left[(0.59)\, (118)^{1.25} \right]$$
$$S_F(f) = (0.05)\, (194.9) / [(0.59)\, (388.9)]$$
$$S_F(f) = (0.05)\, (194.9) / [229.5]$$
$$S_F(f) = 0.0425\ \text{N}^2\text{m}^{-2}\text{s}$$

The maximum mean square of vibration response, $\overline{y^2}$, for the U-bend tube is calculated analogously to Example 10-1. Thus, $\overline{y^2}$ at midspan, $\ell/2$, is,

$$= (0.0425)(2)(1.1) / \left[64\pi^3 (118^3)(0.791^2)(0.0191) \right]$$
$$= (0.0425)(2)(1.1) / \left[64\pi^3 (1.964 \times 10^4) \right]$$
$$= (0.0425)(2)(1.1) / \left[3.896 \times 10^7 \right]$$
$$= 2.40 \times 10^{-9} \text{m}^2$$

Therefore, the rms vibration response at the midspan is,

$$y_{(\ell/2)} = \left(2.40 \times 10^{-9} \right)^{0.5}$$
$$y_{(\ell/2)} = 49 \times 10^{-6} \text{m rms} = 49\ \mu\text{m rms}$$

The rms vibration response at the U-bend tube midspan calculated with the earlier guideline (49 μm rms) is within a factor of two of that calculated with the current guideline (24.4 μm rms) (see Example 10-1).

If every second AVB in the U-bend were ineffective, $L_e = 2 \times \ell = 1.18$ m and $f = f_{1ineffective} = 29.5$ Hz (see Example 4-4). Thus, neglecting the effects of the increased span length and decreased tube vibration frequency on $S_F(f)$, $\overline{y^2}$ at midspan, $\ell/2$, is,

$$= (0.0425)(2)(1.1) / \left[64\pi^3 (29.5^3)(0.791^2)(0.0294) \right]$$
$$= (0.0425)(2)(1.1) / \left[64\pi^3 (472.2) \right]$$
$$= (0.0425)(2)(1.1) / \left[9.371 \times 10^5 \right]$$
$$= 9.98 \times 10^{-8} \text{m}^2$$

Therefore, the rms vibration response at the midspan is.

$$y_{(\ell/2)} = \left(9.98 \times 10^{-8} \right)^{0.5}$$
$$y_{(\ell/2)} = 316 \times 10^{-6} \text{m rms} = 316\ \mu\text{m rms}$$

The calculated rms vibration response at the midspan would be greater (by approximately 70%), if the effects of increased span length and decreased tube vibration frequency on $S_F(f)$ were taken into account.

Example A-2 Fretting-Wear Damage Prediction in a Steam Generator U-Bend

Fretting-wear damage prediction is discussed in Chapter 12. The amount of fretting-wear damage due to random turbulence excitation can be predicted using Eq. (12-29),

$$\dot{W}_N = 16\pi^3 f^3 m \ell y_{(\ell/2)rms}^2 \zeta_S \tag{12-29}$$

where the support damping, ζ_S, is calculated using Eq. (12-6) modified to include two-phase damping,

$$\zeta_S = \zeta_T - \zeta_V - \zeta_{TP}$$

From Example 6-1,

$$\zeta_T = 1.91\%$$
$$\zeta_V = 0.028\%$$
$$\zeta_{TP} = 1.13\%, \text{and}$$
$$\zeta_S = \zeta_{SF} + \zeta_F = 0.68 + 0.072 = 0.752\%$$

for $\varepsilon_g = 85\%$ and effective U-bend supports.

From Example 10-1, the mid-span vibration response for the U-bend tube for $\varepsilon_g = 85\%$ and effective U-bend supports, $y(\ell/2)$, is 24.4 μm rms. For this level of vibration, $\dot{W}_N$ is,

$$\dot{W}_N = 16\pi^3 (118^3)(0.791)(0.59)(24.4 \times 10^{-6})^2 (0.00752)$$
$$\dot{W}_N = 16\pi^3 (1.643 \times 10^6)(0.791)(0.59)(5.950 \times 10^{-10})(0.00752)$$
$$\dot{W}_N = 1.70 \times 10^{-3} \, \text{W} = 1.70 \, \text{mW}$$

where $f = f_1 = 118$ Hz (see Example 4-4), $m = 0.791$ kg/m (see Example 4-2) and $\ell = 0.59$ m.

The fretting-wear damage resulting from this level of work-rate is calculated using Eq. (12-2),

$$\dot{V} = K_{FW}\dot{W}_N \tag{12-2}$$

Where K_{FW} is wear coefficient obtained experimentally as discussed in Chapter 13.

For a wear coefficient of $K_{FW} = 20 \times 10^{-15}$ m^2/N, the volume wear rate, $\dot{V}$ is,

$$\dot{V} = (20 \times 10^{-15})(1.70 \times 10^{-3})$$
$$\dot{V} = 3.40 \times 10^{-17} \, \text{m}^3/\text{s}$$

At this volume wear rate, the total worn volume, V, after a typical component lifetime, $T_s = 20$ years is,

$$V = (3.40 \times 10^{-17})(20)(365)(24)(3600)$$

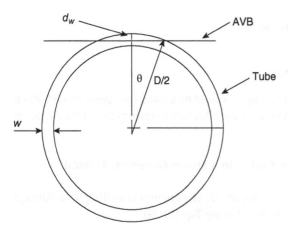

Fig. 12-14a Contact Between a Steam Generator Tube and Flat-Bar-Type (AVB) Support.

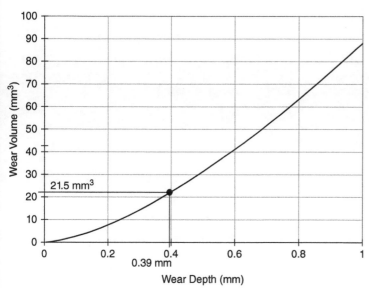

Fig. A-1 Wear Volume versus Wear Depth.

$$V = 2.15 \times 10^{-8} \text{ m}^3 = 21.5 \text{ mm}^3$$

For flat-bar-type supports, it is assumed that wear is occurring on one side and only on the tubes. Thus, the bars and the wear scars remain flat (see Fig. 12-14a). This assumption leads to,

$$V = \frac{LD^2}{8}(2\theta - \sin 2\theta) \qquad (12\text{-}32)$$

where θ is the intersection angle,

$$\theta = arccos\left[(D - 2d_w)/D\right] \qquad (12\text{-}33)$$

Thus, the tube wall fretting-wear depth, d_w, may be estimated using Eq. (12-32) and Eq. (12-33). However, these equations need to be solved iteratively. A curve of V versus d_w for $L = 15$ mm and $D = 20$ mm is shown in Fig. A-1.

The wear volume of 21.5 mm^3 for this geometry corresponds to a wear depth of approximately 390 μm or 39% tube wall reduction for a wall thickness of 1.0 mm, which is marginally acceptable from an operation point of view.

Subject Index

a

acoustic mode 35–36, 356
  frequencies 35, 356
  fundamental 358
acoustic noise 7, 34, 354, 364
acoustic resonance 34–36, Chapter 11
  critical velocities for 356–357
  damping criteria 36, 358–363
  elimination of 364
  lock-in 356–357
  parameter 35–36
  sound pressure level of 361–363
  susceptibility to 35–36
added mass, (see hydrodynamic mass)
anisotropic approach 76
anti-vibration bar (or AVB), (see supports)
Archard's equation 40, 402–403
axial flow-induced vibration 13, 64–65
  fluidelastic instability 212, 220, 264–265
  random turbulence excitation 287, 318–323

b

beam theory 93–98
  free 93–98
  slender 93, 95
bending moment 94, 98–100, 106, 108, 110, 113
boiling water reactor (or BWR), (see nuclear
  reactors)
buckling instability 212–213, 264

c

CANDU 6–7, 47, 49–50, 157, 225, 375, 377, 389,
  399–400, 404, 410, 417, 419, 421–422, 424
cavitating annular flow 287
cavity 352
  acoustic modes in 35, 356, 365
  standing waves in 34–35, 354–356, 365
centrifugal force 75
characteristic length,
  span or tube 20, 22, 131, 139
  void 301
circulation ratio (in steam generator) 53, 446
clearance 6, 9, 37, 43
  diametral 25, 37, 43, 123–124, 132, 387, 389,
  417, 419–420
  effect of (on damping) 20, 24–25, 119–120,
  123–124, 132–133, 136–138, 144
  effect of (on fluidelastic instability)
  185, 210
codes, computer 9, 16, 36–38, 48, 74, 78–79,
  123, 286, 377, 385, 435
  EPRI SG FW 9
  H3DMAP 9
  PIPO1, 38–39
  THIRST 16, 74, 77–78
  VIBIC 9, 123, 377, 385
coherence function 233
confinement 19, 23, 88–89, 133, 144
  coefficient 88–89, 135

Flow-Induced Vibration Handbook for Nuclear and Process Equipment, First Edition.
Michel J. Pettigrew, Colette E. Taylor, and Nigel J. Fisher.
© 2022 John Wiley & Sons, Inc. This Work is a co-publication between ASME Press and John Wiley & Sons, Inc.